Clusters and Superclusters of Galaxies

NATO ASI Series

Advanced Science Institutes Series

A Series presenting the results of activities sponsored by the NATO Science Committee, which aims at the dissemination of advanced scientific and technological knowledge, with a view to strengthening links between scientific communities.

The Series is published by an international board of publishers in conjunction with the NATO Scientific Affairs Division

A Life Sciences	Plenum Publishing Corporation
B Physics	London and New York
C Mathematical	Kluwer Academic Publishers
and Physical Sciences	Dordrecht, Boston and London
D Behavioural and Social Sciences	
E Applied Sciences	
F Computer and Systems Sciences	Springer-Verlag
G Ecological Sciences	Berlin, Heidelberg, New York, London,
H Cell Biology	Paris and Tokyo
I Global Environmental Change	

NATO-PCO-DATA BASE

The electronic index to the NATO ASI Series provides full bibliographical references (with keywords and/or abstracts) to more than 30000 contributions from international scientists published in all sections of the NATO ASI Series.
Access to the NATO-PCO-DATA BASE is possible in two ways:

– via online FILE 128 (NATO-PCO-DATA BASE) hosted by ESRIN,
Via Galileo Galilei, I-00044 Frascati, Italy.

– via CD-ROM "NATO-PCO-DATA BASE" with user-friendly retrieval software in English, French and German (© WTV GmbH and DATAWARE Technologies Inc. 1989).

The CD-ROM can be ordered through any member of the Board of Publishers or through NATO-PCO, Overijse, Belgium.

Series C: Mathematical and Physical Sciences - Vol. 366

Clusters and Superclusters of Galaxies

edited by

A. C. Fabian

Institute of Astronomy,
University of Cambridge,
Cambridge, U.K.

Springer Science+Business Media, B.V.

Proceedings of the NATO Advanced Study Institute on
Clusters and Superclusters of Galaxies
Cambridge, U.K.
July 1–10, 1991

ISBN 978-0-7923-1702-9   ISBN 978-94-011-2482-9 (eBook)
DOI 10.1007/978-94-011-2482-9

Institute of Astronomy
NATO A.S.I. 1991

Clusters and Super-clusters of Galaxies

M. Fukugita M. Hudson P. Anwum P. Stein D. Hngsbry A.C. Baker D.L. Clemens 2.T. Corrigan K. Amum R.H. den Hartog M.C.T. Kruyswyk M.P. van Haarlem J. Butcher M.G. Yates N. Menci S. Colafrancesco S. Bardelli J. Dalcanton C. Canizares D. Woods C. Frenk S. Raychaudhury P.B. Lilje

H.C. Ferguson E. Escalera G. Giuricin H. Bohringer A.M. Dunn E. Schulman C. Cox S. Baum L. Feretti F. Pearce R.E. White P.A. Thomas J. Gr. G. Vettolani J. Gregorini R. Scaramella A.I. Zabludoff G. Soucail R.G. Bower R. Guzman-Llorente P. Coles C. Scharf O. Lahav

S. Schindler R. Croddaiee H. Siddiqui M. Girova G.F. Lewis J.W. Allen M.J. Ward G.B. Dalton C.P. O'Dea S. Colombi R.A. Schwarz B. Whitmore C.S. Crawford C.L. Sarazin R.F. Mushotzky F.N. Owen F.J. Carrera U.G. Briel J.R. Bond H. Andernach L.C. Jafelice A.C. Edge C. Moss

J. Moss H. Ebeling M. Kowalski W. Saslaw P. Gukerlahartia A. Babul N. Katz D. White S. Davies M. Bremer G. Siewert G. Thomm A. Dekel J.A. Peacock M.S. Vogeley R.G. Mann J.A. Eilek P.C. Tribble J. Loveday R. Johnstone A.C.S. Friaça J.R. Lucy W. Forman

H.K.C. Yee J.G. Bartlett M. Dickinson N.A. Bahcall M. Voit M. Donahue I. Gioia W. Jaffe M.W. Wise J.P. Hughes B.R. McNamara J.N. Bregman S. White R. Kirshner A.C. Fabian M. Haas G. Chincarini B. Tully H. Quintana N. Kaiser M. Colless M. West C. Jones A. Evrard J.P. Henry

TABLE OF CONTENTS

viii

PREFACE

Clusters and superclusters of galaxies are the largest clearly–defined objects in the Universe. They are the peaks and mountain ranges of its mass distribution. Clusters are bound by gravity, most of which is due to dark matter. Clusters also contain substantial amounts of diffuse gas in the form of an intracluster medium, with a mass exceeding that in the stars visible in the member galaxies. It is not known yet whether superclusters are gravitationally bound, but they are certainly very massive and must impede the universal expansion in their neighbourhood.

Mapping the content, masses and distribution of clusters and superclusters is an exciting task that is the subject of this book. It contains the reviews from the NATO Advanced Study Institute on 'Clusters and Superclusters of Galaxies' held in Cambridge, England from July 1st – 10th 1991. The field has been rapidly expanding over the past 5 years or so and the ASI provided an excellent opportunity to take stock. X-ray data are revealing substructure, luminosity evolution and cold gas in clusters, radio data show structured magnetic fields and neutral gas, and provide a new measurement of the Hubble constant. Optical studies demonstrate morphological changes in the member galaxies and reveal that clusters act as giant gravitational lenses, amplifying distant galaxies into luminous arcs. Measurements of the distribution and velocities of galaxies and clusters show significant inhomogeneities on supercluster scales and larger, such as the 'Great Wall', the 'Great Attractor' and the 'Bootes Void'. On the very largest measured scales, however, the most recent data suggest that the Universe is relatively smooth.

Distant clusters are studied both directly and by using radio–loud quasars as flags of the central cluster galaxies. They offer much for tackling the challenging problem of the evolution of clusters and clustering, and how the observed large–scale structure originated. This takes us back to the intial spectrum of fluctuations in the Universe, their development and interaction with time.

The topics at the ASI were chosen to emphasise the new developments and to minimize direct overlap with the many conferences on large–scale structure itself. The authors were asked to be as pedagogical as possible; to include the most recent results, but also to set them in perspective. I am pleased that most have complied with this request and hope that the book, right from Martin Rees' excellent introductory chapter to beyond Pat Henry's admonition to 'remember this stuff', is recommendable both to first year graduate students and to experts.

I am very grateful to the NATO Science Committee and Dr L. Da Cunha for funding this Advanced Study Institute, to Professors Martin Rees and Donald Lynden-Bell for permitting it to be held at the Institute of Astronomy, to the Scientific Organizing Committee (H. Böhringer, A. Cavaliere, G.P. Efstathiou, R.S. Ellis, C. Sarazin), to the scientific members of the Local Organizing Committee (A. Babul, M.M. Colless, A.C. Edge, R.M. Johnstone and S. Raychaudhury), who collected, edited and circulated to the participants the contributed talks and poster papers, and in particular to the other members, Michael Ingham and Judith Moss, for much organizational help.

A.C. Fabian
Institute of Astronomy
Madingley Road
Cambridge CB3 0HA
U.K.

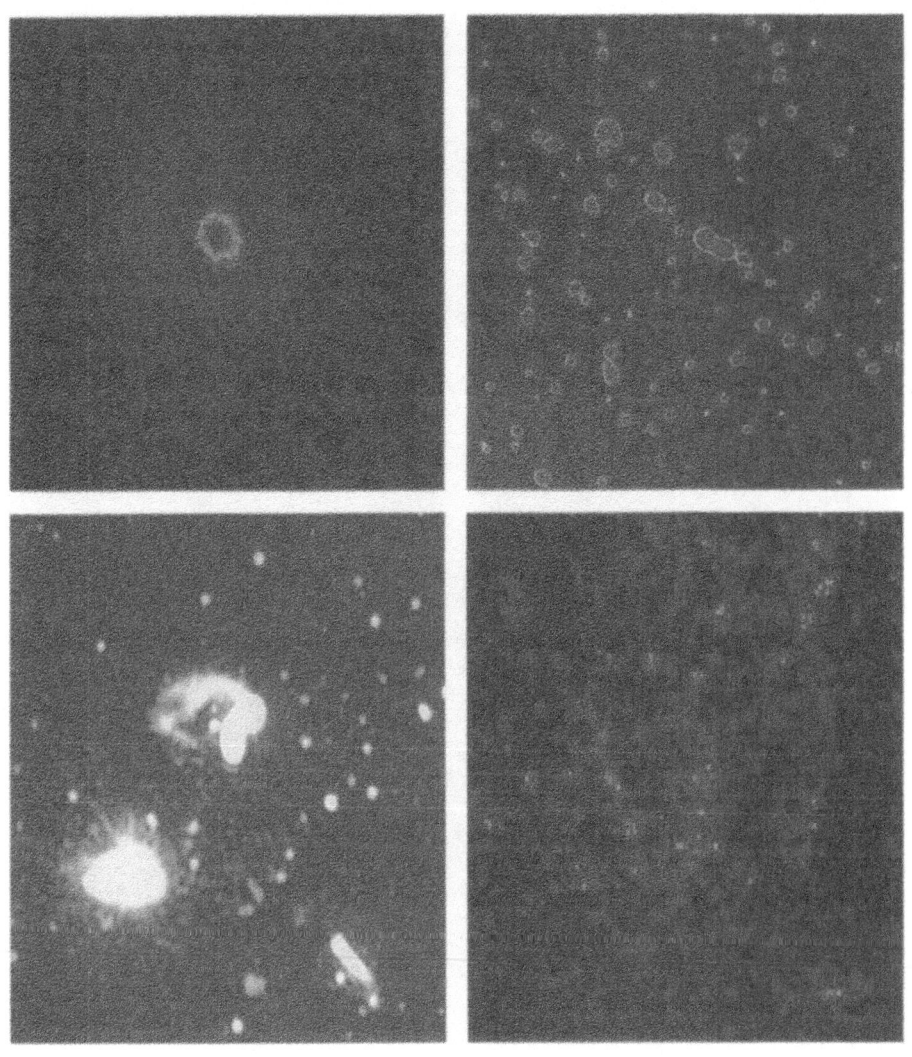

Top left: ROSAT PSPC image of A2199 (G Stewart, Leicester)

Top right: Optical image of Cl0016+16 (I Smail, Durham)

Lower left: Radio image of A426 (G de Bruyn, Dwingeloo)

Bottom right: Optical isopleths of the Shapley Supercluster (S Raychaudhury, Cambridge)

CLUSTERS OF GALAXIES: AN INTRODUCTORY SURVEY

Martin J. Rees
Institute of Astronomy
Madingley Road
Cambridge CB3 0HA
U.K.

1. Introduction

In this introductory talk, I shall try to highlight a number of issues which will be addressed during the conference.To reduce the risk of trespassing on the territory of later speakers, I shall limit myself to general and tentative remarks. I shall try to focus on what we can learn about the *origin* of clusters and superclusters, emphasising that almost all the observations to be discussed at this meeting are relevant to this issue. There are many classes of astronomical objects whose properties can be studied and understood in detail even in ignorance of their origins. To take an extreme example, our understanding of solar physics, and the general structure of the Sun at the present day, is in no way impeded by uncertainties about how it formed, or about star formation in general. This is, at least in part, because the characteristic timescales in the Sun are very short compared to its age: the Sun's dynamical timescale is less than an hour, so the period of solar oscillations is shorter by a factor of around 10^{14} than the Sun's total age. All dynamical memory of how the Sun formed was therefore erased long ago, and is irrelevant to our attempts to understand its present structure. In contrast, the dynamical timescales for clusters and superclusters are not much shorter than the age of the universe. These systems therefore retain an imprint of how they formed. Moreover, they are so bright that we can observe them out to high redshifts, and therefore at earlier cosmic epochs when they were less evolved. Clusters and superclusters tell us about structures in the early universe, and the subject of our conference impinges directly on cosmology.

Our knowledge of clusters and superclusters has expanded greatly within the last decade. But it is salutary to recall that the subject has a rather longer history. Indeed, the classic work of Shapley and Ames in the 1930s had already delineated many of the features of the largescale galactic distribution which exercise us so much today. From their studies of the distribution of galaxies brighter than 13th magnitude, Shapley and Ames were already able, in 1938, to delineate the Virgo cluster, several concentrations of clusters at greater distances, and draw attention to the asymmetry between the northern and southern galactic hemispheres. Further

A. C. Fabian (ed.), *Clusters and Superclusters of Galaxies*, 1–15.

landmarks in the subject are associated with the names of Abell, de Vaucouleurs, Oort, Einasto, and many others. It is on these pioneering foundations that later work, including contributions to be reported at this meeting, has been built.

2. Clusters versus Superclusters

It would be a mistake to get bogged down in definitions right at the start, but one can perhaps make a useful distinction between a cluster and a supercluster on the basis of simple dynamics. Figure 1 depicts how the radii of overdense spheres behave during the expansion of the universe, assuming that pressure effects can be neglected. If the initial overdensity is large, expansion is halted at an early stage, and there is ample time for the system to recollapse and establish a virial equilibrium. If the initial amplitude is rather smaller, the sphere may by now have stopped expanding, and commenced its infall, without yet having virialised. And a sphere with sufficiently small initial perturbation will still be expanding, though it would have suffered an excess deceleration, and its constituent particles would therefore not be moving exactly with the mean Hubble flow. In the simple case of spherical perturbations in an Einstein-de Sitter universe, any system which is already virialised must have a present density more than 200 times the mean. A system which has halted its collapse and is now displaying infall must have more than 5 times the mean density.

The simple dynamics depicted in the figure is relevant in two different, but related, contexts. If we imagine a single sphere, which condenses around a central high density peak, the dashed lines can represent different shells: the inner ones feel a large fractional overdensity, and collapse early; the outer ones feel only a small perturbation, and therefore are merely slightly decelerated. If the early universe contained a spectrum of initial fluctuations, such that the amplitude fell off towards larger scales, then we can also use the same figure to infer that smaller mass systems will tend to have already virialised, whereas larger scales, which initially had much smaller amplitude, would be dynamically younger, and would not yet have achieved dynamical equilibrium. I shall return later to discuss this process in the context of specific cosmogonic models which postulate random fluctuations with a specific spectrum, and can be simulated by N-body calculations.

The general features of three contrasting cosmogonies are summarised in Figure 2 and its caption. The first bound systems to condense are those for which the fractional density perturbation is largest at recombination. In some models, for instance the neutrino-dominated adiabatic scheme, the first bound systems would have a cluster (or even supercluster) mass, and galaxies would form in a 'top down' way. In other models the build up of structure is hierarchical, with small systems condensing first. I shall come back later to discuss the popular Cold Dark Matter model, which is depicted by the central panel of Figure 2. This is a hierarchical scheme, with the feature that the spectrum at low masses is very flat, implying that structure builds up quickly and recently.

It is perhaps helpful to define a *cluster* as a gravitationally-bound system which, at least at its centre, has achieved virial equilibrium; and a *supercluster* as a larger system which, despite having virialised substructure, is overall in a dynamically

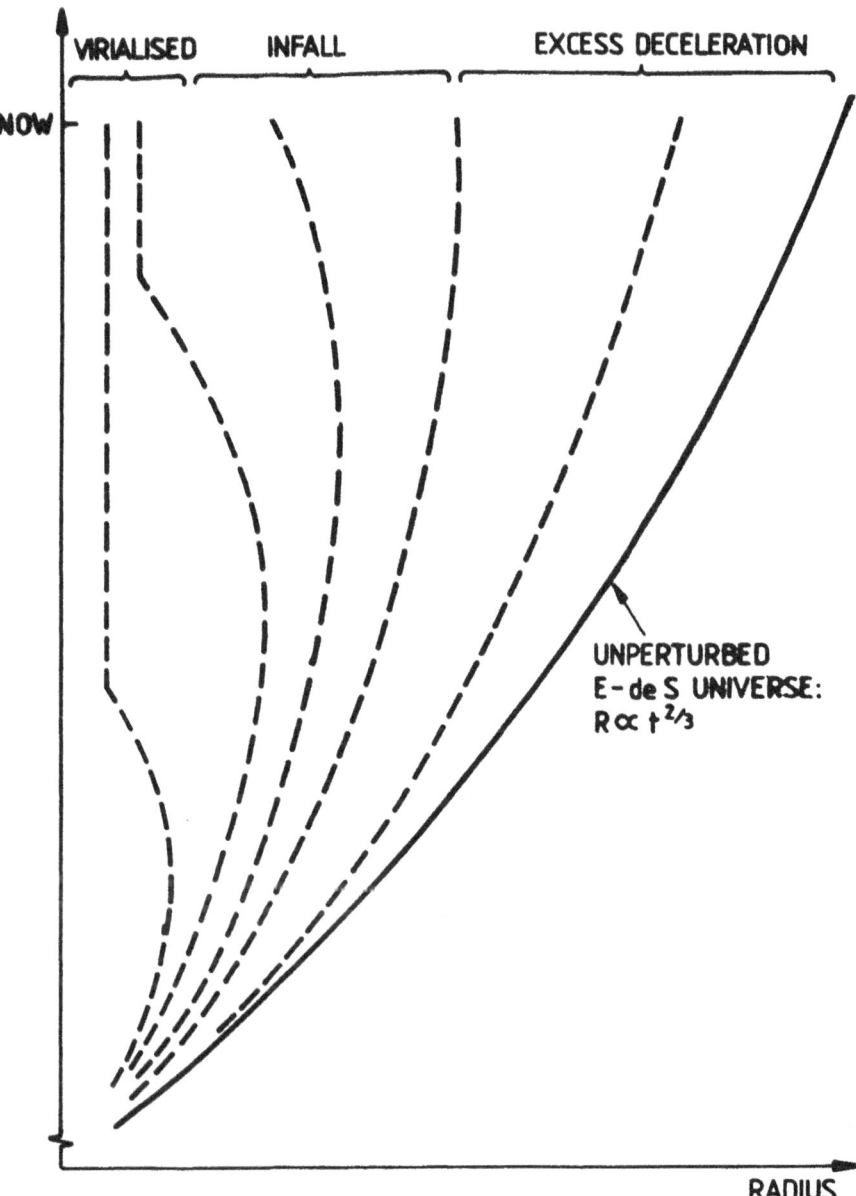

Fig. 1. The dynamics of overdense spheres with $p = 0$ in the expanding universe. The larger the initial overdensity, the earlier the sphere's expansion halts. Systems which are already virialised must have $\rho/\rho_{\rm crit} \gtrsim 200$; those which are now infalling must have $\rho/\rho_{\rm crit} \gtrsim 5$. Note that the overdensity factor of bound or collapsing systems must be even larger in a universe with $\Omega < 1$ and mean density below $\rho_{\rm crit}$.

4

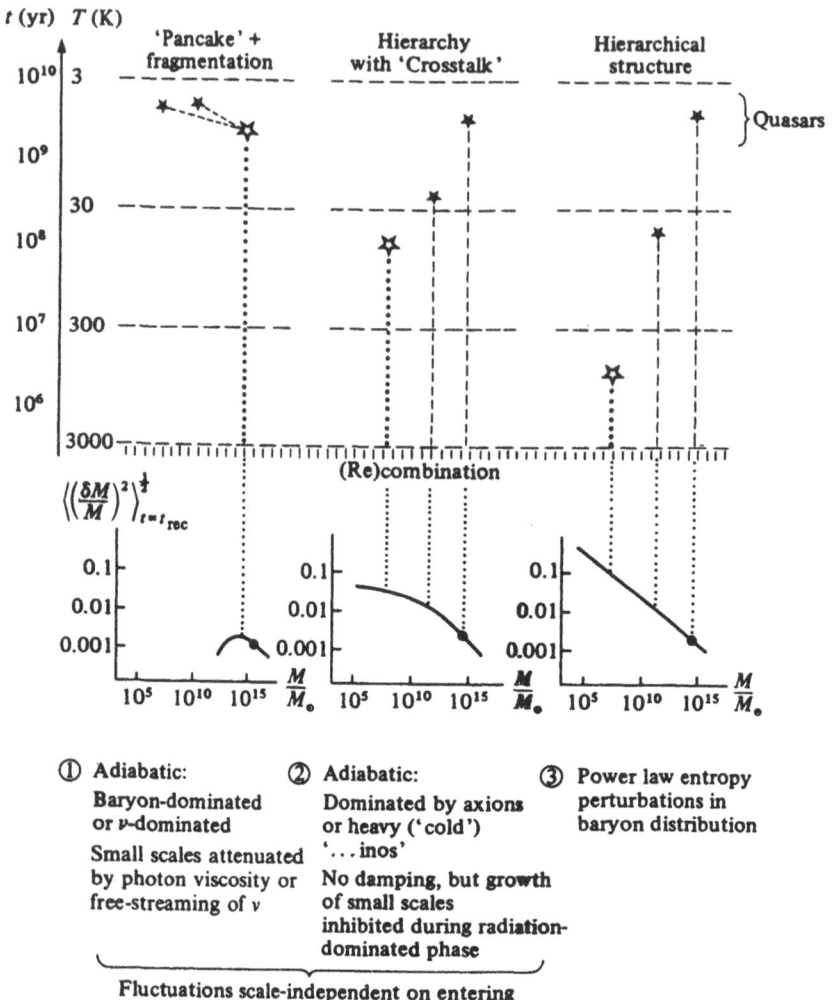

Fig. 2. The cosmogonic processes after (re)combination depend on the spectrum of fluctuations which have survived damping processes, etc., at earlier times. After t_{rec}, linear growth proceeds roughly according to the law $(\delta\rho/\rho) \propto t^{2/3}$, and the first gravitationally bound systems to form will have the mass for which the density contrasts at t_{rec} are biggest. In case (1), (super)clusters are the first systems to condense, and they form quite recently. Indeed, models of this kind run into difficulties because we observe quasars with redshifts z as large as 5, whereas if superclusters had collapsed as early as this, they would now be denser, and display a higher density contrast than is seen. In case (2), baryonic systems of $\sim 10^6\ M_\odot$ condense in potential wells produced by 'cold dark matter' or 'inos', which are presumed to be slow moving, so that they are not homogenised on small scales as neutrinos are (this is the 'CDM' model). The third case shows the spectrum that might arise from primordial entropy perturbations. In cases (2) and (3), subgalactic systems would form *before* galaxies; if these subgalactic systems provided an energy input, they could in principle generate 'secondary' perturbations on larger scales which could swamp the genuinely primordial ones.

younger state, perhaps even still expanding with the universe, albeit at a decelerated rate. Most galaxies are not in large clusters. The scale on which the correlation function is unity is about 8 megaparsecs, and the typical number of galaxies in a sphere of that radius is only a few. Rich clusters, in any gravitational instability scenario, would have evolved from regions where the initial fluctuation amplitude on the relevant scale was exceptionally large.

3. Virialised Clusters

Let us 'home in' now on rich clusters – not forgetting, however, that these are exceptional regions of space that evolved, probably, from 3σ fluctuations. If they formed in a hierarchical way, their internal substructure would be destroyed during collapse and virialisation of the system.

The central cores of rich clusters, being dynamically relaxed, are less likely to retain a memory of the formation process. However, it is these regions about which most is known observationally. The dominant gravitating mass is the dark matter component. This cannot be directly observed, but the galaxies and hot gas can be used to determine the density profile of the dark matter, even though they may not themselves have the same density profile. The motions of the galaxies in principle should allow us to trace out the gravitational potential well. But in practice this is difficult because we do not know whether galactic orbits are isotropic, or else predominantly radial. These orbits may also be affected by dynamical friction on the dark matter.

The gas distribution relative to the dark matter is primarily a function of the parameter β, equal to the ratio of the gas temperature and the virial temperature. Our best evidence on the hot gas comes from X-ray data. However, problems arise because we don't have high spectral resolution combined with high angular resolution. Moreover, the gas may be inhomogeneous, and partially supported by bulk motions rather than by thermal pressure.

A new method of tracing the overall mass distribution in clusters is now becoming feasible, at least for some systems at high redshifts. This is the detection of gravitational lensing of very faint high redshift galaxies by the mass distribution in a foreground cluster. The discovery of the so-called 'arcs', and the realisation in 1987 that these are gravitationally-magnified and distorted images of distant galaxies, has been followed by further evidence that many of the background galaxies along lines of sight to rich clusters are distorted. Data of this kind in principle allow reconstruction of the projected column density of gravitating mass in the cluster. Although this technique is just beginning to be applied, it has already led to the surprising conclusion that clusters are rather more centrally condensed than was believed to be the case.

The gas is an important (and, indeed, generally the dominant) baryonic constituent of clusters. The gas-to-star ratio is of order unity in poor clusters, but up to 5 in rich clusters. It could be even substantially higher if there were a lot of gas at large radii, or in cool cloudlets. (Gas in either of these forms is hard to detect.) Although the dominant mass is the dark matter, it seems that in rich clusters the baryons contribute at least 10 per cent of the total. This is in itself an interesting cosmological result: standard cosmic nucleosynthesis implies that Ω_b is less than 0.1; so, even if the dark matter were entirely non-baryonic, then if Ω were equal to unity there would need to be segregation of baryons on scales as large as clusters. Whereas dissipative effects can readily account for a concentration of baryons in individual galaxies, it is less clear how they might operate on cluster scales.

X-ray data show clear evidence that the intracluster gas contains heavy elements, especially Fe. However, it is unclear whether the inferred abundance applies to the entire cluster, or only to the central regions which dominate the X-ray emission. The total nucleosynthesis requirements would obviously be less stringent in the latter case.

A further uncertainty concerns the homogeneity of intracluster gas. Different bits of gas may have been shock heated on to different adiabats – indeed this is very likely if the cluster results from successive mergers in a hierarchical cosmogony. The density profile of the gas will then be modified by effects of buoyancy and sedimentation. Inhomogeneities are needed in order to trigger thermal instability, which is strongly indicated to be important from the evidence of cooling flows (which will undoubtedly figure prominently in this conference). The fate of the cooled gas is still a mystery. Can clouds or filaments of very cool gas, with $T < 10^4$ K, survive in the cluster environment? Do magnetic fields or plasma processes inhibit thermal conductivity enough to allow the observed temperature gradients, and the inferred inhomogeneities, to develop and persist? Does the cool gas quickly turn into stars? If so, these stars must be of low mass, and could even be a significant part of the dark matter.

Another probe of hot gas in clusters, which has been recognised in principle for many years but is now becoming observationally useful, is the Sunyaev-Zeldovich effect. This is a measure of the product of the Thomson depth through the cluster and the electron temperature (or, equivalently, of the gas pressure if the size of the cluster is known). This effect reduces the microwave background temperature on the Rayleigh-Jeans part of the spectrum. By combining measurements of the profile of this temperature dip with X-ray maps of the cluster, the Hubble constant can in principle be measured.

4. Formation and Evolution of Clusters

Clusters can be observed at sufficiently large redshifts that evolutionary effects are expected. Indeed, an exciting recent result is the discovery that the X-ray emission from clusters is increasing with cosmic time, even in a sample restricted to redshifts less than about 0.5. This discovery is qualitatively consistent with the expectations of hierarchical models, according to which big clusters would have formed only recently as the outcome of mergers of smaller ones. And it has stimulated further discussion of cluster formation, particularly in the context of specific models.

The most popular and best-defined cosmogonic model is the standard Cold Dark Matter scheme. On the assumption that the fluctuations have the same r.m.s. amplitude on all scales when that scale enters the horizon (the 'Harrison-Zeldovich' spectrum), one can then calculate the r.m.s. fluctuations at the recombination epoch as a function of mass scale M. See Figure 3 and its caption for more explanation.

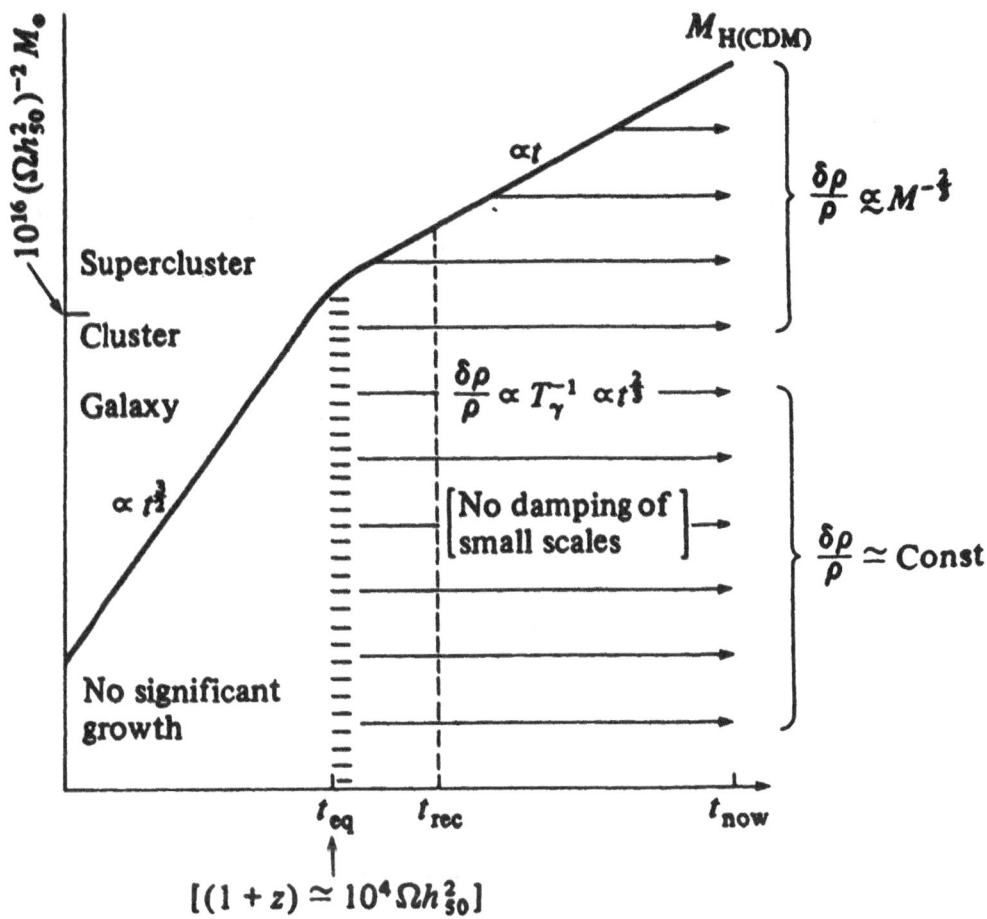

Fig. 3. The growth of adiabatic fluctuations in a universe dominated by 'cold dark matter'. The mass of cold dark matter within the particle horizon is shown as a function of time on a log-log plot. For $t > t_{eq}$ (corresponding to the redshift indicated on the figure), all scales grow at the same rate. Before t_{eq}, when the expansion is radiation-dominated, there is essentially no growth, because the growth timescale much exceeds the expansion timescale. If the fluctuations on all scales enter the horizon with equal amplitude (the 'Harrison-Zeldovich' hypothesis), then the present-day spectrum would have the approximate form written on the right-hand side. The cold dark matter perturbations start to grow at t_{eq}, whereas radiation pressure inhibits growth of baryonic fluctuations on relevant scales until the (later) recombination time t_{rec}. Cold dark matter fluctuations thus have a 'head start' (baryons being able to fall into the resultant potential well after t_{rec}); this permits an acceptable cosmogonic scheme with lower fluctuation amplitude ϵ, and smaller microwave background fluctuations, than in a baryon-dominated universe (cf. Fig 5).

On the further assumption that the amplitude distribution is gaussian, one can set up initial conditions for an N-body simulation which can compute the distribution of the dark matter as a function of cosmic epoch. Work of this kind leads to good quantitative agreement with the properties of individual galactic halos and small groups. Moreover, rich clusters can be interpreted as the outcome of exceptionally high amplitude peaks in the initial noise distribution on mass scales of order $10^{15} M_\odot$.

It is of course the baryons, particularly the hot gas and the galaxies themselves, which are directly observed, so what is really needed is a simulation which takes account of how the gas evolves, and how it condenses into galaxies. This is a much more challenging computational task, where, as we will hear at this meeting, the first steps are now being taken. However, even without detailed calculations, the CDM model can give a convincing answer to a question which arises in any hierarchical model: namely, what determines the demarcation between an individual galaxy and a cluster? Why is something like the Coma cluster not just a single big galaxy?

The answer to this question depends crucially on the cooling time for gas trapped in a potential well. If this is short compared to the dynamical time, no quasi-static equilibrium is possible, and the gas will fragment into stars, perhaps after collapsing to a disc. If the cooling is slower than free fall, but still can occur within the Hubble time, then the gas, though quasi-static, will gradually deflate towards the centre of the potential well. If the cooling is slower still, the shock-heated gas in effect remains at the virial temperature. These three regimes are depicted in Figure 4, in a plot of mass versus radius. From this figure, we can see why there is an upper limit to the mass of galaxies. In a hierarchical build up of structure, as in the CDM model, the bound systems at early epochs are dense, and the potential wells are shallow enough that the gas does not get heated to a very high temperature. In these situations, cooling is rapid and effective. But beyond a certain stage, when the potential wells are deeper and the density lower, cooling is less effective. The critical mass, of order $10^{12} M_\odot$, therefore sets an upper limit to the size of a galaxy. If larger aggregates build up by gravitational clustering, the gas within them will remain hot. This is essentially the distinction between a large galaxy and a system such as Coma.

5. Inhomogeneities on Supercluster (and Larger) Scales

On the larger scales of superclusters, we cannot assume that the systems are virialised, but the gravitational effect of the excess density will cause velocity perturbations which are in principle measurable. For a spherical perturbation where the overdensity is not large enough to cause turnaround, the velocity perturbation is related to the overdensity, and to Ω, by the following equation.

$$\frac{\Delta v}{v_{\text{Hubble}}} = f(\Omega) \frac{\Delta \rho}{\rho}, \tag{1}$$

where $f(\Omega) \underset{\sim}{\propto} \Omega^{0.6}$. The overdensity factor $\Delta \rho / \rho$ may not be equal to the overdensity in luminous galaxies. Indeed it is customary to regard these quantities as related by a bias factor b. Our local peculiar velocity, measurable by our motion through

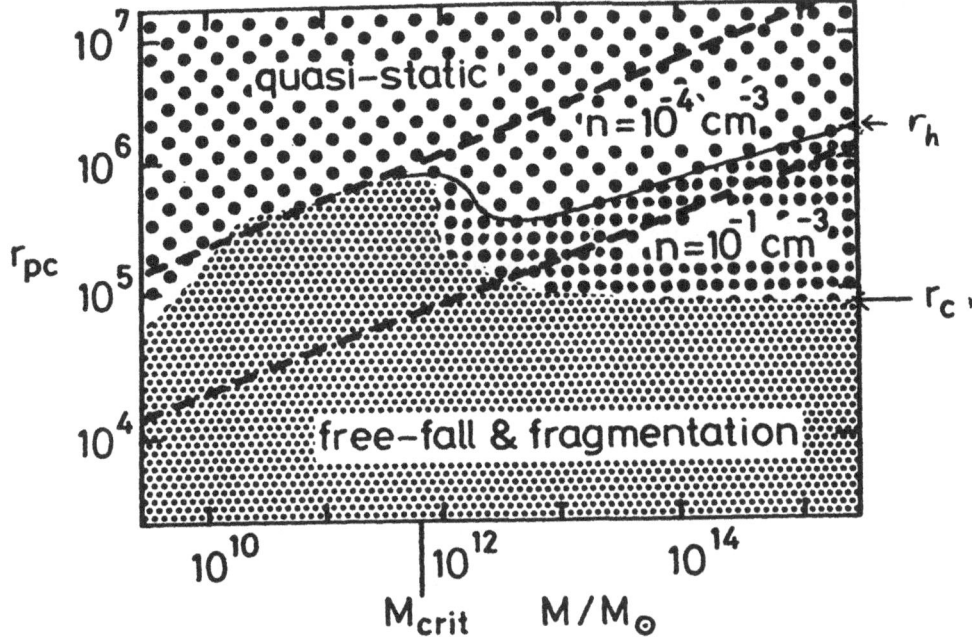

Fig. 4. This diagram delineates the mass-radius relation for a self-gravitating gas cloud whose cooling and dynamical timescales are equal (assuming cooling due only to bremsstrahlung, H and He recombination, and line emission). A cloud of given mass whose radius was initially very large would deflate quasistatically (because $t_{cool} > t_{dyn}$) until it crossed the critical line r_c; it would then collapse in free fall and could fragment into stars. This simple argument (which can readily be modified to allow for non-spherical geometry, a non-baryonic component of mass, etc.) suggests why, irrespective of the cosmological details, no galaxies form with baryonic masses $> 10^{12} M_\odot$ and radii $> 10^5$ pc. There is an intermediate regime, important for clusters and cooling flows, where $r_c < r < r_h$, and the clouds are quasistatic but can deflate in less than the Hubble time.

the microwave background, is generally thought to have been induced by inhomogeneities in the distribution of galaxies around us. Clusters pull, and voids in effect 'push', our Local Group.

We can readily infer, since gravity and light both obey inverse square laws, that the contributions to our peculiar velocity made by any particular galaxy or group is proportional to the light which we receive from it, provided that b is indeed a constant. We note also that the gravitational potential fluctuation ($\propto \Delta M/R$) required to cause a given dv ($\propto \Delta M/R^2$) is larger if the scale is larger.

The above equation can be applied to our own local motion just from two-dimensional data, without knowledge of the redshifts of galaxies. The next step, however, is a much more ambitious one, namely to apply the equation to infer the mass distribution throughout our local region of space, given adequate knowledge of the peculiar velocities of all the galaxies. The peculiar velocities can, of course, only be determined if one has a distance indicator which is sufficiently precise to distinguish between the actual distance of a galaxy and the distance it would have if it was obeying the Hubble flow. The main practical uncertainties in the measurement and interpretation of largescale streaming stem from the problems of calibrating the distance indicators.

In principle, however, knowledge of the largescale velocity field would allow us to infer the mass distribution without any assumption that galaxies trace mass, or that the quantity b is a universal ratio. We need only assume that galaxies are test particles whose motions are gravitationally induced. The main problems are still with the distance measurements, and a lingering uncertainty about whether the luminosity calibrators (velocity dispersion, diameter, etc.) are really independent of location and environment.

The expected deviations from Hubble flow are only a few hundred kilometres per second at most. There is very little hope of ever measuring these reliably by existing methods at distances beyond $\sim$ 100 megaparsecs, because they are then only a very small fraction of the recession velocity, and therefore would require still more accurate distance indicators. There are, however, in principle other quite different methods which could determine peculiar velocities of clusters at still greater distances. For instance, there is an extra contribution to the Sunyaev-Zeldovich effect due to the bulk motion of the cluster. This is equivalent to a simple 'moving mirror' effect, with amplitude equal to the Doppler shift multiplied by the Compton optical depth. (Transverse motions, incidentally, would yield polarisation, but this is a higher order effect and still more challenging for observers.)

What we would really like would be a method of directly delineating the dark matter distribution on large scales. One possibility which has recently been suggested is gravitational distortion of faint background galaxies. A supercluster would not have such a high column density as the core of a rich cluster, and therefore could not give strongly distorted or highly magnified images resembling the 'arcs'. But even if the induced distribution were small the fact that the angular diameter of a supercluster may be of order $1°$ means that up to 10^5 galaxies would share a correlated distortion; so it is by no means hopeless to look for an effect of only a few per cent.

At the moment one cannot conclusively decide between two extreme views regarding the dark matter distribution. On the one hand, it could be that on large scales the overall mass distribution is smooth, and the apparent largescale structure in the galaxy distribution represents simply some largescale spatial modulation of the efficiency of galaxy formation. On the other hand, the density of galaxies could be a reliable tracer of the dark matter distribution on all scales, except for a simple biasing factor b.

We can nevertheless say with great confidence that on scales much larger than 50 megaparsecs the overall mass distribution must be relatively smooth, in the sense that the amplitude of the r.m.s. density fluctuations is $\ll 1$. Our best evidence here comes from the microwave background anisotropies. For scales exceeding the horizon scale at the epoch of recombination, the dominant contribution to $\Delta T/T$ comes from the gravitational potential or metric fluctuations on the last scattering surface. This is the so-called 'Sachs-Wolfe' effect. The upper limits are of order

10^{-5}, and these imply that on the corresponding scale L ($\sim \theta ct_H$) we must have

$$\left(\frac{\Delta\rho}{\rho}\right)_{\text{r.m.s.}} \lesssim 10^{-5}\left(\frac{L}{ct_H}\right)^2 \qquad (2)$$

Even a mass concentration as large as the Great Attractor would, if placed on the last scattering surface, yield a fluctuation little more than 10^{-5}.

The microwave background anisotropies also, as (2) implies, constrain the potential fluctuation on mass scales much larger than 100 megaparsecs. This is of interest because it allows us to test whether the initial fluctuation spectrum does indeed have a power law form. The assumption of a power law is generally made because it seems unnatural to feed the mass scale appropriate to clusters or galaxies into the conditions of the early universe. Figure 5 summarises the various constraints on the fluctuation amplitude on scales stretching all the way from sub-stellar masses up to the Hubble radius and beyond. We see that if the initial fluctuation spectrum is a power law, its slope is well constrained. Moreover, if it is a power law with the scale-independent Harrison-Zeldovich form, the amplitude is well constrained to be a bit less than 10^{-5}. It is perhaps worth noting, however, that the case for favouring a Harrison-Zeldovich spectrum would be very much weaker if the universe were not, as theorists like to believe, flat with exactly the critical density. If Ω differs significantly from unity, there *is* a preferred length scale in the universe, essentially the Robertson-Walker curvature radius. Given that this scale, a substantial fraction of the present Hubble radius, would have to have been imprinted at an early time, it would then be by no means implausible to suppose that other scales, related to it by some modest numerical factor, might also have been singled out.

It is interesting to ask whether a nearby cluster or supercluster would perturb the microwave background. The Sachs-Wolfe effect due to the precursor of a cluster on the last scattering surface causes a temperature perturbation of order $\Delta\Phi/c^2$, where Φ is the Newtonian potential; this is of order $\frac{v^2}{c^2}$ where v is the velocity dispersion in the cluster after virialisation.

But a similar cluster with redshift $z < 1$ produces a much smaller effect on the microwave background. The effect of the cluster's potential well on the microwave photons is, to first order, zero: the photons gain energy falling in but lose it again climbing out the other side. The only non-zero effects are of higher order: they are due to the extra time delay experienced by photons going through the potential well of the cluster, and the deepening of the potential well as the photons go through. For a linear perturbation of radius r, these effects, the first a blueshift and the second a redshift, are both smaller than the Sachs-Wolfe effect by $\frac{r}{ct_H}$. For a nonlinear collapsing perturbation the second effect may exceed the first by the ratio of the collapse rate to the Hubble expansion rate, but it is still generally much smaller than the Sachs-Wolfe effect that would be associated with a similar cluster observed in its embryonic state on the surface of last scattering.

6. Clusters and Protoclusters at High z

It is clear that we shall learn a great deal from more extensive X-ray data on clusters at very high redshifts. Also, quasars surveys have now discovered enough objects to

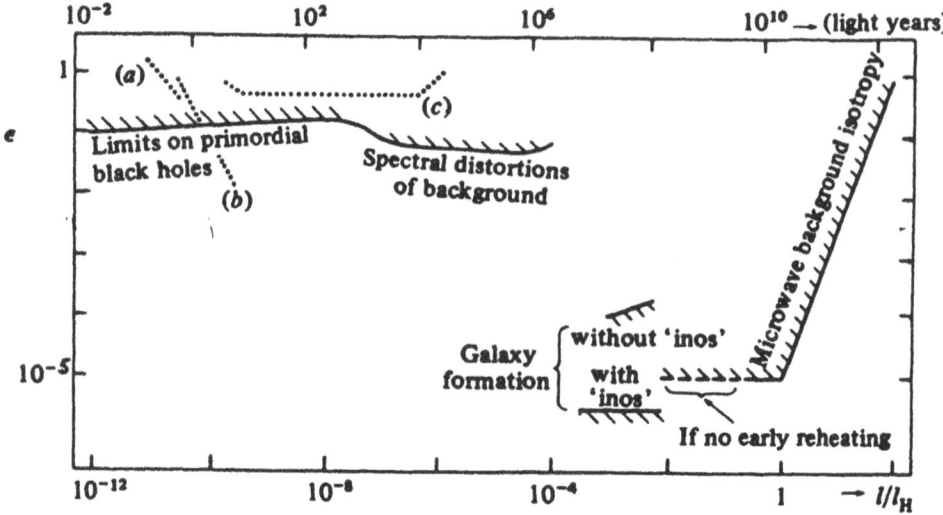

Fig. 5. This diagram depicts the astrophysical limits on the amplitude ϵ of adiabatic metric fluctuations, on various scales l. On large scales, the microwave background isotropy offers stringent upper limits. On very small scales, ϵ must merely be not so large that too much of the universe collapses as the relevant scales enter the horizon; the absence of distortion in the microwave background sets a slightly better limit for mass-scales $10^4 - 10^{13}\ M_\odot$. The requirement that bound systems have 'turned around' by the present epoch gives *lower* limits (also plotted). A spectrum with $\epsilon \simeq 10^{-5}$ on all scales is acceptable if the universe is dominated by non-baryonic matter which can start clustering before t_{rec}. If ϵ has a power-law dependence on l and is $\sim 10^{-5}$ on the scales relevant to galaxy formation, then it cannot fall off more steeply than $\propto l_{-0.15}$ without causing excessive production of primordial mini-holes. Also plotted (dotted lines) are the amplitudes ϵ_g of primordial gravitational waves that can, within the next few years, be probed by: (a) doppler tracking of spacecraft, (b) timing of 'quiet' pulsars, and (c) timing of the orbit of the binary pulsar. This diagram is drawn assuming a 'flat' background universe with $\Omega_0 = 1$. If $\Omega_0 < 1$ the limits on large scales are modified. Indeed the appropriate definition of ϵ is then somewhat ambiguous on scales exceeding the Robertson-Walker curvature radius.

yield quantitative data on how quasars are clustered. There is tantalising evidence that clustering exists, but does not depend steeply on redshift. The rate of evolution of clustering is a diagnostic for the cosmological density parameter, and also for the biasing factor. Clustering evolves more rapidly at later epochs, under the action of gravity, if Ω is high. On the other hand, if the objects being studied, galaxies or quasars, are strongly biased in their distribution relative to the mass, then the apparent clustering will develop less dramatically with redshift.

There is a possibility of using quasars to study incipient clustering right back to redshifts of order 5. I should like, however, briefly to mention a technique that offers the chance of probing largescale structure at even higher redshifts, perhaps even before the first galaxies and quasars 'switched on' and reheated the primordial plasma. This technique depends on studying the 21cm line expected from diffuse neutral hydrogen. In terms of brightness temperature, this line contributes much less than the 2.7° which we get from the microwave background. Its temperature is also much less than the non-thermal radio background due to synchrotron emission from extragalactic sources. It may nonetheless be possible to pick out the 21cm contribution, because of its characteristic angular structure, combined with fine structure in frequency space. See Figure 6 and its caption. The contribution to the radio background temperature at 1420 MHz due to uniformly distributed hydrogen at redshift z is easily calculated to be

$$J_{HI} \simeq 0.1(1+z)^{\frac{1}{2}}\Omega_{HI}f \text{ K.} \tag{3}$$

The factor f is unity if the spin temperature is much higher than the radiation temperature. If there had been no heat input into the primordial gas before the relevant epoch, then the spin temperature would be lower than the radiation temperature and the intergalactic gas would show up in absorption. If a region lies along our line-of-sight which has a higher density than average, or has an expansion rate slower than the mean Hubble flow, the contribution from the 21cm line would be enhanced. For a linear fluctuation, the enhancement is five thirds the amplitude of the density perturbation, the extra two thirds coming from the reduced expansion rate in an overdense growing perturbation, which increases the hydrogen column density per unit redshift interval.

If neutral hydrogen has largescale inhomogeneities, or its velocities are perturbed from the mean Hubble flow, there will therefore be spatial and spectral structure in the radio background we receive from it. Though small compared to the continuum radio background, this may be detectable by difference measurements, switching between nearby frequencies and directions. The expected fine structure in frequency space would allow the signal from non-uniform neutral hydrogen to be distinguished from patchiness in the non-thermal synchrotron background.

The most hopeful possibility for carrying out this kind of 'tomography' on protoclusters involves the proposed GMRT in India. This instrument is planned to comprise an array of 34 dishes, each 45m in diameter. The dishes will not be sufficiently well-surfaced to be effective at high frequencies. However, the array will be eight times more sensitive than the VLA at 327 MHz. It will also operate at lower frequencies in the range of 150 – 250 MHz, where the artificial background, particularly at its proposed location, is particularly low. This corresponds to redshifts between 6 and 8.5. The sensitivity is such that protoclusters would be detected, if they have the properties predicted by some theories. Furthermore, the precursors of specially large systems resembling the Great Attractor or the Shapley Concentration should reveal themselves in this way.

7. Concluding Comments

Let me conclude by looking forward into the distant cosmic future. As the universe continues to expand, the scale of clustering will grow. In a simple model with $\Omega = 1$ and scale independent initial perturbations, the typical mass scale of clustering rises

14

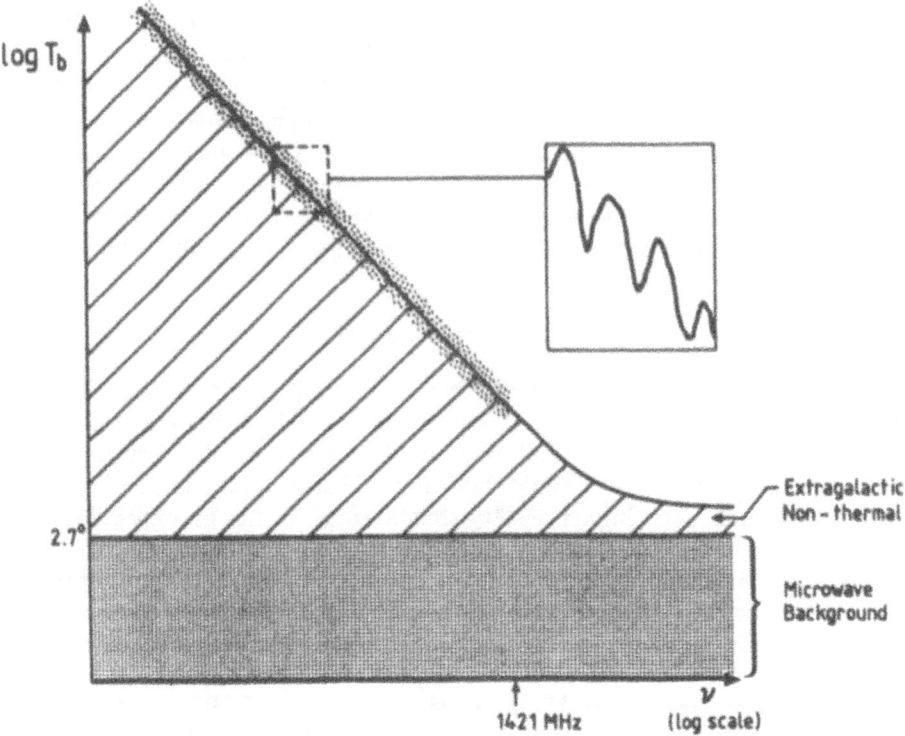

Fig. 6. The dominant extragalactic backgrounds in the radio bands are the primordial 2.7 K black body radiation, and the non-thermal synchrotron background, whose brightness temperature goes as $\sim v^{-2.7}$. Intergalactic HI emits and/or absorbs via the 21 cm transition, and in consequence changes the background temperature. Although this effect would be undetectably small if the HI were smoothly distributed, any 'clumping' of the gas would create spectral and angular structure in the background. By scanning in angle using a narrow bandwidth of frequencies, structures in the high-z neutral hydrogen could be detected. By comparing the angular structures seen in two 'maps' made at slightly different frequencies, one could distinguish between effects due to discrete non-thermal sources (for which the two maps would correlate) and those due to HI (where the maps would not correlate), and thereby detect incipient large scale structure at redshifts $z > 5$.

in direct proportion to the age of the universe. In this particular model, the virial velocities and temperatures of typical clusters are independent of time. The typical number of galaxies in a rich cluster is now about 10^3. At a much later time, this number will, instead, be

$$\sim 10^3 \left(\frac{t}{10^{10} \text{ yrs}} \right). \tag{4}$$

The typical number of galaxies in the mean correlation scale is now only about 10. It is the smallness of this number which makes it hard to study clusters; for instance, we cannot map out their shapes, isodensity contours, etc., with any precision because of discreteness effects. But at later times this number will also have risen in proportion to the size of rich clusters. Moreover, the characteristic cooling time for the gas in a cluster increases as t^3, even if Ω_{gas} does not decrease. Therefore, one will not need to worry about the complications of cooling flows in the remote future.

In a cosmic perspective and time-frame we may be observing clusters at a confusing stage in their development, but if we view astronomical advances on a human timescale we can feel very encouraged. Progress in delineating largescale structures, streaming velocities, and the optical and X-ray properties of individual clusters, not only locally but far enough away to see evolution, is rapid and encouraging. Now is therefore the best of times for a conference like this.

NOTE The topics mentioned in this introductory survey are all treated more fully in later chapters. For this reason, and to avoid duplication, literature references are not given here.

THE STRUCTURE OF GALAXY CLUSTERS

Simon D.M. White
Institute of Astronomy
Madingley Road
Cambridge CB3 0HA
U.K.

ABSTRACT. I discuss both equilibrium and nonequilibrium methods for determining the masses of galaxy clusters, and point out the difficulties encountered when applying these methods to real optical and X-ray data. The best data sets now give quite precise determinations of the mass in the inner regions of relaxed clusters. The observed ratio of baryonic to total mass in these regions is at least three times larger than that predicted by the standard model for cosmological nucleosynthesis in a flat universe. This suggests either (i) that the universe is open, (ii) that the theory of light element production needs modification, or (iii) that some unknown process has led to a substantial concentration of baryonic material in clusters relative to the rest of the matter in the Universe. The degree of substructure in clusters is an indicator of their mode of formation. Significant substructure is evident in many clusters at both optical and X-ray wavelengths. Although its strength and prevalence are hard to quantify, it clearly causes a number of biases both in cluster definition and in the measurement of cluster properties. I illustrate the problems using clusters selected from simulations of a Cold Dark Matter universe according to standard observational criteria. Such a model provides a good qualitative and quantitative description of observed cluster samples.

1. Introduction

The Coma cluster was the object for which Fritz Zwicky (1933) first noticed that dynamical mass determinations required very large amounts of unseen material. Since that time rich galaxy clusters have been studied in great detail and at new wavelengths, but his original conclusion has remained unchanged. Clusters are largest objects which appear to be at least approximately in equilibrium, and, as I discuss below, application of Newton's laws of dynamics and gravitation then require about 80% of their mass to be in unidentified "dark matter" of some kind. Observations of the lensing effect of clusters on background galaxies confirm this conclusion while dispensing with the hypothesis of equilibrium (*e.g.* Soucail, this volume). The improvement of X-ray data on galaxy clusters has not only greatly reduced the uncertainties in such mass determinations, it has also provided a very direct means of measuring the mass of gas in the intergalactic medium. In general this turns out to be comparable to or somewhat larger than the mass of the observed stellar populations in the galaxies. Together, these two components make up 10 – 20% of

17

A. C. Fabian (ed.), Clusters and Superclusters of Galaxies, 17–28.
© 1992 *Kluwer Academic Publishers*.

the total mass in clusters. This in itself is an interesting result, because current calculations of the cosmological synthesis of light elements indicate that only a much smaller fraction of the mass in a flat universe can be in baryonic form.

Rich clusters form in different ways in different theories for the origin and evolution of structure. Thus, in the standard model for a neutrino-dominated universe (e.g. Doroshkevich et al. 1980) clusters are expected to collapse asymmetrically, but nevertheless as coherent units. Again, if structure has been imposed on the universe by "cosmic texture", then clusters form coherently by almost spherical inflow onto a site of texture collapse and annihilation (Turok 1989). On the other hand, hierarchical clustering theories such as the Cold Dark Matter model (CDM: Peebles 1982; Blumenthal et al. 1984; Davis et al. 1985) predict clusters to build up by the merging of preexisting nonlinear subunits. One might hope to discriminate between these models by studying the internal structure of clusters. In particular, it seems natural to expect substantially more internal irregularity in clusters in the latter model than in the first two. However, it is difficult to be quantitative about this statement without detailed modelling of the formation both of the galaxies and of the clusters. Considerable progress has been made for the CDM model, finding fair agreement with a wide range of observations (see Frenk 1991 for a review), but a similar amount of effort has yet to be invested in the other models. In the case of the neutrino-dominated (Hot Dark Matter) model this is in part because it appears to be in serious conflict with observation (White et al. 1983, 1984; Zeng and White 1991; but cf Centrella et al. 1988). The observational evidence for substructure in clusters is also controversial. This is a consequence of the difficulty in assigning significance to an apparent subgrouping in the distribution of a few hundred galaxies. This is another area where X-ray images have had a substantial impact, and the large field of view and high S/N attainable with ROSAT will undoubtedly bring major improvements over the next two or three years.

In the next section I discuss mass determinations from optical data, both for the inner equilibrium regions of clusters, and from the infall pattern at large radii. Section 3 then notes how X-ray data can provide considerably more precise results, and shows the extent of agreement with optical measurements for the Coma cluster, perhaps the best observed cluster in the sky. The Coma data provide a good opportunity to assess the extent of the conflict between the baryon fraction in clusters and that expected in a flat universe. I go through the numbers in Section 4. Finally, Section 5 discusses the extent of substructure in observed clusters, and its effect on the measurement of cluster velocity dispersions, masses, and abundances. The effects to be expected are illustrated using artificial clusters obtained by applying standard observational selection criteria to "galaxy" catalogues generated from N-body simulations of a CDM universe.

2. Optical mass determinations

For an idealised spherical galaxy cluster in equilibrium, the equation of hydrostatic equilibrium for the galaxy population takes the form,

$$\frac{1}{\rho_{gal}}\frac{d(\rho_{gal}\sigma_r^2)}{dr} + 2\frac{\sigma_r^2 - \sigma_t^2}{r} = -\frac{GM(r)}{r^2}, \tag{1}$$

where $\rho_{gal}(r)$ is the number density of galaxies, $\sigma_r(r)$ and $\sigma_t(r)$ are the galaxy velocity dispersions at each radius in the radial and tangential directions, and $M(r)$ is the *total* mass enclosed within radius r. Observations of galaxies allow us to estimate the projected galaxy density, which can, in principle, be inverted to give $\rho_{gal}(r)$, and a single line-of-sight velocity dispersion. The two functions σ_r and σ_t cannot both, therefore, be determined from the observations, and it is impossible to derive $M(r)$ without further assumptions even in this idealised situation. Assumptions which have often been adopted in the past to produce a closed system include an isotropic velocity distribution ($\sigma_r = \sigma_t$ at all radii) and light tracing mass ($\rho \propto \rho_{gal}$). Note that these two assumptions are not equivalent, and neither has an *a priori* physical justification. The and White (1986) and Merritt (1987) use data for the Coma cluster to study the uncertainties introduced by this indeterminacy if no additional assumptions are introduced. They found a wide variety of behaviour to be consistent with the observational data. Models with near-isotropic velocity distributions require mass density profiles similar to the observed galaxy density profile. Models with predominantly radial orbits require a less centrally concentrated mass distribution, while near-circular orbits require a more concentrated distribution.

Merritt (1987) points out that the data, in principle, contain more information that can be brought to bear on this problem. He shows that models with different orbital eccentricities predict different shapes for the *distribution* of line-of-sight velocity at intermediate radii. Circular orbits produce a flat-topped distribution, while radial orbits produce distributions with extended wings. Unfortunately, real clusters rarely have enough measured redshifts to apply this test, and in the best case, Coma, the velocity histogram is strongly skewed, suggesting that the cluster cannot be considered as a spherical equilibrium system. This difficulty is clearly linked closely to the question of substructure discussed in Section 5.

Although many very different mass profiles are consistent with the optical data in Coma, the allowed $M(r)$ all cross each other at $r = 0.5\text{--}1.0h^{-1}\text{Mpc}$ (h is Hubble's constant in units of 100 km/s/Mpc), roughly in the middle of the logarithmic range covered by data. Thus the total mass within $1.0h^{-1}\text{Mpc}$ is quite well constrained and lies within a factor of about 1.5 of $6 \times 10^{14}h^{-1}\text{M}_\odot$. The masses contained both by smaller radii (*i.e.* the central or core mass density) and by larger radii (*i.e.* the total virialised mass of the cluster) are much more uncertain. It is worth noting that the period of a circular orbit becomes equal to the age of the Universe at about $1.7h^{-1}\text{Mpc}$, so that this marks the approximate boundary of the region of the cluster which can be considered in equilibrium. Indeed, fitting the detailed infall solution of Bertschinger (1985) to the cluster implies that the standoff accretion shock should be at about $2.3h^{-1}\text{Mpc}$. In a spherical model, material at larger radii is cold and on purely radial Keplerian orbits.

This simple infall model provides an alternative means of determining cluster masses at large radii. The mass interior to a shell is related to its radius and infall velocity through the requirement that a radial orbit about mass, $M(r)$, should leave $r = 0$ at time $t = 0$ and be at distance, r, with velocity, v_r, at time $t = t_0$, the present age of the Universe. Between the shock radius and the turnround radius (where $v_r = 0$, $r_{ta} \sim 7h^{-1}\text{Mpc}$ for Coma) the infall velocity, v_r, is negative.

Along any line-of sight passing through this region the redshift-distance relation has a very characteristic pattern. Well in the foreground of the cluster, redshifts are proportional to distance from the observer and are less than that of the cluster. When the line-of-sight crosses the turnround sphere the redshift becomes equal to that of the cluster, and from there until the distance of the cluster the observed redshift exceeds the cluster mean because of the infall. Beyond the cluster distance the pattern is reversed. The observed redshift is first below the cluster mean because of infall, it is equal to it again on the turnround sphere, and at large radii it tends again to the unperturbed Hubble flow. As demonstrated by Kaiser (1987), the boundaries of the region where three different distances correspond to a given redshift are expected to show up as caustics in the galaxy density in a plot of redshift against projected distance from cluster centre. Notice that the position of these caustics depends neither on the radial density distribution of galaxies (which merely determines how they are populated) nor on the mean density of the Universe, Ω_0, (since the motion of each shell is determined by the mass interior to it alone).

Caustic patterns in real cluster data have been studied by Regös and Geller (1989). They are quite difficult to see, and their detection is not completely convincing. This is partly due to the fact that at most a few hundred galaxies have redshifts in the best cluster fields, and the statistics are therefore rather poor. A more important problem, however, is the existence of substantial subclustering in the outer regions of clusters which breaks the spherical symmetry and blurs out the caustics (see Section 5 below). The test would probably work best if applied to a statistical ensemble of clusters suitably scaled by a measure of their central velocity dispersion (which should be proportional to turnround radius). For the objects they study, Regös and Geller find a mean mass-to-light ratio somewhat larger than those obtained for the virialised regions of clusters. Under the additional assumption that the galaxy-to-mass ratio within these infall regions is typical of the Universe as a whole, they conclude that $\Omega_0 \lesssim 0.5$.

3. X-ray mass determinations

The hot intergalactic medium in rich clusters provides and alternative means of measuring their mass. For a locally homogeneous gas in hydrostatic equilibrium within a spherical potential well the equation of hydrostatic equilibrium can be written in the form,

$$M(r) = \frac{kT(r)r}{G\mu}\left[-\frac{d\ln\rho_g}{d\ln r} - \frac{d\ln T}{d\ln r}\right], \tag{2}$$

where $\rho_g(r)$ and $T(r)$ are the density and temperature of the gas, μ is its mean molecular weight, and k is Boltzmann's constant. As was the case for the galaxy distribution (equation 1) this is a purely local equation for the total mass within r. However, whereas the observable data are insufficient in practice to determine all the variables in equation (1), and so measure a mass, the X-ray emission from intracluster gas allows both its density and temperature to be determined. Furthermore, use of sufficient integration time, or of a large enough telescope, permits the measurement of these properties with arbitrarily high S/N. This contrasts with the optical case where one is limited by the finite and relatively small number of galaxies. The X-ray surface brightness of clusters typically drops below the extragalactic background at $0.5 - 1.0h^{-1}$Mpc, so measurement of the surface brightness and particularly of the temperature becomes difficult at much larger radii. Temperature measurements so far have mostly come from observations with a wide field of view. These primarily give information about the inner regions since rather little of the integrated flux of a cluster comes from $r > 1.0h^{-1}$Mpc.

There has been considerable disagreement between the masses derived by different groups using these methods. In general this is a result of extrapolation to regions well beyond those for which the data are reliable in an attempt to approximate *total* cluster masses by values within some large radius. This is a mistake. Clusters do not have well defined total masses, and the large uncertainty introduced by the extrapolation masks the inherent precision of the observations. The difficulty is that equation (2) requires the temperature at the radius within which the mass is measured. For large r this is very poorly constrained by the data since almost all the flux in the best spectral measurements comes from much smaller radii. It cannot be overemphasised that an X-ray mass measurement within, say, $r = 2.0h^{-1}$Mpc is only as good as the *measurement* of the gas temperature at that radius. (The polytropic equations of state often used to infer the temperature at large radius have no *a priori* physical justification and are very likely to be misleading.) In all clusters observed to date the uncertainty in this quantity is probably at least a factor of two. One is on much safer ground quoting cluster masses within some smaller radius (0.5 or at most $1.0h^{-1}$Mpc) where the temperature is much better constrained.

The Coma cluster provides a good illustration of these difficulties. In a provocative paper Cowie *et al.* (1987) argued that the integrated X-ray spectrum of the cluster requires multiple temperature components. In their preferred model the hottest gas is at cluster centre and the temperature drops rapidly outside the cluster core. The mass then converges rapidly at $r \gtrsim 0.5h^{-1}$Mpc leading to total values considerably below those normally quoted. The and White (1988) countered this by noting that even if the model of Cowie *et al.* is accepted for the inner regions of the cluster, it can be modified beyond $r = 0.5h^{-1}$Mpc in a way which has no detectable effect on the X-ray surface brightness profile or on the integrated spectrum, yet implies masses at large radii in agreement with the conventional values. The point is simply that the mass at large radii depends on the temperature at large radii which is not significantly constrained by the integrated spectrum. Hughes (1989) came back to this problem and carried out a thorough analysis of all the available spectral and imaging data. He concluded that there is no observational support for a strong temperature gradient over the range 0.0–$0.5h^{-1}$Mpc. Indeed his preferred model has a mass density profile which parallels the galaxies and has a normalisation in perfect agreement with that inferred from the optical data if light traces mass. Since this model also requires the galaxy velocity distribution to be approximately isotropic, the simplest assumptions about cluster structure seem in excellent agreement with all the available data. From fitting a wide variety of models Hughes concluded that $M(0.5h^{-1}\text{Mpc}) = 3 \times 10^{14}h^{-1}\text{M}_\odot$ with a total allowed range of $\pm 30\%$.

New data from long pointed observations of Coma with ROSAT suggest that the cluster structure is considerably more complex than these results might lead one to suppose, and that this story may therefore change again. Both the gas emissivity and the gas temperature show substantial large-scale asymmetries with clear evidence of substructure on a variety of scales; the hottest gas does not appear to be at the cluster centre and a considerable drop in temperature is seen at large radii (White *et al.* , in preparation).

4. Cosmic nucleosynthesis and the baryon content of clusters

The observational data imply directly that a substantial fraction of the mass of rich clusters is in baryonic form. This is an interesting observation because current calculations of the formation of the light elements suggest that only a much smaller fraction of the cosmic closure density can be baryonic. Again it is useful to use the Coma cluster in order to quantify the discrepancy. However, data on other clusters suggest that Coma is in no way exceptional.

The optical and X-ray analyses discussed in the last two sections agree well and give a mass within $1.0h^{-1}$Mpc near

$$M(1.0h^{-1}\text{Mpc}) = 6 \times 10^{14}h^{-1}\text{M}_\odot, \tag{3}$$

with a total allowed range of $\pm 30\%$ for the various models tested so far. This radius is the one at which the optical data give the best constraint, and is an extrapolation of perhaps a factor of two beyond the corresponding optimal radius for the X-ray data. The latter also allow a direct measurement of the gas density as a function of radius, and so of the total gas mass in this same region:

$$M_{gas}(1.0h^{-1}\text{Mpc}) = 2.7 \times 10^{13}h^{-2.5}\text{M}_\odot. \tag{4}$$

The range of uncertainty in this case is about $\pm 25\%$. Notice that the Hubble constant dependence of this expression differs from that in equation (3) because this mass estimate is not derived from dynamical considerations. The and White (1988) use photoelectric measurements of the *total* light from the central regions of the cluster (which is, in fact, mostly due to the galaxies) to derive a direct estimate of $1.76 \times 10^{12}h^{-2}\text{L}_\odot$ for the visual luminosity within this same region. Multiplying this by a mean visual M/L ratio of $15h$, taken from the elliptical galaxies in Table 4.2 of Binney and Tremaine (1987), gives an estimate for the mass of stars:

$$M_*(1.0h^{-1}\text{Mpc}) = 2.6 \times 10^{13}h^{-1}\text{M}_\odot. \tag{5}$$

The uncertainty in this number is formally small, but note that it implicitly assumes that the stellar populations are everywhere identical to those in the cores of bright elliptical galaxies. Comparing the sum of (4) and (5) to (3) now gives a lower limit to the fraction of the Coma cluster (within $1.0h^{-1}$Mpc) which is baryonic,

$$F_b \geq \frac{M_{gas} + M_*}{M} = 0.043 + 0.045h^{-1.5}. \tag{6}$$

This may be compared with limits on the fraction of the closure density which can be baryonic derived from comparison of the observed light element abundances with the predictions of cosmological nucleosynthesis. According to Olive *et al.* (1990) this gives to a constraint,

$$\Omega_b \leq 0.015h^{-2}. \tag{7}$$

For $h = 1$ the discrepancy between these two limits is a factor of almost 6, and even for $h = 0.5$ it is nearly a factor of 3. This seems much too large to attribute to uncertainties in the cluster data.

The simplest way to reconcile these numbers is to assume that the total cosmic matter density is much less than the closure density. One can then take $F_b = \Omega_b/\Omega_0$ so that the mix of material in the cluster is indeed representative; the theoretical prejudice in favour of a flat universe could be saved, if necessary, by the rather inelegant introduction of a cosmological constant. A second possibility is to argue that the theory of nucleosynthesis should be modified in a way that relaxes the stringency of its constraint on Ω_b. This has become less plausible as improving lattice gauge theory models have shown that the quark-hadron phase transition almost certainly cannot be of the kind needed to generate inhomogeneities at nucleosynthesis (Fukugita and Hogan 1991). A third alternative is that baryons are preferentially concentrated to the inner regions of clusters. Recent data from the ROSAT survey (Briel *et al.* 1991) show that the gas mass in Coma increases approximately in proportion to radius out to at least $2.0h^{-1}\text{Mpc}$. This increase is considerably faster than the increase in the number of galaxies and is probably faster than that of the mass distribution. Thus it seems likely that the discrepancy at the nominal radius of the accretion shock ($r_s \sim 2.3h^{-1}\text{Mpc}$) is at least as large as in the region where we have measured it directly. Since this radius separates material which has already fallen into the cluster from material which has yet to reach it, it is difficult to see where the additional gas needed to increase F_b could have come from. Overall, this discrepancy seems a strong argument in favour of an open universe

5. Substructure in clusters

Most of the analysis of clusters in the previous sections has approximated them as spherically symmetric objects in equilibrium. As I have noted repeatedly, there is plenty of evidence that this assumption is seriously in error for many systems. It is difficult to know how to quantify structural irregularities in clusters, or how to assess the statistical and physical significance of any substructure which is found. A wide range of techniques has been tried, and different authors have often come to opposing conclusions about the same cluster. For example, Geller and Beers (1982) set up a scheme for drawing surface density contours from a set of galaxy positions on the sky, and proposed a technique for measuring the significance of any local peak in the resulting map. They found significant substructure in about 40% of the sample of 65 clusters studied by Dressler (1980). In contrast, West *et al.* (1987) found rather little substructure in this same sample and concluded that many of the "significant" concentrations in the maps of Geller and Beers were in fact due to Poisson noise. This kind of difference is probably best settled by applying the same analysis to real data and to Monte Carlo simulations. West *et al.* carried out a variety of N-body simulations to provide such artificial clusters and concluded that substructure in the inner regions of clusters agrees with Dressler's data and is weak in all theoretical models, but that it becomes strong at radii of order $5h^{-1}$Mpc in hierarchical clustering theories.

X-ray images provide another means to find substructure and have the advantage that S/N can be increased "at will" by increasing the exposure time. Early in the Einstein mission a number of double clusters were found where two comparable mass concentrations are clearly present (Forman et al. 1981). In other cases quite complex X-ray structure turned out to correspond well to structure in the galaxy distribution (Gioia et al. 1982). ROSAT can be expected to increase greatly the sample of clusters for which detailed analysis is possible. The Coma cluster is an interesting example. Although the archetype of a rich, concentrated, and relaxed cluster, there have been persistent claims of substructure, in particular around each of the two supergiant galaxies in the cluster core. By using all the available galaxy photometry in the field, Mellier et al. (1988) were able to draw galaxy surface density contours with considerably better resolution than earlier workers. Their map showed "at least nine significant peaks". They support the reality of these peaks by showing that each is almost coincident with one of the ten brightest galaxies in the cluster, and, furthermore, that nearby objects often have surprisingly small velocities relative to these bright galaxies. Although Einstein data had provided no significant evidence for substructure in Coma, a long pointed observation with ROSAT shows substantial extended emission, as well as emission from the dominant galaxy itself, for all four of the Mellier et al. peaks in the field of view (White et al. in preparation). Thus their conclusion that the cluster is irregular and dynamically young seems to be dramatically confirmed by the X-ray data.

The presence of substructure in clusters can significantly bias the statistical properties of observed samples, because selecting the highest density, richest systems in projection favours cases where neighbouring independent lumps are seen projected on top of each other. It is often very difficult to tell that this has happened, even with a large amount of redshift data, and as a result the high richness and high velocity dispersion tails of the cluster distribution can be artificially extended. Such effects depend sensitively on the interaction between the observational criteria used to define rich clusters and the detailed three-dimensional clustering properties of the galaxy distribution. They can only be assessed quantitatively by careful and realistic simulation both of the clustering and of the observing process. An attempt to carry out such an assessment for clusters in a Cold Dark Matter universe was carried out by Frenk et al. (1990) and I have reported further results from this work in White (1991). We constructed catalogues of "galaxies" from simulations of a flat, CDM-dominated universe, projected them onto the "sky" and then selected clusters using criteria patterned as closely as possible on those used to construct the Abell catalogue (Abell 1958). Cluster membership and a velocity dispersion for the objects seen within one Abell radius were then determined by applying standard observational techniques to the "redshift" histogram. The velocity dispersions obtained in this way were often significantly larger that the velocity dispersion of the largest true clump along the cluster line-of-sight. This is illustrated in Figure 1 for the 39 clusters found in one particular simulation. Although only two of the

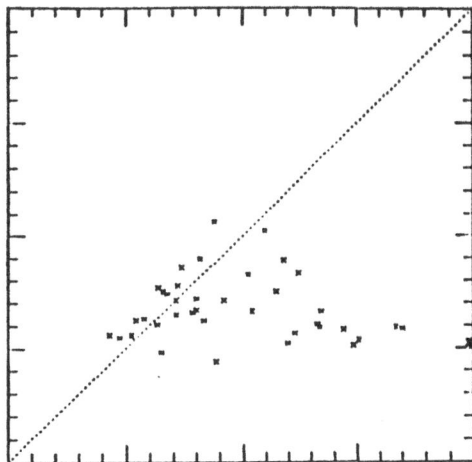

Fig. 1. True versus apparent velocity dispersion for a set of 39 clusters selected according to Abell's criteria from the projection of a CDM simulation. The apparent dispersion (x-axis) is measured using standard observational methods from the redshift histogram of apparent cluster members within $1.5 h^{-1}$ Mpc. The true velocity dispersion (y-axis) is that of the most massive clump in the simulation along the line-of-sight to the cluster. Both axes run from 0 to 2000 km/s with tickmarks each 100 km/s.

lumps in this model had true velocity dispersions (slightly) in excess of 1000 km/s, 17 clusters had apparent dispersions larger than this, including one with a dispersion of almost 2000 km/s. The result of this effect is to give the distribution of cluster velocity dispersions a tail extending out beyond 2000 km/s when in reality the simulations contain no clusters with velocity dispersions much above 1000 km/s. This has a relatively modest effect on the typical masses and M/L ratios estimated for clusters, but it can seriously prejudice the use of cluster abundances to estimate the amplitude of primordial fluctuations. Frenk *et al.* (1990) show that these contamination problems are significantly reduced if clusters are defined and their velocity dispersions measured at higher density contrast (and hence in smaller regions) than in the Abell catalogue.

One might hope to uncover projection problems by looking in more detail for substructure in the joint position-redshift distribution of galaxies. Tests with our simulated clusters suggest that this should often be possible, but that many clusters with multiple components will remain undetected. Figures 2 and 3 show two examples. In the cluster of Figure 2 the redshift histogram shows a bimodality which, although only marginally significant, looks highly suspicious. The cluster membership criteria of Colless (1987; Colless and Hewett 1987) would, however, include all the galaxies in the shaded part of the histogram and give a cluster velocity dispersion of 1258 km/s. The distribution of line-of-sight distances reveals that there are indeed three well seperated components in this cluster. However, the

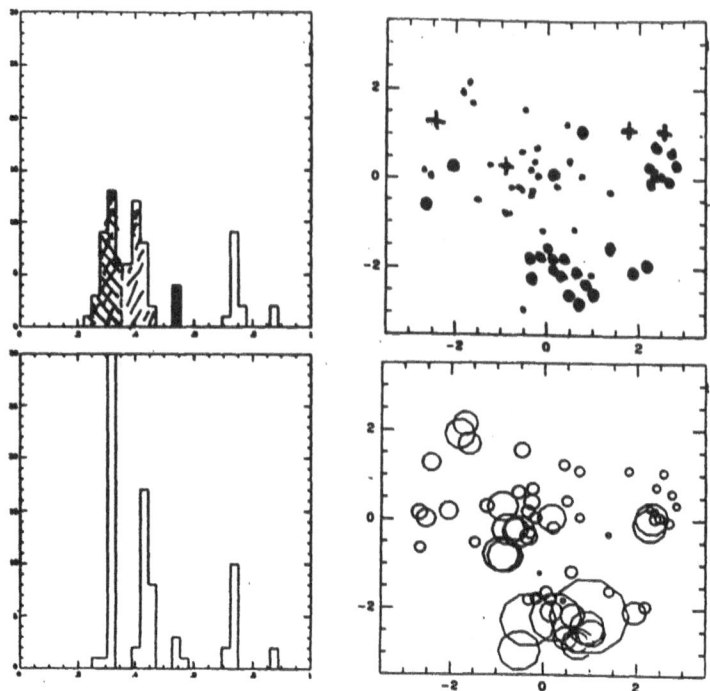

Fig. 2. The top left panel is a redshift histogram for all galaxies within the Abell radius of a cluster found in the projection of a CDM simulation. The box is 18000 km/s wide, and each bin corresponds to 500 km/s. Only the galaxies in the shaded part of the histogram are accepted as cluster members by Colless's (1987) membership criteria. The lower left panel gives the histogram of true distances for these same galaxies. The cluster is a superposition of three well separated clumps. The upper right panel shows the distribution of "cluster members" on the sky; crosses correspond to the filled histogram and heavy dots to the cross-hatched histogram. The lower right panel shows the substructure test of Dressler and Shectman applied to these data. Large circles occur in regions where the local kinematics do not match those of the cluster as a whole.

galaxies on the two sides of the main redshift peak are quite well separated on the sky, and give a strong signal in the test devised by Dressler and Shectman (1988) to search for correlations of the local velocity distribution with position on the sky. Note, by contrast that the four galaxies of the small disjoint group are scattered relatively uniformly across the face of the cluster. The case in Figure 3 is rather different. The velocity histogram is unimodal and only slightly skew, and it gives a "reasonable" velocity dispersion of 865 km/s. Nevertheless, the distance histogram shows that a large disjoint subclump is superposed on the main cluster. There is no clear separation on the sky between the distribution of the subclump galaxies and that of the true cluster members. As a result the Dressler-Shectman test gives no significant signal. The appearance of three distinct density maxima on the sky is,

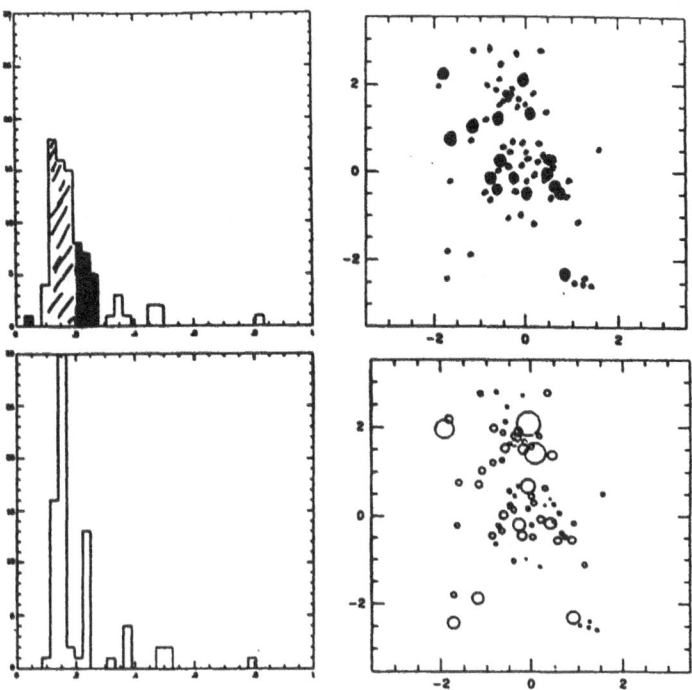

Fig. 3. A second simulated cluster displayed as in Figure 2. In this case the heavy dots correspond to the filled part of the redshift histogram, and the light dots to the other "cluster members" in the hatched part of the histogram.

in fact, misleading, because all three contain galaxies from each of the two spatially separated subgroupings in roughly equal proportions. In this cluster the inflation of the velocity dispersion by the superposed clump is impossible to detect from the observationally accessible information.

High S/N X-ray data will undoubtedly improve our ability to detect clusters with this kind of substructure. The case of Coma is interesting in this context also. Although substructure is seen in the galaxy distribution on the sky and is confirmed by the X-ray map, the distribution of velocities shows no structure at all, and the cluster gives no signal in the Dressler-Shectman test (see the original paper). Thus substructure, when present, can show up in the sky distribution only (Coma), in the redshift distribution only (small clump in Figure 2), in both (large clumps in Figure 2), or in neither (Figure 3). In general, substructure in the cluster sample studied by Dressler and Shectman and in the newer sample presented by Zabludoff at this meeting seems very similar to that in our CDM model. Frenk *et al.* (1990) found that this same model gives quantitative agreement with observed cluster abundances as a function of velocity dispersion, luminosity, or X-ray temperature. Taken at face value these results seem to give strong support for the idea that clusters formed by hierarchical clustering from CDM-like initial conditions.

28

Acknowledgements

Some of the original research reported in this paper was supported by grants from the NSF and NASA to the University of Arizona.

References

Abell, G., 1958. ApJS, 3, 211.
Bertschinger, E., 1985. ApJS, 58,39.
Binney, J., & Tremaine, S.D., 1987. Galactic Dynamics, Princeton Univ. Press.
Blumenthal, G.R., Faber, S.M., Primack, J.R., & Rees, M.J., 1984. Nature, 311, 517.
Briel, U.G., Henry, J.P., & Böhringer, H., 1991. MPE preprint.
Centrella, J., Gallagher, J.S., Melott, A.L., & Bushouse, H., 1988. ApJ, 333, 24.
Colless, M.M., 1987. Ph.D. thesis, Cambridge University.
Colless, M.M., & Hewett, P.C., 1987. MNRAS, 224, 453.
Cowie, L.L., Henriksen, M., & Mushotzky, R., 1987. ApJ, 317, 593.
Davis, M., Efstathiou, G., Frenk, C.S., & White, S.D.M., 1985. ApJ, 292, 371.
Doroshkevich, A.G. et al. , 1980. Ann.N.Y.Acad.Sci., 373, 32.
Dressler, A., 1980. ApJ, 236, 351.
Dressler, A., & Shectman, S., 1988. AJ, 95, 985.
Forman, W., et al. 1981. ApJ, 243, 114.
Frenk, C.S., White, S.D.M., Efstathiou, G., & Davis, M., 1990. ApJ, 351, 10.
Frenk, C.S., 1991. Physica Scripta T36, 70.
Fukugita, M., & Hogan, C.J., 1991. Nature, 354, 17.
Geller, M.J., & Beers, T.C., 1982. PASP, 94, 421.
Gioia, I.M., et al. 1982. ApJ, 255, L17.
Hughes, J.P., 1989 ApJ, 337, 21.
Kaiser, N., 1987. MNRAS, 227, 1.
Mellier, Y., et al. 1988. AA, 199, 67.
Merritt, D., 1987. ApJ, 313, 121.
Olive, K.A., Schramm, D.N., Steigman, G., & Walker, T.P., 1990. Phys.Lett.B, 236, 454.
Peebles, P.J.E., 1982. ApJ, 258, 415.
Regős, E., & Geller, M.J., 1989. AJ, 98, 735.
The, L.S. & White, S.D.M., 1986. AJ, 92, 1248.
The, L.S. & White, S.D.M., 1988. AJ, 95, 15.
Turok, N., 1989. PRL, 63, 2625.
West, M.J., Oemler, A., & Dekel, A., 1988. ApJ, 327, 1.
White, S.D.M., 1991 Large-Scale Structures and Peculiar Motions in the Universe (eds. Latham, D.W., & Da Costa, L.N.) ASP, p285.
White, S.D.M., Davis, M., & Frenk, C.S., 1984. MNRAS, 209, 27P.
White, S.D.M., Frenk, C.S., & Davis, M., 1983. ApJ, 274, L1.
Zeng, N., & White, S.D.M. 1991. ApJ, 374, 1.
Zwicky, F., 1933. Helvetica Physica Acta, 6, 110.

ENVIRONMENTAL INFLUENCES ON GALAXY MORPHOLOGY

Augustus Oemler, Jr.
Department of Astronomy
Yale University
New Haven CT 06511
U.S.A.

ABSTRACT. Populations of galaxies show systematic variations with environment, which reach an extreme in the cores of rich clusters. This review describes the nature of these variations and attempts to identify those factors in the galaxies' environment which are responsible for them. There is evidence that the environment exerts its effect over the entire history of the galaxy, rather than just at the time of formation, but which of many candidate processes are most important is not yet clear.

1. Introduction

It has been recognized, since early in this century, that some galaxies are clustered, and that the galaxies which inhabit clusters are systematically different from those outside. Galaxies in clusters are predominantly ellipticals and S0's, while those outside, in what is conventionally called the *field*, are mostly spirals and irregulars. In the 40 years since Baade & Spitzer's (1951) first attempt to understand this phenomenon, there has been a growing recognition that these environmental variations hold an important clue to the origin of all galaxy types. If one could identify the conditions in the environment which were responsible for population variations, and if one could understand the means by which these conditions effected the changes seen, one would make significant progress towards understanding the formation and evolution of all galaxies.

In this review I shall describe what is now known about the variations of the properties of galaxies with environment. I shall not restrict myself to clusters, since an understanding of the properties of cluster galaxies may only be possible in a broader context, considering the entire range of conditions from the lowest density voids to the cores of clusters. The discussion will however, be confined to three related properties of galaxies: their stellar content, gas content, and location along the Hubble Sequence, which, for the sake of brevity, will be referred to collectively as *morphology*. I shall ignore the subject of environmental variations in the mass and luminosity distributions within galaxies, the galaxy luminosity function, and galactic dynamics. Although partly dictated by limitations of space, this is not an entirely arbitrary division, as the former properties are all aspects of the internal

A. C. Fabian (ed.), Clusters and Superclusters of Galaxies, 29–47.
© *1992 Kluwer Academic Publishers.*

TABLE I
Morphological Mix vs Environment

Environment	% Type		
	E	S0	Sp
Field	10	20	70
Poor Group	10	20	70
Rich Group	10	30	60
Cluster	20	40	40

populations within galaxies, and may be more intimitely linked to each other than they are to the latter properties.

2. The Morphology-Environment Relation Today

2.1. THE ELLIPTICAL/S0/SPIRAL DISTRIBUTION

The best-established environmental variation is that of the relative proportions of ellipticals, S0's and spirals. Table I illustrates the phenomenon. These values are taken from a number of sources, including field redshift surveys, group catalogs, and studies of rich clusters. Since populations vary within each category, the numbers should be regarded as representative rather than universally applicable. As the number of galaxies in a grouping increases, the fraction of spirals decreases. At first, most of the lost spirals are replaced by S0's, whose numbers increase. However, in the richest associations, the fraction of ellipticals increases, in some cases overtaking that of the S0's.

Oemler (1974, Fig. 9) showed that similar variations occur within individual rich clusters. Excluding the small number of clusters of irregular, unrelaxed appearance, one finds strong radial gradients in the galaxy mix, with the fraction of S0's increasing towards the center, and the fraction of E's rising even more rapidly. The fall in the surface density of spirals towards the cluster center is consistent with a central space density of zero. Dressler (1980) recast the same phenomenon in terms of the local surface density of galaxies, S_{gal}. He showed that the fraction of spirals is a decreasing function, and the fraction of S0's an increasing function of S_{gal}. In a manner similar to that evident in Table I, the fraction of E's varies only slowly with density until S_{gal} becomes quite high, after which it increases rapidly.

The observations cited above appear quite robust. Each of the three population trends- with group richness, with location within a group, and with local surface density of galaxies- has been demonstrated by a number of workers, with consistent results, and all three appear to be harmonious, suggesting a picture in which the S0 abundance increases steadily with the richness and/or density of the environment, and the elliptical population increases in a more nonlinear manner. Some caution is needed, however. The classification of galaxy morphology is as much art as science. The dividing line between S0's and spirals is not sharp, and moves as the quality of the images used to do the classification changes. With images that are less than

TABLE II
Spiral Colors vs Environment

Type	B-V Field	Group	Cluster
Sa	0.77	0.78 ± .01	0.94 ± .02
Sb	0.69	0.68 ± .04	0.87 ± .03
Sc	0.57	0.57 ± .02	0.78 ± .04
Irr	0.45	0.49 ± .08	0.61 ± .07

perfect (as all images are) an increasing fraction of weak-featured spirals might be read as an increasing fraction of S0's. As I shall discuss later, similar or worse errors are likely in distinguishing ellipticals from S0's.

2.2. THE DISTRIBUTION OF SPIRAL SUB-TYPES

There is some evidence that the relative frequency of different spiral types varies with environment in a way similar to the frequencies of broader morphological classes. Gisler (1980) determined the cluster membership of a sample of UGC galaxies (Nilsen 1973) with measured velocities, using the Catalog of Galaxies and Clusters of Galaxies (Zwicky et al. 1961-68). Ordering the environments in a sequence of increasing density, from the field, through open clusters, medium compact clusters, to compact clusters, Gisler found that the relative frequency of Sc's fell rapidly with density. Sb's fell less rapidly, Sa's were of roughly constant frequency, and S0's rose rapidly.

Similar results have been found by Giovanelli, Haynes & Chincarini (1986) from a study of galaxies in the Perseus-Pisces supercluster. When the relative frequency of Hubble types was plotted against projected galaxy density, late-type spirals fell fastest with S_{gal}, intermediate types were of constant frequency, and the earliest spiral types increased slightly in frequency in the denser environments.

Since the mean disk-to-bulge ratio (D/B) decreases towards the early (E/S0) end of the Hubble sequence, these results are also consistent with Dressler's (1980) finding that D/B of spirals decreases in denser environments. Solanes, Salvador-Solé & Sanromà (1989) have shown that, in Dressler's data, there is no trend of D/B with environment within one spiral type, so that the overall trend of D/B with S_{gal} within Dressler's data must be due to a shift among the relative frequency of different spiral types.

2.3. GALAXY COLORS

Table II summarizes the colors of spiral and irregular galaxies of each Hubble type as a function of environment. The values for field galaxies are the mean for all objects with photometry presented in the *Second Reference Catalogue of Bright Galaxies* (de Vaucouleurs, de Vaucouleurs, & Corwin, 1976).The values for group members are derived from that subset of the above galaxies which are identified by Geller & Huchra (1983) as being members of groups. The values for cluster members are taken from the photometry of Butcher & Oemler (1985), transformed to B-V. The colors of galaxies in all three data sets have been corrected for Galactic reddening, but not for reddening within the individual galaxies. There is a striking uniformity in the trend for all Hubble types: group galaxies are indistinguishable from field galaxies, but cluster members are substantially redder.

Color variations among E/S0 galaxies are, if present, likely to be more subtle, and the evidence is ambiguous. Sandage and Visvanathan claimed to see no difference in the colors of E/S0's in the field, groups, and clusters, but that conclusion was contested by Larson, Tinsley & Caldwell (1980). Caldwell & Oemler (1981) saw some evidence that the colors of cluster E/S0's were bluer in environments rich in spirals than elsewhere, but the result can only be called suggestive, as can Eder's (1990) finding that Coma Cluster S0's are slightly redder than field S0's. Colors are only sensitive to differences in the amount of recent star formation, but Bower, Ellis, Rose & Sharples (1990) have found that narrow-band spectral indices of ellipticals in the field, the Virgo Cluster, and the Coma Cluster suggest that galaxies in the latter environment (the richest of the three) had less extended histories of star formation than the others.

The cumulative effect of the three trends described above- variations with environment in the fraction of spirals, the fraction of spirals that are of late-type, and the mean color of each morphological type- results in a strong trend of galaxy color with environment. Butcher & Oemler (1984) define the parameter f_B to be the fraction of galaxies in a population with rest frame B-V colors at least 0.2 mag bluer than the color of the E/S0 sequence at the same absolute magnitude. f_B is a rough measure of the fraction of galaxies with significant amounts of recent star formation. Figure 1, derived from data in Allington-Smith et al. (1992, hereafter AEZO) summarizes the trend of f_B with system richness (L_{cl} is the integrated luminosity, in solar units, of all galaxies in the system). Filled circles are data on groups and clusters from AEZO, Geller & Huchra (1983), and Butcher & Oemler (1984); the hatched area at the left represents the field. The smooth curve is an eye fit to the trend of blue fraction with richness. This is hardly a unique solution; a sharper, more step-like line would be an equally good fit. The qualitative result is, however, unambiguous: clusters and rich groups have many fewer blue galaxies than poor groups and the field.

2.4. THE GAS CONTENT OF SPIRALS

After some early controversy, it is now well established that there are systematic differences in the gas content of field and cluster galaxies. The state of our knowledge of the phenomenon has been recently reviewed by Haynes (1990) and Kenney (1990). Using the surface density of HI gas within an optical isophote as a measure of atomic gas content, one finds that at least some galaxies in some clusters are depleted in gas relative to galaxies of the same morphological type found in the field. Many spirals in the core of the Coma cluster and other similarly rich, compact clusters have no more than one tenth as much atomic hydrogen as field galaxies of the same type. The fraction of galaxies with significant depletion, and the mean

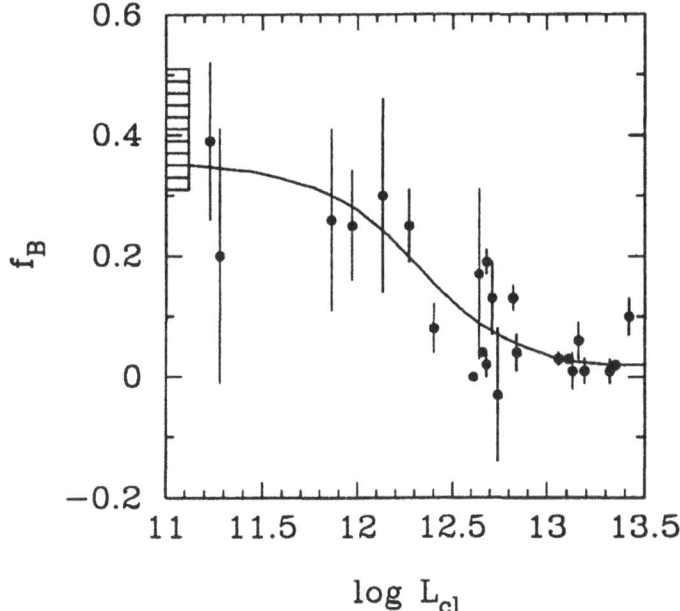

Fig. 1. Fraction of blue galaxies in groupings of galaxies versus their total galaxy population, measured in units of solar luminosity.

amount of depletion increases towards the centers of clusters. Beyond a distance of two or three times the Abell radius ($1.5h^{-1}Mpc$) the amount of depletion is undetectable.

Dressler (1986) has demonstrated, using the run of galaxy velocity dispersion with radius within the cluster, that the gas deficient cluster members tend to be on radial orbits, which take them close to the cluster center, while those galaxies which are not deficient are on more circular orbits, keeping them far from the cluster core. Giovanelli & Haynes (1985) have found that the mean gas deficiency in a cluster is proportional to the total hot gas content of the cluster, as measured by its X Ray luminosity. The mean depletion is a function of Hubble type (vid Kenney 1990, Table 1): S0's and early-type spirals show the most, and late-type spirals and irregulars the least depletion.

The gas depletion seems to be confined to the atomic gas, and to the outer parts of the galaxy. Kenney & Young (1989) demonstrated that, even in galaxies with severely depleted HI, the molecular gas content was generally unchanged. Thus, whatever removes the atomic gas does not, apparently, affect the molecular gas. Conversely, the constancy of the molecular gas implies that the deficiency in HI cannot be due to the conversion of gas from atomic to molecular forms. High resolution mapping of HI deficient galaxies in Virgo by a number of workers has shown that the gas disks are truncated: gas tends to be removed from the outside in, rather than simultaneously over the entire disk.

2.5. IDENTIFICATION OF THE BASIC PHYSICAL VARIABLES?

What is the relationship between the observable quantities which we have described above and the underlying physical properties of the galaxy populations? There are two points to keep in mind during the following discussion. Firstly, in discussing the role of environment in influencing galaxy populations, no prejudgment is being made as to whether the process is the transformation of the properties of individual galaxies during their lifetime, or a change in the relative birthrates of galaxies with different properties. Secondly, even in the case of the transformation of a preexisting galaxy population, it may not be necessary for individual galaxies to change their type. All the populations under discussion are *observed* populations, usually those whose luminosities are higher than some minimum value. Therefore, for example, a change in the ratio of two types of galaxies might be due not to a transformation between types but rather to the fading of galaxies of one type below the minimum luminosity of the observed population.

How then, is the environment affecting galaxy populations? There appear to be four main effects:

1) *In the central regions of many clusters, the extent of the atomic gas disks of spirals is truncated.*

2) *The star formation rate of many cluster spirals (as inferred from their colors) is lower than that of field spirals of the same type.* Indeed, the observations may significantly underestimate the amount by which star formation has been diminished, if the reduced star formation shifts the galaxy towards earlier Hubble type (see point 3, below). The decreased star formation may or may not be caused by the truncation of the gas disks. In the Virgo cluster, gas deficient galaxies seem to have lower than normal star formation rates (Kennicutt 1983), but there seems to be no such trend in several other clusters (Kennicutt, Bothun & Schommer 1984).

3) *In richer and/or denser environments, the distribution of the morphological types of disk galaxies is skewed towards the early (S0) end.* It seems clear that a total end of star formation in a spiral will, in time, transform the galaxy into an S0. There has been some controversy over whether the structural properties of the S0 galaxies in clusters are consistent with those expected from a population of dead spirals (Burstein 1979a,b, Dressler 1980), but recent studies suggest that they are (Dressler 1986, Eder 1990). Whether a partial reduction in the star formation rate will create a normal spiral of earlier type rather than an anaemic spiral (van den Bergh 1976) is less clear, because the Hubble type of a disk galaxy depends on the amount of star formation, the disk to bulge ratio, and the shape of the arms. A decrease in the star formation rate will certainly affect the first two characteristics, but it is not clear whether the effect will be sufficient to cause all of the observed shift in Hubble type.

4) *The increasing fraction of ellipticals and the decreasing mean D/B of disk galaxies indicate that the relative luminosity functions of disks and spheroids shifts, in the latter's favor, in dense and/or rich environments.* Because the mechanisms available to alter disks and spheroids are different, it is important to decide whether

this shift is due to an enhancement of spheroids or a diminution of disks. Dressler (1980) demonstrates that, in his sample, the bulges of disk galaxies become more luminous in denser regions. However, Larson, Tinsley & Caldwell (1980) show that, in a data set limited by total (disk plus bulge) luminosity, a not implausibly large diminution of disk luminosity can increase the average bulge luminosity of the galaxies remaining above the luminosity limit by an amount comparable to that seen by Dressler. Recently, Solanes, Salvador-Solé & Sanromà (1989) have presented evidence that this is, indeed, occurring in Dressler's sample, and that all of the variation in D/B which can be seen is due to a fading of the disks in the denser cluster regions.

The above argument cannot explain the rising frequency of elliptical galaxies in cluster cores. However, if classifying disk galaxies is often an art, distinguishing S0's from E's is a black art. The characteristic visual features which allow one to distinguish between spheroid and face-on disk are, at best subtle, and, at worst, illusory. There exists little confirming evidence to demonstrate that the classifications of face-on S0's and E's are not systematically in error. Even if systematically correct, many individual galaxies of each type must inevitably be misclassified. Thus, if the true abundance of S0's rises, the apparent observed abundance of E's will also rise. There is also, undoubtedly, a minimum disk-to-bulge ratio below which S0's will be systematically misclassified as ellipticals. Therefore, a decrease in the mean disk-to-bulge ratio of S0's in the cores of clusters would inevitably result in an increase in the perceived E to S0 ratio. I would not go so far as to claim that these classification problems are responsible for all of the observed variation in the fraction of ellipticals. However, it is not out of the question, and this possibility should be kept in mind while judging the ability of various astrophysical processes to produced the observed population variations.

3. The Nature of the Environmental Dependence

I have, so far, been deliberately vague in specifying whether the observed population variations are a function of the richness and/or other properties of the association in which the galaxy lives, or of the local galaxy density. However, since the answer is of major importance to understanding the underlying physical cause of the morphology-environment relation, we must establish precisely what characteristic of the environment is most correlated with galaxy morphology.

Each point of view has been advocated by a number of workers, and for each there is considerable supporting evidence. The galaxy populations of clusters are well correlated with the overall dynamical state of the cluster (Abell 1965, Butcher & Oemler 1979). Irregular clusters with little central concentration, which appear to be dynamically young, have galaxy populations much closer to those of the field than to those found in regular, centrally concentrated, relaxed-appearing clusters. Within the latter, but not the former, there is a strong radial gradient in the population (Oemler 1974). Figure 1 demonstrates that the population is also a strong function of cluster richness, perhaps even a step-function. These observations suggest that it is the cluster as a coherent system which is responsible for the population variations within it.

36

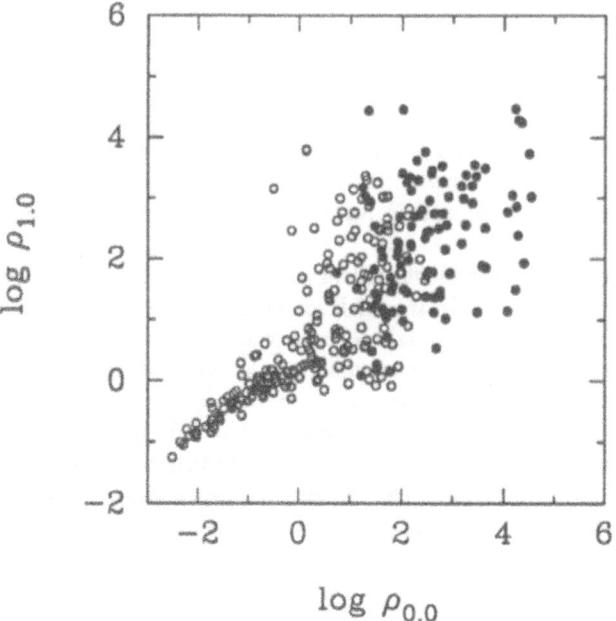

Fig. 2. Correlation between local galaxy densities near galaxies today and at an earlier epoch corresponding to $z = 1.0$. Filled circles are those galaxies now within one Abell radius of the center of a rich cluster.

The opposite view has been advocated by Dressler (1980), who demonstrated that the morphological mix within a cluster is well-correlated with the very local surface density of neighboring galaxies, even in those irregular clusters with no ordered structure or radial population segregation. Powerful support for this view is provided by Postman & Geller's (1984) finding that the population is equally well correlated with the local space density of galaxies, and that the correlation extends beyond the confines of rich clusters, to galaxies in smaller groups. Giovanelli, Haynes & Chincarini (1986) have pushed even further, demonstrating comparable trends in the general field.

The arguments for a local effect appear quite compelling, but there is one significant problem. The crossing times in the central regions of relaxed clusters are much shorter than the Hubble time, and the galaxy populations there should be rather well mixed. This may be demonstrated with the cluster models calculated by West, Dekel & Oemler (1987). Figure 2 plots, for galaxies in a cubic region of size $45h^{-1}Mpc$ centered on a typical rich cluster, the space density near each galaxy at the present epoch versus the density at the epoch which would be observed at a redshift of 1.0. The overall correlation is rather good. However, if one confines one's attention to the filled circles, which represent galaxies whose present locations are within one Abell radius of the cluster center, the correlation almost disappears.

Correlations between present environment and that at even earlier times are even poorer.

Thus, the present local density near a galaxy in a rich cluster is generally very different from the local density at earlier epochs. Unless the morphological types of galaxies were determined in the *very* recent past, the local density cannot be the key parameter in the cluster morphology-environment relation. In a recent paper, Sanromà & Salvador-Solé (1990) have presented additional evidence that it is not. They have reanalyzed the morphology-local density relation in Dressler's data, reproducing Dressler's result. However, if the positions of the galaxies in Dressler's clusters are azimuthally scrambled, so that local information is lost and only the radial position is preserved, the resulting morphology-local density relation is *still* identical to Dressler's. This suggests that the original correlation of morphology with local density is only an indirect one, due to the fact that the cluster population and local galaxy density are both correlated with radius.

I conclude from this that, within rich, relaxed clusters, global cluster effects dominate the morphology-environment relation. However, outside clusters, as Figure 2 demonstrates, populations are *not* well mixed, and *all* effects are local. Thus, if the morphology-environment relation extends beyond rich clusters, the processes giving rise to it and the conditions under which they operate may differ significantly from those prevailing within clusters. Both the Postman & Geller, and the Giovanelli, Haynes, & Chincarini results suggest that it does, but there are troublesome ambiguities. Both samples contain systems of a wide range of richnesses, from very poor groups to rich clusters such as Virgo and Coma. Galaxies in all systems may contribute to the populations at each value of the density, but all need not contribute to the sensitivity to density displayed by the total population. Postman and Geller find that morphology depends of density only above a threshold value, which they identify with systems whose crossing times are equal to the Hubble time. It is striking, however, that this value is also equal to the lowest value seen in rich clusters. Could the few richest systems in their sample be providing all of the signal for an environmental dependence?

This problem could be dealt with easily by a reanalysis of the two data sets, subdivided by the richness of the associations in which the galaxies dwell. However, there are two other, almost unavoidable effects, each of which will tend to smear any sharp transition in population inside and outside of clusters into a gradual trend with local environment: projection effects (we can never be sure of a galaxy's position along the line of sight) and galaxy orbits (Figure 2 shows that even in low density regions there is some mixing.) I want to stress that I am not arguing against the existence of a morphology-environment relation outside of clusters; I am only pointing out that establishing the existence and nature of one is a very difficult task. However, the answer is very important, and additional work to provide it is clearly needed.

4. The Morphology-Environment Relation at Earlier Times

The phenomena described above are open to many interpretations, including the fundamental alternatives of whether environment affects the formation or subsequent evolution of galaxies. In principle, observations of systems at earlier epochs can discriminate between these possibilities. For this reason, Butcher and I began, fifteen years ago, a program of photometry of high-redshift clusters (Butcher and Oemler 1978, 1984). Since then, a number of other workers have contributed substantial amounts of additional photometry, all of which is consistent with the conclusion that there has been significant evolution of the colors of galaxies in clusters since the epoch observed at $z = 0.5$. On average, the mean f_B of populations in the cores of centrally concentrated galaxies grows from about 0.02 today to 0.20 at $z = 0.5$.

Although more abundant, these blue galaxies have a radial distribution similar to that of spirals in clusters today. Their distributions in color and absolute magnitude also resemble those of today's spirals. This suggests that these blue objects are the spiral pregenitors of the red spirals and S0's which are overabundant in today's clusters. If true, there are two alternate pictures which are consistent with these observations. If the color evolution were confined to the cores of rich clusters, the morphology-environment relation would have steepened with time, implying that its origin was in environmentally-driven galaxy evolution. If the evolution seen in cluster galaxies were typical of galaxies in all environments, recent evolution would not have enhanced the morphology-environment relation, and its origin would have to be sought at earlier times, perhaps the time of formation.

A resolution of this question rests on the behaviour of galaxies in other environments. It has long been known that counts of faint field galaxies increase too rapidly towards faint magnitudes, and their mean colors are too blue, to be consistent with the simplest models for a non-evolving population (vid Tyson 1988). The most straightforward interpretation of this behavior has been that the field galaxies were undergoing evolution of a kind and degree similar to that seen in clusters. However, recent redshift surveys of faint field galaxies by Broadhurst, Ellis & Shanks (1988) and Colless et al. (1990) have made such an interpretation untenable. The observed redshift distributions are consistent with no evolution at all, at least among the luminous galaxies which are found by such a survey. Indeed, at $z = 0.4$, I calculate that the Colless et al. sample had $f_B \approx 0.42$, indistinguishable from that characterizing the field today. Whatever is responsible for the behaviour of the faint galaxy counts (starbursts and mergers in fainter, nearer galaxies have been suggested by the authors), it is not, apparently, indicative of general evolution of luminous field galaxies.

Such a conclusion is supported by the recent study, by AEZO, of galaxies in small groups, whose properties span the difference between the field and rich clusters. This study suggests that the rate of evolution is a strongly increasing function of group richness. The morphology-environment relation appears to be almost flat at $z = 0.4$, implying that its present form is, indeed, the result of environmentally-driven evolution, most of which has occurred in the last third of the Hubble time.

Colors provide, however, only a limited view of the stellar populations in a galaxy, much more limited than that provided by spectra. Dressler & Gunn (1982, 1985, 1992) and others have, as of the time of this writing, obtained spectra for over 400 galaxies in about 15 clusters with redshifts greater than 0.2. These spectra have demonstrated that, contrary to the conclusions drawn from the galaxy colors, most of the blue galaxies do not have the spectral energy distributions of normal spirals. Dressler and Gunn identified three types: active galactic nuclei (AGN), galaxies undergoing starbursts (SB), and galaxies which appear to be in a post-starburst phase (PSB). Detailed spectral synthesis of blue galaxies in three clusters by Couch & Sharples (1987) suggests that, while there are a few blue galaxies with spectra consistent with a normal history of star formation, most are in some phase of a starburst or its aftermath. Newberry, Boroson & Kirshner (1990), in a similar analysis, find that the PSB spectra can equally well be explained as the aftermath of a sudden truncation of star formation in a normal late-type spiral.

It is clear, then, that these galaxies have had unusual histories of star formation. It is less clear what conclusions about the evolution of galaxy morphology one should draw from this fact. One possible view is that there is nothing particularly remarkable about these spectral peculiarities. Faint field redshift surveys have shown that many galaxies in all environments exhibited somewhat greater signs of activity at earlier epochs. More to the point, it should be noted that the blue cluster galaxies inhabited a region of galaxy parameter space which is not occupied today: gas-rich galaxies in regions of high density, high relative velocities, and high intergalactic medium density. Given the potential for disturbance of the galactic interstellar medium which this environment presented, it is not surprising that their histories of star formation were less placid than those of spirals today. I will discuss later several mechanisms which could have produced such disturbances. All, significantly, would also contribute to the observed evolution in the galactic star formation rate. In this view, the spectral peculiarities discovered by Dressler and Gunn can be viewed as natural, and, indeed, as signatures of the very processes which produced the morphology-environment relation observed today.

This naive interpretation, is not, however, the only possible view. Dressler and Gunn propose that the spectral peculiarities are an indication that the blue cluster galaxies are the product of unusual conditions which existed only in clusters at earlier epochs, and are not, therefore, relevant to the origin of the morphology-environment relation, or to the evolution of galaxies in general. Building on the impression that the blue galaxies tend to form a ring about the cluster center and tend to have substantially higher velocity dispersions than the red galaxies, they suggest that they are infalling field galaxies, whose interstellar medium has been strongly perturbed when it encounters a shock front in the intracluster medium. This shock causes a burst of *ram pressure induced star formation*, after which the remaining interstellar medium is swept from the galaxy. The cluster models of Evrard (1990) predict such a shock in clusters at earlier epochs, and predict that it would be much weaker today. A detailed modelling of the Dressler-Gunn process by Evrard (1991) shows that, assuming that ram pressure induced star formation does work, the expected number of blue post-shock galaxies would evolve in a manner similar to the observed evolution of blue cluster members.

This is a very nice model, which can explain the observed spectral peculiarities of the blue cluster galaxies. Because it attributes all of the observed star formation to infalling galaxies undergoing a unique encounter with the intracluster medium, it requires no evolution of galaxies in the long-term cluster population. However, this model is not without problems. For one thing, no one has ever seen ram pressure induced star formation. There is little or no evidence from nearby clusters that ram pressure can affect the molecular gas or induce star formation. Nevertheless, many of the alternate mechanisms, which the naive model needs to explain the starbursts, are equally speculative, so this is not, by itself, a fatal flaw.

The model does make quantitative predictions, which can be tested with existing data. Because the blue galaxies are weakly-bound infalling objects, their velocity dispersion should be much higher than that of the other cluster members: Evrard's model predicts $\sigma_{blue}/\sigma_{red} = 1.7$. On the other hand, the naive model predicts only a modest increase in velocity dispersion for the blue galaxies. Using velocities from the literature for 151 blue and 248 red galaxies in 14 clusters, AEZO find $\sigma_{blue}/\sigma_{red} = 1.21 \pm .09$. Such a low ratio is not consistent with that expected for infallers.

The infaller model also predicts that the starbursts will occur at a well-defined site, the spherical shock which, at the epoch observed, is expected to be at a radius of about 1 Mpc. The galaxies are on radial orbits, falling towards the center at velocities of order 2000 kms^{-1}. The time needed for them to reach the center after being shocked is about the same as the evolution time of the stars in the burst. Thus, even in projection, there should be a strong radial gradient in the colors and spectral properties of the blue galaxies, with the SB galaxies in a ring at ≈ 1 Mpc, and increasingly red PSB galaxies towards the center. Figure 3, from AEZO, summarizes the data on spectroscopically-confirmed cluster members in four clusters (Abell 370, Cl1447+2619, Cl0024+1654, and 3C295) with $0.37 \leq z \leq 0.47$. Δ(B-V) is the color difference between the galaxy and the mean of E/S0's. It is clear that the blue galaxies are less concentrated towards the cluster centers than the red galaxies. However, among the blue galaxies (those with Δ(B-V)≥ 0.15) there is no perceptible radial color gradient.

The ram pressure induced star formation model appears, then, to be in imperfect agreement with the extant data on blue cluster galaxies. Those data suggest that the blue galaxies are not infallers, but rather long-term cluster members. Until it is proved untenable, I prefer the naive conclusion that their blue colors and spectral peculiarities are signs of evolutionary phenomena among normal galaxies, which have led, by the present day, to the observed morphology-environment relation.

5. Possible Mechanisms for the Morphology-Environment Effect

I think that it is necessary to begin the consideration of possible causes of the phenomena described above with a warning. All but one of the galaxy properties under discussion (B/D, gas content, spectral energy distribution) are linked to the other, the Hubble Type. All effects operate in the same sense: in richer/denser environments, galaxy properties move towards those at the early (E/S0) end of the Hubble Sequence. Because of this, there is a very strong temptation towards imposing a unitary framework on the phenomena, in two ways. The data are very incomplete, with observations of only a subset of the many possible correlations of galaxy property with environment and epoch. One is tempted to fill in the missing correlations in the obvious way, assuming, for example that, because the colors of cluster galaxies vary with epoch, the Hubble Type, gas content and B/D do as

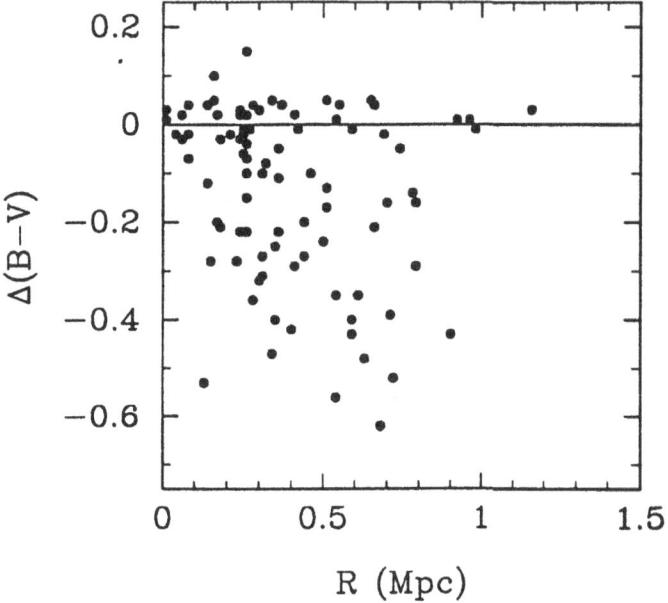

Fig. 3. Radial location versus color of spectroscopically-confirmed members of 4 high-redshift clusters.

well. Equally, it is tempting to assume that one process must be sought which can explain all of the observed effects.

There is no justification for either assumption. There are, as we shall see, many processes which might produce the observed variations in galaxy properties. Each operates under different conditions and each will produce somewhat different effects on the galaxies. Simplicity prefers that one process be responsible for all of the variations of properties in all environments and epochs, but the universe is not necessarily that simple.

5.1. CONDITIONS AT FORMATION

The first set of alternatives to be considered is that of nature versus nurture: whether the variations in galaxy populations were established at the time of formation, or have arisen through evolutionary processes since then. The idea of primordial effects has been very popular. For example, Faber (1981), Dekel & Silk (1986), and others have argued that many of the properties of a galaxy are determined by the initial amplitude of the density fluctuation which led to its formation. The phenomenon of biased galaxy formation (Kaiser 1986) then provides a means of coupling the amplitudes of the galactic scale perturbations to larger scale density variations, thus linking galaxy morphology to large-scale environment. (It is this link which is the relevant factor here. Few would dispute that the properties of a galaxy today must be at least partially determined by those of the protogalactic lump from which it formed. What is important in establishing the origin of the morphology-environment relation is whether the present spatial correlations in galaxy properties are due to spatial correlations in the initial conditions of the protogalaxies, or to spatially coherent effects acting on the galaxies since formation.)

The claimed relation between morphology and local density has often been invoked as an argument for primordial effects. Evrard, Silk & Szalay (1990) have modeled in detail the morphology-local density relation expected in a cold dark matter cosmological model, with morphological type assumed to be a function of initial peak amplitude; and they find results very similar to Dressler's. Unfortunately, the problems, discussed in Section 3, with the interpretation of the cluster data in terms of a morphology-local density relation throw this entire enterprise into doubt. Evrard et al.'s result seems to contradict the conclusion I reached from Figure 2, that mixing has erased the connection of present environment to initial conditions. But, in fact, the results are consistent, because their simulation covers a volume ($8h^{-1}$ Mpc on a side) much larger than the well-mixed cluster core, and should, therefore, produce a correlation more like that of the open circles in Figure 2 than the filled circles.

The observed evolution of the f_B vs Richness relation also limits the scope for initial conditions, but here I must heed my own earlier warnings. The only aspect of the morphology-environment relation which has been observed to evolve is the distribution of galaxy colors and star formation rates. It is not yet known whether the uniformity in f_B observed among galaxy populations at $z = 0.5$ also applies to the bulge-to-disk ratio, Hubble type, and other galaxy characteristics. Nevertheless, the preponderance of the evidence suggests that cluster galaxies are unpromising candidates for primordial population differences.

Outside of clusters there is more scope for primordial differences. Indeed, in low velocity dispersion environments, in which there has been little or no mixing, the differences between nature and nuture are only a matter of timescales, since a galaxy has seen a similar environment since birth. This is particularly true of galactic disks, which probably build up slowly. In their case, the distinction between initial conditions and subsequent environmental effects is merely semantic. Given this fact, and given that cluster galaxies, at the very least, have been substantially modified by the environment over long periods of time, I would suggest the following hypothesis: The spatial coherence in the properties of galaxies is a function of the integral of all the environmental effects which have acted upon them since formation. In clusters the integral is determined by the global properties of the cluster and the orbit of the galaxy within it. Outside of clusters it is determined by local conditions.

5.2. ENVIRONMENTAL EFFECTS

A number of processes have been proposed which could influence the evolution of a galaxy. It will be useful, while considering them, to keep in mind a few environmental trends. Clusters (and possibly groups) observe the following scaling relations (West, Oemler, & Dekel 1989, Koopman, Tucker, & Oemler 1992):

$$R_{cl} \propto L_{cl}^{1/2} \quad \sigma_{cl} \propto L_{cl}^{1/3} \quad M_{gas} \propto L_{cl}^2,$$

where R_{cl} is any cluster scale, σ_{cl} is the cluster velocity dispersion, M_{gas} is the mass of X Ray emitting intracluster gas, and L_{cl} is the total cluster luminosity. From the virtual universality of the cluster galaxy luminosity function, it follows that the cluster richness, $N_{gal} \propto L_{cl}$. A first conclusion, then, is that, in clusters and groups, $\rho_{gal} \propto L_{cl}^{-1/2}$. Thus, contrary to intuition, richer clusters are *less* dense than poorer ones. $\Delta\rho/\rho \propto N^{-1/2}$ is, of course, what one would expect for a Poissonian model for clustering.

5.2.1. Galaxy-Intracluster Medium Interactions. The most popular mechanism for the environmentally-induced transformation of cluster galaxies has been the stripping of the galactic interstellar medium due to interractions with the intracluster medium through which the galaxies move (vid Kenney 1991). Although other processes such as evaporation, viscous drag, and turbulent flow may contribute, most attention has been given to the effects of the *ram pressure* of the intracluster medium (Gunn & Gott 1972). Ram pressure stripping provides a natural means of transforming spirals into S0's. It has the great virtue (lacking by many) that it can be seen to work. There are several galaxies in nearby clusters whose ISM show clear signs of disturbance like that expected from ram pressure stripping.

Because $P_{ram} \propto \rho_{gas} v_{gal}^2$, it is most effective in cluster centers, and therefore can explain the observed dependence of galaxy morphology and gas depletion on the distance from the cluster center, and the dependence of gas depletion on the shapes of galaxy orbits. The density dependence is also consistent with the observed dependence of gas depletion on cluster X ray luminosity. It has been shown (Gisler 1979) that the stripping efficiency has a threshold which is dependent on the amount of ISM and the rate at which it is replenished. This sensitivity provides a natural explanation for the fact that early-type spirals, with less ISM and a lower replenishment rate, are more depleted than late types. From the scaling relations summarized above, we find $\rho_{gas} \propto L_{cl}^{1/2}$ and, thus, $P_{ram} \propto L_{cl}^{7/6}$, consistent with the observed morphology-cluster richness trends.

Ram pressure stripping can, therefore, explain much; but it cannot explain all. Although stripping can, apparently, deplete the atomic gas content of a galaxy, it is not at all obvious that this has further effects. The persistence of the molecular gas, and the poor correlation between gas deficiency and star formation rate, casts doubt on stripping's role in fundamentally transforming galaxies. Although the amount of gas deficiency is well correlated with cluster X ray luminosity, a good indicator of cluster gas mass, the overall galaxy content is not. Figure 4 presents the relation between cluster spiral fraction and X Ray luminosity. Spiral fractions are taken from Oemler (1974), Dressler (1980), and Butcher & Oemler (1979). Following Butcher & Oemler (1985) they are determined within R_{30}, the radius containing 30 percent of the projected galaxy distribution of the cluster. X ray luminosities are taken from Jones & Forman (1984) and Edge (1989), and are on Edge's system. (Open circles represent irregular, presumably unrelaxed clusters which may not have the high random velocities necessary for stipping to work.) The correlation is clearly poor for all clusters.

44

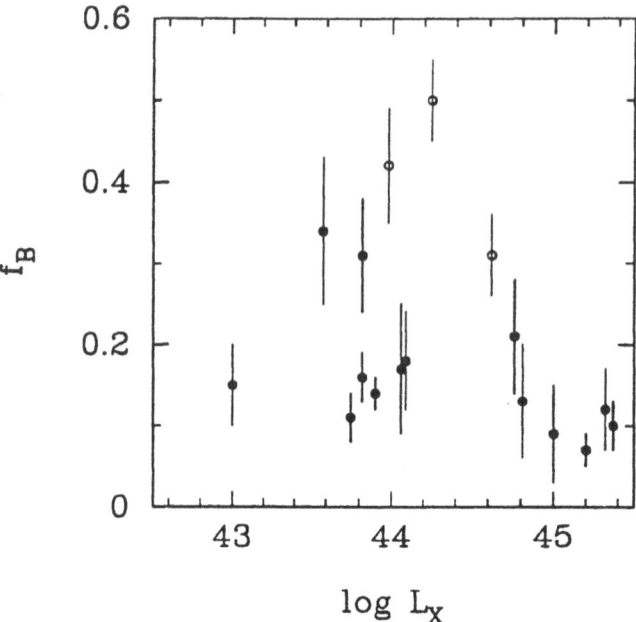

Fig. 4. Correlation of the spiral content of clusters with their X Ray luminosity. Open circles represent irregular clusters.

Gisler (1980) pointed out an even more serious problem. Ram pressure is most effective on early-type spirals, as the gas depletion data shows. However, as environments become richer and/or denser, it is the *latest*-type spirals which disappear first, and the early-types last. Also, ram pressure stripping can only work in dense, high velocity dispersion environments. If the morphology-environment relation does, indeed, extend beyond cluster cores, other processes must be responsible. Finally, stripping cannot explain the increased fraction of ellipticals in cluster cores, unless my hypothesis of misclassified S0's is correct. All of these problems suggest that, while stripping's role in affecting the atomic gas content of cluster galaxies may be significant, it plays only a minor role in the morphology-environment relation.

5.2.2. Galaxy-Galaxy Interactions. Galaxies interact not only with the intracluster medium; they also interact with each other. Galaxy-galaxy collisions were originally hypothesized by Baade & Spitzer (1951) as the mechanism for transforming cluster spirals into S0's. Barnes & Hernquist (1991) and others have shown that mergers between disk galaxies can drive gas from the disk into the center, producing S0's. In some cases the disks may be completely disrupted to create an elliptical. Interpenetrating encounters between gas-rich galaxies that do not lead to mergers are expected to be very effective at removing the gas, including, probably, the molecular clouds which are unaffected by encounters with the more tenuous intracluster medium. Tidal encounters between galaxies can lead to bursts of star formation, the best example of which is the M51 system. If sufficiently vigorous, such bursts can be effective in depleting the gas content of a galaxy.

The encounter rate between group and cluster galaxies is proportional to $\rho\sigma \propto L_{cl}^{-1/6}$, and thus is comparable in all environments. The expected number of such encounters is not negligible. The average cluster member might expect about one interpenetrating collision over its lifetime, and about 10 significant tidal events. The main problem is one of velocities. Mergers are only practical, and tidal encounters only effective, if the relative velocities of the galaxies are comparable to their internal velocities, Because these are a factor of about 5 to 10 smaller than typical velocities within rich clusters, gravitational encounters should only be important in small groups.

However, this reasoning, while plausible, is not totally consonant with observations. NGC4438, in the Virgo Cluster, seems to be an example of severe damage from a high-velocity gravitational encounter (Combes et al. 1988). Lavery & Henry (1988) and Lavery(1991), have presented observations which appear to demonstrate that a significant fraction of the high redshift blue cluster galaxies are undergoing gravitational encounters. It may be that tidal encounters (though probably not mergers) are more effective at high velocities than the simplest reasoning would suggest.

5.2.3. Other Gravitational Processes. Another intriguing, and very promising, process has been proposed by Byrd & Valtonen (1990). In rich, centrally concentrated clusters of galaxies, the cluster potential well can itself exert a significant tidal force on galaxies passing through the core. Because the scale of the cluster core is over an order of magnitude larger than that of individual galaxies, the tidal encounter lasts proportionably longer than galaxy-galaxy encounters do. Thus, even in the highest velocity dispersion clusters, the integrated tidal effect can be as large or larger than that of quite slow galaxy-galaxy encounters. This process has the required sense of proportionality: it is stronger in richer, more concentrated clusters. Its initial effect will be to produce a centrally concentrated population of starburst galaxies. However, as starbursts consume the gas in these galaxies, the cluster population will evolve towards one in which the central galaxies are all red, with a shell of blue starburst and poststarburst galaxies surrounding them, very much like what is observed in high redshift clusters.

5.2.4. Death by Starvation. One of the most attractive mechanisms for morphological segregation was proposed by Larson, Tinsley & Caldwell (1980). In the typical spiral galaxy, calculations show that the gas consumption timescale is much less than the Hubble time. Unless we are living in the last epoch of active star formation in the universe, galaxies must have means of replenishing their internal gas supply from external sources. Such sources might include primordial gas clouds and gas-rich dwarf satellite galaxies. The amount of gas in such systems is sufficient to maintain star formation in the universe for a quite long time, and many examples exist which are suggestive of mergers of such gas-rich systems with larger galaxies.

A Galaxy's hold on this external supply is much more tenuous than its hold on its own interstellar medium. One would expect environments which are much too gentle to remove the ISM would be capable of disrupting a galaxy's external gas supply. Therefore, this process will be effective over a much wider range of environments that ram pressure stripping, but it will become increasingly effective in denser, higher velocity dispersion regions. Its effects are delayed: once the external supply is cut off, a galaxy will continue to form stars from its internal supply until that is exhausted. The one shortcoming of this process is its gentleness: it will not create starbursts.

5.3. CONCLUSIONS

It should be clear, from the above discusion, that there are many possible means by which the properties of galaxies can be transformed after birth. All of those which I have listed operate in the needed sense: they drive galaxies toward the early end of the Hubble sequence. All are probably effective at some level and under some conditions. However, none appears to be sufficient to explain all of the observed population variations by itself.

My guess is that all of these processes, and perhaps others, play some role in creating population variations. Unfortunately, such a catholic hypothesis is very difficult to verify (and almost impossible to disprove). The most promising approach may be to study the environment in which galaxy transformation is most active. At earlier epochs, clusters were that environment, and a detailed study of the blue galaxies in high-redshift clusters should reveal much. At the present epoch, the corresponding environment, that at which the morphology-environment relation has the steepest gradient, is found in groups of middling richness. A search for galaxy transformations in such groups may be very rewarding.

Is there any room left for primordial population variations? I think that there is, particularly in explaining the variations in numbers of ellipticals. The most effective way to prove the need for primordial effects will be a negative one. If one can demonstrate that no evolutionary process can account for spatial variations in spheroid populations, one will have to invoke conditions at formation. Conversely, a demonstration that all variations in the relative luminosity functions of disks and spheroids are due to evolutionary processes would be an almost fatal blow to the concept of biased galaxy formation.

I am grateful to Jeff Kenney for several very illuminating discussions about the effects of environment on the interstellar medium of galaxies.

References

Abell, G.O. 1965, ARA&A 3, 1.
Allington-Smith, J.R., Ellis, R.S., Zirbel, E., and Oemler, A. 1992, in preparation.
Baade, W., and Spitzer, L. 1951, ApJ 113, 413.
Barnes, J.E., and Hernquist, L.E. 1991, ApJ Lett. 370, L65.
Bower, R.G., Ellis, R.S., Rose, J.A., and Sharples, R.M. 1990, AJ 99, 530.
Broadhurst, T.J., Ellis, R.S., and Shanks, T. 1988, MNRAS 235, 827.
Burstein, D.H. 1979a, ApJ 234, 435.
Burstein, D.H. 1979b, ApJ 234, 829.

Butcher, H.R., and Oemler, A. 1978, ApJ 219, 18.
Butcher, H.R., and Oemler, A. 1979, ApJ 226, 559.
Butcher, H.R., and Oemler, A. 1984, ApJ 285, 426.
Butcher, H.R., and Oemler, A. 1985, ApJS 57, 665.
Byrd, G., and Valtonen, M. 1990, ApJ 350, 89.
Caldwell, C.N., and Oemler, A. 1981, AJ 86, 1424.
Colless, M., Ellis, R.S., Taylor, K., and Hook, R.N. 1990, MNRAS 244, 408.
Combes, f., Dupraz, C., Casoli, F., and Pagnai, L. 1988, A&A 203, L9.
Couch, W.J., and Sharples, R.M. 1987, MNRAS 229, 423.
Dekel, A., and Silk, J. 1986, ApJ 303, 39.
de Vaucouleurs, G., de Vaucouleurs, A., and Corwin, H.G. 1976, *Second Reference Catalogue of Bright Galaxies* (Austin: Univ. of Texas Press).
Dressler, A. 1980, ApJ 236, 351.
Dressler, A. 1986, ApJ 301, 35.
Dressler, A., and Gunn, J.E. 1982, ApJ 263, 533.
Dressler, A., and Gunn, J.E. 1985, ApJ 294, 70.
Dressler, A., and Gunn, J.E. 1992, preprint.
Eder, J.A. 1990, PhD thesis, Yale University.
Edge, A.C. 1989, PhD thesis, University of Leicester
Evrard, A.E. 1990, ApJ 363, 349.
Evrard, A.E. 1991, MNRAS 248, 8p.
Evrard, A.E., Silk, J., and Szalay, A.S. 1990, ApJ 365, 13.
Faber, S.M. 1982, in *Astrophysical Cosmology* ed. H.A. Bruek, G.V. Coyne, and M.S. Longair (Pontif. Acad. Sci.), p. 191.
Geller, M.J., and Huchra, J.P. 1983, ApJS 52, 61.
Giovanelli, R., and Haynes, M.P. 1985, ApJ 292, 404.
Giovanelli, R., Haynes, M.P., and Chincarini, G.L. 1986, ApJ 300, 77.
Gisler, G.R. 1979, ApJ 228, 385.
Gisler, G.R. 1980, AJ 85, 623.
Gunn, J.E., and Gott, J.R. 1972, ApJ 176, 1.
Haynes, M.P. 1990, in *Clusters of Galaxies* ed. W.R. Oergerle, M.J. Fitchett, and L. Danly (Cambridge: Camb. Univ. Press), p. 177.
Jones, C., and Forman, W. 1984, ApJ 276, 38.
Kaiser, N. 1986, in *Inner Space, Outer Space*, ed. E.W. Kolb, M.S. Turner, D, Lindley, K. Olive, and D. Sekel (Chicago: Univ. Chicago Press), p. 287.
Kenney, J.D.P. 1990, in *The Interstellar Medium in Galaxies*, ed. H.A. Thronson and J.M. Shull (Dordrecht: Kluwer), p. 151.
Kenney, J.D.P., and Young, J.S. 1988, ApJ 326, 588.
Kennicutt, R.C. 1983, AJ 88, 483.
Kennicutt, R.C., Bothun, G.D., and Schommer, R.A. 1984, AJ 89, 1279.
Koopman, S., Tucker, D.L., and Oemler, A. 1992, in preparation.
Larson, R.B., Tinsley, B.M., and Caldwell, C.N. 1980, ApJ 237, 692.
Lavery, R.J., and Henry, J.P. 1988, ApJ 330, 596.
Lavery, R.J. 1991, presentation at this conference.
Newberry, M.V., Boroson, T.A., and Kirshner, R.P. 1990, ApJ 350, 585.
Nilsen, P.N. 1973, Uppsala General Catalog of Galaxies, Uppsala Obs. Ann., Vol. 6.
Oemler, A. 1974, ApJ 194, 1.
Postman, M., and Geller, M.J. 1984, ApJ 281, 95.
Sanromà, M., and Salvador-Solé, E. 1990, ApJ 360, 16.
Solanes, J.M., Salvador-Solé, E., and Sanromà, M. 1989, AJ 98, 798.
Tyson, J.A. 1988, AJ 96, 1.
van den Bergh, S. 1976, ApJ 206, 883.
West, M.J., Dekel, A., and Oemler, A. 1987, ApJ 316, 1.
West, M.J., Oemler, A., and Dekel, A. 1989, ApJ 346, 539.
Zwicky, F., Herzog, E., Wild, P., Karpowicz, M., and Kowal, C.T. 1961-1968, *The Catalog of Galaxies and Clusters of Galaxies* (Pasadena: Calif. Inst. of Tech.)

IMAGING THE HOT INTRACLUSTER MEDIUM

C. Jones and W. Forman
Harvard-Smithsonian Center for Astrophysics
60 Garden Street
Cambridge MA 02138
U.S.A.

ABSTRACT. Clusters of galaxies contain large amounts of hot, diffuse x-ray emitting gas. Observations of this gas can be used to measure the gas mass, the cluster virial mass, and the cluster morphology. From these measurements. a lower limit to the baryonic mass in clusters, the formation efficiency of galaxies in different environments, and the presence and scale of cluster substructure can be determined.

1. An Introduction to the X-ray Properties of Groups and Clusters

The luminous material in clusters of galaxies falls primarily into two forms – the visible galaxies and the hot, x-ray emitting intracluster medium. The richest, densest clusters contain predominantly elliptical and lenticular (S0) galaxies, while in less dense clusters, up to half the galaxies are spirals. Observations of the relative velocities of galaxies (the velocity dispersion) in rich clusters result in mass-to-light ratios in solar units of ~ 250 (in V, or ~ 325 in B for $H_0 = 50$ km sec^{-1} Mpc^{-1}). Thus, with a mass-to-light ratio of ~ 8 for individual galaxies, only about 3% of the total cluster mass is contained within the visible galaxies. X-ray observations of clusters provide a different view from that obtained at visible wavelengths. Although x-ray emission from individual galaxies in the cluster is sometimes seen, the dominant source of x-ray emission is thermal bremsstrahlung from a hot, intracluster medium (ICM) whose mass is comparable to or greater than that of the galaxies.

Rich (Abell-like) clusters have x-ray luminosities ranging from as low as those of individual bright galaxies up to 1000 times brighter — $10^{42} - 10^{45}$ ergs sec^{-1} (Jones *et al.* 1979; Abramopoulos and Ku 1983; Jones and Forman 1984, 1991). Gas temperatures range from a few 10^7 to 10^8 K (Mushotzky *et al.* 1978, Edge 1989) and are comparable to the equivalent temperatures as measured by the velocity dispersions for the galaxies in the cluster. The gas densities in the cores of rich clusters lie in the range $10^{-2} - 10^{-3}$ cm^{-3} and the inferred cooling times of the gas can be as small as 10^9 years (Fabian, Nulsen, and Canizares 1984, 1991). The gas mass is typically a few 10^{14} M$_\odot$ within the central few Mpc of rich clusters

A. C. Fabian (ed.), Clusters and Superclusters of Galaxies, 49–70.
© 1992 *Kluwer Academic Publishers.*

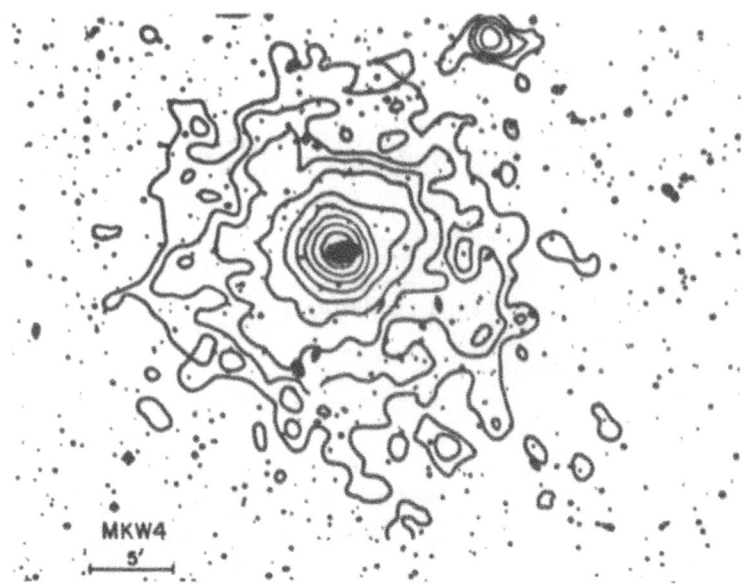

Figure 1 – X-ray contours are shown superposed on the optical photograph of the MKW4 group of galaxies (North is up, East to the left). The group has a redshift of 0.0196 and has an x-ray luminosity of 2×10^{44} ergs sec. An unresolved source (unrelated to the group) lies to the northwest of the extended emission.

or $\sim 10 - -30\%$ of the total cluster mass. Compact (dense) groups of galaxies (notably those selected by Morgan, Kaiser, and White 1975 and Albert, White, and Morgan 1977) have predominantly early-type galaxy populations and are also bright in x-rays (Schwartz, Schwarz, and Tucker 1980; Kriss, Cioffi, and Canizares 1983). These groups have gas masses of $\sim 10^{12} - 10^{13}$ M$_\odot$ (within ~ 0.5 Mpc) and gas temperatures up to a few 10^7 K. Luminosities range up to 10^{44} ergs sec^{-1}, well into the regime populated by rich clusters. An example of the x-ray emission from a compact group is shown in Figure 1. In general, and as the image illustrates for MKW4, the x-ray emission from compact groups is azimuthally symmetric and is centered on the bright D galaxy which dominates the group.

While the visible galaxies and the x-ray emitting hot gas are important components of clusters, most of the mass in rich clusters is "dark matter." Although this material has not been directly observed at any wavelength and its nature remains unknown, as discussed in Section 2, x-ray and visible light observations can determine the amount and distribution of this dark matter in clusters.

Present epoch clusters display a wide variety of properties, both in visible light and at x-ray wavelengths, which can be understood in a framework of cluster dynamical evolution. Gunn and Gott (1972) noted that, while the dynamical timescale of the Coma cluster (a rich, relaxed system) was less than a Hubble time, most other less dense clusters would have dynamical timescales greater than a Hubble time and, hence, could not be fully relaxed. The more evolved, more x-ray luminous clusters have a hotter intracluster medium (Mushotzky et $al.$ 1978, Edge 1989). Optically richer clusters tend to be more x-ray luminous (Jones and Forman 1978, Henry *et al.* 1983) (and have hotter gas), although as shown in Giacconi and Burg (1990), even within an Abell richness class, clusters exhibit a wide range in x-ray luminosity.

2. Cluster Virial Mass Determinations

The existence of extensive amounts of dark matter was first inferred from studies of the dynamics of the cluster galaxies (Zwicky 1933). Early estimates of cluster masses were based on the standard virial theorem or one of its variants. For a spherical system, the virial theorem can be used to derive an expression for the mass of the system:

$$M_{total} = <v^2>/G < F(r)r^{-1}> \tag{1}$$

where $<v^2>$ is the cluster velocity dispersion and $F(r)$ is the fraction of the total mass within a radius r. If one assumes that the visible light traces the total mass, then the cluster mass can be estimated by determining the radial distribution of the galaxies in the cluster and their velocity dispersion. However, The and White (1986) and Merritt (1987) have shown that such estimates may be considerably in error if the visible light does not trace the mass. Merritt showed that for the Coma cluster, relaxing the light-follows-mass constraint allowed the total mass to range from 1.5 to $11 \times 10^{15} h_{50}^{-1}$ solar masses. Thus, even for Coma, perhaps the best studied cluster, the total mass derived from optical observations alone remains uncertain by nearly an order of magnitude. For clusters in general, the optical determination of the mass is hampered by several effects. First, many clusters have not yet virialized and therefore, the application of the virial theorem is inappropriate. Related to cluster dynamics is the problem of internal substructure which can be difficult to identify in optical galaxy counts. Second, the nature of the galaxy orbits in clusters is unknown and hence it cannot be assumed that the galaxy velocity distribution is isotropic. Finally, there are a limited number of bright galaxies, especially in poorer clusters, whose properties (position and velocity) can be measured.

The x-ray emitting gas, while itself an important mass component of clusters is also an ideal tracer – through the hydrostatic equation – of the total underlying mass, without the problems inherent in the optical determinations. The advantages of the x-ray determinations are fourfold. First, the gas is in hydrostatic equilibrium. The applicability of the hydrostatic equation to the gas requires that the sound crossing time of the system be short over the size scale of interest. The sound crossing time in a cluster is given by $\tau_{sound} = 7 \times 10^8 D_{Mpc}/T_8^{1/2}$ years where D_{Mpc} is the size of the region in Mpc and T_8 is the gas temperature in units of 10^8 K. Also, since the sound crossing time is comparable to the dynamical timescale, the gas is generally in equilibrium with the gravitational potential (Cavaliere 1980; Sarazin 1988). Thus, for most hot (x-ray luminous) clusters where $T_8 \sim 0.5 - 1$, the hydrostatic condition is generally applicable over the central few Mpc. Second, the mean free path of protons in the cluster, which can be expressed as $\lambda = 23 T_8^2/n_{-3}$ kpc where n_{-3} is the gas density in units of 10^{-3} cm^{-3} is relatively short. This short mean free path assures that the velocity distribution in the gas is isotropic and that the uncertainty regarding tangential and line-of-sight velocity components which hampers optical analyses, does not apply to the x-ray observations. Third, imaging x-ray spectroscopy provides a means to distinguish background/foreground projections and superposed substructure since one can measure both the morphology of the system and the energy of the x-ray photons. Structures with different mass are characterized by different gas temperatures and are apparent in x-ray images with moderate spectral resolution. Thus, a small group superposed on a rich cluster, while probably apparent in the surface brightness distribution, also appears as a region of cooler gas superposed on the hotter medium of the rich cluster. Finally, there is no physical limit to the statistical precision of the x-ray measurements. While the number of optically bright galaxies is limited, longer x-ray observations provide better determined gas parameters needed in the mass determinations (up to the limit where systematic effects become important).

The hydrostatic equation for a spherical system yields an expression for the gravitating mass, $M(r)$, interior to the radius r

$$M(r) = \frac{-kT_{gas}(r)}{G\mu m_p} \left(\frac{d\ln\rho}{d\ln r} + \frac{d\ln T_{gas}}{d\ln r} \right) r \qquad (2)$$

where $\rho(r)$ is the gas density, G is the gravitational constant, μ = mean molecular weight, m_p = hydrogen mass, k = Boltzmann's constant, and $T_{gas}(r)$ is the gas temperature. Given the isotropy of the gas, the ability to distinguish superpositions, and the validity of the hydrostatic assumption, this expression can be confidently applied to clusters or subclusters having reasonable symmetry. Thus, in principle, for simple geometries, the gravitating mass can be determined as a function of radius using x-ray parameters alone. In addition, Fabricant, Rybicki, and Gorenstein (1984) showed that even for systems with considerable ellipticity (either prolate or oblate) the spherical assumption gives masses which are in error by no more than about 20%.

As shown in equation 2, this measurement of the cluster mass requires the determination of both the gas density distribution, which can be derived from the observed x-ray surface brightness, and the gas temperature distribution. Although overall emission-weighted gas temperatures have now been determined for about 100 clusters (see David 1991), spatially resolved x-ray spectral measurements have been made for only a small but growing number of systems. The *Einstein* Observatory had little effective area above 4 keV, so could map the temperature structure of only cool gas clusters. For M87, the central galaxy in the Virgo cluster, whose temperature is roughly 3×10^7K, Fabricant and Gorenstein (1983) used the *Einstein* IPC observations to determine the gas temperature and density distributions and used these to calculate the radial distribution of the total mass. Similarly, Matilsky *et al.* (1985) mapped the mass around N4696, the central galaxy in the Centaurus cluster. Both analyses reported large amounts of dark matter with the total mass increasing approximately proportional to the radius. Figure 2 shows the determination of the virial mass for M87. Recently Koyama *et al.* (1991) measured the temperature distribution at large distances from M87 using north–south scanning observations from *Ginga*. Their results show an apparently isothermal distribution from M87 towards the north, but an increase in the temperature about three degrees south of M87. If the temperature increase in the south can be at-

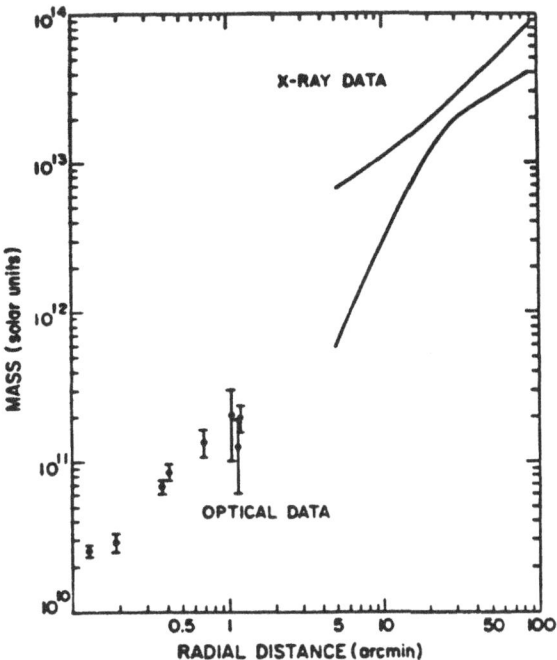

Figure 2 shows constraints on the mass of M87 as a function of radius. The inner points are based on optical velocity dispersion measurements (Sargent et al 1978). Intermediate points are based on globular cluster velocities. The inferred mass in the outer region is determined by Fabricant and Gorenstein (1983) from the x-ray observations.

tributed to the bright point source seen near this location in the ROSAT survey (Bohringer, these proceedings), these results would suggest that the gas around M87 is isothermal to a radius of about 3 degrees which corresponds to a linear distance of nearly 1 Mpc. ROSAT, whose effective area declines rapidly above ~ 2 keV, provides high quality spectroscopic observations of cool clusters, groups and individual elliptical galaxies. Such observations are presently being used to determine virial masses to large radii with high precision. To constrain the virial mass distribution in the hotter Coma cluster, Hughes and his colleagues (Hughes 1989, Hughes *et al.* 1988a, 1988b) have used integrated x-ray spectra from the Tenma, EXOSAT and *Ginga* satellites. Hughes (1991) recently analyzed *Ginga* scanning observations to measure the temperature distribution in Coma at large radii. His results (see Figure 3) show an approximately isothermal core region with the gas temperature declining beyond several core radii.

In general, x-ray spectral observations of both cool and hot clusters suggest a temperature distribution which is fairly isothermal, within a few core radii of the center (with sometimes the addition of a central cooling core). Beyond this region, the temperature declines. David, Hughes and Tucker (1991) have recently modelled clusters assuming either an initial adiabatic temperature distribution or an evolving

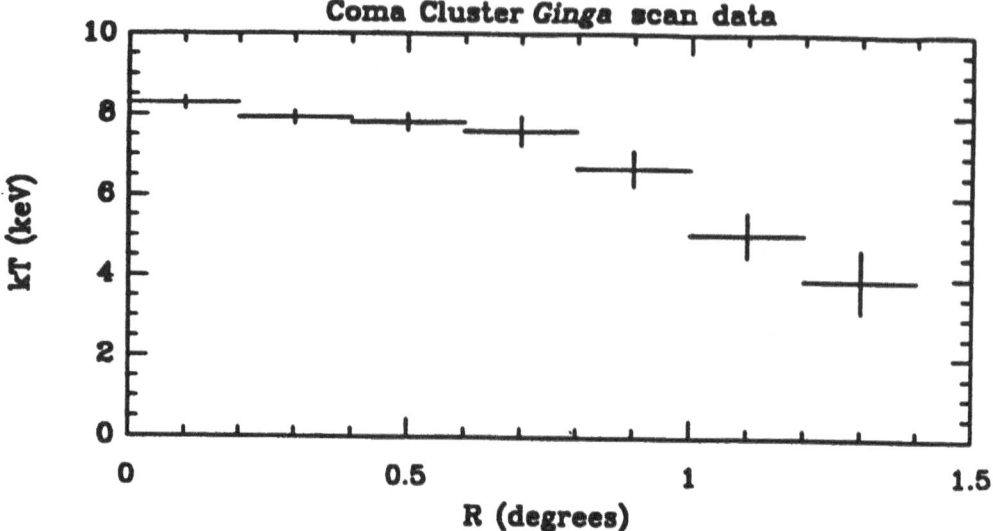

Figure 3 shows the results of thermal fits for the Coma cluster derived from *Ginga* scanning observations by Hughes (1991). The results are shown radially binned around the cluster center, although the entire cluster was scanned. Different directions from the cluster center showed the same temperature behavior, but with less statistical precision.

cluster potential with accreting gas. They found that thermal conduction in the hot gas (with 0.1 or more of the full Spitzer conductivity) transports and redistributes the thermal energy of the gas and, in the central few core radii, transforms the initial non-isothermal temperature distribution into an isothermal one. Furthermore, thermal conduction does not suppress the growth of a cooling flow in the cluster core, but actually enhances the conditions required for the formation of a flow.

For clusters with hot intracluster gas temperatures, new capabilities are required to take full advantage of the promise of x-ray studies for the determination of cluster masses. BBXRT and in 1993, ASTRO-D, a joint Japanese/US mission, have adequate spatial resolution and high energy response to determine spatially resolved temperatures of nearby clusters. For example, at the distance of Coma, the spatial resolution of ASTRO-D ($\sim$ 2 arcminutes) corresponds to less than 100 kpc. Recent BBXRT observations (see Mushotzky, these proceedings), have been made of several nearby high temperature clusters.

3. The Cluster Gas Mass

One can use the x-ray surface brightness to determine physical cluster parameters. For a spherically symmetric cluster or subcluster, the radial x-ray surface brightness distribution is accurately described by the expression

$$S(r) = S(0)(1 + (r/a_x)^2)^{-3\beta+1/2} \tag{3}$$

where $S(0)$ is the central surface brightness, a_x is the cluster core radius and, in the hydrostatic-isothermal β model (Cavaliere and Fusco-Femiano 1976), β is the ratio of the energy per unit mass in the galaxies to that in the gas. With the exception of the central cusp in cooling flow clusters, this form for the surface brightness distribution has been found to be an adequate description to the extent that gas has been traced in clusters, out to 8-10 core radii (see Jones and Forman 1984; Henriksen and Mushotzky 1986; Briel *et al.* 1991).

The expression in equation 3 for the cluster surface brightness can readily be inverted to give the gas density distribution

$$n(r) = n(0)(1 + (r/a_x)^2)^{-3\beta/2} \tag{4}$$

Formally, this inversion requires the assumption that the gas be isothermal. However, the count rate from a fixed mass (and volume) of gas in the *Einstein* IPC energy band (0.5-4.5 keV) changes by less than 10% over the full range of cluster temperatures from 2 to 15 keV (see Figure 4 of Fabricant, Gorenstein, and Lecar 1980). Therefore, the surface brightness distribution can be used to derive the gas density distribution to an accuracy of 3%. For the ROSAT PSPC, the sensitivity of the count rate to the gas temperature is even smaller. Hence, the cluster surface brightness as measured by either *Einstein* or ROSAT is nearly a direct measure of the gas density profile. In addition, measurement of the gas density (and gas mass) relies only on our basic understanding of the physics of thermal radiation from hot gas and does not depend on assumptions regarding the gas distribution. In particular, measuring the gas mass does not require assumptions that the gas distribution is adiabatic, or isothermal, or even that the gas is in hydrostatic equilibrium (see David *et al.* 1990).

The gas mass in clusters is in the range of a few $\times 10^{14}$ solar masses within the central few Mpc. For clusters with measured gas temperatures and measured surface brightness profiles, Figure 4 shows the fraction of the virial mass which is hot gas. In general as the virial mass of the cluster increases, so does the gas mass, so that the fraction of the virial mass which is hot gas remains relatively constant. The intracluster medium is typically 10 to 30% of the cluster virial mass, although for a few clusters, the fraction reaches beyond 40% (see also Henriksen and Mushotzky, 1985). In an $\Omega = 1$ universe, for standard Big Bang nucleosynthesis models, the baryonic fraction of the total mass is less than about 6%. The higher baryonic fraction in clusters implies that for these models, the baryonic mass must be segregated on cluster scales.

To address the questions of the origin of the ICM and the efficiency of galaxy formation in different environments, the mass in intracluster gas can be compared with the cluster galaxy mass. As shown in Figure 5, for a small sample of clusters, the ratio of the mass in gas to that in galaxies varies systematically from poor to rich clusters (David *et al.* 1990). In particular, in groups and poor clusters the gas mass (as measured within five core radii) is comparable to the galaxy mass, while in

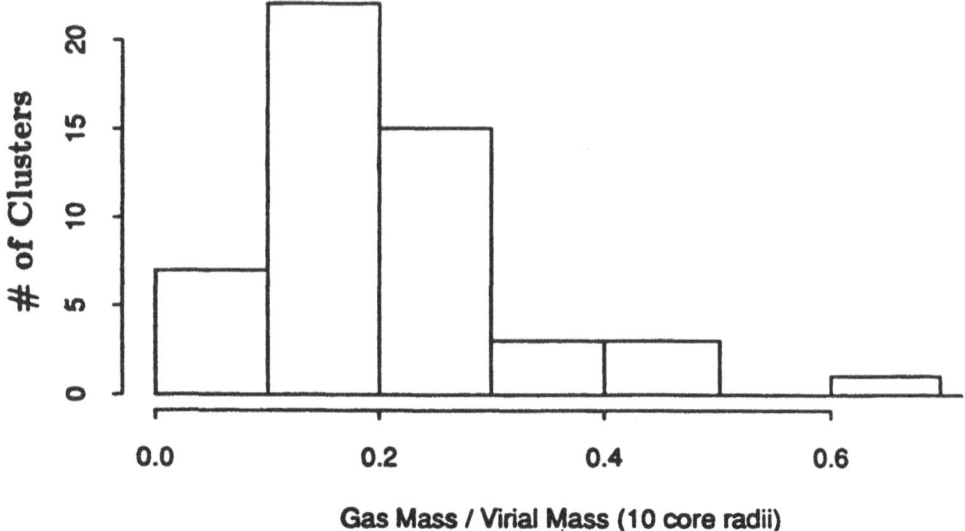

Figure 4 shows a histogram of the fraction of the virial mass which is hot gas. The cluster virial mass is measured within ten cluster core radii (derived from the measured gas temperature assuming that the gas is isothermal) The gas mass of the ICM also is measured within ten core radii.

the richest, hottest clusters, the gas mass is up to six times the mass in the galaxies. For a larger sample of clusters, Arnaud *et al.* (1991) compared the stellar mass and gas mass as measured within 3 Mpc radii. As shown in Figure 6, the gas mass increases as $M_{gas} \propto M_{stellar}^{1.9}$. It is well known that the mass-to-light ratio increases with the size of the system (e.g., Blumenthal *et al.* 1984). However, from poor to rich clusters, the ratio of x-ray emitting gas to virial mass remains relatively constant (Figure 4; Abramopoulos and Ku 1983). Thus, one can understand qualitatively why the ratio of gas mass to stellar mass should increase from poor to rich clusters.

These correlations of gas to galaxy mass can be used to compute the efficiency of galaxy formation and can provide important constraints on formation scenarios. The galaxy formation efficiency—the conversion of baryons from gas to stars in galaxies— can be written as $\epsilon = M_{stellar}/(M_{stellar} + M_{gas})$ or equivalently as $\epsilon = (1 + M_{gas}/M_{stellar})^{-1}$ if we make three assumptions: 1) that clusters are closed systems (no gas is gained or lost), 2) that the gas mass expelled from galaxies is small compared to that presently seen in clusters, and 3) that the two luminous components – galaxies and gas – comprise the total baryonic complement of clusters. Thus, by measuring $M_{gas}/M_{stellar}$, we determine that the efficiency of star (and galaxy) formation ranges from 50% for groups and poor clusters ($M_{gas}/M_{stellar} \approx 1$) to about 15% for the richest clusters ($M_{gas}/M_{stellar} \approx 6$). Alternatively, if we assume that considerable mass is ejected by galaxies and we estimate this ejected mass per unit stellar mass by making the extreme assumption that all the gas mass observed in poor clusters was originally expelled from galaxies, then the galaxy

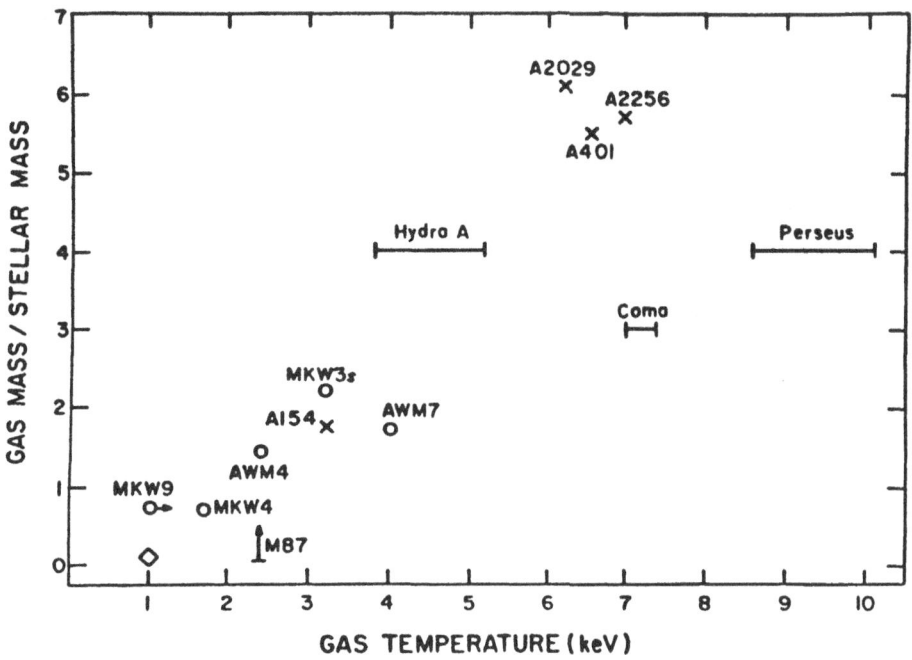

Figure 5 shows the ratio of the gas mass to the stellar mass plotted against the temperature of the ICM. The gas and stellar masses are evaluated within five core radii (from David *et al.* 1990)

formation efficiency in groups is 100%, while that in the richest clusters remains essentially unchanged. As an intermediate example, we can use the results of David, Forman and Jones (1991) who numerically simulated the dynamic evolution of the interstellar medium for elliptical galaxies in which Type II supernova driven winds enrich the surrounding ICM. In their model which best matched the observed ICM abundances, galaxies eject 25% of their initial baryonic mass into the ICM. For this example, the galaxy formation efficiency in groups would increase from 50% to 66%, while that in rich clusters again changes insignificantly. Thus, although the ratio of luminous material (gas+stars) to total mass remains relatively constant for all clusters (Blumenthal *et al.* 1984), the efficiency of galaxy formation decreases as one moves to richer systems. In other words, although the richest systems obviously produced more galaxies, they produced them less efficiently.

David and Blumenthal (1991) carried out calculations of galaxy formation efficiency within the context of cold dark matter scenarios with biasing (Blumenthal et al 1984 and Bardeen *et al.* 1986). In such scenarios, the initial fluctuations which will become groups are larger in amplitude than those which become clusters, since there is more power in the initial fluctuations on smaller linear scales. Biasing requires that galaxies form only in perturbations whose density exceeds a critical density threshold. Since the form of the power spectrum of density perturbations implies that smaller mass perturbations are generally of larger amplitude, galaxy

58

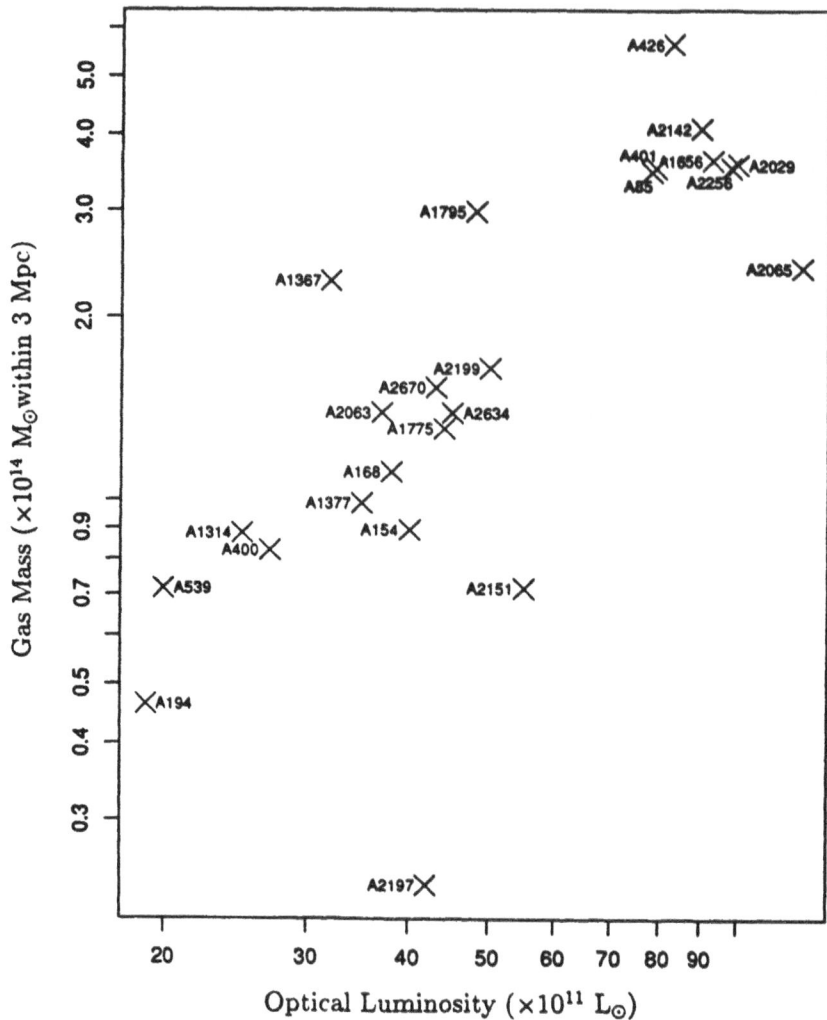

Figure 6 shows the cluster gas mass (from Jones and Forman, 1991) plotted against the stellar luminosity (from Arnaud *et al.* 1991). In this figure, both the gas mass and the stellar luminosity are determined within 3 Mpc of the cluster center. The greater increase in gas mass compared to stellar luminosity between groups and rich clusters suggests a decreasing efficiency of galaxy formation in rich cluster environments.

perturbations in groups are more likely to exceed the critical threshold than are galaxy perturbations in rich clusters. Quantitatively, the predictions derived from CDM models agree with the observations, so long as clusters correspond to 2-3σ fluctuations, which implies that the biasing parameter be close to unity.

4. The Origin of the Intracluster Medium

Since, the ICM mass is equal to or greater than that in stars, it is of particular importance to determine the origin of this large fraction of the known baryonic mass. One of the most important results related to the origin of the ICM was the discovery of emission lines from iron and other heavy elements in the energy spectrum of the hot gas. Cluster x-ray spectra show both that the x-ray emission is thermal in origin (Serlemitsos et al. 1977, Lea et al. 1982) and that the heavy element abundances are between 20 and 50% of the solar value (Mushotzky 1984; Edge 1989; Hatsukade 1989). Since heavy elements can be produced only through thermonuclear reactions in stars or by supernovae, the discovery that the intracluster medium was enriched in iron requires that material, processed through stars, be ejected into the ICM.

The possible origins for the ICM ⹁re that either the gas is primordial or that it has been ejected from galaxies. In an entirely primordial origin for the ICM, the gas has not undergone nucleosynthesis in stellar systems, but has fallen into the gravitational potential of the cluster (e.g., Gunn and Gott 1972). Alternatively, galaxy formation could have been 100% efficient and the observed ICM would then have been ejected or stripped from galaxies comprising the cluster. The near solar abundance of the ICM measured by early x-ray experiments (see Mushotzky 1984 for an early summary and Edge 1989 and Arnaud et al. 1991 for recent compilations) led to the suggestion that a large fraction of the material in the ICM was ejected from galaxies (e.g., DeYoung 1978).

While the enriched material must come from the galaxies (in the absence of a very early stellar Population III outside of galaxies), recent studies suggested that the bulk of the ICM in a rich cluster could not have originated within the galaxies because the ICM mass is up to several times larger than the mass of the galactic stellar component. In particular, numerical modelling of the hot gaseous coronae around elliptical galaxies shows that over a Hubble time these galaxies can enrich the ICM in heavy elements, but can contribute only a fraction (less than $\sim$50%) of their stellar mass to the ICM (David et al. 1991; White 1991). Thus, a considerable fraction of the ICM must be "left over" from the formation of galaxies.

We also can use the ratios of the stellar mass to gas mass in groups and clusters to limit both the mass lost to the ICM from present epoch galaxies as well as any contribution from early, Population III stars. Specifically, so long as the IMF's and the Population III component of groups and clusters are similar and no gas is lost from the systems, then we would expect the stellar contribution to the ICM per unit stellar mass to be the same in all groups and clusters. The approximate equality of gas mass and stellar mass in the low x-ray luminosity Morgan groups (MKW4, MKW9, and AWM4) limits the contribution to the ICM by all stars to no more than the present stellar mass. Thus, in the richest clusters, where the gas mass is three to six times the stellar mass, only a small fraction of the ICM could have been produced in stars. The large fraction of the luminous baryonic matter in the ICM of rich clusters requires a dominant primordial component (in most standard scenarios), while the abundances of heavy elements (25% to 50% of the solar value) in the ICM require injection of enriched material that has undergone nucleosynthesis in stellar systems.

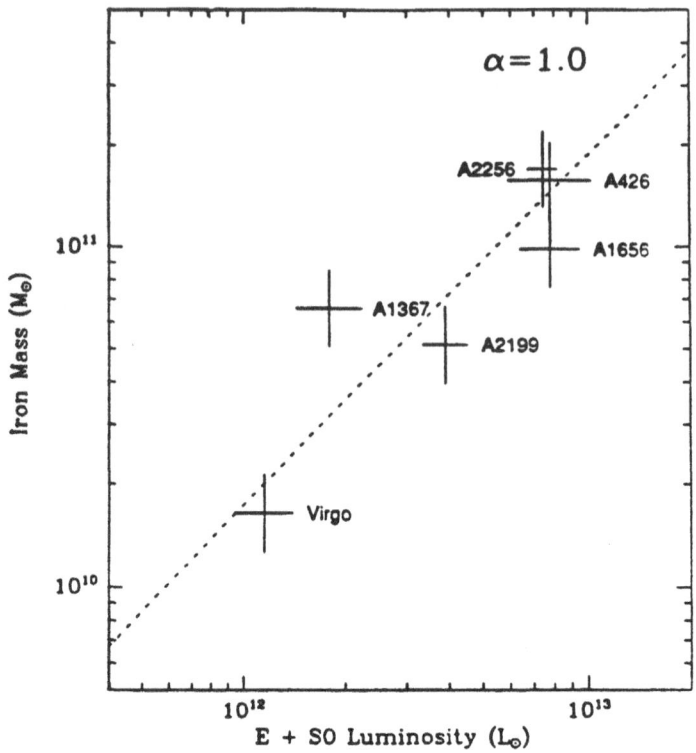

Figure 7 shows the mass of iron in clusters (computed assuming that the iron is uniformly distributed in the ICM) against the integrated stellar luminosity of cluster galaxies (Arnaud *et al.* 1991).

4.1. CORRELATION OF IRON ABUNDANCE WITH T_{GAS}

The increasing ratio of gas mass to stellar mass with cluster mass combined with an understanding of the production of heavy elements, predicts a correlation of heavy element abundance with gas temperature so long as the galaxy populations of these systems are similar (David *et al.* 1991; White 1991).

The groups which are luminous x-ray sources are dense systems and have galaxy populations comparable to rich clusters (Morgan *et al.* 1975). Also, the correlation of galaxy population with local density (Dressler 1980, and Postman and Geller 1984) requires the similarity of the galaxy populations in the groups and clusters. Therefore, the production of heavy elements should be directly proportional to the stellar light, or equivalently stellar mass, since comparable populations will have similar mass-to-light ratios. In fact, Arnaud *et al.* (1991) recently compared the iron mass in clusters (assuming a uniform distribution of iron in the ICM) with the cluster galaxies' optical luminosities. Their results (Figure 7) show that the iron mass in clusters increases linearly with the luminosity (mass) of the early type galaxies in clusters. Thus, as discussed above, although the total gas mass increases with cluster mass faster than does the stellar mass, these results suggest that the production of iron is directly related to the present epoch stellar luminosity. Hence,

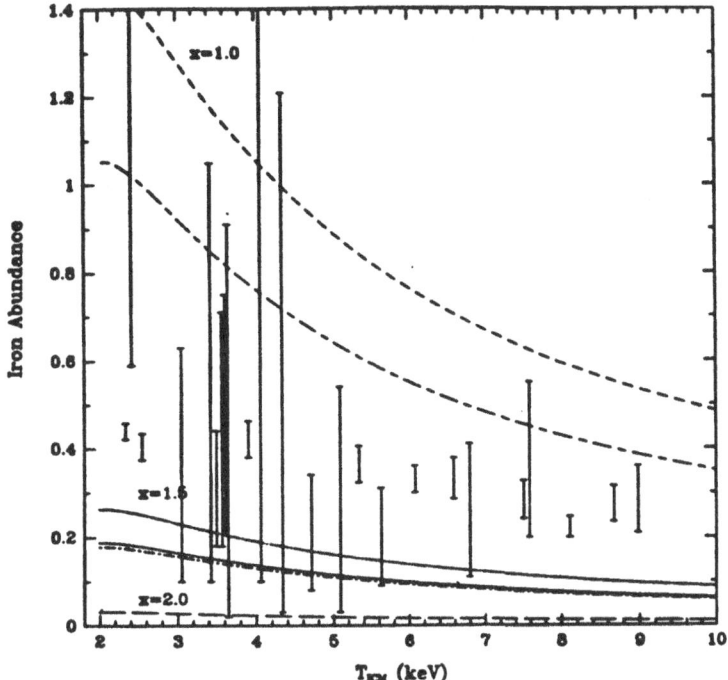

Figure 8. The iron abundance (as a fraction of the solar value; from Arnaud *et al.* 1991) is plotted against gas temperature. The smooth curves (from David *et al.* 1991) are predictions based on galactic wind models with different exponents (x) of the initial mass functions and a parameterization of the relation between $M_{gas}/M_{stellar}$ and T_{gas}. The long dash–dot curve has x=1.5 and Tammann's supernova rate. The short dash–dot curve has x=1.5 and 0.1 of Tammann's rate. The other curves are for 0.25 of Tammann's rate. The long dash–short dash curve is for x=1.0 and assumes that the galaxies retain a solar abundance of iron.

the larger the ratio of gas mass to stellar mass, the more dilute the stellar products like iron. Since $M_{gas}/M_{stellar}$ increases with increasing T_{gas}, hotter clusters (those with larger $M_{gas}/M_{stellar}$) should have lower iron abundances than cooler clusters. This prediction assumes that the clusters and groups are closed systems, i.e., no significant amount of gas is expelled from or accreted into the cluster.

Figure 8 shows quantitative predictions for the correlation of iron abundance with gas temperature and the presently measured iron abundances. (see David *et al.* 1991). The gas in the ICM is enriched both during an early phase of massive star supernovae (type II) and continuing through the present with primarily type I supernovae and mass loss from older stars. As Figure 8 shows, present estimates of supernova yields can explain the observed heavy element abundances in the intracluster gas. These results require that the first generation of stars formed with a relatively flat IMF (with preferentially high masses). The ejected gas is highly enriched and is diluted to the observed values by mixing with the predominantly primordial component of the ICM. Since the gravitational potential of poor clusters is sufficient to bind the enriched material ejected in supernova winds driven from the galaxies, none of the material injected into the ICM should be lost from the cluster as a whole. Furthermore, based on enrichment rates, extensive amounts of matter could not have been expelled by the galaxies and entirely lost from the rich clusters, if their ICM's are to have the observed heavy-element abundances. Therefore, the change in the ratio of gas mass to stellar mass with cluster richness cannot be explained by a loss of hot intracluster material from the groups and poor clusters. The relative constancy over rich and poor clusters of the fraction of the cluster virial mass made up by luminous material (stars and gas) also supports the notion of a "closed" system.

The assumption that groups and clusters are closed systems (i.e., gas is not expelled or accreted in significant quantities) can be tested by observing clusters with progressively lower temperatures. If ejection from the group or cluster becomes important below some temperature, T_{crit}, then one would observe an increasing heavy element abundance from the hottest clusters down to those with temperatures equal to T_{crit}. Below T_{crit}, the winds would serve to expel enriched material and the abundance would decline (or remain constant) as the gas temperature decreases further.

The question of the actual distribution of iron within the intracluster medium is only now beginning to be addressed with observations. (See Mushotzky's contribution to these proceedings for recent BBXRT results.) In the Perseus cluster, Ponman *et al.* (1990) reported that the iron emission was not uniform, but was concentrated around NGC1275 at the center of the cluster. However, this result conflicts with the BBXRT analysis which suggests a more uniform distribution. In the Virgo cluster, Koyama *et al.* (1991) found an iron abundance of half solar within 300 kpc of M87 and a lower iron abundance (0.1 to 0.2 of solar) in the outer cluster region. From *Ginga* scanning observations of the Coma cluster, Hughes (1991) has found a nearly uniform iron abundance of 0.2 solar to a radius of 2 Mpc (see Figure 9).

4.2. THE ENERGY CONTENT OF THE ICM

The changing ratio of gas mass to stellar mass also will affect the energy (or temperature) of the ICM. By measuring the surface brightness profiles and independently by measuring the ratio of the velocity dispersion to the gas temperature, one can estimate β, the energy per unit mass of the galaxies compared to that of the gas, if the cluster ICM (or at least the central core) can be characterized by the isothermal hydrostatic model (Cavaliere and Fusco–Femiano 1976). From the surface brightness profiles, β is generally $\sim 2/3$ for rich clusters. The β values calculated from the measured velocity dispersions (v) and gas temperatures ($\beta = \mu m_h v^2 / 3kT_{gas}$) have a broad range (but see Edge 1989). By comparison to rich clusters, the surface brightness profiles for hot gas around single dominant cluster galaxies such as M87 and the cD groups such as AWM7 yield a somewhat smaller β value $\sim 1/2$ (Kriss *et al.* 1983; their parameter $\alpha = -3\beta$). This implies that the groups and individual central galaxies have more energy per unit mass in gas compared to the constituent galaxies than do rich clusters. For the groups and poor clusters where the stellar

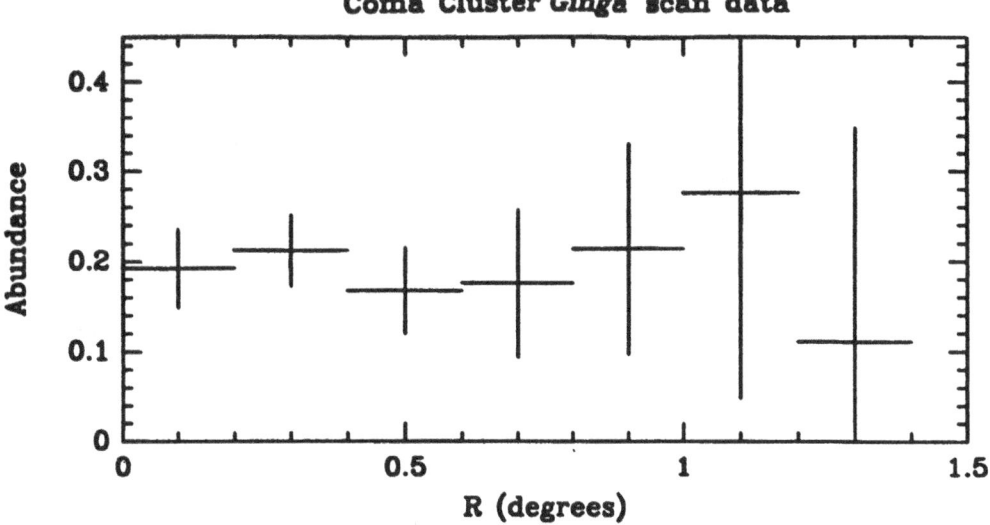

Figure 9. The radial abundance of iron for the Coma cluster derived from *Ginga* scanning observations by Hughes (1991) shows a nearly uniform iron distribution to nearly 2 Mpc (0.8° from the cluster center.

mass is comparable to the gas mass, David *et al.* (1991) and White (1991) have shown that the same material, ejected into the ICM by the galaxies, which provides the requisite heavy element enrichment, can also provide significant heating to account for the observed difference in typical β values between the groups and the clusters.

5. Substructure in Clusters

One of the indicators of the dynamical state of a cluster is the degree of substructure as demonstrated by either the galaxy or the gas distribution. The x-ray observations have been found to be particularly suited to studies of the structure of the cluster potential. *Einstein* images show clear evidence for cluster substructure (e.g. Forman *et al.* 1981; Gioia *et al.* 1982). Figure 10 shows x-ray isointensity contours of four clusters – SC0627-54, A98, A1750, and A2151 – each of which had been selected optically as a single system. The figure shows that each cluster consists of two separate structures. Furthermore, each of the subclusters had a redshift consistent with being members of a single collapsing system (see Forman *et al.* 1981, Henry *et al.* 1981, Dressler and Shectman 1988). The projected separations of the subclusters are typically less than 1.5 Mpc (for $H_0 = 50$ km sec^{-1} Mpc^{-1}). Detailed optical observations of the galaxy distributions confirmed the bimodal nature of these clusters (Beers, Huchra, and Geller 1983). For A98 and A1750, Beers, Geller,

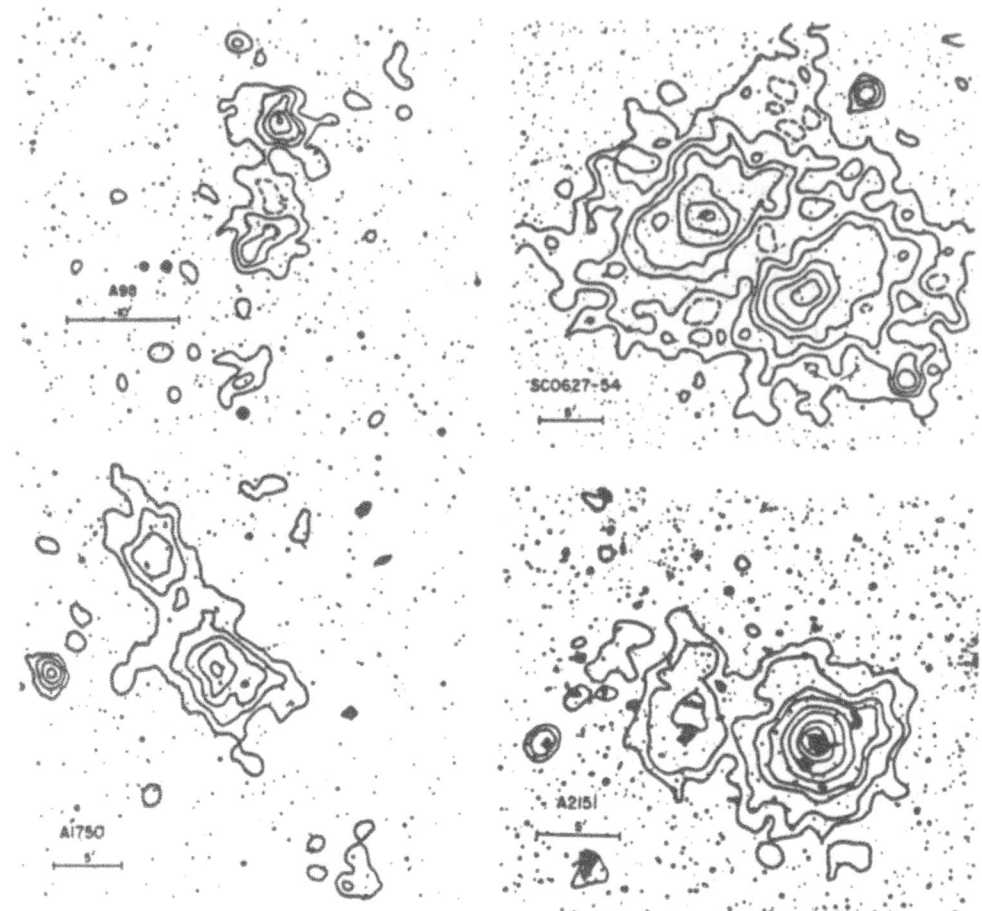

Figure 10. The x-ray iso-intensity contours of four double clusters are shown superposed on the optical photographs. Each subcluster in the pair is at a comparable redshift. These images emphasize the dynamically young state of many rich clusters of galaxies.

and Huchra (1982) and Beers *et al.* (1991) used extensive spectroscopic observations and argued that in each cluster, the two subclusters are gravitationally bound and infalling.

Other x-ray cluster images show still richer substructure. For example, A514, shown in Figure 11, has at least three mass condensations. The separate structures have x-ray luminosities ranging from 6×10^{42} to 6×10^{43} ergs sec^{-1} (Jones and Forman 1991).

The morphological classification of clusters is closely related to the study of substructure (see Bahcall 1977 for a review). The earliest *Einstein* imaging observations suggested the existence of two families of clusters – those with small

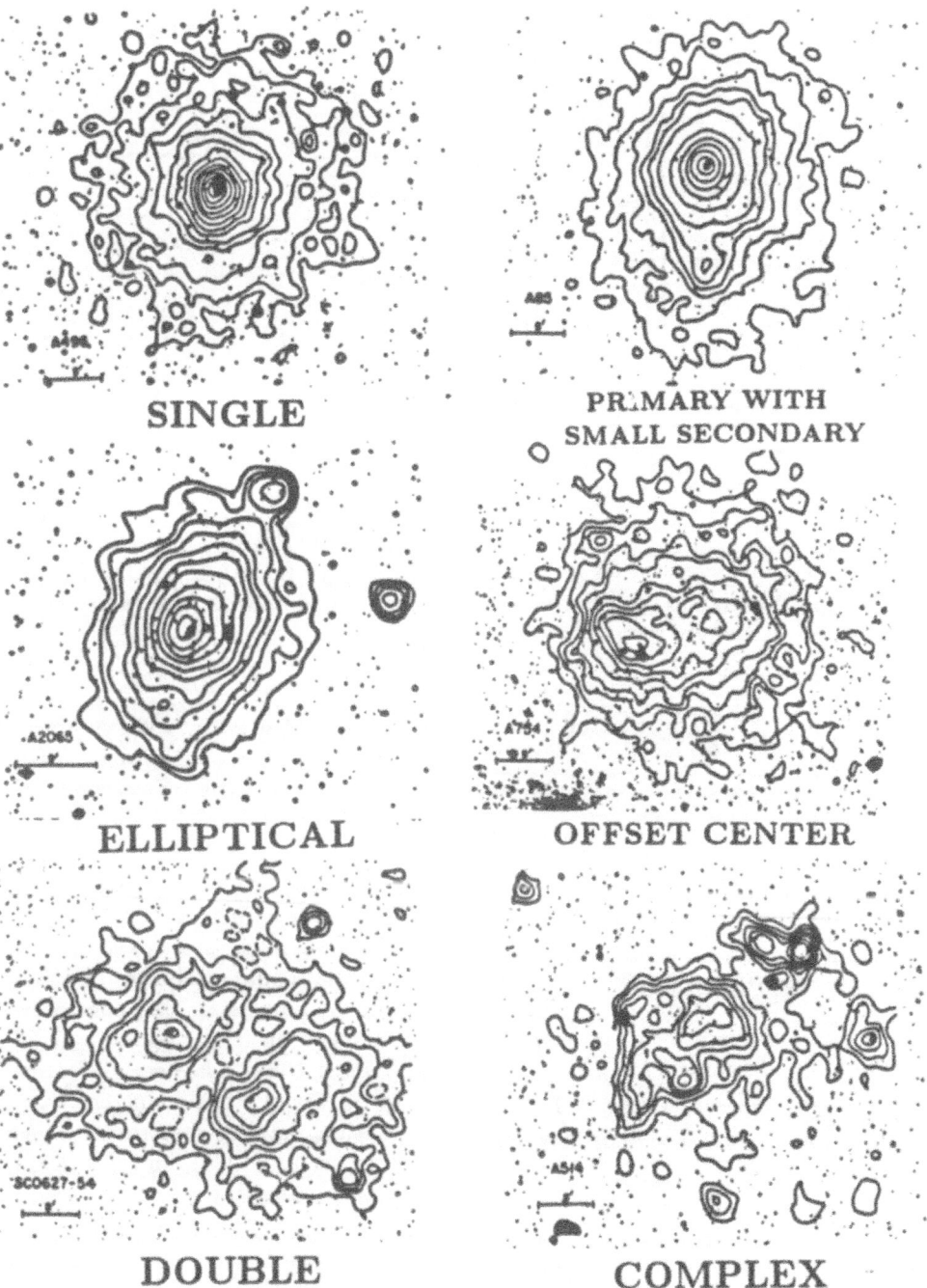

SINGLE

PRIMARY WITH SMALL SECONDARY

ELLIPTICAL

OFFSET CENTER

DOUBLE

COMPLEX

Figure 11 shows x-ray isointensity contours superposed on optical sky prints for six x-ray defined cluster morphological classes.

66

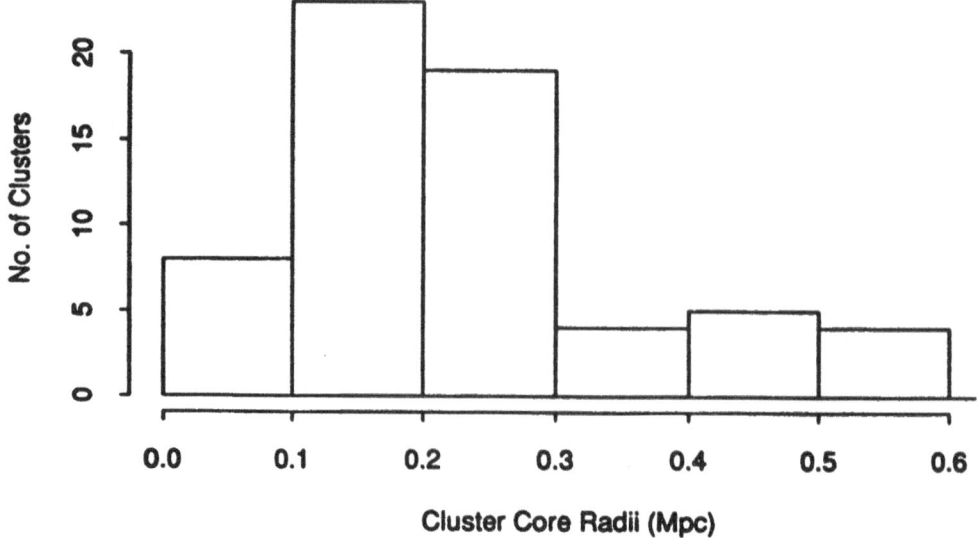

Figure 12 is a histogram of 63 core radii determined from fitting *Einstein* surface brightness profiles (Jones and Forman 1984; 1991). Clusters with core radii greater than about 0.4 Mpc are merger candidates.

core radii having a bright centrally located galaxy and those with larger core radii and no central bright galaxy (Forman and Jones 1982; Jones and Forman 1984). Figure 12, a histogram of well-determined core radii measured from *Einstein* observations, shows that most clusters have core radii between 0.1 and 0.3 Mpc although the distribution extends to 0.6 Mpc.

However, as more detailed analyses have been made, the question has arisen whether the class of large core radius clusters exists or whether they are, in fact, multiple systems in the process of merging and are seen as single systems as a result of overlapping x-ray surface brightness distributions that could not be easily distinguished with the *Einstein* images. For example, the Coma cluster, long considered the prototype of a large core radius, relaxed system, has been suggested to be a multiple system undergoing a merger (Fitchett and Webster 1987). The nearly identical looking cluster A2256 recently has been shown to consist of two subclusters based on ROSAT PSPC images (Briel *et al.* 1991). In addition, detailed spectroscopy such as that by Dressler and Shectman (1988), Fabricant *et al.* (1986), and Fabricant, Kent, and Kurtz (1989) has shown evidence for substructure in previously "single" clusters.

Extensive studies of substructure in clusters have been based on optical galaxy counts and spectroscopy. Geller and Beers (1982), Baier (1978), Dressler and Shectman (1988), Colless (1987), and Fitchett and Webster (1987), among others, found from analysis of galaxy counts and velocities that evidence for substructure is strong and often found in rich clusters. Based on their simulations, West, Oemler, and Dekel (1988) argued that from the optical observations, there was little evidence

Table 1 – Frequencies of Cluster Morphological Classes

X-ray Morphological Class	Example	Number	Percent	Mean L_x
S – single symmetric peak	A401	120	56	29.3×10^{43}
O – offset center	A2319	10	5	44.5
E – elliptical	A2256	31	14	30.8
C – complex, multiple structures	A514	27	13	9.6
D – double (roughly equal components)	A98	13	6	22.1
P – primary with a small secondary	A85	7	3	19.9
G – primarily galaxy emission	A2666	7	3	0.4

for substructure in the inner 1-2 Mpc of clusters (see also West and Bothun 1990). X-ray observations can help clarify the scale and frequency of cluster substructure.

We analyzed *Einstein* observations of a sample of $\sim$ 400 clusters with redshifts less than about 0.2 (Jones and Forman 1991). Of these clusters, 208 are x-ray bright enough to "classify". We defined seven cluster morphological classes. Six of these are illustrated in Figure 11. The seventh class is one in which the x-ray emission is relatively weak and arises primarily from galaxies – usually from a single bright galaxy at the cluster center. Table 1 gives the classes we defined, the name of a member of the class, the number of clusters in each class, the percent of the sample in that class, and the mean x-ray luminosity (within 1 Mpc radius) for the class.

As Table 1 shows, substructure is common. Those classes clearly exhibiting substructure (C, D, and P), comprise 22% of the classifiable sample. These multiple-peaked structures are evidence for a still evolving cluster potential. As noted above, the elliptical (type E; e.g. A2256) and offset center clusters (type O; e.g. A2319), which correspond to large core radii clusters, also may be clusters with substructure. If the large core radius systems are presently undergoing mergers, the percentage of surveyed clusters with substructure rises to about 40%. (If "large core radius" clusters are actually systems undergoing mergers, this also would mean that clusters have only a relatively small range in their core radii – see Figure 12 – all clusters form with about the same characteristic size.)

The detection of subclustering in x-ray images depends on the cluster brightness, the observation length and the spatial resolution of the detector. Our results on the frequency of subclustering are based on 208 clusters (with $z \leq 0.2$) whose *Einstein* IPC observations contain sufficient source counts both for detection of subclusters containing 10% of the total cluster luminosity and for verification that the substructure is extended (rather than a foreground/background point source). Therefore it is easier to detect substructuring in x-ray brighter sources. One also might expect that as clusters evolve and become more x-ray luminous, that the fraction of clusters showing substructure might change. Thus for both scientific reasons and to examine biases in detecting substructure, it is useful to determine the fraction of clusters with substructure for different x-ray luminosity intervals. Table 2 gives these results for our *Einstein* survey. In defining clusters with substructure, we selected classes O, E, C, D, and P.

Although the x-ray observations show considerable substructure for all cluster

Table 2 – Distribution of Clusters with Substructure
Total Clusters 208 [‡]

X-ray Luminosity ($\times 10^{44}$)	Number of Clusters		% With Structure
	With Structure	Total	
< 0.5	17	47	46 [†]
0.5-1.0	22	40	55
1.0-2.0	22	47	47
2.0-4.0	12	36	33
> 4.0	15	38	39

[‡] – Cluster morphological class G (primarily galaxy emission) is omitted from this table.
[†] – This value represents a lower limit since the weakest clusters tend to be classified as having no structure due to a lack of statistically significant structure.

luminosity classes, one should consider the type and scale of subclustering. As West and Bothun (1990) and West (1990) have emphasized, it is important to separate clusters whose substructure is a true vestige of smaller subsystems that have recently merged from others such as those in which a small group lies outside or is falling into a virialized cluster. The "complex" and "double" systems, which are 20% of this sample, are clusters which are undergoing mergers, while some of the x-ray luminous, hot clusters which are types "elliptical"(A2256), "offset center"(A2319), or those with a small secondary subcluster (A85) may contain a dominant virialized cluster.

Extensive simulations have been performed for various cosmological scenarios (e.g., Aarseth, Gott, and Turner 1979; Efstathiou and Eastwood 1981; Frenk, White, and Davis 1983; Klypin and Shandarin 1983; Davis *et al.* 1985; Albrecht and Turok 1985 and others). These simulations are becoming more and more realistic and make theoretical predictions for observed cluster properties. One important property is the presence (and amount) of substructure in and around clusters which can be used to derive the small-scale power in the initial perturbation spectrum. Different gravitational instability scenarios predict different amounts of power. For example, pancake scenarios predict little small scale structure, since small scale perturbations are erased at early epochs. Alternatively, in hierarchical models where large structures are formed by mergers of smaller systems, one would predict considerable amounts of small scale structure. In comparing these models to the x-ray observations, one can begin to use the ensemble of substructure properties, including the scale and percentage of subclustering as a function of x-ray luminosity or gas temperature, to constrain the models. Richstone, Loeb, and Turner (1991) recently used the presence of substructure in clusters to place a lower limit on the density of the universe. Since at late times, the growth of structure is sharply curtailed in a low density universe, they found that the significant fraction of present epoch clusters with substructure implies a dense ($\Omega \gtrsim 0.5$) universe.

6. Conclusions

Since the discovery that clusters of galaxies, as a class, were luminous x-ray sources, x-ray images of the hot intracluster gas have provided unique insights into a variety of problems relating directly to clusters of galaxies. As examples, the mass of gas in the ICM, the presence of heavy elements, and the prevalence of substructure have implications for the origin of the ICM, the interactions of galaxies with the ICM, and the dynamical evolution of clusters. X-ray observations of clusters also address the broader questions of cluster cosmological evolution, the predicted level of CMBR fluctuations (due primarily to the Sunyaev-Zeldovitch effect), galaxy formation efficiency in different environments, tests of cosmological models and limits on cosmological parameters. Future observations promise definitive results on some questions, in particular measurement of the gas mass at large cluster radii, better determination of virial masses and the distribution of dark matter in clusters, measurement of the distribution in the ICM of heavy elements, and conclusive determinations of cluster evolution.

Acknowledgements

We thank Andy Fabian and the organizing committees for making possible the NATO ASI on Clusters and Superclusters of Galaxies. This work was supported through NASA Contracts NAS8-39073 and NAS8-30751.

References

Aarseth, S, Gott, J., and Turner, E. 1979, Ap. J., 236, 43.

Abramopoulos, F. and Ku, W. 1983, Ap. J., 271, 446.

Albert, C., White, R., and Morgan, W. 1977, Ap. J., 211, 309.

Albrecht, A. and Turok, N. 1985, Phys. Rev. Lett., 54, 1868.

Arnaud, M., Rothenflug, R., Boulade, O., Vigroux, L., and Vangioni-Flan, E. 1991, A & A, in press.

Bahcall, N. 1977, Rev. Astr. and Astrophys. 15, 505.

Baier, F. 1978, Astron. Nach. 299, 311.

Bardeen, J. M., Bond, J., Kaiser, N., and Szalay, A. 1986, Ap. J., 304, 15.

Beers, T.C., Geller, M.J., and Huchra, J.P., 1982, Ap. J. 257, 23.

Beers, T.C., Geller, M.J., and Huchra, J.P., 1983, Ap. J. 264, 356.

Blumenthal, G., Faber, S., Primack, J., and Rees, M. 1984, Nature, 311, 517.

Briel, U., Ebeling, H., Edge, A., Hartner, G., Henry, J., Schwarz, R., and Voges, W. 1991, A & A.

Briel, U., Henry, J.P., and Bohringer, H., 1991 in "Clusters and Superclusters of Galaxies," NATO Advanced Studies Institute, Cambridge, England.

Cavaliere, A. and Fusco-Femiano, R. 1976, A. & A., 49, 137.

Cavaliere, A, 1980 in X-ray Astronomy ed. R. Giacconi and G. Setti, Dordrecht Reidel, 20.

Colless, M. 1987, Thesis, University of Cambridge.

David, L. 1991, private communication.

David, L. and Blumenthal, G. 1991, almost submitted to Ap. J.

David, L., Hughes, J., and Tucker, W. 1991, submitted to Ap. J.

David, L., Arnaud, K., Forman, W., and Jones, C. 1990, Ap. J., 356, 32.

David, L., Forman, W., and Jones, C. 1991, Ap. J., in press.

Davis, M., Efstathiou, G., Frenk, C., and White, S. 1985, Ap. J. 292, 371.

DeYoung, D. 1978, Ap. J., 223, 47.

Edge, A. 1989, Ph. D. thesis, Leicester University.

Efstathiou, G. and Eastwood, J. 1981, MNRAS, 194, 503.

Dressler, A. 1980, Ap. J., 236, 351.

70

Dressler, A. and Shectman, S., 1988, A.J. 95, 985.

Fabian, A., Nulsen, P., and Canizares, C. 1984, Nature, 310, 733.

Fabian, A.C., Nulsen, P.E.J., and Canizares, C. 1991, A & A Rev., 2, 191.

Fabricant, D. and Gorenstein, P. 1983, Ap. J., 267, 535.

Fabricant, D., Rybicki, G., and Gorenstein, P. 1984, Ap. J., 286, 186.

Fabricant, D., Lecar, M., and Gorenstein, P. 1981, Ap. J., 241, 552.

Fabricant, D., Beers, T., Geller, M., Gorenstein, P., Huchra, J., and Kurtz, M. 1986, Ap. J., 308, 530.

Fabricant, D., Kent, S., and Kurtz, M. 1989, Ap. J., 336, 77.

Fitchett, M. and Webster, R. 1987, *Ap. J.* 317, 653.

Forman, W. and Jones, C. 1982, Ann. Rev. Astron. Astrophys., 20, 547.

Forman, W. Bechtold, J., Blair, W., Giacconi, R., Van Speybroeck, L., and Jones, C. 1981, Ap. J. (Letters), 243, L133.

Frenk, C., White, S., and Davis, M. 1983, Ap. J., 271, 417.

Geller, M.J., and Beers, T.C. 1982. PASP, 94, 421.

Gioia I. M., Geller, M., Huchra, J., Maccacaro, T., Steiner, J., and Stocke, J. 1982, Ap. J. (Letters), 255, L17.

Gunn, J.E. and Gott, J.R. 1972, *Ap.J.*, 176, 1.

Hatsukade, I. 1989, Ph. D. thesis, Osaka University.

Henriksen, M. and Mushotzky, R. 1985, Ap. J., 292, 441.

Henry, J. P., Henriksen. M., Charles, P., and Thorstensèn, J. 1981, Ap. J. (Letters), 243, L137.

Hughes, J. 1991, private communication.

Hughes, J. 1989, Ap. J., 337, 21.

Hughes. J., Yamashita, K., Okumura, Y., Tsunemi, H., and Matsuoka, M. 1988a, Ap. J., 327, 615.

Hughes, J., Gorenstein, P., and Fabricant, D. 1988b, Ap. J., 329, 82.

Jones, C., Mandel, E., Schwarz, J., Forman, W., Murray, S., and Harnden, R. 1979, Ap. J., 234, L21.

Jones, C., and Forman, W., 1978, Ap. J., 224, 1.

Jones, C. and Forman, W. 1984, Ap. J., 276, 38.

Jones, C. and Forman, W. 1991 submitted to Ap. J.

Klypin, A. and Shandarin, S. 1983, MNRAS, 204, 891.

Koyama, K., Takano, S., and Tawara, Y. 1991, Nature, 350, 135.

Kriss, G., Cioffi, G., and Canizares, C. 1983, Ap. J., 272, 439.

Lea, S., Mushotzky, R., and Holt, S. 1982, Ap. J. 262, 24.

Matilsky, T., Jones, C., and Forman, W. 1985, Ap.J., 291, 621.

Merritt, D. 1987, Ap. J., 313, 121.

Morgan, W.W., Kayser, S., and White, R.A. 1975, Ap.J., 199, 545.

Mushotzky, R., Serlemitsos, P., Smith, B., Boldt, E., and Holt, S. 1978, Ap. J. 225, 21.

Mushotzky, R. 1984, Physica Scripta, 77, 157.

Ponman, T., Bertram, D., Church, M., Eyles, M., Watt, M., Skinner, G., Willmore, A. 1990, Nature, 347, 450.

Postman, M. and Geller, M. 1984, Ap. J., 281, 95.

Richstone, D., Loeb, A, and Turner, E. 1991, submitted to Ap. J. Letters.

Sarazin, C. 1988, *X-ray Emission from Clusters of Galaxies*, Cambridge University Press

Sargent, W., Young, P., Boksenberg, A., Shortridge, K., Lynds, C., and Hartwick, F. 1978, Ap. J., 221, 731.

Schwartz, D., Schwarz, J., Tucker, W. 1980, Ap. J. (Lett.), 238, L59.

Serlemitsos, P., Smith, B., Boldt, E., Holt, S., and Swank, J. 1977, Ap.J. (Letters), 211, L63.

Soltan, A. and Henry, J. P. 1983, Ap. J., 271, 442.

The, L. and White, S. 1986, *AJ*, 92, 1248.

West, M. and Bothun, G. 1990, Ap. J., 350, 36.

West, M., Oemler, A., and Dekel, A. 1988, Ap. J., 327, 1.

White, R. 1991, Ap. J., 367, 69.

Zwicky, F. 1933, Helv. Phys. Acta 6, 110.

ROSAT Observations of Clusters of Galaxies

H. Böhringer, R.A. Schwarz, U.G. Briel, W. Voges, H. Ebeling, G. Hartner,
R.G. Cruddace
Max-Planck-Institut für Extraterrestrische Physik
D-8046 Garching
Germany

ABSTRACT. The ROSAT observatory with its high spatial resolution X-ray telescope is an ideal
instrument for the study of clusters of galaxies. In the first part of the mission an All Sky X-ray
Survey was conducted with ROSAT. In this paper we discuss first results of ROSAT observations
of galaxy clusters from the All Sky Survey and some early pointed observations.

Results from detailed morphological studies on some prominent nearby clusters with high
enough photon yield in the survey like Virgo, Perseus, nad AWM7 are presented to illustrate the
capabilities of ROSAT in galaxy cluster studies. In addition we discuss the prospects for detections
of clusters of galaxies in the All Sky Survey and their possible implication for cosmological research.

1. Introduction

The German X-ray observatory ROSAT (Trümper *et al.*, 1991) which was launched
on June 1st, 1990 has a higher sensitivity and spatial resolution than previous X-
ray telescopes. The soft X-ray energy window of the ROSAT X-ray telescope (0.1
to 2.4 keV) covers part of the wavelength region where the hot intracluster gas of
galaxy clusters has its radiative emission maximum. ROSAT is therefore an ideal
instrument for the detection and morphological study of clusters of galaxies.

The ROSAT mission is divided in two parts. Half a year at the beginning of the
mission was devoted to an All Sky Survey that lasted from August 1990 to January
1991. The survey is followed by pointed observations which will also cover detailed
studies of individual clusters. There is therefore a twofold interest in research on
clusters of galaxies conducted with the ROSAT X-ray Observatory: The pointed
observations with longer exposure times will allow detailed studies of the morphology, dynamical state, and other physical properties of individual clusters, while the
survey results mainly offer the unique possibility to study a large statistical all sky
sample of X-ray selected clusters.

The X-ray emitting hot intracluster gas of clusters of galaxies is optically thin
and has a temperature very close to the virial temperature corresponding to the
depth of the gravitational potential well of the cluster. Thus the expected and
actually observed intracluster gas temperatures range from 2 to 10 keV. In that
temperature range the X-ray flux observed in the ROSAT band depends directly

A. C. Fabian (ed.), Clusters and Superclusters of Galaxies, 71–90.

on the emission meassure of the gas with very little variation with temperature. With ROSAT images one can therefore directly trace the projected gas density distribution. The hot intracluster gas fills the gravitational potential well nearly in hydrostatic equilibrium and thus illuminates the shape of the potential of the cluster. The mass and the depth of the gravitational potential can be deduced from the gas distribution within the limits to which the gas temperature is known. The main application of ROSAT studies of individual cluster will therefore be concerned with the determination of the gas mass and distribution, the energy radiated by the gas and the existence of cooling flows, the determination of cluster masses and the study of the distribution and nature of the "dark matter", the evolutionary state of the cluster, the morphology of the gravitational potential well in connection with gravitational lenses, and the study of the Sunyaev-Zel'dovich effect in combination with radio observations.

In the All Sky Survey thousands of clusters of galaxies should be detected. First statistical results indicate that most of these clusters should be new detections. A effort in optical follow-up investigations will be required to positively identify the cluster candidates selected from the All Sky Survey. The final sample of X-ray selected clusters of galaxies will offer a unique possibility for statistical studies. Up to now there is no clear and unique concept what properties characterize a cluster of galaxies and what selection rules should be applied to find clusters observationally. A comparison of the properties of the sample of X-ray selected clusters with those compiled in optical catalogues (e.g. Abell, 1958; Zwicky et al., 1964; Abell, Corwin, and Olowin, 1990) should give some very interesting information how the nature of the galaxy clusters found changes with the applied selection rules. One may than be able to use different pragmatic cluster definitions for different applications.

Despite this ambiguity in the exact definition of a cluster of galaxies, most of the observed clusters can still be characterized quite well by being close to the idealized equilibrium state given for example by the King model (King, 1962). In that sense clusters of galaxies are the largest units in the Universe that can be characterized by a proper constitution of their own in contrast to larger features as for example walls, voids, and filaments.

The ROSAT All Sky Sample of X-ray selected clusters of galaxies will therefore be a good basis to study the large scale structure of the Universe. Using the mass or luminosity function of clusters of galaxies one can probe for length scales of the order of 10 Mpc (see article by Henry in this volume). Larger scales can be studied through a correlation analysis of the spatial distribution of the clusters up to the extent of the survey which could easily exceed 1000 Mpc (A Hubble constant of 50 km s^{-1} Mpc^{-1} is used here and throughout the text).

This paper provides some details about the ROSAT observatory and the observing mission in section 2. Results on the X-ray morphology and physical structure of clusters and cooling flows are given in section 3. Section 4 provides a dicussion of the first statistical results of cluster detections in the Rosat All Sky Survey, and section 5 is a summary.

2. The ROSAT Mission

A summary of first results obtained with ROSAT which provide a good impression of the capabilities of the observatory were published by Trümper et al.(1991). A detailed description of the instument parameters relevant to observers can be found in the "Rosat AO2 Call for Proposals" (MPE, 1991). The main instrument on board of the ROSAT observatory is the X-ray telescope that consists of a fourfold nested Wolter type mirror configuration. It covers an energy range from 0.07 to 2.4 keV. The second instrument is the Wide Field EUV Camera with a spectral window from 20 to 300 eV. Due to the strong absorption of the interstellar medium of the Galaxy in the EUV band the Wide Field Camera is only of limited application for extragalactic objects such as clusters and we will here concentrate on the results from the X-ray telescope. The first observations of a cooling flow cluster, A2199, with the ROSAT Wide Field Camera has been reported by G. Stewart at this conference.

The X-ray telescope carries two types of focal plane detectors. The Position Sensitive Proportional Counter (PSPC) has a spatial resolution better than 30 arcsec within a radius of 20 arcmin around the telescope axis. It has a limited energy resolution which is roughly about 45 % FWHM at 1 keV and the relative resolution varies roughly inversely proportional to the square root of the energy. The second type of detector, the high resolution imager (HRI), has a better spatial resolution of less than 5 arcsec near the telescope axis. This detector has no energy discrimination, however, and is about half as sensitive. Compared to the EINSTEIN IPC the ROSAT PSPC has an increased sensitivity that is roughly a factor of two to three higher (depending very much on the type of X-ray spectrum; for most cluster observations the sensitivity increase is about a factor of two). Fig. 1 shows the effective photon collecting area of the X-ray telescope with the PSPC detector as a function of energy. The energy band is devided into two parts due to photoabsorption of carbon in the plastic window of the PSPC.

X-ray spectra of optically thin, hot thermal gas as they would be observed by the ROSAT PSPC instrument are given in Fig. 2 for four different gas temperatures. The spectra are normalized to the same emission measure. The observed temperatures for the intracluster gas range from about 2 to $12 \cdot 10^7$ K. One notes that in this temperature range the spectra are quite similar. The differences are becoming larger only for temperatures below 10^6 K. The ROSAT energy window is unfortunately too small to observe the exponential cutoff in the Bremsstrahlung spectrum for the high cluster temperatures. Also the absolute energy flux integrated over the ROSAT window (defined as 0.1 to 2.4 keV) is not very sensitive for the gas temperature. This has the disadvantage that cluster gas temperatures are very difficult to measure with ROSAT but it has the advantage on the other hand that the X-ray flux is directly related to the emission measure and thus gas densities can be determined without having precise temperature information.

The small variation of the spectral shape shown in Fig. 2 can be used to determine the temperatures for spectra with very good photon statistics. There are two major features that characterize the change of the PSPC X-ray spectrum for temperatures between 10^7 K and 10^8 K which can been seen in the Figure. The

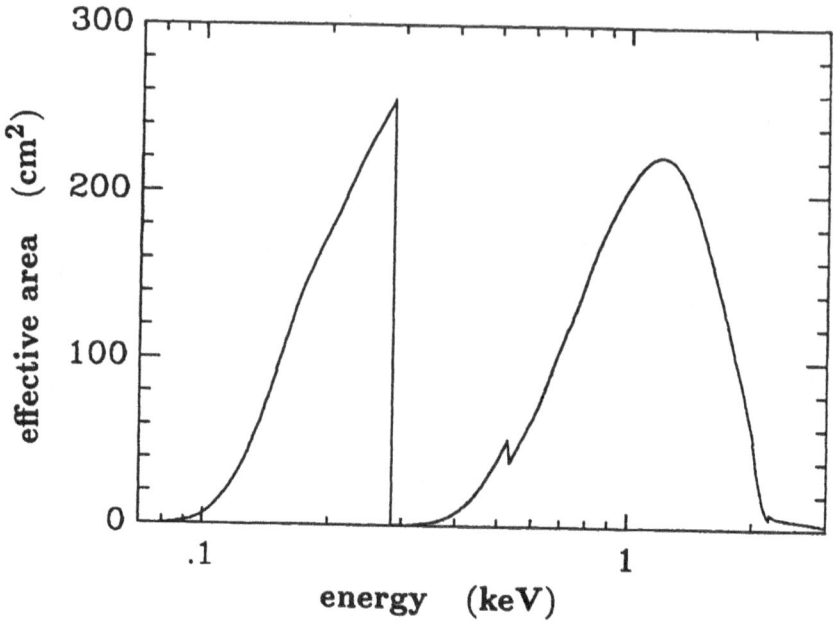

Fig. 1. Effective photon collecting area of the ROSAT X-ray telescope with the PSPC detector as a function of energy

temperature dependence of the gaunt factor causes a change of the slope of the free-free radiation spectrum which is reflected in the total spectrum. The second feature is the peak around 1 keV which is produced mainly by a blend of iron L-shell lines. These lines only become prominent at temperatures below about 2 keV. The height of this line emission peak depends therefore not only on the gas temperature but also on the metallicity of the gas.

The energy discrimination of the ROSAT PSPC detector is also very useful to partly discriminate between the foreground and the cluster X-ray emission. Most of the X-ray background radiation comes from hot gas within our galaxy which has a temperature of the order of one million degree. It is even slightly softer than the radiation for $3 \cdot 10^6$ K gas shown in Fig. 2. By filtering out the photons with energies below 0.4 keV for example one can reduce the X-ray background to about one fourth while of the order of 75% of the cluster emission is retained. This method is frequently used to obtain a good signal to noise ratio for images of clusters. The cut at 0.4 keV has been adopted in the standard analysis of the ROSAT data to divide the ROSAT energy window in two parts and to determine hardness ratios of discovered X-ray sources.

After two month of calibration and verification observations half a year of the ROSAT mission was devoted to an All Sky Survey which was almost completed at the end of January 1991. A strip covering about 5% of the sky was missed at

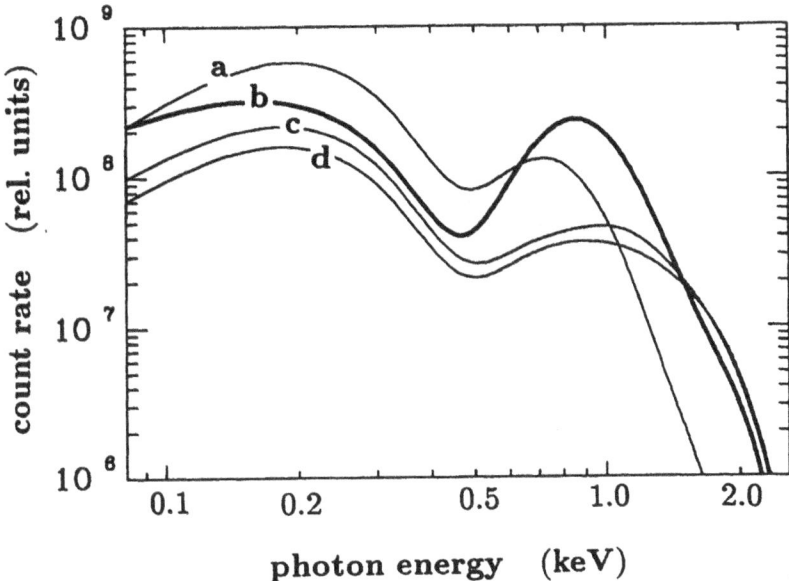

Fig. 2. PSPC X-ray spectra for optically thin thermal plasma at temperatures of $3 \cdot 10^6$ K, (a), $1 \cdot 10^7$ K, (b), $3 \cdot 10^7$ K, (c), and $1 \cdot 10^8$ K, (d). The abundances of the heavy elements used in the calculations had half their solar values.

the time due to an instrument failure and was successfully scanned recently, thus completing the All Sky Survey. During the Survey the sky was scanned by the ROSAT telescope in great circles in a plane perpendicular to the solar direction. Following the sun the whole sky is thus covered in half a year. Due to the overlap of the scan circles at the ecliptic poles the exposure is much greater at these poles. The instrument has to be shut down frequently when the orbit penetrates the radiation belts. This affects the southern sky more severely than the north due to the South Atlantic Anomaly in the Earth's magnetic field. The minimum exposure time in the equatorial regions is about 400 sec while exposure times of several 10 000 sec are reached at the ecliptic poles. Fig. 3 gives an exposure time histogram for the entire sky as predicted from preflight simulations.

In the first round of the standard analysis of the ROSAT All Sky Survey the data are organized in 90 two degree wide strips following great ecliptic circles. The first processing is almost completed covering 85 strips at present. If the current source detection rate is extrapolated to the whole sky one finds an estimated number of about 60,000 detected sources. First tests with the All Sky Survey data show that a sensitivity limit of about 10^{-12} erg s^{-1} cm^{-2} (for the energy band 0.1 to 2.4 keV) is easily reached for most of the sky for objects with cluster type spectra.

In the current standard analysis program possible X-ray sources are found in a

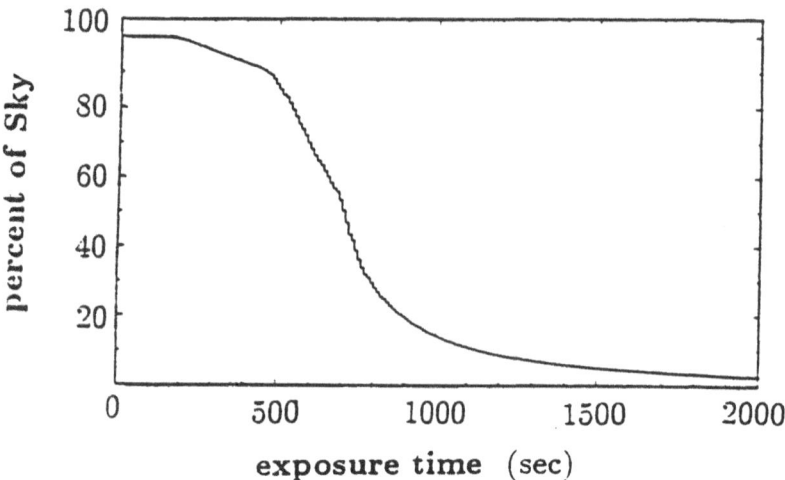

Fig. 3. Exposure time histogram for the ROSAT All Sky Survey as predicted from preflight simulations

first screening step by a sliding window technique using square windows of different sizes. In a second step a maximum likelihood analysis for each source candidate is performed and data below a threshold corresponding to about a 5 σ significance are rejected. Preflight simulations showed that this leads to less than about 1% false detections. During the maximum likelihood analysis the data are also inspected for a possible extent of the source by fitting varying Gaussian profiles. Thus the output of the source detection in the standard analysis provides the parameters: count rate, hardness ratio, extent parameters, and a test for time variability. A statistical analysis of these data in search for detections of galaxy clusters is described in section 4.

3. X-ray Morphology of some Nearby Clusters

In this section we present results on observations of nearby clusters in the ROSAT All Sky Survey and from an early pointed observation on A 2256.

A prominent cluster in our immediate neighbourhood is the Virgo cluster of galaxies. In optical images the cluster extends over an area of at least 10 degrees diameter. Due to its large size the cluster has so far never been imaged as a whole in X-rays. Scans of the cluster region by the GINGA collimated detector indicate that the X-ray emission is very extended (Takano *et al.*, 1989).

In the ROSAT All Sky Survey the Virgo cluster region was scanned with an average exposure time of about 460 sec. Fig. 4 shows a contour plot of the Survey data. For this plot only the hard photons have been used to enhance the signal to noise ratio and the image has been smoothed with a Gaussian filter. The most prominent features are the giant X-ray halo of M87 and the much smaller halo

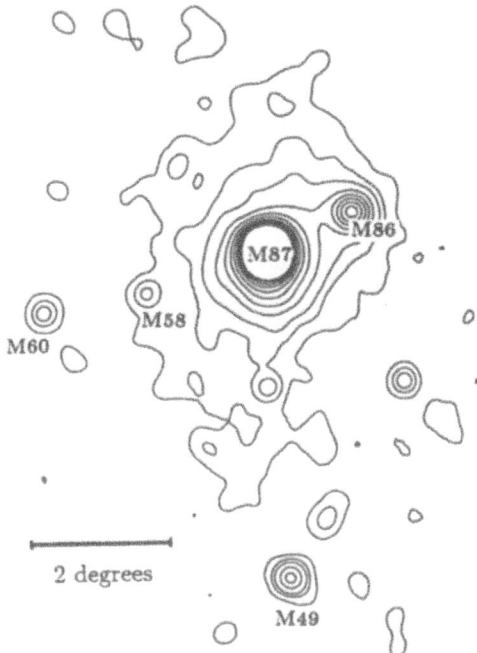

Fig. 4. Contour plot of an X-ray images of the Virgo cluster from the ROSAT All Sky Survey. Only hard photons ($E \geq 0.4$ keV) and a Gaussian filter with a width of 20 arcmin were used to construct this image. The most prominent elliptical galaxies with observed X-ray emission are labelled in the plot.

of M49 (NGC 4472). Besides the almost spherically symmetric halos there is also diffuse emission between the two halos that is observed clearly for the first time. The bright elliptical galaxiey M58 - and possibly M60 - which are labelled in the Figure, are also enclosed by the diffuse X-ray emission region. It is interesting that the diffuse emission is extended asymmetrically to the east between the two halos, because a similar shape is visible in the contour plots of the galaxy densities from the optical survey by Binggeli et al.(1987).

The lowest contour in Fig. 4 has been chosen such that it lies well above the observed variations of the X-ray background in the ROSAT Survey. A deeper look at the cluster which would allow us to determine wether the two X-ray halos of M87 and M49 are really connected by diffuse emission can only be performed when more is known about the variations of the X-ray background in the Survey on scales of about a degree.

In the following we will concentrate on the analysis of the two X-ray halos. The azimuthally averaged surface brightness profile looks very much like that from the EINSTEIN IPC observations (Fabricant et al., 1980), but in the present data the X-ray emission can be traced to larger radii. One can therefore determine that the gas mass contained in the halo of M87 within a radius of 0.63 Mpc (2 degrees) is

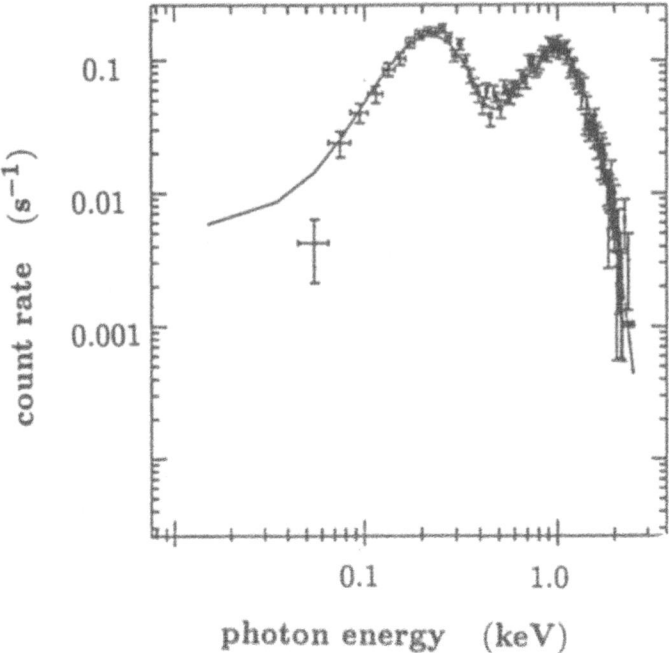

Fig. 5. ROSAT PSPC X-ray spectrum of the innermost region of the X-ray halo of M87 within a radius of 7 arcmin. A spectrum calculated by means of the Raymond and Smith code has been fitted to the data yielding a best fit at 1.86 keV which is shown in the Figure as straight line.

about $8 \cdot 10^{12}$ M$_\odot$.

The image of the M87 X-ray halo contains more than 25,000 photons within a radius of 60 arcmin. This allows us to split the large data sample into several radial bins and to obtain a good X-ray spectrum for each of the regions separately. We have chosen here a binning with radial boundaries at 7, 15, 30, and 60 arcmin. Fig. 5 shows for example the background subtracted X-ray spectrum of the innermost bin together with the best fitting theoretical spectrum obtained with the radiation code of Raymond and Smith (1977). A region outside the $1°$ radius, where the cluster emission is very faint, has been used to determine the background spectrum that was subtracted from the data.

χ^2 fits of theoretical spectra with various temperatures to the same X-ray spectrum, where the normalization factor, the interstellar absorption column density, and the metallicity of the gas have been treated as free fitting parameters, show that the temperature in the innermost bin is quite well confined to the range 1.7 - 2.1 keV. The number of photons is lower in the outer bins while the temperature is increasing. Therefore the determination of the temperature becomes more and more difficult. Fig. 6 shows for example the χ^2 fits as a function of temperature for the second bin (7 to 15 arcmin). The error bars are asymmetric with a smaller

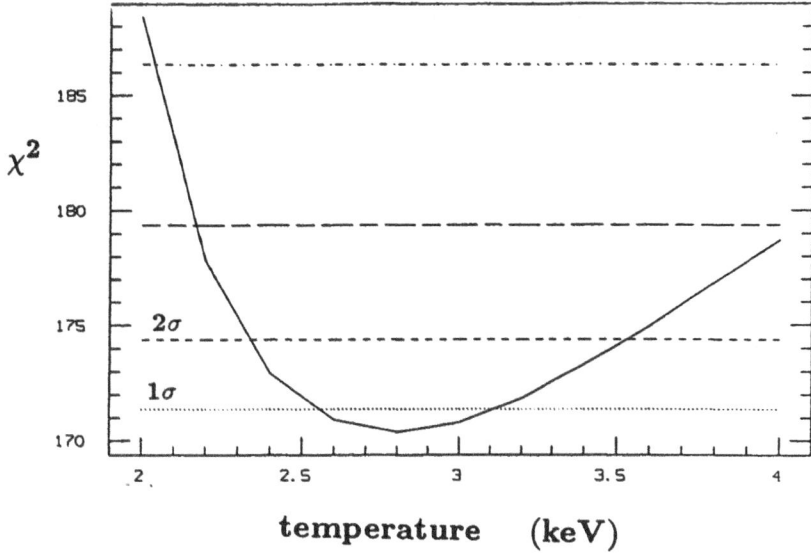

temperature   (keV)

Fig. 6. χ^2 fit of theoretical spectra as a function of temperature to the X-ray spectrum of the gas in the halo of M87 in the region between 7 and 15 arcmin radius.

uncertainty towards the lower temperatures, which is also a result of the fact that the variation of the spectrum is larger at lower temperatures.

Fig. 7 summarizes these results in the form of a temperature profile for the M87 halo. If the average temperature is determined by χ^2 fitting over the whole selected region one obtains a value of $2.4\left[^{+0.35}_{-0.2}\right]$ keV which is in good agreement with the results of the GINGA satellite by Koyama et al.(1991) who found a temperature of 2.19 ($\pm$0.04) keV. Only in the innermost part of the cluster, at radii $\leq$ 15 arcmin, is a significant temperature drop observed.

A deprojection analysis of the surface brightness profile (following Fabian et al., 1981) yields a cooling flow radius of about 120 kpc inside which the cooling time is shorter than the Hubble time. This cooling flow radius of $\sim$ 20 arcmin is consistent with the observed temperature drop inside a 15 arcmin radius and also its magnitude is within the limits of the predictions of a one-phase cooling flow model. The flow rate in the cooling flow that is determined from the surface brightness profile has a value of 20 ($\pm$4) $M_\odot$ y^{-1} in good accord with the results by Stewart et al. (1984).

One can also try to determine the metallicity of the gas through spectral fitting. If this is done for the coldest, innermost region one finds a value of 0.44 ($\pm$0.13) of the solar value which is in surprisingly good agreement with the GINGA results obtained by Koyama et al. (1991) of 0.44 ($\pm$0.018). But one has to be very careful

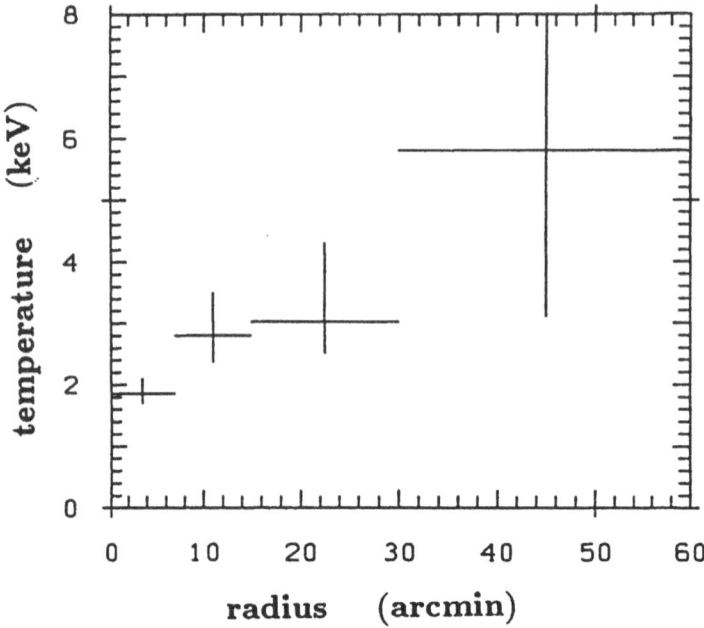

Fig. 7. Temperature profile of the gas in the X-ray halo of M87. The vertical error bars show the 2σ uncertainties.

in the interpretation of the ROSAT spectra. The gas in the cooling flow region should be expected to have a broad temperature distribution. The main spectral feature that has a dependence on the metallicity is the peak in the spectrum at 1 keV which is mainly produced by iron L-shell lines. These lines have their emission maximum at temperatures which lie mostly below the bulk temperature of the gas (see e.g. Canizares et al., 1983). Therefore gas colder than about 10^7 K can give a more than proportional contribution to the 1 keV peak resulting in erroneously high metallicities. A correct analysis can only be performed through a multiphase cooling flow model where calculated spectra for the proper temperature distribution are fitted to the observations.

The X-ray halo around M49 is much fainter and smaller than the one of M87. Fig. 8 shows the azimuthally averaged surface brightness profile of the M49 halo. The core radius of M49 can probably not be resolved in the Survey image due to the limited resolution of the PSPC and the poor photon statistics. Fitting a β-model (Jones and Forman, 1984) to the data one obtains upper limits to the core radius of about 2.5 arcmin and the β parameter of 0.8. Earlier EINSTEIN HRI observations give lower values for these parameters with a core radius around 7 arcsec and $\beta \sim 0.4$ (e.g. Thomas, 1986). A spectral fit to the data within a radius of 10 arcmin gives temperature values in the range 1.25 (±0.3) keV, but the photon statistics are much poorer than for the analysis of the M87 data. A much better

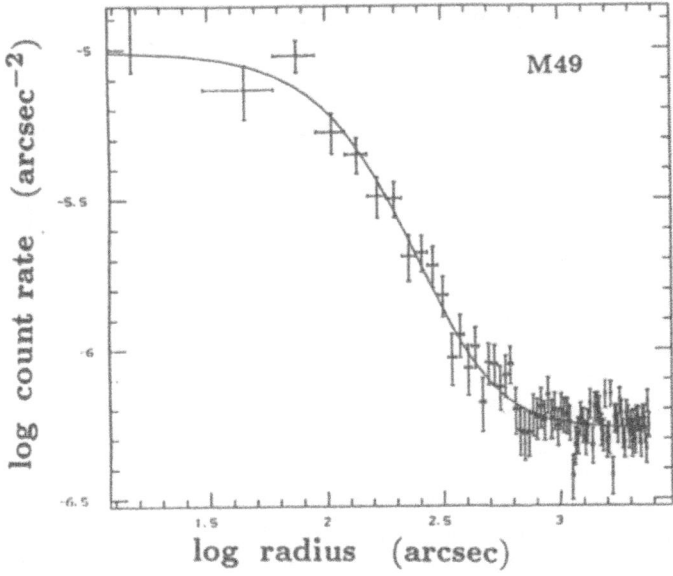

Fig. 8. Azimuthally averaged surface brightness profile of the X-ray halo of M49 (NGC 4472) from the ROSAT All Sky Survey. A comparison with EINSTEIN HRI observations indicate the the central maximum of the halo is not yet well resolved in this image.

result will be achieved with the scheduled pointed observation. The temperature determined with GINGA of 1.9 ($\pm$0.4) is higher than our result, but due to the large field of view GINGA may have seen some additional harder background emission from the cluster. Results from an EINSTEIN IPC analysis by Forman and Jones (referenced in Thomas, 1986) yield temperatures in the range 0.9 to 1.38 keV within a radius of about 5 arcmin, in good agreement with the present result.

The brightest cluster in the Survey is the Perseus cluster of galaxies. Some data have already been presented by Schwarz *et al.* in the book of contributed talks. Fig. 9 shows a contour plot of the Perseus cluster from the ROSAT Survey. Again the area of the image is larger than has been explored previously in X-rays and a trace of the prominent optical feature, the bright chain of galaxies extending from NGC 1275 to IC 310, is for the first time clearly observed at X-ray wavelength. IC 310 is in fact situated in a local X-ray emission maximum. Collimator experiments on rockets (Cash *et al.*, 1976) and on SPARTAN (Snyder *et al.*, 1990) had earlier found some indication of X-ray emission from the bright line of galaxies. The total ROSAT PSPC count rate of Perseus integrated out to a radius of 42 arcmin is about 33.6 cts s^{-1} corresponding to a luminosity of $1.4 \cdot 10^{45}$ erg s^{-1}. The luminosity of the substructure containing IC 310 is only about 3.5% of that.

One of the puzzles about the Perseus cluster is the fact that reasonable models reproducing the observed distribution of the X-ray emitting gas have to assume a shallower gravitational potential than is suggested by the optically observed galaxy velocity distribution (e.g. Fabian *et al.*, 1981; Kent and Sargent, 1983). Analysing the X-ray spectra for different parts of the X-ray image of Perseus one notes a temperature gradient across the cluster at radii outside the cooling flow region. The temperature values determined for the different sectors marked in Fig. 9 are: $2.06\left[^{+0.71}_{-0.36}\right]$ keV for (1), $2.15\left[^{+0.9}_{-0.25}\right]$ keV for (2), $3.61\left[^{+6.1}_{-1.3}\right]$ keV for (3), $3.0\left[^{+2.15}_{-0.7}\right]$ keV for (4), and ≥ 5.2 keV for (5), respectively. The errors quoted correspond to 2σ deviations. Thus the cluster is hotter in the western part. An indication of an east-west temperature gradient was also found in the SPARTAN observations by Snyder *et al.* (1990). The average temperature determined for the Perseus cluster image from the ROSAT Survey data is $5.27\left[^{+1.33}_{-1.22}\right]$ keV in good accord with previous results from collimated proportional counter spectrometers. Such a strong temperature gradient could have for example been caused by a recent infall of a major subcluster. The results of the hydrodynamic simulations of a cluster merger event presented by Schindler at the conference did show such an asymmetric heating of the cluster leaving the side of the infall hotter. This would imply that Perseus is a very unsettled cluster and therefore may explain the discrepancy between the mass estimates from velocity and gas distribution data.

In addition to the large scale temperature gradient a significant temperature drop is observed towards the centre of the cluster, which is connected to the presence of a strong cooling flow as in the case of M87. The temperature in the innermost bin for which X-ray spectra were analysed (with a radius of 3.3 arcmin) is $2.63\left[^{+1.0}_{-0.6}\right]$ keV.

The cluster Abell 2256, which has a size comparable to Coma and is located at a redshift of about z=0.06, was studied extensively in X-rays with EINSTEIN and in the optical (115 known galaxy redshifts) and was considered as a well relaxed cluster with a somewhat elliptical shape (Fabricant *et al.*, 1986, 1989). A2256 was observed with ROSAT in the early calibration phase in the pointing mode for 17,323 sec. Analysing the X-ray image of A2256, shown in Fig. 10, Briel *et al.*(1991) discovered that the data can best be described as a composition of two subclusters. The main body appears to be a very symmetric cluster while the second component, which is quite compact, is distorted on the side pointing towards the main cluster.

This seems to imply that in A2256 two subclusters are merging. A reinspection of the galaxy velocity data supports this interpretation. If the galaxy velocities are averaged in a sector ranging from p.a. 220° to 310°, which is the region covered by the infalling smaller component, one finds an average velocity of 16977 km s^{-1}. The average galaxy velocity in the remaining sector is 17,817 km s^{-1} implying that the smaller component with the lower velocity is falling in from behind. The velocity difference is rather high and if it is caused purely by the gravitational attraction of the two subsystems they must have approached to within 1 to 2 Mpc of each other, if the mass of the total system is about 10^{15} M$_\odot$. This is consistent with the fact that the two gas halos overlap and that the smaller component is already distorted.

A2256 is also one of the few strong and extended cluster radio halos discovered so far (Hanisch, 1982). The shape of the radio halo is rather peculiar (e.g. Bridle and Fomalont, 1976; Bridle *et al.*, 1979) and has been a puzzle to observers. Fig. 11

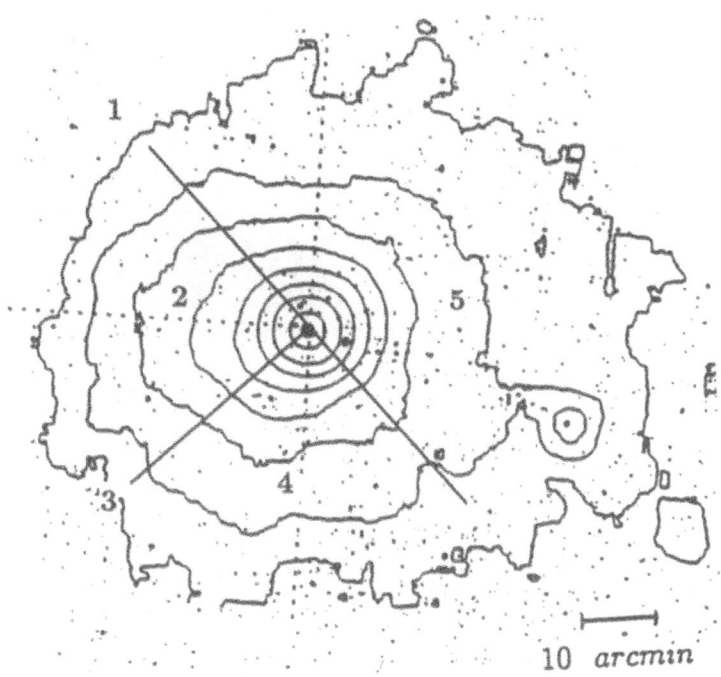

10 arcmin

Fig. 9. X-ray image of the Perseus cluster from the ROSAT All Sky Survey. The contour levels are logarithmically spaced. The X-ray map is superposed onto the POSS plate. The main maximum coincides with NGC 1275 while the small western maximum falls ontop of the galaxy IC 310. The gas temperatures for the sectors inticated by the numbers are given in the text.

shows the 610 MHz Westerboork map of A2256 from Bridle and Fomalont (1976). The locations of the two maxima in the X-ray surface brightness are marked on the radio map to show the correspondance of the X-ray and radio features. The gas cloud of the second component lies behind the three prominent head-tail radio galaxies. Their tails are swept backwards probably tracing the infall direction of the merging clump (as remarked also by Fabian and Daines in the proceedings of the contributed papers and posters). Taken together there is observational evidence in the X-ray, optical, and radio wavelength region that A2256 is a merging cluster.

In addition to the three head-tail radio galaxies there is a very extended diffuse emission region on the side of the infalling clump from which more than half of the total radio flux of the A2256 radio halo is received. From the radio spectrum of A2256 published by Bridle et $al.$ (1979) one can calculate the energy content of the relativistic electrons in the radio halo responsible for the observed synchrotron emission to be about $5 \cdot 10^{58}$ erg $B^{-2.5}$ h_{50}^{-2}, where B is the magnetic field strength in μGauss. In steady state the electron population needs an energy supply to com-

84

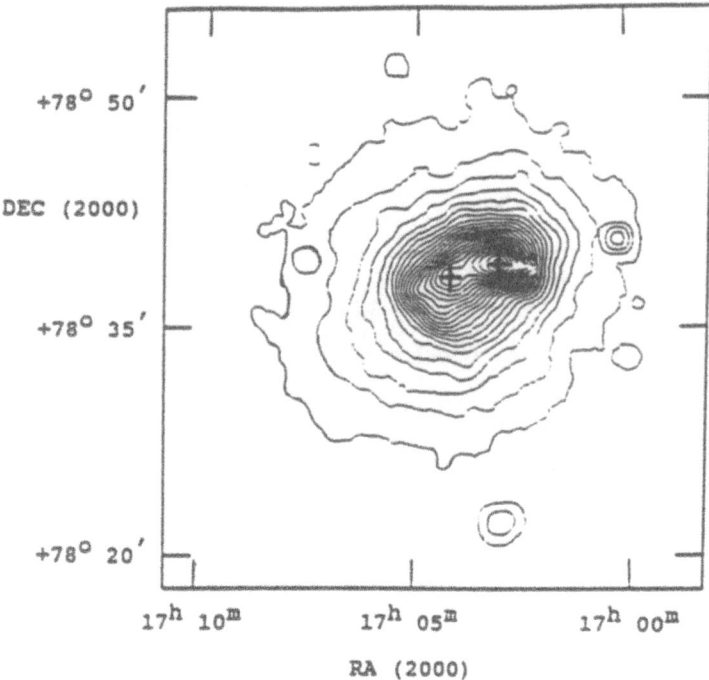

Fig. 10. X-ray image of the cluster A2256 taken with the ROSAT PSPC. The contours are in intervals of 2 PSPC counts s^{-1} per 8×8 arcsec^2 pixel and the image has been smoothed with a Gaussian with a width of 48 arcsec. Two crosses that mark the maxima reappear in Fig. 11 and allow to correlate this X-ray image with a radio map.

pensate the Compton losses of about $2 \cdot 10^{42}$ erg s^{-1} B^{-2} h_{50}^{-2}. It has been proposed previously that radio halos may be powered by relativistic electrons or other relativistic particles leaking out from the radio lobes of the active cluster galaxies or alternatively that the synchrotron electrons are produced by *in situ* acceleration in the turbulent wakes of galaxies in the intracluster medium (e.g. Hanish, 1982). All these mechanisms have to be extremly effective and in the first case the particle transport has to be very rapid to explain the energetics deduced from the observations. A merging cluster like A2256 offers another very attractive solution. The intracluster medium shock wave resulting from the merger event could be an ideal site for cosmic ray acceleration. An analysis of the X-ray images of the two subsystems of A2256 implies that the gas density at the possible contact surface of the two systems, at a distance of about 1 Mpc from each of the subcluster centres, is of the order of 10^{-4} cm^{-3}. For a velocity difference of about 1500 km s^{-1} the energy released in the shock wave is of the order of $3 \cdot 10^{45}$ erg s^{-1}, enough to power the cloud of relativistic electrons. The diffuse radio halo may thus mark the trail of the incoming clump where the shock wave acceleration has been operative in the

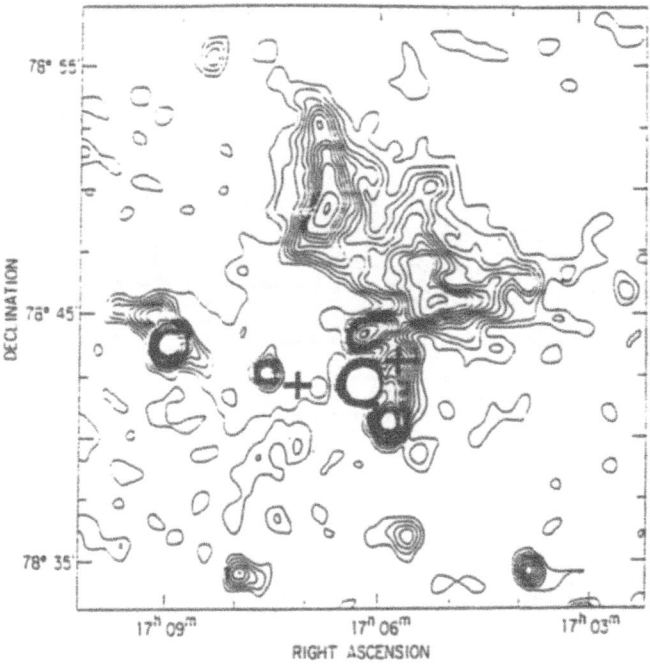

Fig. 11. Radio map of Abell 2256 at 610 MHz obtained with the Westerbork Synthesis Radio Telescope by Bridle and Fomalont (1976). The two crosses mark the location of the two maxima in the X-ray image. The three prominent head-tail galaxies seem to be associated with the smaller infalling subcluster. While the intracluster gas of this subcluster is being stopped by the main cluster gas, the galaxies are falling ahead of the gas component.

recent past.

One can in turn calculate the inverse Compton radiation that is produced by the observed synchrotron electrons and obtains an X-ray flux at 1 keV of about $1 \cdot 10^{-13}$ erg s^{-1} cm^{-2} keV^{-1} B^{-2} which is neglegible in comparison to the observed X-ray flux from the second component of about $6 \cdot 10^{-12}$ erg s^{-1} cm^{-2} in the ROSAT band (for magnetic fields above or around 1 μG). This is consistent with the X-ray observations which imply that the radiation must come mostly from thermal emission of gas at around 2 keV (see Briel et al., 1991).

AWM 7 was found as a group of galaxies around the prominent cD galaxy NGC 1129 (Albert et al., 1977). This object was observed for about 420 sec in the ROSAT All Sky Survey and detected with a count rate of $\sim$ 7.8 cts s^{-1} (inside a cluster radius of 30 arcmin, corresponding to $\sim$ 0.9 Mpc). The cD-group has a galactic latitude below 20° and thus did not enter the ACO catalogue, but it is at least comparable to a richness class 0 Abell cluster. Fig. 12 shows the X-ray image of

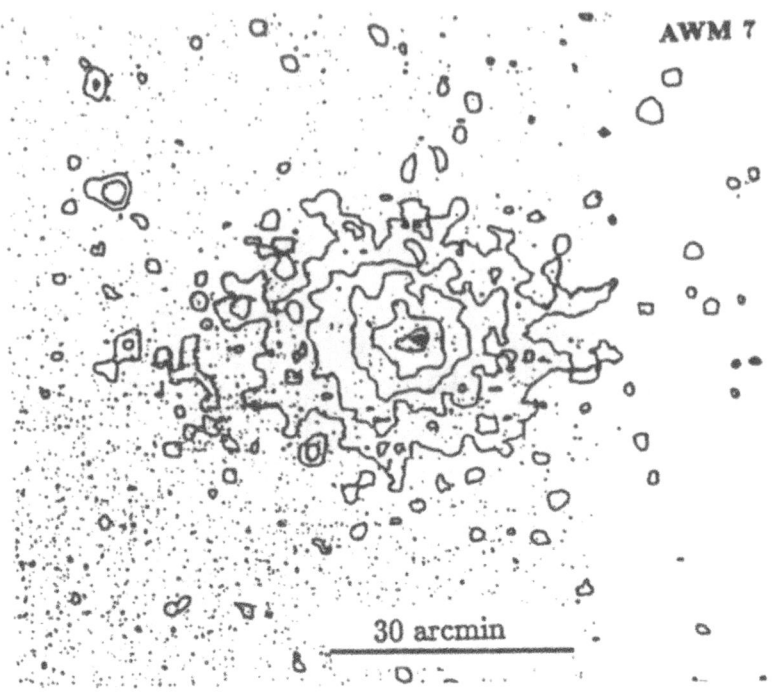

Fig. 12. X-ray image of the galaxy cluster AWM 7 from the ROSAT All Sky Survey. The X-ray emission is almost centred on the central dominant cD galaxy NGC 1129. The x-ray emission can be traced out to a radius of ~ 1 Mpc.

AWM 7 from the All Sky Survey. The central cD galaxy NGC 1129 is found to have a small offset to the X-ray centre determined at intermediate radii.

An azimuthally averaged X-ray surface brightness profile for the cD-group AWM 7 is shown in Fig. 13. An asymmetric component of the X-ray surface brightness located at the position of NGC 1129 has been removed before the azimuthal average was taken. A β model fit to the surface brightness profile yields the parameters $\beta = 0.6$ and the core radius $r_c = 5.1$ arcmin (~ 150 kpc). The total gas mass out to a radius of 0.9 Mpc is about $2.5 \cdot 10^{13}$ $M_\odot$. Thus the gas to virial mass ratio out to this radius in the cluster is roughly 7 to 20%.

4. Cluster Detections in the All Sky Survey

While in the previous sections results on observations of single clusters were reviewed, this section provides a discussion of the prospects of cluster detections in the ROSAT All Sky Survey in general. Using the known exposure distribution over the sky which for the most part is about 400 sec, and assuming that 15 to 25 photons are sufficient to detect a cluster source, one can roughly calculate a flux limit for sources with cluster type spectra. For the areas of the sky where the interstellar hydrogen column density is less than $4 \cdot 10^{20}$ cm^{-2} one finds a limiting flux of 0.6 to $1 \cdot 10^{-12}$ erg cm^{-2} s^{-1}. In an area of ~ 30 deg^2 at the ecliptic poles the flux

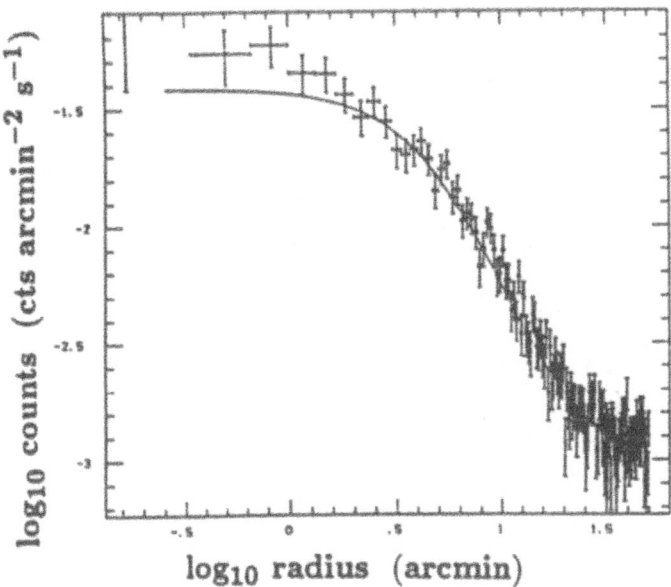

Fig. 13. X-ray surface brightness profile of the cD-cluster AWM7 obtained from the data shown in Fig. 12. An asymmetric peak centred on NGC 1129 near the centre of the cluster has been cut out from the image for the determination of the surface brightness profile.

limit is a good order of magnitude lower. A first analysis of the Survey data using logN/logS statistics of known clusters as well as inspecting the detection limit of a sample of EINSTEIN cluster X-ray sources shows that the detection limit lies close to 10^{-12} erg cm^{-2} s^{-1} for most of the sky.

These estimates do not take into account that some of the faint extended cluster sources with a total flux above the quoted detection limit will be missed in the current analysis. Some of these types of clusters have been found by eye inspection of the ROSAT Survey sky maps. Currently the detection software is being improved so that some of these sources may be detected automatically.

Using the known X-ray luminosity function from clusters detected in previous X-ray surveys one can estimate the number of clusters that should be detected in the ROSAT All Sky Survey. The results are shown in Fig. 14 which gives the expected number of cluster detections in redshift bins for a flux limit of $5 \cdot 10^{-13}$ erg cm^{-2} s^{-1} for an area of 8 ster of the sky (excluding a 40 deg wide strip around the galactic plane) and for a flux limit of $5 \cdot 10^{-14}$ cm^{-2} s^{-1} in the area of 30 deg^2 around the ecliptic poles. For the calculations an X-ray luminosity function for the 2 - 10 keV energy band of

$$n(L_x) = 3 \cdot 10^{-7} \ exp(L_x/8.2) \ L_x^{-1.6} \qquad (1)$$

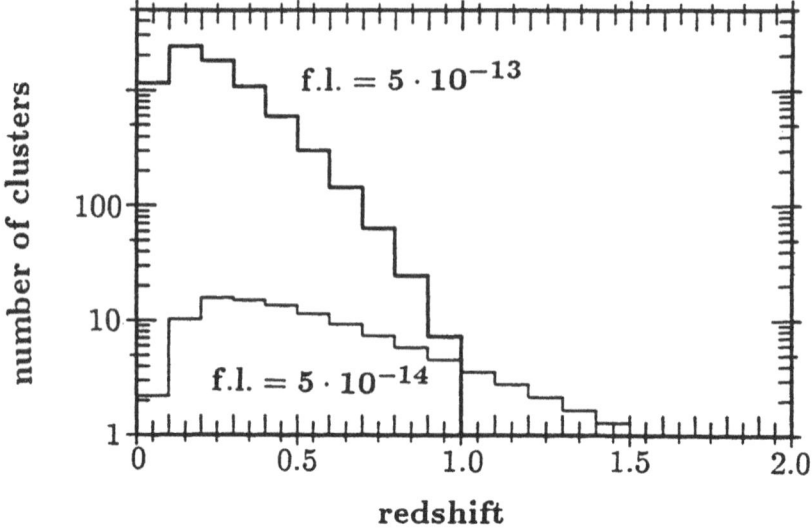

Fig. 14. Histogram of the expected number of cluster detections in the All Sky Survey as a function of redshift for a flux limit of $5 \cdot 10^{-13}$ erg cm^{-2} s^{-1} for the entire sky and for a flux limit of $5 \cdot 10^{-14}$ erg cm^{-2} s^{-1} for an area of 30 deg^2 around the ecliptic poles.

has been used where L_x is in units of 10^{44} erg s^{-1} and $n(L_x)$ in units of Mpc^{-3}. Eq. (1) is a good fit to the combined results for HEAO 1 of Kowalski et al. (1984) and the EXOSAT/HEAO 1 results of Edge et al. (1990). A temperature - flux relation of $L_x = 24 \cdot (T/10^8 K)^{2.7}$ was used to convert these fluxes into fluxes for the ROSAT band. Standard cosmological parameters ($H_o = 50$ km Mpc^{-1} s^{-1}, $\Omega = 1$, and $\Lambda = 0$) were used for the calculations. The results show that about 4000 to 8000 clusters of galaxies should be found in the All Sky Survey for a flux limit between $5 \cdot 10^{-13}$ and 10^{-12} erg cm^{-2} s^{-1}.

The above calculations do not allow for evolutionary effects of clusters of galaxies. If such effects are taken into account for the rich clusters as they were found in the EINSTEIN Medium Sensitivity Survey (EMSS; Gioia et al., 1990 a,b) and in the HEAO 1/EXOSAT data (Edge et al., 1990) the histograms have a steep cutoff at around z $\sim$ 0.5 and z $\sim$ 0.9, respectively. The estimated total number of cluster detections is not very much affected by that, however. Most of the clusters detected in the survey will have redshifts below 0.3 but a few rich clusters out to z $\sim$ 1 should also be found.

From the experience with the EINSTEIN Medium Sensitivity Survey which is only slightly deeper than the ROSAT Survey one would expect to find clusters of galaxies for about 10% of the X-ray sources. With the present detection rate this gives an estimate of about 6000 cluster detections in good agreement with the above prediction.

Searching for clusters from the Abell, Corvin, and Olowin catalog (ACO, 1989) among the ROSAT X-ray sources from 38 strips (45 % of the sky) 479 ACO clusters were detected with a confidence of 90% that the sample is not contaminated by other X-ray emitting objects (as described in more detail by Ebeling in the proceedings of the contributed papers). This list includes the supplementary galaxy groups in the ACO compilation. Lowering the certainty threshold the sample can be increased to slightly more than 500. Compared to the total number of 2686 ACO clusters in the study area one finds that only around 18% of the ACO objects are detected in the Survey.

Of the order of 300 additional clusters from the Zwicky catalogue and other published compilations were found in the same study region. Extrapolation of these detection rates to the entire sky gives a number of only 1500 to 2000 catalogued clusters of galaxies that should be found in the ROSAT All Sky Survey. Compared to the expected number of cluster detections the known clusters would make up only about one third of the final X-ray selected cluster sample.

In addition to published astronomical catalogues a catalogue of objects derived from COSMOS scans of the UK Schmidt plates of the southern sky (Heydon-Dumbleton et al., 1989) is used for the ROSAT source identification (in collaboration with the Royal Observatory, Edinburgh and the Naval Research Laboratory, Washington). The object catalogue distinguishes between objects that appear starlike on the plates and possible galaxies. The regions around X-ray sources were searched for galaxy overdensities and a first statistical analysis shows that the number of clusters found is commensurate with the above estimates and the pre-flight simulations. Samples of these clusters have been at La Silla and CTIO and clusters were found in the redshift range 0.1 to 0.25. At the moment studies are being conducted to quantify the detection statistics of clusters by this X-ray survey-optical plate correlation. It seems that most of the missing clusters in the estimates above can be supplied by this identification method.

5. Summary

The results from ROSAT observations of individual clusters of galaxies have shown that far more details in the cluster structures can be resolved, using the higher spatial resolution and the energy discrimination of the ROSAT PSPC detector, than possible up to now. The few ROSAT HRI images which have been analysed so far promise even higher spatial resolution. The program of long pointed observations of clusters, part completed and part still in planning, promise a far more detailed determination of cluster masses, gas content, and dynamical states of clusters.

An extremely interesting result from the ROSAT All Sky Survey is the fact that probably only one third of the X-ray detected clusters of galaxies are known optically and only a fraction of the known optical clusters are found in the Survey. Surveying in X-rays one is more sensitive for the deepest parts of the potential wells of clusters while the optical detection depends more on the global morphology. It is interesting that both detection techniques lead to quite different cluster samples.

The X-ray selected all sky sample of clusters of galaxies will therefore be very important for cosmological studies such as a spatial correlation analysis, analysis of the cluster mass function, and evolution studies.

Acknowledgements

We would like to thank the ROSAT team for the help in the data analysis and for making these unique observations possible. H.B. acknowledges support from the Deutsche Forschungsgemeinschaft.

References

Abell, G.O., 1958,*Astrophys. J. Suppl.*, **3**,211.
Abell, G.O., Corwin, H.G., Olowin, R.P., 1989,*Astrophys. J. Suppl.*, **70**, 1.
Albert, C.E., White, R.A., Morgan, W.W., 1977, *Astrophys. J.*, **211**, 309.
Binggeli, B., Tammann, G.A., Sandage, A., 1987, *Astron. J.*, **94**, 251.
Bridle, A.H., Fomalont, E.B., 1976, *Astron. Astrophys.*, **52**, 107.
Bridle, A.H., Fomalont, E.B., Miley, G.K., Valentijn, E.A., 1979, *Astron. Astrophys.*, **80**, 201.
Briel, U.G., Henry, J.P., Schwarz, R.A., Böhringer, H., Ebeling, H., Edge, A.C., Hartner, G.D., Schindler, S., & Voges, W., 1991, *Astron. Astrophys.*, **246** L10.
Canizares, C.R., Clark, G.W., Jernigen, J.G., Markert, T.H., *Astrophys. J.*, **262**, 33.
Cash, W., Malina, R.F., Wolf, R.S., 1976, *Astrophys. J.*, **209**, L111.
Edge, A.C., Stewart, G.C., Fabian, A.C., & Arnaud, K.A., 1990, *Mon. Not. R. astr. Soc.*, **245**, 559.
Fabian, A.C., Hu, E.M., Cowie, L.L., Grindley, J., 1981, *Astrophys. J.*, **248**, 47.
Fabricant, D., Lecar, M., Gorenstein, P., 1980, *Astrophys. J.*, **241**, 552.
Fabricant, D., Kybicki, G., & Gorenstein, P., 1984, *Astrophys. J.*, **286**, 186.
Fabricant, D., Kent, S., & Kurtz, M., 1989, *Astrophys. J.*, bf 336, 77.
Gioia, I.M., Henry, J.P., Maccacaro, T., Morris, S.L., Stocke, J.T., & Wolter, A., 1990a, *Astrophys. J.*, **356**, L35.
Gioia, I.M., Maccacaro, T., Schild, R.E., Wolter, A., Stocke, J.T., Morris, S.L., & Henry, J.P., 1990b, *Astrophys. J., Suppl.*, **72**, 567.
Hanisch, R.J., 1982, *Astron. Astrophys.*, **116**, 137.
Heydon-Dumbleton, N.H., Collins, C.A., & MacGillivray, H.T., 1989, *Mon. Not. R. astr. Soc.*, **238**, 379.
Jones, C., Forman, W., 1984, *Astrophys. J.*, **276**, 38.
Kent, S.M., and Sarget, W.L.W., 1983, *Astron. J.*, **88**, 697.
King, I., 1962, *Astron. J.*, **67**, 471.
Koyama, K., Takano, S., and Tawara, Y., 1991, *Nature*, **350**, 135.
Raymond, J.C., & Smith, B.W., 1977, *Astrophys. J. Suppl.*, **35**, 419.
Snyder, W.A., Kowalski, M.P., Cruddace, R.G., Fritz, G.G., 1990, *Astrophys. J.*, **365**, 460.
Stewart, G.C., Canizares, C.R., Fabian, A.C., Nulsen, P.E.J., 1984, *Astrophys. J.*, **278**, 536.
Takano, S., Awaki, H., Koyama, K., Kunieda, H., Tawara, Y., Yamauchi, S., Makisahima, K., and Ohashi, T., 1989, *Nature*, **340**, 289.
Thomas, P.A., 1986, *Mon. Not. R. astr. Soc.*, **220**, 949.
Trümper, J., *et al.*, 1991, *Astron. Astrophys.*, **246** L1.

X-RAY SPECTRAL IMAGES OF CLUSTERS

Richard Mushotzky
Laboratory for High Energy Astrophysics
NASA Goddard Space Flight Center
Greenbelt MD 20771
USA

ABSTRACT. We briefly review the scientific requirements for spatially resolved X-ray spectroscopy in order to determine the mass of the cluster, its metallicity, its state of evolution and the detailed physics of cooling flows. We discuss the available pre-Rosat and BBXRT observations of Perseus, Coma and Virgo and conclude that there is no simple obvious pattern to the temperature or elemental abundance distribution. We discuss the capability of Rosat for spatially resolved spectroscopy and present preliminary BBXRT results on several clusters. The BBXRT results are in good agreement with simple cooling flow theory, in 3 cases confirm the existence of cold material in the cluster, find only shallow abundance gradients and constrain the mass profile in the central regions of the cluster.

1. Introduction

The X-ray emission from clusters of galaxies is due, primarily, to thermal bremsstrahlung from a hot, optically thin, gas enriched with heavy elements and is contained in the cluster by its potential well (Sarazin 1986 for a comprehensive review). Spatially resolved X-ray spectroscopy, the subject of this paper, gives directly the density law, $\rho(r)$, the temperature profile, $T(r)$, and the distribution of the heavy elements (O, Ne, Mg, Si, S, Ca, Ar, Fe). One must stress that the X-ray emission in the cluster is a dominant process. Most of the visible baryons (Edge and Stewart 1991, Jones this symposium) are in the hot gas with typical values inside 1–3 Mpc being $M_{gas}/M_{stars} \sim 2-10$ and $M_{gas}/M_{total} \sim 0.1-0.2$. X-ray spectra of optically thin plasmas can be roughly divided into two regime: $T < 2 \times 10^7$ degrees where line cooling dominates and $T > 2 \times 10^7$ K where bremsstrahlung cooling dominates (Figure 1). Thus the emission from 'low' kT plasmas is dominated by lines while hotter plasmas are dominated by the continuum. The sensitivity of the strength of various lines of different elements produced in low temperature plasmas to the temperature is illustrated in Figure 2. Because one expects variation in the temperature of the cluster IGM with position the lines produced in the gas will have strongly variable equivalent widths. We shall only consider a subset of the topics for which spatially resolved X-ray spectroscopy is necessary : 1) determining the total mass of the cluster 2) measuring the distribution and emission measure of cooling

A. C. Fabian (ed.), Clusters and Superclusters of Galaxies, 91–108.
© 1992 *Kluwer Academic Publishers.*

gas 3) searching for cold gas in the cluster 4) determining the distribution of heavy elements 5) looking for dynamical evidence of mergers

1.1. MASS DETERMINATION

It has been shown in detail (Fabricant, Rybicki and Gorenstein 1984) that if the ICM is in hydrostatic equilibrium and the external pressure is negligible the total mass of the cluster can be expressed as

$$M(< r) = (kT_g(r)/Gm_H)[(d\log\rho/d\log r) + (d\log T/d\log r)]r$$

where $T(r)$ is the gas temperature and ρ the gas density. Small deviations from hydrostatic equilibrium might be expected in clusters due to mergers but detailed numerical simulations show that the effect on the derived cluster mass is not strong.

Thus one needs to determine the temperature at some fiducial radius and its distribution with radius. The determination of the mass using this method does not require assumptions about the form of the potential or a particular scale length and is numerically stable. Inspection of the Rosat data shows that the X-ray gas profile frequently extends to $\sim 3\,\mathrm{Mpc}$. Thus this method can be used to determine the mass profile out to radii where the assumption of collisional equilibrium no longer applies.

1.2. COOLING FLOWS

It has been realized since the first detection of large increases in the X-ray central surface brightness of clusters (Fabian these proceedings) that the crucial information for determining the actual parameters in the cooling flow is the measurement of the distribution of gas temperature with position in the cluster. Results from the Einstein Solid State Spectrometer (SSS; Mushotzky and Szymkowiak 1987) and Focal Plane Crystal Spectrometer (FPCS; Canizares et al. 1987) show the presence of emission lines in the center of several clusters which can only arise in gas whose temperature is considerably less than the average temperature determined by non-imaging large beam proportional counter data. However it is clear from Figure 1 that these low sensitivity spectrometers, which could only detect the strongest spectral lines, could not detect gas in the critical temperature regime between $7 \times 10^7\,\mathrm{K}$ (the mean average cluster temperature) and $2 \times 10^7\,\mathrm{K}$ which is the temperature range over which the gas loses most of its energy. What is required is a measurement of the continuum temperature and the line strengths as a function of position in the cluster with sufficient spectral and spatial resolution to determine where the gas is losing most of its energy. Such information can determine the mass accretion rate and the physical status of the gas and test, in detail, the cooling flow hypothesis.

1.3. EXISTENCE OF COLD MATTER IN CLUSTERS

If somehow the gas in the cooling flow does not collapse into condensed objects, or if there is an inflow of cold material (such as the ISM of galaxies) into the cluster one might expect considerable amounts of cold material in the cluster. There are

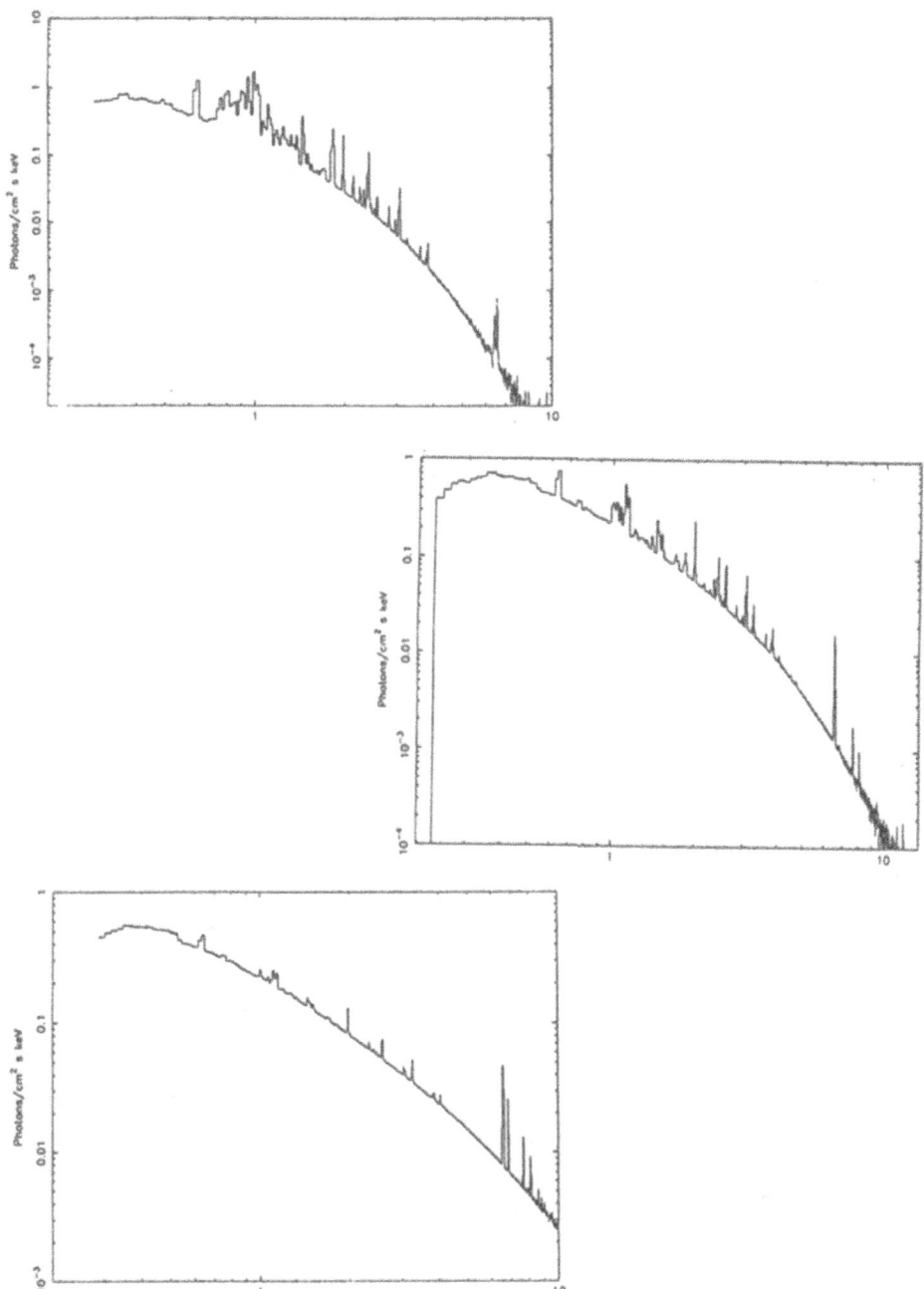

Fig. 1. The $0.3 - 10\,\mathrm{keV}$ emission from a coronal plasma of $1/2$ solar abundance at $kT = 1\,\mathrm{keV}$ (top left), 2 keV (top right) and 6 keV. Notice the strong changes in the spectrum as a function of temperature.

94

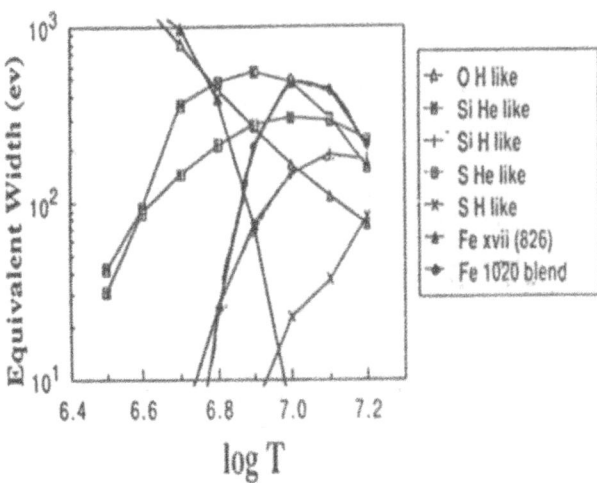

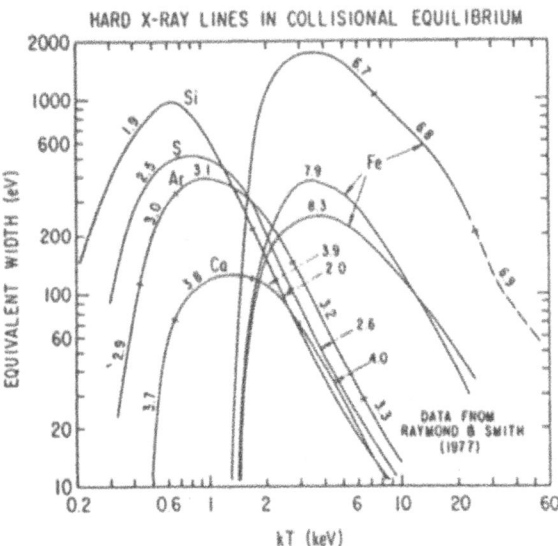

Fig. 2. Top panel; the equivalent width plotted versus temperature for some of the strongest lines in the 0.6 – 3 keV band from a collisional plasma. Note the strong temperature dependences. Lower panel; the equivalent width plotted versus temperature for some of the strongest lines in the 1.8 – 7 keV band. The 'line blends' are weighted by the energy that would be measured by a proportional counter. Note that the H– and He–like '6.7' keV line is the strongest line in the spectrum for $kT > 2\,\mathrm{keV}$.

indications from optical reddening data and HI measurements that this might be the case. Recently (White *et al.* 1991 and White this symposium) there has been strong evidence presented from Einstein SSS spectra that there is an excess of cold material along the line of sight to the center of the cluster. And that it was detected preferentially (?) in clusters with strong cooling flows. To further study this phenomenon requires an increase in sensitivity to determine if this effect is seen in clusters with lower central surface brightness, an increase in the sample size to determine how this material correlates with other cluster properties and the ability to map its distribution and determine its ionization state.

1.4. ORIGIN OF HEAVY ELEMENTS

The origin of the heavy elements in the IGM of the cluster is not completely understood at present (David , Jones and Forman 1990, Renzini *et al.* 1991, Jones this symposium). If the distribution of the heavy elements is consistent with the distribution of gas (Ponman *et al.* this symposium, Kowalksi *et al.* 1991) the total mass of, iron, in for example the Perseus cluster is $2.5 \times 10^{11} \, M_{\odot}$. The 'best bet' model presently is one in which an early stage of rapid star formation in cluster ellipticals resulted in a wind which ejected the material from the galaxy. However other models such as ram pressure stripping of galaxies falling into the cluster, galactic winds earlier (or later) in the life of the cluster (galaxy) are also reasonable. What is required is a map of the distribution of the heavy elements, to determine the actual mass of iron and oxygen, a determination of the O/Fe ratio to measure the fraction of metals produced by massive stars, and a measure of the evolution of the metal abundance in the IGM with time.

1.5. CLUSTER EVOLUTION

As shown in several detailed numerical simulations at this symposium (G. Evrard, P. Thomas, S. Schindler) there are expected to be strong variations in the cluster IGM temperature during and shortly after a merger. The gas is strongly shocked and may not be in equilibrium. One might anticipate that the spectral evidence for a merger would be the presence of strong spatial/spectral structure with, perhaps, line strengths that indicate non-equilibrium conditions (similar to supernova remnants). Given the Rosat and Einstein images of several clusters which clearly show that mergers are frequent this is an effect which should be searched for.

2. Summary of Non-imaging Cluster Spectral Data

The vast bulk of cluster X-ray spectroscopy has been obtained with non-imaging collimated proportional counters sensitive in the 2-20 keV range on the HEAO-1 (3x1.5 FWHM), EXOSAT (45' FWHM) and Ginga (1x2° FWHM) satellites. There has been a smaller data set obtained with with the Einstein SSS and FPCS in the 0.5-4 keV band with a ~ 3 arcmin radius field of view. All of the clusters for which sufficient signal has been obtained have shown emission lines from heavy elements (see Butcher this symposium). The best data have been obtained for Fe because it produces the strongest lines in the 2 – 20 keV band which is observable with a proportional counter (figures 1 and 2). The average Fe abundance is $\sim 1/3$ solar but there seems to be a pattern with the hotter more luminous X-ray clusters having somewhat lower abundances. Because of the relatively short exposures of the SSS it was able to detect emission lines from lower Z elements (Figure 1) in only those clusters which have strong 'cooling flows'. Thus the determination of the relative abundance of Mg, Si, S are somewhat model dependent. However it is clear that they are within a factor of 3 of solar. There has been an indication in M87 and Perseus from the FPCS data that the O/Fe ratio is larger than solar. The FPCS and the SSS data show the presence of emission lines due to low temperature gas in the central regions of several clusters, thus giving strong impetus to the cooling flow scenario. There has been no high quality spectroscopy in the 0.3–2.5 keV band for the total cluster emission but results from Rosat should soon be forthcoming.

There is a strong correlation (Mushotzky 1984, 1987; Edge and Stewart 1991) between cluster luminosity and temperature, $L \sim T^{5/2}$, and luminosity and central surface brightness, $S(0) \sim L^{0.8}$. These correlations show that the abundances of lower Z species are difficult to determine in high luminosity clusters because they are hot and the equivalent widths (Figure 1) will be very low while imaging detectors have difficultly in detecting low luminosity clusters far from the center relative to high luminosity clusters. The relatively low surface brightness of low luminosity clusters also indicates that spatially resolved spectra of these objects at large distances from the center will be difficult. A wide variety of correlations between optical and X-ray properties is discussed by Mushotzky (1984, 1987, 1990) and Edge and Stewart (1991).

3. Previous Results On Spatially Resolved X-ray Spectroscopy

Because until 1990 an imaging X-ray spectrometer had not been flown into space and the technical difficulty of making an imaging telescope with large collecting area at $E > 4$ keV, the bulk of the available results have depended on rather indirect means to infer the spatially dependent quantities in the cluster gas. These can be briefly characterized as no imaging at all, lowest quality imaging and brute force imaging.

3.1. NO IMAGING CAPABILITY

The integrated spectra of cluster are clearly a convolution of the temperature and emission measure profile. Very high quality spectra of clusters such as those obtained for a few objects by HEAO-1 A-2 (Henriksen 1987) show a deviation from an isothermal model (see also Hatsukade 1991). An ansatz which connects the density and the temperature distribution (Cowie, Henriksen and Mushotzky 1987, Hughes 1989, The and White 1989) allows an inference of the temperature profile by using the imaging data which constrains the density profile. Because one cannot determine which of several possible ansatzs is 'true', this method, while potentially very powerful, has considerable systematic uncertainty (The and White 1988). In particular for Coma, the best measured cluster, at least two fundamentally different ansatzs, a polytropic model and one which is isothermal in the center and adiabatic at the outside are equally consistent with the data and have very different physical consequences with respect to the distribution of cluster mass.

3.2. VERY LOW ANGULAR RESOLUTION

X-ray proportional counters with 45' (Exosat and the Einstein MPC), 60 × 120' (Ginga) and 180' (Tenma) fields of view have been used to obtain constrains on the temperature profiles and Fe abundance gradients in the 3 largest clusters (Perseus, Coma and Virgo). These data have been obtained either by direct mapping or by comparing detectors with different fields of view.

3.3. THE VIRGO CLUSTER

Ginga scanning data (Koyama *et al.* 1990) show that there is no temperature gradient within a few degrees ($\sim 1\,\mathrm{Mpc}$) of M87. However, these authors claim that there exists a strong gradient in the Fe abundance with the abundance peaking at 1/2 solar on M87 and dropping to 0.2 solar, 1 degree (0.35 Mpc) away. However, because of the strong dependence on Fe equivalent line strength versus temperature in the 2-3 keV temperature range of the Virgo cluster this measurement is difficult with the relatively low energy resolution Ginga data. Exosat results (Edge 1990) show no strong temperature or Fe gradient in the central degree field of view. Einstein solid state detector data (Lea, Mushotzky and Holt 1982) imply a greater than solar abundance in the central regions where the cooling flow dominates. Fitting a cooling flow model to these data does not change this result significantly giving a best fit overall abundance of 1.3 ± 0.3 solar. It is interesting, however, that the best fit isothermal model gives 0.54 solar consistent with the Ginga proportional counter results.

3.4. THE COMA CLUSTER

Independent analysis of the Exosat ME data (Hughes *et al.* 1989, Edge 1990) show that the temperature is hotter in the center than in the outer regions. A wide range of models can describe the temperature distribution and the range of temperatures seen by the Ginga and Tenma detectors (see above) with Hughes (1989) favoring an isothermal core model. However, there is no evidence for an Fe abundance gradient within a radius of 1.5 degrees ($3h_{50}^{-1}\,\mathrm{Mpc}$). The Fe can be distributed either like the gas or the galaxies and the minimum Fe abundance averaged over the central 3 Mpc region is 0.19 solar, compared to the average value of 0.24, implying that the gas is well mixed (Hughes 1991).

3.5. THE PERSEUS CLUSTER

Edge (1990) using Exosat data detected no temperature gradient and no obvious abundance gradient within the central 40' but his best fit temperature of 4.8 keV is less than that seen by the big beam detectors and the abundance determined for the central regions $A \sim 0.3 \pm 0.05$ solar is consistent (but systematically less than) the average value obtained by larger field counters such as Ginga (Allen *et al.* 1991),

HEAO-1 A-2 and Tenma and the central regions of the cluster by the Einstein SSS arguing that no strong gradients exist in the outer regions.

3.6. BRUTE FORCE IMAGING

Strip scans with a collimator narrow in one direction (Ulmer et al. 1987, Snyder et al. 1990) have been performed by the Spartan I space shuttle experiment on the Perseus cluster. This technique allows a 'simple' image to be synthesized. These data show that the cluster is cooler in the central regions with an average temperature of 4.3×10^7 K in the central 6' and is more or less isothermal in the outer regions with an average temperature of 7.2×10^7 K with an upper limit on the effective polytropic index in the angular range from 6-20' of 1.14 (Snyder et al. 1990). The Fe abundance data (Kowalski et al. 1991) show that the Fe is centrally condensed with an abundance of 0.8 ± 0.15 in the central 4' ring dropping to 0.2 ± 0.15 at $\sim 15'$. There is no evidence of Fe line emission further out than 22' ($0.7 h_{50}^{-1}$ Mpc).

The other imaging technique was a Dicke shadow mask used by the Birmingham group in the SL2 XRT flown on the space shuttle. In this technology a solid plate with a 'random' pattern of holes is used to cast shadows on a imaging detector. Under a wide variety of likely source distributions a direct image can be inferred. This technique works very well for a distribution of point sources but is more difficult for the diffuse low surface brightness objects such as clusters. The results for the Perseus cluster reported by Ponman et al. (1990) and Eyles et al. (1991) show a strong central concentration of Fe with the abundance in the central 6' being 1.3 ± 0.2 solar in the best fitting model dropping to < 0.2 solar at $r \sim 25'$ ($0.8 h_{50}^{-1}$ Mpc). The temperature profile is complex with a central minimum rising to a maximum at $r \sim 20'$ and dropping slowly with radius. The derived temperature profiles are in approximate agreement with the analysis of Cowie, Henriksen and Mushotzky (1986) using 'non-imaging' data and a polytropic assumption with an effective polytropic index of 1.23 (+0.09,-0.03). The disagreement in this index with the Spartan data of Snyder et al. 1990 may indicate that a polytrope is not a good description of the temperature profile or that the experiments are in disagreement. These data (and those from Rosat for A2199 and A2142) confirm that the X-ray emitting gas is a substantial fraction of the total mass of the system at $r > 1$ Mpc.

These authors have also reported data on the Coma and Ophiuchus clusters (Ponman et al. 1991) and find no strong evidence for an Fe gradient. However, the data are of poorer quality than for the Perseus cluster.

3.7. SUMMARY

There is no clear pattern in the spatially resolved data obtained to date and there is disagreement for the Perseus cluster. Inside 1 Mpc Virgo is isothermal (excluding the cooling flow at the center) while Coma clearly shows a drop in temperature at $r > 2$ Mpc. Coma shows no abundance gradient, Virgo a strong negative gradient and the situation is uncertain for Perseus. While both the Spartan and SL2 XRT data show an Fe abundance gradient there are considerable differences in the central value. Clearly these data need confirmation and a larger sample size.

4. Direct Imaging

4.1. ROSAT

The Rosat PSPC offers fair energy resolution, $E/\Delta E \sim 2.5$ in the 0.1-2 keV band, combined with good spatial resolution, low background, large field of view, and good counting statistics and the opportunity to observe a large number of targets. There have already been several clusters analyzed (A2199 and A2319 presented by Stewart and co-workers at this conference and A2256 by Briel *et al.* 1991). For clusters whose effective temperature is less than 3 keV it is relatively straightforward to measure a color temperature in a set of annular rings or to determine a spectral hardness ratio as a function of position in the cluster. For hotter clusters considerably greater signal is required since the determination of the temperature depends on measuring the slope of the temperature dependent gaunt function rather then determining the high energy cut off in the bremsstrahlung continuum. For clusters whose temperature is less than 2 keV (see introduction) the fact that the spectrum is dominated by the Fe L line complex gives the Rosat PSPC a very fine vernier on the temperature since the peak of the Fe L emission line complex is very temperature sensitive (Figures 1, 2). We should expect a large number of results in the next year as the pointed data are analyzed. The preliminary results shown in this conference for A2199 confirm that the center is indeed cooler than the outer regions, as implied by the 'cooling flow' hypothesis, and shows that the temperature rises to a peak at 1 Mpc and drops further out, consistent with the results inferred by Henriksen (1987).

4.2. BBXRT

BBXRT (the Broad Band X-ray Telescope), a 7 day shuttle flight flown in Dec 1990 has a fair image quality (Figure 3) with a central pixel (called A0 below) of 2' radius and outer pixels (A1-A4) of radius 8', good spectral resolution, $E/\Delta E \sim 50$ at 6 keV, broad band coverage (0.3-12 keV), low background and large collecting area (Serlemitsos *et al.* 1991). However BBXRT suffered severely from a very short flight and a poorly functioning pointer. Only 7 clusters had exposure times of more than 1000 seconds (Perseus, A496, A262, A2256, A665, A754 and Virgo). In the next section we will be presenting preliminary BBXRT results on selected clusters.

5. Preliminary BBXRT Results

The BBXRT production data were delivered in May 1991 and thus many of our results on clusters must be considered preliminary. Because of their great interest and impact on the subject some of them are presented in this section. We will not discuss the BBXRT results for A754, Virgo and A665. The BBXRT observations of individual clusters belong to different individuals : in particular K. Arnaud has done much of the work on Perseus and T. Miyaji on A2256.

5.1. THE PERSEUS CLUSTER

Much of this work has been done by K. Arnaud and is presented in Arnaud *et al.* (1991). In one of our observations (Figure 3) BBXRT was more or less centered

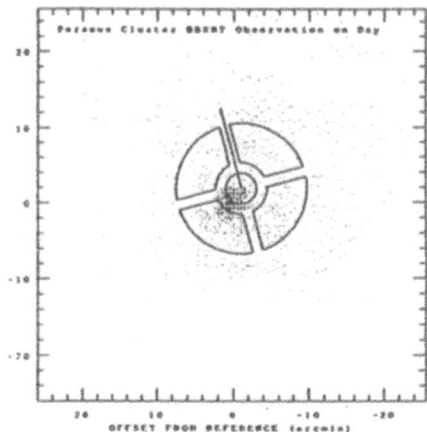

Fig. 3. One of 3 BBXRT observations of the Perseus cluster. The IPC image is presented as a gray scale over the BBXRT pixels. The central pixel is A0 and in counter–clockwise order from the top right the quadrants are A1 – A4. Note that in this observation, NGC1275, the site of the brightest X-ray emission, lies on one of the detector ribs.

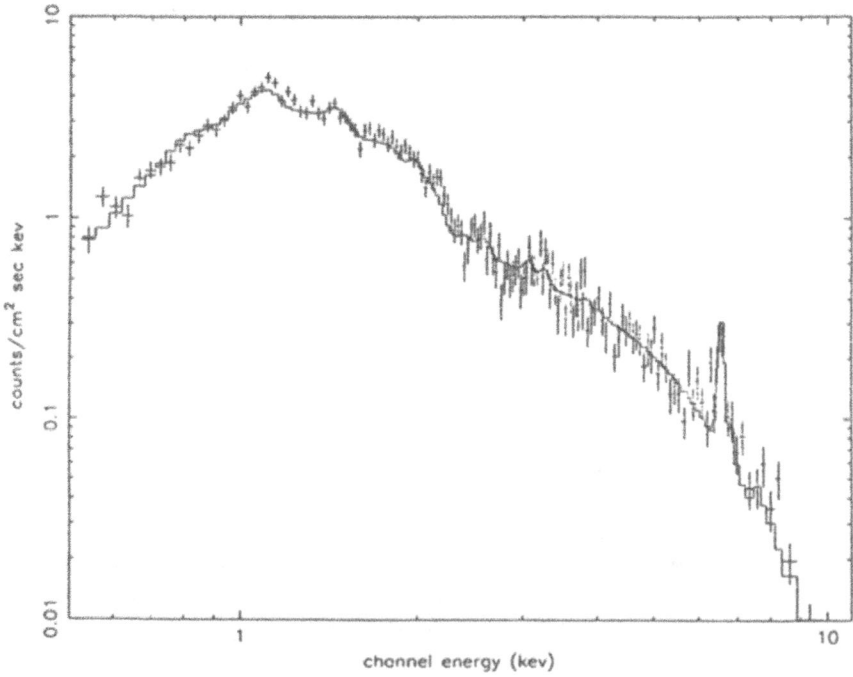

Fig. 4. A cooling flow model fitted to the spectrum of the central region of the Perseus cluster. The fit requires a column density in excess of the galactic value, consistent with the SSS result of White *et al.* (1991) and an Fe abundance of 0.65 solar.

on the core of the cluster and two other observations allowed determination of cluster properties out to ~ 20'. Unfortunately much of the flux from NGC 1275 fell on the detector ribs. Pixels A0 and A3 more or less bracketed the central core and show evidence for multi- temperature gas within 4' of the center. The best fit abundance is $\sim 0.6 \pm 0.15$ (90% confidence errors) solar (Figure 4), in disagreement with the results of Ponman et al.. The abundance determined from just the Fe-K He line at 6.67 keV and from a global fit to a cooling flow model are in agreement for pixel A0. Further out in the cluster pixels A1 and A2 ~ 10' from the center show little evidence for multi- temperature structure and have $kT \sim 6$ keV and abundance ~ 0.5, slightly larger than the best fit 'big' beam proportional counter abundance of 0.42 ± 0.03 (Allen et al. 1991). At an intermediate radius, pixel A4 ~ 4' from the center, has weak ($\delta\chi^2 \sim 20$) evidence for cool gas with a ratio of emission measure of cool to hot gas $15\times$ less then in A3, while slightly further out in an other observation of Perseus ~ 6' from center we find $kT \sim 6.9(+2.5, -1.7)$ keV (at 90% confidence) with $N(H)$ and $A < 0.75$ consistent with the cluster average. Thus the cooling component seems confined to the central 10' and is consistent with the cooling radius of ~ 6' derived from imaging data (Fabian et al. 1981). Other data obtained at ~ 15' (~ 0.5 Mpc) from center are consistent with galactic absorption and $5.8 < kT < 10$ keV. The abundance is consistent with $\sim 1/2$ solar but errors are large. The temperatures derived from BBXRT are in good agreement with those of Eyles et al. within the region of overlap. With good agreement with the BBXRT differential data and the Ginga integral data the Eyles et al. results are the best to date on mapping the cluster mass distribution.

The redshift of the Fe XXV He-like line at a variety of positions in the central 10' of the cluster is consistent with the optical redshift to ± 300 km s^{-1}. Thus the galaxies are roughly stationary in the frame of reference of the X-ray emitting gas, ruling out a recent high velocity merger as the origin of the high Perseus cluster central velocity dispersion.

We thus conclude that in Perseus that there is strong evidence for cooler gas confined to the central 10' and a weak abundance gradient with all the BBXRT points in the central 10' having consistent abundances but lying systematically slightly higher then the big beam averages.

Using the equation of hydrostatic equilibrium, the mass inside 0.3 Mpc is 1×10^{14} M$_\odot$ in good agreement with the Eyles et al. data, consistent with the minimum mass estimator but inconsistent with the virial estimator $M \sim 10^{15}\sigma^2 r_{\text{Mpc}}$ of $2 - 4 \times 10^{14}$ M$_\odot$. Thus in the central regions, as pointed out by Eyles et al., the mass and M/L ratio derived from X- ray imaging spectroscopy is $1/2$ of that derived using the virial theorem. The M/L ratio for Perseus derived from spatially resolved X-ray spectroscopy is ~ 150, in good agreement with the average for other clusters (Mushotzky 1991).

5.2. A262

The central 2' ($100h_{50}^{-1}$ kpc) regions of the cluster are not well fit by a isothermal model. The 2-10 keV data have $kT \sim 2.3 \pm 0.4$ keV in excellent agreement with the Ginga data (Butcher et al. 1991) $kT = 2.23 \pm 0.1$ keV. However when fitting the full

BBXRT bandpass, down to 0.4 keV, the mean temperature found is lower, $kT \sim$ 1.6 keV and the fit (Figure 5) is poor. A two-component spectral decomposition, while unphysical, is adequate to account for the present data. The presence of emission lines due to $kT < 1$ keV gas in the central regions and a 'cooler' component is direct evidence for cool gas at $r < 2$'. There is a strong indication of excess absorption in the central regions with a best fit $\Delta N_H \sim 2.2 \times 10^{21}$ cm^{-2} (Figure 5b), similar to that seen in several other cooling flow clusters. The mean abundance in the center, determined with a two component model, is $A = 0.44(0.25, 0.80)_{90\% confidence}$, is also consistent with the Ginga global value of 0.51 +/-0.13.

We have two determinations of the abundance and temperature in the outer regions of the cluster. At $r \sim 5$' and 10' there is no evidence of cooler gas with $kT = 2.4$ (1.87, 3.0) and 2.4 (1.8, 3.0) and the abundance of 0.4 $(0.1, 1.1)_{90\% confidence}$ and 0.73 (0.2, 1.75) solar, respectively. The mass estimate at $r \sim 6$ ' (.17 Mpc) is 3.5×10^{13} M$_\odot$ which is $\sim 75\%$ of the virial estimate.

5.3. A2256

Much of this work was done by T. Miyaji. We confirm the Rosat detection of a cool component located near the Rosat 'lump' (Briel et al. 1991). The temperature in the central 10' is consistent with Ginga average temperature but there is weak evidence for a temperature gradient with the cluster being hotter in the center. The abundance in the central 10' is consistent with Ginga average and thus there is no strong evidence for an abundance gradient

There is a oxygen absorption edge in the 'east' bump and in the 'cool' Rosat lump (Figure 6). The energy of the edge is consistent with cool material with the ionization state of oxygen being O^{+++} or less, with an optical depth in Oxygen of $\tau \sim 1 - 1.5$. This feature is not present in other locations in the cluster at this level indicating that the absorption is localized. This is the first detection of X-ray absorption in a cluster other than in the central regions (White et al. 1991).

5.4. A496

The central regions of this cluster show a cooler central temperature (Figure 7) and the presence of emission lines due to species that are only present at even lower temperatures and thus gives direct evidence for cooler gas in the central region. The Fe abundance in central region is $\sim 0.7 \pm 0.3$ solar consistent with the Ginga value of 0.5 ± 0.05, as is the data at $r \sim 10$' $(0.56 h_{50}^{-1}$ Mpc) from the center where the temperature and Fe abundance is consistent with the Ginga cluster average data.

The mass inside $\sim 1/2$ Mpc derived from the equation of hydrostatic equilibrium is is 1.9×10^{14} M$_\odot$ while use of virial theorem gives $M \sim 2.5 \times 10^{14}$ M$_\odot$ ($\sigma = 705$ km s^{-1}). Thus the situation is similar to Perseus and A262 where $M_X \sim 0.7 M_{opt}$. There is no strong evidence for an Fe abundance gradient but their could have been a factor of 2 change over central 10'. As in the Perseus cluster the measured X-ray redshift $z_{Xray} = 0.035 \pm 0.01$ is consistent within 1000 km s^{-1} of the optical redshift $z_{opt} = 0.0316$.

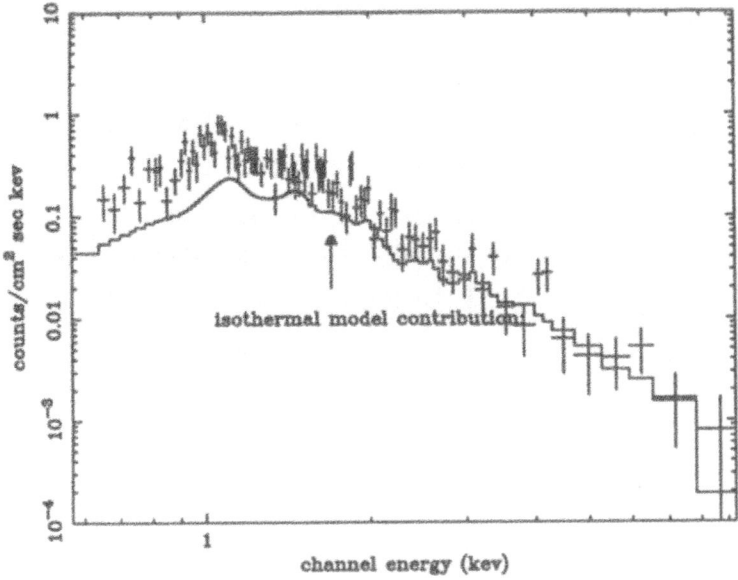

Fig. 5. The spectrum of the central regions of A262. The contribution of the cooling gas to the model spectrum has been removed to emphasize its contribution to the total flux.

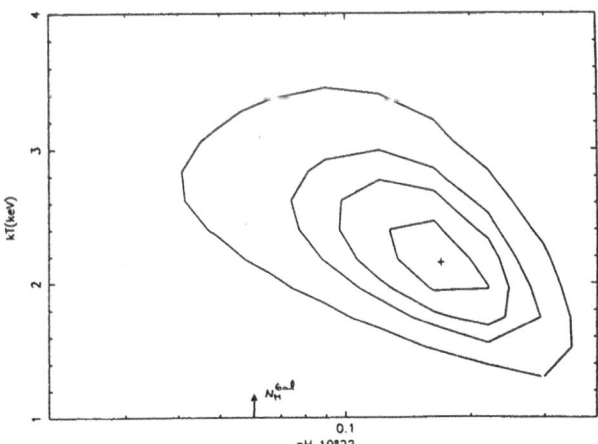

Fig. 6. Probability contours for column density plotted versus central temperature in A262. The contours correspond to 40, 60 and 90% confidence for 2 parameters of interest. The galactic value of 5×10^{20} cm^{-2} (indicated by the arrow on the x axis) is excluded at 95% confidence.

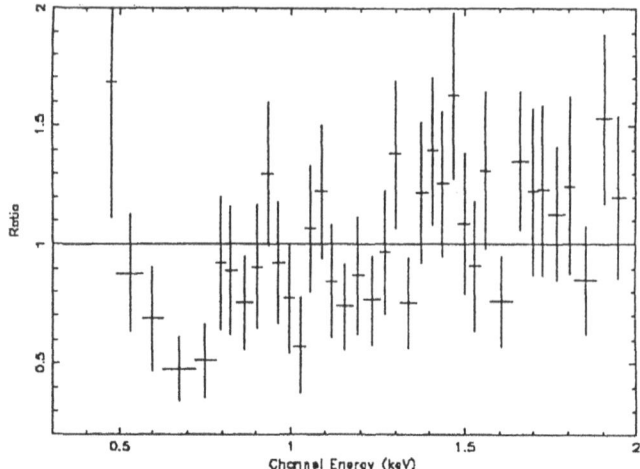

Fig. 7. The ratio of a model with no oxygen absorption to the data from one of the central regions of A2256. Note the sharp dip at $E \sim 600\,\mathrm{eV}$.

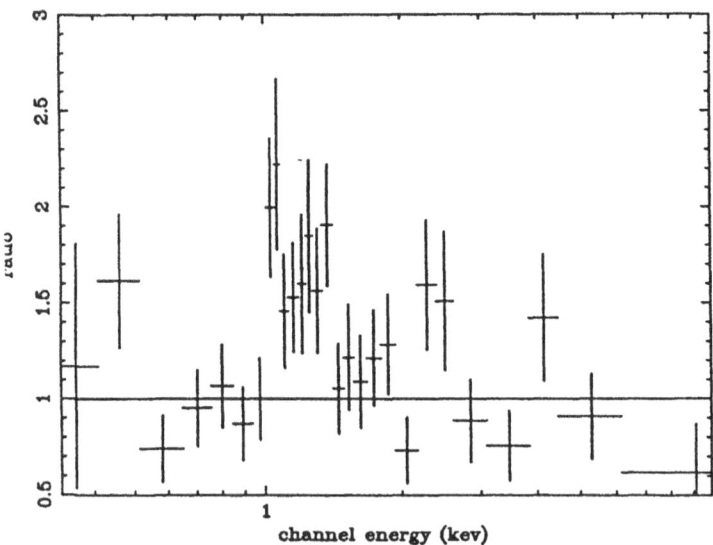

Fig. 8. Residuals to the spectrum in the central region of A496 when fitted with a 4 keV continuum model. These residuals represent line radiation from gas cooler than the average cluster temperature and are primarily due to Fe L lines.

5.5. BBXRT DATA SUMMARY

The Fe abundances obtained at effective radii of 2 and 10' are in good agreement with the Ginga abundances obtained with a much larger effective radius. This indicates that abundance gradients are shallow or absent within the radius where most of the X-ray flux originates. However, we are not able to comment on Fe abundance gradients at larger radii. We are able to 'confirm' that the simplest cooling flow models are roughly correct and are able to detect the emission lines from $T < 10^7$ K gas in the center as well as the cooler continuum temperature predicted by this model. The gas is observed to cool at radii which in good agreement with the results of K. Arnaud (1987) from the IPC data. The temperatures measured at $\sim 10'$ are in good agreement with the Ginga average temperature. This indicates that the temperature gradient in general is rather shallower than adiabatic.

We have rather spectacular new results on A2256 and are able to confirm the presence of the 'cool' clump first detected by Rosat (Briel *et al.* 1991). However, in addition we detect the strong signature of absorption by oxygen at two positions in the cluster. We have been able to estimate for the first time the binding mass in the central regions in a relatively model independent way and confirm that the minimum M/L ratio is > 80 for A262, A496 and Perseus. The existence of cool absorbing material in the central regions of A262 and Perseus strengthens the conclusions drawn from the SSS analysis of cooling flow clusters. We are trying to derive the distribution of the cold material from the BBXRT imaging data. What is perhaps most important is that the limited BBXRT results show that the future of X-ray imaging spectroscopy of clusters is very bright.

6. Preliminary Results of Spatially Resolved Spectroscopy

6.1. MASS AND MASS DISTRIBUTION

The BBXRT and SL2 XRT (Eyles *et al.* 1991) data for Perseus are in good agreement and show that the mass inside 0.5 Mpc is less then implied by the virial theorem (Kent and Gunn) and that the best estimate is $M/L \sim 100h_{50}$. This value is consistent with that of well-measured clusters using optical data of $< M/L > \sim 140h_{50}$ (Mushotzky 1990). While these central values inside ~ 1 Mpc seem relatively stable, data at larger radii where much of the mass of the systems lies, is required to come to firm conclusions about the total mass. However it is clear that dark matter is necessary. However, as pointed out by Cowie, Henriksen and Mushotzky (1986), Eyles *et al.* 1991 and Stewart (this symposium) a large fraction of the total binding mass $\sim 0.2 - 0.3$ within $3h_{50}^{-1}$ Mpc, lies in the X-ray emitting hot gas and that this ratio appears to increase with cluster radius. Thus with 'X-ray eyes' the dark matter appears concentrated in the center and the ratio of baryonic matter to dark matter rises with increasing distance from the center. If true, and so far all available data sets are consistent with this scenario, it implies a fundamental change in our view of the formation and evolution of clusters.

6.2. COOLING FLOWS

The BBXRT data confirm the simplest predictions of the cooling flow models. The presence of a range of temperatures in the central regions and the onset of cooling at roughly the predicted radius. The Rosat, Spartan and SL2 XRT data are also in good agreement. The next level of analysis, the distribution of emission measure with temperature needs to be done with the BBXRT data. The influence of X-ray absorption (White *et al.* 1991) on the cooling flow analysis needs to be evaluated since it will suppress the signal of low temperature gas, particularly He and H- like oxygen lines, in the central regions.

6.3. ELEMENTAL ABUNDANCES

The BBXRT data have the potential for determining the abundances of O, Mg, Si, S and Fe in the cooling flow clusters. For A262, which has a low ($kT \sim 2.5\,\mathrm{keV}$) average temperature there is the potential for determining for the first time the abundances of these elements in a non-cooling flow situation. The BBXRT data show that in the central $\sim 1\,\mathrm{Mpc}$ that the abundance of Fe does not vary strongly with radius. In particular, for Perseus the two pixels closest to the center have Fe abundance estimates of 0.65 ± 0.13 and 0.69 ± 0.1 solar respectively. However, it is only in conjunction with the Ginga data, which sample the entire cluster, that the inference that the abundance gradients at larger radii are not strong can be made. It thus appears from the BBXRT data that clusters are well mixed. However, the disagreement with the SL2 XRT abundance results for Perseus are not understood at present.

6.4. ABSORPTION IN THE CENTRAL REGIONS

The Einstein SSS results and preliminary BBXRT and Rosat results indicate the presence of cold material in the central regions of clusters. The relationship of this cold material to cooling flows is not certain and the BBXRT detection of absorption due to such material in A2256, a non-cooling flow cluster, may indicate that is more widespread. Further BBXRT analysis of the other clusters in the sample should indicate whether the cool material is concentrated in the central regions.

6.5. CLUSTER EVOLUTION

The Rosat and BBXRT data for A2256 show that there is a strong spectral signal of merging clusters. The numerical simulations of cluster evolution shown in this meeting also indicate such a signature. It is unclear if the models and data agree. It will be possible with X-ray spectroscopy to examine the dynamical state of the IGM and compare it with the indications of dynamical age shown by the galaxies. Since the gas relaxes much faster then the galaxies, spatially resolveD x-ray spectroscopy will give unequivocal evidence of 'only' recent mergers. As indicated by Evrard (this symposium) the expectation is that, in recent mergers, there should be an appreciable velocity difference, of the order of the infall velocity $\sim 2000\,\mathrm{km\,s^{-1}}$. The absence of such a velocity difference in Perseus and A262, at least in the center, constrains such ideas. The ability of X-ray spectroscopy to determine the redshift of the cluster gas with an accuracy similar to that of the optical galaxy data will allow a search for velocity difference and determination of the relative velocities of the two cluster components.

7. Conclusion and Prospects

Even the relatively few results from spatially resolved X-ray spectroscopy obtained prior to 1992 have given us a fundamentally new picture of the IGM in clusters, the distribution of dark matter and the fate of the gas in cooling flows and cluster mergers. The future in the next 2 years is very bright with the acquisition of many new deep Rosat cluster exposures and the advent of Astro-D data. Analysis of deep Rosat images should determine the evolution of the cluster IGM and provide detailed information about the general evolution of clusters (Henry, this symposium). Astro-D with roughly the same band pass and spectroscopic capability as BBXRT combined with a true imaging capability ($\sim$ 1' spatial resolution with both the X-ray CCDs and imaging gas scintillator proportional counters) will vastly increase our knowledge of low, $z < 0.3$, clusters. By the end of 1993 we should know the distribution of the binding mass in several clusters of different 'types', the distribution of heavy elements and their abundance ratios, the ratio of baryonic to total mass and either a confirmation or refutation of the 'strong' cooling flow predictions. Preliminary results will be obtained on the evolution of the cluster potential in depth and form and the evolution of the metal abundance in the IGM.

I hope we will meet again in 1994 to digest this new feast of data and re-formulate our picture of clusters. I predict that many of the ideas discussed in detail at this meeting will be strongly tested.

Acknowledgements

First of all I would like to thank Peter Serlemitsos and the rest of the BBXRT team for developing, building and flying the BBXRT and for allowing me to present preliminary results at this conference. Special thanks are due to K. Arnaud and T. Miyaji for the preliminary results on the Perseus and A2256 clusters. I thank A. Fabian for his kind invitation and hospitality in Cambridge and the meeting organizers for a most productive occasion.

References

(References to papers in this symposium have been suppressed)
Allen, S. *et al.*, 1991. *Mon. Not. R. astr. Soc.*, in press.
Arnaud, K., 1987. PhD. thesis, University of Cambridge.
Arnaud, K. and the BBXRT Team, 1991. Proceedings of the Yamada Meeting on Frontiers of High Energy Astrophysics in press.
Briel, U. *et al.*, 1991. *Astr. Astrophys.*, L10.
Canizares, C. Markert,T., and Donahue, M., 1987. in NATO Workshop 'Cooling Flows in Clusters of Galaxies' edited by A. Fabian, Kluwer, p 63.
Cowie,L. Henriksen, M. and Mushotzky, R., 1987. *Astrophys. J.*, 337, 593.
David, L., Jones, C. and Forman, W., 1990. *Astrophys. J.*
Edge, A., 1990. PhD Thesis, University of Leicester.

Edge, A., and Stewart, G., 1991. *Mon. Not. R. astr. Soc.*, 252, 414.

Eyles,C.J. *et al.*, 1991. *Astrophys. J.*, 376, 23.

Fabricant, D, Rybicki, G and Gorenstein, P., 1984. *Astrophys. J.*, 286, 186.

Henriksen, M., 1987. PhD. Thesis, University of Maryland.

Fabian, A.C., Hu, E, Cowie, E. and Grindlay, J., 1981. *Astrophys. J.*, 248, 47.

Hatsukade, X., 1991. PhD Thesis, University of Osaka.

Hughes, J. Yamashita ,K. Okumura,Y. Tsunemi, H. and Matsuoka, M., 1988. *Astrophys. J.*, 327, 615.

Hughes, J., 1989. *Astrophys. J.*, 337, 21.

Hughes, J., 1991. in Iron Line diagnostics in X-ray Sources pg 80 eds A.Treves, G. Perola, L. Stella Springer-Verlag.

Kent, S. and Sargent, W., 1983. *Astr. J.*, 88, 697.

Kowalski, M.P. *et al.*, 1991. in Iron Line diagnostics in X-ray Sources pg 72 eds A.Treves, G. Perola, L. Stella, Springer-Verlag.

Koyama,K. *et al.*, 1991. Nature.

Lea, S., Mushotzky, R.F. & Holt, S., 1982. *Astrophys. J.*, 262, 24.

Mushotzky, R., 1984. Physica Scripta T7 157.

Mushotzky , R., 1988. Hot Thin Plasmas in Astrophysics, ed R. Pallavicini, Kluwer, p.273.

Mushotzky , R., 1990, in 'After the First 3 Minutes' eds S.Holt and C. Bennet AIP Press.

Mushotzky, R and Szymkowiak, A., 1987. in NATO Workshop 'Cooling Flows in Clusters of Galaxies' edited by A. Fabian, Kluwer, p 53.

Ponman, T. *et al.*, 1990. *Nature*, 347, 450.

Ponman, T. J. *et al.*, 1991. in Iron Line diagnostics in X-ray Sources pg 76 eds A.Treves, G. Perola, L. Stella Springer-Verlag.

Renzini *et al.*, 1991. *Astrophys. J.*,

Sarazin, C., 1989. Rev Mod Phys. 58,1.

Serlemitos, P. and the BBXRT Team, 1991. Proceedings of the Yamada Meeting on Frontiers of High Energy Astrophysics in press.

Snyder, W. *et al.*, 1990. *Astrophys. J.*, 365, 460.

The, L.S. and White S.D.M., 1988. *Astr. J.*, 95, 15.

Ulmer, M. Cruddace, R. Fenmore, E., Fritz, G. and Snyder, W., 1987. *Astrophys. J.*, 319, 118.

White, D.A., Johnstone, R.M., Fabian, A.C., Arnaud, K.A. and Mushotzky, R.F., 1991. *Mon. Not. R. astr. Soc.*, 252, 72.

RADIO STUDIES OF CLUSTERS

Walter Jaffe
Leiden Observatory
P.B. 9513
Leiden, 2300 RA
Netherlands

ABSTRACT. We consider those aspects of the intracluster diffuse medium which can be profitably studied with radio techniques. The continuum morphology of the extended cluster radio sources suggests the existence of substructure in the medium and of energy input into the medium from massive binary galaxies and infalling material. The Faraday rotation in radio halos, and near radio sources, yields estimates of the magnetic field strength in these regions, and suggests that the field is created in turbulent dynamos. Observations at 21 cm and 2.6mm of neutral gas in absorption and emission show the existence of $\sim 10^{10} M_{\odot}$ of such gas in some clusters and limit, but do not exclude the possibility of much higher masses of very cold ($\sim 4K$) gas. Finally, we show that the radio galaxies present in the centers of many cooling flow clusters must dramatically affect the flow conditions in their vicinity.

1. Prelude

It is somewhat awkward to discuss "Radio Studies of Clusters" since we would rather discuss physical aspects of an interesting astronomical phenomenon (clusters) than a specific observing technology (radio astronomy). We do it anyway because there are important phases of the diffuse medium in clusters which are difficult to observe with any other technique: the magnetic field, the neutral gas at 10 to 100 K, and the *very* hot component ($10^{13}K$, the cosmic ray gas). The interpretation of radio results within the cluster context is most satisfactory when the physical processes in the radio emitting regions interact directly with those in other regions, for example when the equipartition magnetic field pressure of a radio lobe can be compared to the thermal pressure of the X-ray medium. It is less satisfactory when the radio phenomena cannot be coupled to situations where the phsyics are better understood. When, for instance, we discuss the action of radio jets on the medium, our ignorance of the pressures, densities, magnetic fields *etc.* in jets makes for uncertain conclusions.

This said, we also note that radio observations often provide superlative resolution and dynamic range, thus revealing structures inaccessable with other observing techniques.

2. Morphology

Figure 1 shows a 1400 MHz continuum map of the Perseus (A 426) cluster made with the Westerbork telescope (Sijbreng and de Bruyn, 1991).

It, and a similar map of A 2256 by De Jong and Rottgering (1991) show a variety of features

A. C. Fabian (ed.), Clusters and Superclusters of Galaxies, 109–118.
© 1992 *Kluwer Academic Publishers.*

110

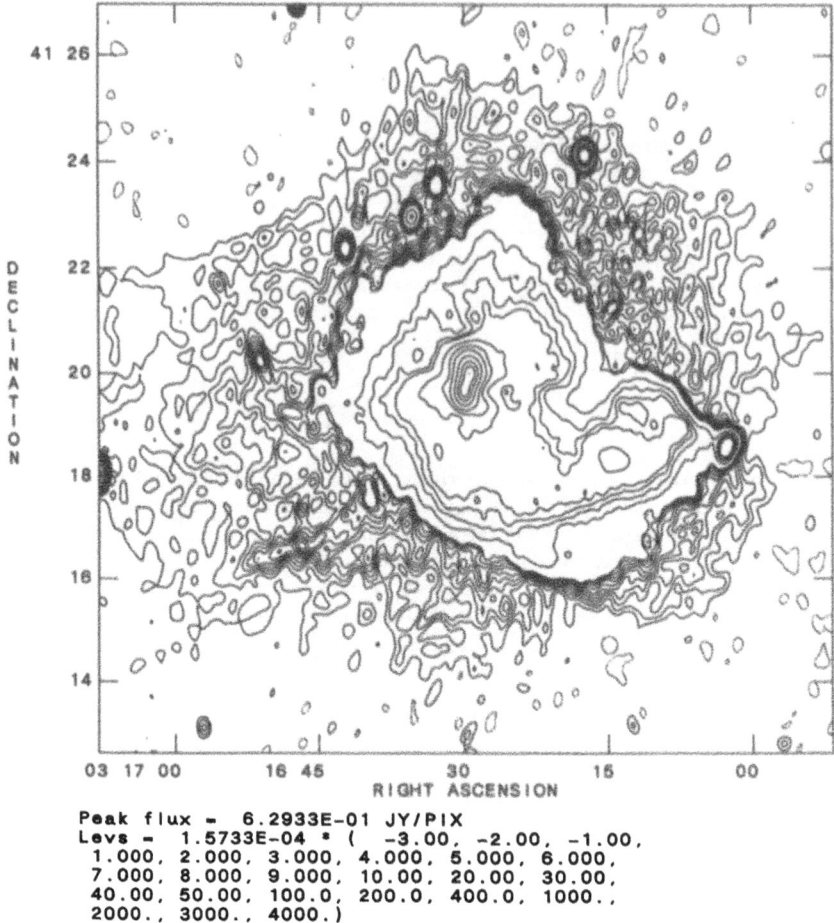

Peak flux = 6.2933E-01 JY/PIX
Levs = 1.5733E-04 * (-3.00, -2.00, -1.00,
1.000, 2.000, 3.000, 4.000, 5.000, 6.000,
7.000, 8.000, 9.000, 10.00, 20.00, 30.00,
40.00, 50.00, 100.0, 200.0, 400.0, 1000.,
2000., 3000., 4000.)

Figure 1. 1400 MHz map of the Perseus Cluster made with the Westerbork Synthesis Radio Telescope by D. Sijbreng (Groningen) and A.G. de Bruyn (Dwingeloo). A point source containing the flux of the milliarcsec source at the nucleus of NGC 1275 has been subtracted.

that suggest specific processes, but also emphsize the complexity of the intracluster medium. In these maps we see clumpy extended halos, some hundreds of kiloparsecs in size. These indicate that the processes controling acceleration and diffusion of cosmic ray electrons in the cluster are quite inhomogeneous. In the Perseus map, for instance, we can identify a small halo about 20 kpc across immediately surrounding the central active galaxy NGC 1275, and a larger complex halo associated with the bright elliptical NGC 1272 to the southwest. This morphology suggests that the acceleration and diffusion of particles is influenced by turbulent motions in the cluster medium generated by the relative motion of these two massive galaxies.

The Perseus map also shows a sharp concave edge in the continuum emission to the northeast. This suggests a contact discontinuity or shock there through which the relativistic electrons do not flow. Perhaps, as the X–ray measurements presented at this meeting imply, clumps of gas remaining from cluster formation are now hitting the main cluster gas and creating such shocks. A similar emission edge is visible in A 2256.

The tailed radio sources offer contradictary insights into the medium. In general the radio-determinated equipartition pressures in the tails are comfortingly close to the thermal pressures derived from X–ray data, but the most recent interpretation (Feretti *et al.*, 1990, Feretti, Perola, and Fanti, 1991) indicate that the equipartition pressures are systematically low. The small scale straightness of some tails (*e.g.* 3C 129, Jaffe & Perola, 1973) indicates that the medium is quiescent, with turbulent velocities below 100 km s^{-1}, while the as yet unexplained sharp bends in others (Burns, *et al.*, 1986) suggest high velocity, large scale turbulence.

3. Halos, Turbulence, and Magnetic Fields

The halos in A 2256 and A 426, and the more diffuse, well known halo in the Coma cluster (*c.f.* Giovannini, 1991) present a number of difficulties, and opportunities for interpretation. Some of these problems have been discussed for years, while others are just becoming current. Characteristic of the halos is their large size (up to a megaparsec), indeterminate form (associated/not associated with identified radio galaxies), and generally steep radio spectrum.

The now well worn but unresolved discussions on halos center on the relativistic electrons. Where do they come from? How do they move about? Why do the appear in only a small fraction of clusters?

The most obvious sources of electrons are the radio galaxies in the clusters. On the other hand, with the exception of Perseus, the halos are neither associated with specific radio galaxies, nor are they entirely smooth. This implies they spread about to cluster size (say 300 kpc) on a time scale commensurate with their synchrotron lifetimes (about 10^8 years for electrons radiating at 1 GHz). The implied speed is 3000 km s^{-1}. Electrons streaming this fast along field lines have an anisotropic distribution in pitch angle; there are more electrons moving in the "foward" hemisphere than in the "backward" hemisphere. Electrons whose pitch angle distribution is strongly anisotropic create Alfvén waves by a type of plasma Cerenkov radiation which scatters the streaming electrons, bringing their streaming spead down to the Alfvén speed. This is probably not above 100 km s^{-1}in clusters. Holman, Ionson and Scott (1979) suggested that damping of short wavelength Alfvén waves by thermal protons prevents electrons from being scattered through pitch angle 0. This would restrict all electrons to forward streaming, which is equivalent to a streaming speed of c/2. This is plenty fast enough, in fact too fast, since the electrons would escape from the cluster. On the other hand this process requires an exquisitely constant mean magnetic field to prevent mirroring of electrons into the backward direction by small field fluctuations.

A more fruitful line of arguement is that the electrons are generated locally by shocks (as we believe to happen in radio sources), or by betatron acceleration when the magnetic field in amplified in turbulence, or by collisions between cosmic ray protons, which have long radiative lifetimes. These ideas are pleasantly unquantitative, since our theories of shock acceleration and turbulent amplification are incomplete, while the cosmic ray protons are essentially unobservable. The turbulent/shock models are interesting because they may connect observations of the halos to other measures of turbulent energy input such as X–ray substructure, massive binary galaxies, or strong radio galaxies.

More recent discussion has moved from the electrons to the magnetic field necessary for synchrotron emission. How strong is it, what is its scale size, and where does it come from? The only hard limit on the field strength is a lower limit of about 0.2 μG derived from the lack of inverse Compton X–ray emission from clusters (Perola and Reinhart, 1972). If the field were lower than this, a high cosmic ray density would be necessary to explain the radio halos, and these cosmic rays would generate too many X–rays by scattering the cosmic microwave background. More, but difficult to interpret, data comes from Faraday rotation measurements. In a magnetized, ionized gas, the two circularly polarized modes of electromagnetic radiation propagate at slightly different speeds. This effect causes the position angle of *linearly* polarized radiation to rotate through an angle proportional to the square of the wavelength:

$$\chi \;=\; RM \times \lambda^2$$

where RM, the *Rotation Measure* has the dimensions of radians m^{-2} and has the value:

$$RM \;=\; 800n_e\, B_\parallel l$$

where n_e is the density of thermal electrons, $B_\parallel$ the average line of sight magnetic field in μG and l the length of the line of sight in kpc. Faraday rotation has been measured for radio galaxies within and behind the Coma cluster halo (Kim *et al.*, 1990) and in the strong radio emitting regions surrounding the cooling flow radio galaxies 3C 274 a.k.a. Virgo A a.k.a.M 87, (Owen, Eilek, and Keel, 1990), and 3C 405 a.k.a. Cygnus A (Dreher, *et al.* 1987). In the Coma cluster the rotation measures are small (< 50 rad m^{-2}) and compatible with forground Galactic rotation, except near the cluster center, where they reach values of ~ 50 rad m^{-2}. If we assume the magnetic field is chaotic with a scale size of about 20 kpc (see below) we can estimate a halo field strength of about 2 μG. This interpretation is somewhat uncertain because most of the measurements near the cluster center are in fact cluster tailed radio galaxies, which might have intrinsic rotation measures of about this order.

The rotation measures near strong radio galaxies are much larger, up to ~ 6000 rad m^{-2}, and the measurements are well resolved. These show that the Faraday rotation is definitely associated with the material immediately surrounding the radio source, and that the scale of the magnetic field variations is about 1 kpc in these regions. If we estimate n_e from X–ray maps of these regions, we find the magnetic fields to range from 20 to 100 μG. These value are so high that the magnetic energy density is comparable to the thermal energy density; the field must be dynamically important in the motions near the sources.

The high field value near sources of turbulent energy suggests that the field is generated by a turbulent dynamo (Jaffe, 1980; Roland, 1981; Ruzmaikin, Sokoloff, and Shukurov, 1989). A simple minded theory then suggests that the value of the turbulent field should be:

$$B_T \;\simeq\; (8\pi)^{\frac{1}{2}}\frac{(P_T R_T)^{\frac{1}{3}}}{D\rho^{\frac{1}{6}}}$$

where P_T is the input turbulent power, R_T the scale size of the energy source, D the size of the turbulent region, and ρ the gas density. The field scale size will be slightly larger than R_T.

For the general cluster medium we can try $\rho \sim 10^{-27}$g cm^{-3}, $D \sim 300$ kpc, $P_T \sim 10^{43}$erg s^{-1}, and $R_T \sim 10$ kpc, which yields $B_T \sim 1\mu$G. Near a cooling flow radio source we use $\rho \sim 10^{-25}$cm^{-3},

$D \sim 30$ kpc, $P_T \sim 10^{45}$, and $R_T \sim 1$ kpc, and we find $B_T \sim 50\mu G$. These numbers are thus at least reasonable.

An alternative explanation for the field near the central galaxy is that the general cluster field is compressed by the cooling flow (Soker and Sarazin, 1990). In this scenario, the initially chaotic field is compressed azmuthally and stretched radially until it is almost entirely radial. Its strength then increases inwards as predicted by flux conservation ($B \sim r^{-2}$) down to a radius where the magnetic energy density equals the thermal energy density (or equivalently, the sound speed equals the Alfvén speed). At this point the magnetic pressure prevents the gas from sinking farther until field line recombination (known as the Parker instability) decouples the gas from the field and allows it to fall. This model predicts field strengths of about 20 μG at the point where the material decouples, about 30 kpc from the center; this is of the right order of magnitude. We would expect the near-center halo radio emission to be radially polarized. This is not observed, but the Faraday rotation of the near central halo is often so high that it is difficult to say what the intrinsic polarization angle is. At high frequencies the Faraday rotation is small, but the halos are weak emitters at high frequencies, so that question is hard to answer. On the other hand the compression of field and cosmic rays in the flow should give a sharply peaked power law emission profile toward the center, perhaps going as r^{-7}. This we don't see in cooling flow clusters like Perseus. There the halo shows a fairly constant plateau in the whole region between NGC 1272 and NGC 1275. So in this case the turbulent dynamo model seems more likely.

A curious footnote to the discussion of halos is their seeming anti–correlation with cooling flows. While Perseus has a radio halo, it is relatively weak, and confined to the cooling flow region. The better known, brighter halos like those in Coma, A 2319, and A 2256 are in clusters without strong cooling flows. This effect might be a direct result of the cooling flow (*e.g.* the flow reduces diffusion away from the central radio galaxy), or the result of a common cause (*e.g.* in clusters with strong subclustering or binary structure the turbulent energy generated by the relative motions of the subclusters both generates the halo magnetic field and cosmic rays, and also breaks up by quiescent conditions necessary to form a cooling flow).

4. Cold Gas

In cooling flows we see plenty of X–rays from gas at $10^6 K$ to $10^8 K$ and optical line emission from gas at $10^4 K$, but what happens to the gas after this? If the flow is truly cooling we expect cool (or in some cases warm or tepid) neutral gas at $10K$ to $1000K$. As long as this gas is diffuse and not too thick we should be able to see its radio emission at 21 cm—if it is in atomic form—or 2.6 mm—if it is molecular and the physics of of CO molecule formation are not too different from that in our Galaxy. This last is a major question.

We can estimate the expected mass in cool gas by multiplying the mass cooling rate, say $100 M_\odot$/yr, by the length of time the material is visible as such. A minimum time scale is that to form stars, at least 10^7yr, so there should be at least $10^9 M_\odot$ in cool gas. On the other hand, if as suggested by White, *et al.*, (1991), the cool gas never condenses out, the appropriate time scale is the Hubble time, and the expected mass is $10^{12} M_\odot$.

Atomic hydrogen can be seen in emission, and, if there is a background continuum source, in absorption. From standard theory, the 21 cm optical depth of an HI cloud is:

$$\tau = 10^{-18.3} N_H/(T_s \Delta V),$$

where N_H is the column density of atomic hydrogen, T_s the spin temperature (usually assumed equal to the kinetic temperature), and ΔV the velocity spread in the gas in km s^{-1}. In emission the observed brightness temperature T_B will be $T_s(1 - e^{-\tau})$. Combining various geometric effects and assuming $\tau \ll 1$ we can then estimate the mass of HI as:

$$M_{HI}/M_\odot \quad = \quad 10^{2.4} S_e \Delta V D_{\rm Mpc}^2,$$

where S_e is the received emission flux in mJy and $D_{\rm Mpc}$ the distance in Mpc. Using the Hubble's law, we can convert the radial velocity V_{1000}, measured in thousand km s^{-1}, to $D_{\rm Mpc}$. With $H_o = 75$ we get

$$M_{HI}/M_\odot \quad = \quad 10^{4.6} S_e V_{1000}^2 \Delta V.$$

The equivalent formula for absorption is:

$$M_{HI}/M_\odot \quad = \quad 10^{3.8} T_s \Omega V_{1000}^2 \Delta V (S_a/S_{cont}).$$

Here Ω is the source solid angle in arcsec^2, and S_a/S_{cont} is the fractional decrease in the continuum flux. Clever readers can take the ratio of these two formulae and prove that:

$$S_a/S_e \quad = \quad T_{B(cont)}/T_S.$$

In other words, absorption exceeds emission when the brightness temperature of the background continuum source exceeds the spin temperature of the gas. Current radio observations have sensitivity limits of $S_e \sim 1$ mJy, resolutions of about 10 arcsec, and dynamic ranges of $S_a/S \sim 10^{-4}$. Thus for $V_{1000} \sim 7$ and $\Delta V \sim 400$, which might be typical for the nearest rich clusters, the limiting detectable emission mass would be about $10^9 M_\odot$. The limiting detectable absorbing mass is a factor of 10 lower, providing a bright background source of $\sim 10 Jy$ is available, and $T_s \sim 100K$, which is often assumed.

In fact 21 cm absorption has been seen in at least 4 clusters: Perseus (Jaffe, 1990), A0335+096, MKW3s (McNamara et al. 1990) and Virgo (this paper), while emission has been seen only in Virgo to date. Making a mass estimate from absorption data requires the angular size Ω of the absorbing region, which generally requires interferometric measurements. These are only available for Perseus and Virgo. In Perseus where we find about $10^{10} M_\odot$ of HI, assuming optical thinness and $T_s = 100K$. CO emission has also been seen in this cluster, (Mirabel,1989) implying a similar mass of H_2 gas. The Virgo maps (Figure 2) show both absorption and emission with masses of about $10^9 M_\odot$ The other clusters looked at by McNamara et al. (1990) typically show no emission above $10^9 M_\odot$.

The predominance of absorption over emission implies that T_s is lower than the brightness temperatures in these sources, which are typically about $100K$ (Jaffe, 1991). The masses are about right for material on its way to becoming stars. The morphologies seen in the high resolution maps of Perseus and Virgo are strange however. The HI is patchy and shows a high velocity dispersion, $\Delta V \sim 400$ km s^{-1}, even when we only see the "front" side in absorption. In a smooth cooling flow we expect to see only fairly narrow–band redshifted material in front of the galactic nucleus. This last seems to be the case for the low resolution observations of A0335+096 and MKW3s. Perhaps these are "normal" cooling flows while the material in Perseus and Virgo comes from a gas rich cloud or spiral galaxy being "eaten" by the central galaxy. If not in fact transient, the Perseus cloud indicates that the cool gas is an important dynamic component in the cooling flow. It contains a large mass ($10^{10} M_\odot$) and a large kinetic energy (10^{58} erg), while its dynamic time scale (size/velocity) is roughly 10^7y. Thus this configuration would require an energy input of 10^{44}erg s^{-1} to sustain; this is similar to the X–ray luminosity. In this scenario we can consider the turbulent neutral gas to be the parent population of the "warm" HII gas; colliding neutral clouds would be shocked up to $10^7 K$ but would rapidly cool off to $10^4 K$ where they are visible in optical line emission. After that they would recombine and rejoin the neutral population , albeit at a lower turbulent velocity. Ultimately, after several collisions, they would sink to the galaxy center and perhaps form stars. Clearly a lot more HI detections are necessary to make any of these scenarios plausible.

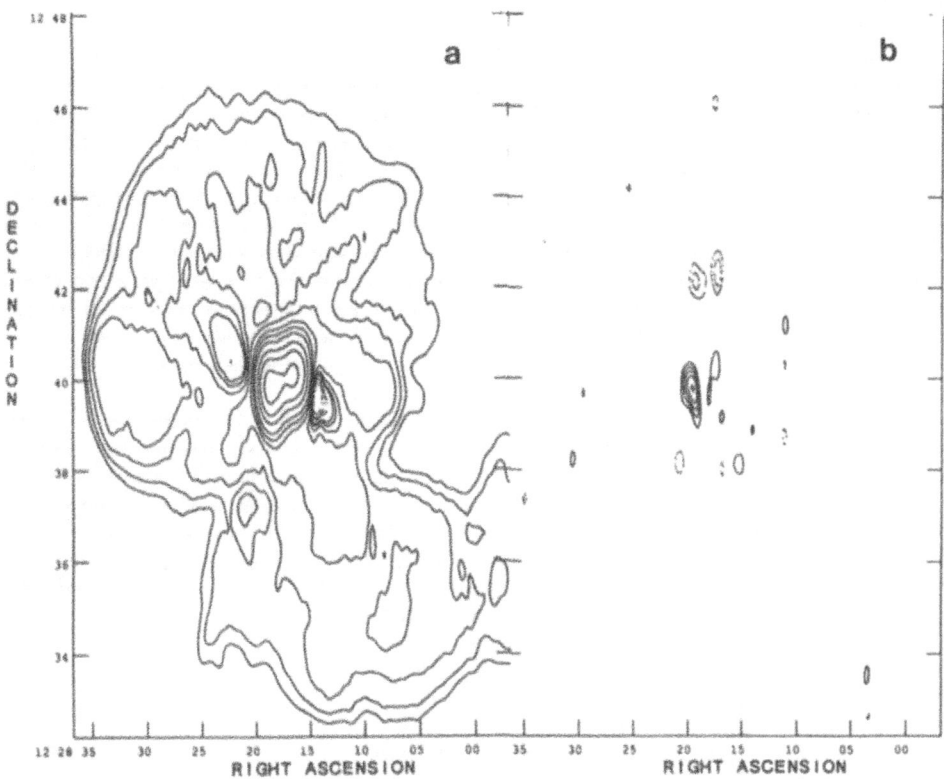

Figure 2 (a) A 1420 MHz continuum map from the Westerbork Synthesis Radio Telescope of the region around Virgo A = M 87. In the center the two bright peaks are the nucleus of M87 to the east and the jet to the west. (b) A single HI channel, 60 km s^{-1} wide, from which the continuum has been subtracted. Visibile are emission to the east of the nucleus, and absorption to the north.

What about the $10^{12} M_\odot$ of neutral material postulated to cause the absorption of soft X-rays from cooling flow clusters? From the above emission limits, this can't be present in an optically thin form of HI, but must be in thick clouds, with optical thicknesses of 100 to 1000 so that than 99% or more of each cloud is shielded from view by the material in front of it. If they are so thick, why don't we see them more clearly in absorption? To get the required optical depth from the surface density implied by the X-ray data, $N_H \sim 10^{21} \text{cm}^{-2}$, requires both a low spin temperature, T_s, and low velocity spread ΔV. In fact the two are related; the velocity spread must be greater than the thermal velocity $\Delta V_t \sim 0.1\sqrt{T_s}$. Inserting this into the above expression for optical depth yields $T_s < 4K$ (*very cold!*), and $\Delta V < 200$ m s^{-1}. Furthermore, the clouds must be quite small, a parsec or less, otherwise the virial velocity of a cloud will exceed the limit on ΔV. In this case our ponderous kiloparsec sized radio beam picks up millions of individual clouds: one or two along each of a million lines of sight. Each cloud absorbs *all* the continuum emission in a very narrow velocity range, 200 m s^{-1}, in a total ensemble velocity spread of 500 km s^{-1}; that is, only about 0.04% of the radiation passing through it. When observed with a low (spatial) resolution, and low (spectral) resolution radio telescope, we see this only fraction of the continuum absorbed. This is be difficult to detect.

Is there any way to test this theory? In some radio galaxies much of the continuum is radiated by a nuclear VLBI source of subparsec dimension. Then the effective spatial resolution of the radio observations far exceeds the actual resolution of the telescope. This is indeed the case in Perseus. On the way to us this radiation passes through at most a few clouds, each of which would block all the 21cm radiation in a narrow velocity band. We (Jaffe and D. Hartmann at Leiden) have observed Perseus with the Dwingeloo radio telescope with a 1 km s^{-1} resolution and find no narrow absorption; there seem to be no clouds along this line. More such cases need to be observed to decide whether this is a statistical fluke or a general result.

5. Energy and Momentum

Let us now turn from using radio telescopes to study cluster environments to considering the role of radio galaxies in modifying these environments. Since we don't understand many of the details of radio source physics, or cooling flow physics for that matter, we will only calculate the contribution of the central radio sources to the overall energy and momentum balance of their surroundings.

The total thermal energy in the intracluster medium is of course very large, about 10^{63}ergs, and altering this energy, even on a Hubble time scale of $\sim 10^{17}$ s requires a very large power, $\sim 10^{46}$erg s^{-1} or so. This is larger than any input or output from X-ray sources, radio sources or whatever, although curiously enough it is about the same level as the total optical power output of the cluster. This large energy content is the reason that the heating of the whole cluster medium is a problem only once, at cluster formation time. Thus the radio galaxies cannot affect the global configuration of the intracluster medium.

The power associated with the *cooling flow* material–that with cooling time less than the Hubble time– is rather smaller. By definition the cooling power is:

$$
\begin{aligned}
P_{cool} &= \dot{M}_{cool} u^2 \\
&\sim 10^{43.5} (\dot{M}_{cool}/100) \, \text{erg s}^{-1},
\end{aligned}
$$

where $\dot{M}_{cool}$ is the mass cooling rate in $M_\odot$/yr, and u the sound speed in the hot gas. The observed *radio* power of the central source in Perseus, 3C 84, is approximately $10^{44.0}$ erg s^{-1} and it is conventional to express the *kinetic* luminosity as

$$
P_{kinetic} = P_{radio} \times IC/\epsilon.
$$

ϵ is the efficiency of conversion of kinetic power into radio power and is variously estimated as 1 to 10 %. IC is an "isotropic correction" for the fact that the radio emission from a compact source or jet may be beamed. In that case our estimate of the total source luminosity based on the observed flux would be an overestimate. 3C 84 is probably beamed but not highly so (the nuclear components are not superluminal). If we take $\epsilon \times IC \sim 1$ we find the kinetic power of the AGN in Perseus to be about equal to the cooling flow luminosity.

3C 84 is an somewhat atypical cluster radio galaxy; its spectrum is flat or rising to mm wavelengths, and this contributes to its high value of P_{radio}. The more common steep spectrum radio galaxies have $P_{radio} \sim 10^{42}$ erg s^{-1}, but these more mundane sources probably have $IC \sim 1$ so $P_{kinetic} \sim 10^{43}$ erg s^{-1}, still comparable to the cooling luminosity.

The radio source should then disrupt or strongly modify the cooling flow in its immediate surroundings, out to 30 kpc or so, and to a lesser extent out to the initial cooling radius, ~ 100 kpc. Turbulence derived from this energy input may help ionize the HII filaments seen in the inner parts of the flow. Heckman et al., (1989), and Stefi Baum's presentation at this conference show that the occurance and spatial extent of the HII filaments correlate better with the radio properties of the central galaxy than with any other obvious cooling flow statistic.

Interestingly, the directed momentum flux in radio jets is itself large enough to affect the flow in the circumgalactic region. We can estimate the total force of a radio jet as:

$$F \quad = \quad \frac{dp}{dt} \quad = \quad P_{kinetic}/v_{jet} \quad \sim 10^{33}/\epsilon \, \text{dynes}.$$

Here we have assumed that for typical radio galaxies $v_{jet} \sim 0.1\,c$. The inward momentum rate of material passing through the cooling radius is $\dot{M}_{cool} v_{flow} \sim 10^{33}$ dynes so that the jet can in principle arrest the flow coming in along its axis. If we make a crude time dependent calculation of what happens when we turn a jet on, we find that the first effect is to blow out the material along the source axis. The outward flow then slows down quickly as surrounding material is entrained, and doesn't get much beyond the about 30 kpc. On a longer time scale the energy input is more important; the gas heated by the jet expands, becomes less dense, and rises through the flow (unless it can reradiate its energy quickly).

We might expect the flow pattern near the galaxy to show large scale bipolar circulation. Material flows out along the radio axes and returns at the equator. We should then see some relation between indicators of the flow (X-radiation, Hα radiation) and the radio axes. We believe in fact that there is some evidence (see the illustrations in Baum et al., 1988) that the Hα regions in cluster radio galaxies avoid the radio axes.

6. Conclusions

1. Continuum radio maps of the extended radio sources in clusters are very rich in detail. The features seen suggest various large scale processes in the intracluster medium.

2. The magnetic field in clusters and near cluster radio galaxies is probably generated by turbulent dynamos. The field strength is $\sim 2\mu G$ in the general field and $\sim 20 \rightarrow 100\mu G$ near active galaxies. The field scale length is ~ 10 kpc in the first case, and as small as 100 pc in the second.

3. Cold gas with a mass of $\sim 10^{10} M_{\odot}$ is present in isolated clumps in several clusters. Larger masses of cold gas, if present, must exist as very small, very cold clouds in order to have avoided detection so far.

4. Radio galaxies dominate the cooling flow pattern in their immediate neighborhood, and perhaps to larger scales.

References

Baum, S.A., Heckman, T., Bridle, A., van Breugel, W.J.M, Miley, G: 1988, *Ap.J.Supp.* **68**, 643

Burns, J.O., O'Dea, C.P., Gregory, S.A., Balonek, T.J.: 1986, *Ap.J.* **307**, 73

de Jong, J.P., Rottgering, H.:1991, *in preparation*

Feretti, L., Spazzoli, O., Gioia, I.M., Giovannini, G., Gregorini, L.: 1990, *A&A* **233**, 325

Feretti, L., Perola, G.C., Fanti, R.: 1991, *this conference*

Giovannini, G., Feretti, L., Venturi, T., Kim, K.–T., Kronberg, P.P.: 1991, *this conference*

Heckman, T.M., Baum, S.A., van Breugel, W.J.M., McCarthy, P.: 1989, *Ap.J.* **338**, 48

Holman, G.D., Ionson, J.A., Scott, J.S.: 1979, *Ap.J.* **228**, 576

Jaffe, W.: 1980, *Ap.J.* **241**, 925

Jaffe, W.: 1990, *A&A* **240**, 254

Jaffe, W.: 1991 *A&A in press*

Jaffe, W., Perola, G.C.: 1973, *A&A* **26**, 423

Kim, K.T., Kronberg, P.P., Dewdney, P.E., Landecker, T.L.: 1990, *Ap.J.* **355**, 29

McNamara, B.R., Bregman, J.N., O'Connell, R.W.: 1990, *Ap.J.* **360**, 20

Mirabel, I.F., Sanders, D.B., Kazes, I.: 1989, *Ap.J.* **340**, L9

Owen, F.N., Eilek, J.A., Keel, W.: 1990, *Ap.J.* **362**, 449

Perola, G.C., Reinhardt, M.: 1972, *A&A* **17**, 432

Roland, J.: 1981, *A&A* **93**, 407

Ruzmaikin, A., Sokoloff, D., Shukurov, A.: 1989, *M.N.R.A.S.* **241**, 1

Sijbreng, D., de Bruyn, A.G.: 1991 *in preparation*

Soker N., Sarazin, C.L.: 1990, *Ap.J.* **348**, 73

White, D. A., Fabian, A.C., Johnstone, R.M., Mushotsky, R.F., Arnaud, K.A.: 1991, *this conference*

FAR INFRARED EMISSION FROM CLUSTERS AND WARMING FLOWS

JOEL N. BREGMAN
Astronomy Department
University of Michigan
Ann Arbor, MI 48109
USA

ABSTRACT. Clusters of galaxies are sources of far infrared radiation and have been detected with the IRAS observatory at 60μm and 100μm. In a complete sample of rich clusters with cD galaxies, far infrared emission is detected toward the cD galaxy in about 12% of the clusters; the average properties of detections are $<z> = 0.06$, $<L_{FIR}> = 3\times10^{44}$ erg/sec, and $<T> = 25\text{-}30$ K. This emission probably arises from warm dust heated by electron collisions from the hot gas. Within the cooling radius of the cluster (central 100-200 kpc), $L_{FIR} \approx 10L_X$, so the dominant cooling of the hot gas may be carried by the dust emission; mass cooling rates of ≈500 M$_\odot$/yr are implied. A search for extended far infrared emission from 56 clusters reveals weak diffuse emission in a small fraction of the systems. This emission, which is unlikely to arise from cluster point sources, is about ten times greater than the compact far infrared luminosity toward the cD galaxies in rich clusters.

Stellar mass loss from galaxies and radiative cooling by dust could produce cool cluster gas at a rate approaching 1000 M$_\odot$/yr unless balanced by a heating mechanism. Thermal electron conduction can be an effective heating agent and we examine whether the X-ray properties are consistent with a large amount of cool gas being evaporated rather than hot gas radiatively cooling. The Fe XVII lines, observed in two clusters to date, provide powerful evidence for cooling gas, although it does not eliminate the possibility that cool gas is being evaporated.

1. Introduction

Far infrared emission from clusters of galaxies is a relatively new field, but of potential importance because the energy radiated can be comparable to or greater than the X-ray luminosity. However, before discussing these observations and their implications, it is valuable to review the information that can be obtained from far infrared observations.

The observations that we will discuss are those obtained with the IRAS satellite, in which fluxes were obtained in four broad bands centered at 12 μm, 25 μm, 60 μm, and 100 μm (Neugebauer et al. 1984). Dust is probably the emitting agent of this radiation and its mass may be determined by adopting a model for the dust, which includes the dust composition, size, opacity, and temperature. The dust

A. C. Fabian (ed.), Clusters and Superclusters of Galaxies, 119–130.
© 1992 *Kluwer Academic Publishers.*

is often assumed to have a power-law size distribution, with grains no larger than about a = 0.25 μm (Mathis, Rumpl, and Nordsieck 1977), so emission occurs in the long wavelength limit ($\lambda >> a$). In this limit, the total emission from an ensemble of dust grains at a single temperature is independent of the size distribution so the conversion from flux to mass is simple under some conditions. However, dust is not at a single temperature and the dust grain opacity is not well understood at all wavelengths, both of which complicate the interpretation of IRAS data (see Draine 1990).

The opacity has been calculated for local Galactic dust by several authors (Hildebrand 1983; Draine and Lee 1984; Rowan-Robinson 1986; Mathis and Whiffen 1989) who are in agreement at wavelengths $\lambda < 40$ μm, but the discrepancy can be considerable at longer wavelengths. For these authors, the range in the calculated opacities is a factor of two at 60 μm, three and a half at 100 μm, a factor of eight at 200 μm and a factor of 30 at 1000 μm. At wavelengths beyond 200 μm, the dust is transparent, so conversion of flux to mass is relatively simple, save for the large uncertainty in the opacity and the difficulty of obtaining data at these wavelengths. At shorter wavelengths where the opacities are better understood, the dust is opaque, so its flux is very sensitive to temperature ($\propto T^4$). When a range of dust temperatures exists, the weak emission from cool dust can be overwhelmed by the higher emissivity from warmer dust, leading to a considerable underestimate in the total dust mass. Dust at a range of temperatures is the rule rather than the exception because of the size distribution and variations in the intensity of the heating agent. When dust grains are small, < 50 Å, the timescale for them to radiate can be shorter than the timescale between absorbing heating photons. Because the heat capacity of small grains is low, they are heated briefly to temperatures that can exceed 1000 K (for the $\lambda < 5$ Å grains), whereupon they cool rapidly and spend most of their time at lower temperatures. Small grains not only dominate the emission in the 12 μm and 25 μm, but contribute to the 60 μm emission as well. The 100 μm emission, and usually the 60 μm flux is determined by the presence of larger, thermally stable dust grains in the temperature range 20-50 K. The few IRAS bands do not provide enough information to determine accurately the temperature distribution. As an illustration of this difficulty, Draine (1990) finds that equally good fits are made to the IRAS observations of NGC 6240 for models that differ in total dust mass by more than an order of magnitude (with the same opacity law).

Uncertainties in dust masses also can be obtained by comparing the IRAS dust mass to that in galaxies where the dust mass can be estimated from other methods, such as by optical extinction or from the neutral gas mass (by assuming a dust to gas ratio). For early-type galaxies, the IRAS dust mass leads to a gas to dust ratio that is nearly an order of magnitude higher than the likely value of 100, suggesting that a considerable amount of cool dust is being overlooked (Young 1990). To summarize, the uncertainties in converting an IRAS flux to a dust mass can be an order of magnitude, and if early type galaxies are a suitable guide, then the dust masses determined in clusters may significantly underestimate their true values.

2. Far Infrared Emission From cD Galaxies In Rich Clusters

The original motivation in searching for diffuse cluster emission arose from the cooling flow model, which suggests that the ~ 100 M$_\odot$/yr of cooling material eventually forms stars (e.g., review by Fabian, Nulsen, and Canizares 1991). If this star formation leads to dust formation and heating, far infrared emission might be visible with the IRAS telescope. The natural location of this emission would be the central region of the cooling flow, which is also the site of the dominant galaxy in a cluster.

The detection of thermal dust emission from cD galaxies was reported upon by three groups in the same year. The well-known cD galaxy in the Centaurs cluster, NGC 4696, was detected at 100 μm as a point source and at the 3σ level at 60μm (de Jong et al. 1990). Also, two groups examined the few dozen brightest X-ray clusters and found 60 μm and/or 100 μm emission toward the central dominant galaxy in the center of the X-ray emission (Bregman, McNamara, and O'Connell 1990; Grabelsky and Ulmer 1990). Approximately 1/3 - 1/2 of the objects were detected above the 3σ level in either of these two energy bands. These detections were compared to those found from ordinary early-type galaxies, which are detected at about the same rate and in the same IRAS bands. The typical flux densities for the normal early-type galaxies are 0.2 mJy (60 μm) and 0.5 mJy (100 μm), and if the far infrared to optical flux density ratio is the same in the massive central dominant galaxies, which are much fainter in apparent optical magnitude, then no cluster emission would be detected. The positive detection indicates that different conditions prevail in these central dominant galaxies for them to be overluminous in the far infrared relative to normal early-type galaxies.

Different models for the source of the dust mass and the heat source predict significant correlations between observable quantities. For example, if the dust is a byproduct of the cooling flow model, its mass should be related to M as is the case for emission line gas. If the dust is heated by electron collisions, there would be a relationship between the infrared emission and the electron density of the hot gas. Despite the small number of detected objects, the samples were examined for correlations between the presence of infrared emission and X-ray or optical properties. However, the only relationship evident was between infrared detectability and distance, with nearer objects more commonly detected. Based on the need for a larger sample and the ability to observe central dominant galaxies at far infrared wavelengths, Cox, Bregman, and Schombert (1992) began a new study of a large, well-defined sample of objects.

For this sample, we chose all Abell clusters of richness class ≥ 0 with a cD galaxy in the center. Unlike previous samples, these objects were selected by optical rather than by X-ray properties, and in most cases, X-ray observations do not exist yet. Of the 167 clusters that meet these criteria, we requested IRAS ADDSCANs at the location of the cD. For those not familiar with this obscure nomenclature, an ADDSCAN is a strip scan across the location of an object, in this case, the cD galaxy. The IRAS instrument has detectors that are 4.5 wide (the "cross scan" direction) and it is diffraction limited along the scan line (the "in-scan" direction), so the resolution is about 0.25 and 0.5 at 60 μm and 100 μm, respectively. Each region of the sky was scanned about 5-15 times at various angles, and to make an ADDSCAN, these separate scans have a baseline removed from them (avoiding regions where sources are thought to exist) and are then added in different ways to form the final product (mean, weighted mean, and median addition of individual scans). Although the standard IPAC software automatically identifies sources, we have chosen to use other analysis software (DRAWSPEC; Liszt 1989) to extract sources and to determine their uncertainty. The use of ADDSCANs is thought to be the most sensitive and accurate method of identifying sources with an rms about 2-3 times lower than in images.

Two important considerations in using the IRAS data are poor angular resolution and the poorly understood noise characteristics of the ADDSCANs, at least partly caused by the ubiquitous cirrus emission. At the redshift of a typical detection ($<z> = 0.06$), 1' corresponds to 50 kpc ($H_o = 50$ km/sec/Mpc is used throughout this paper), so that the observation cannot determine where in the cD the emission occurs, although the beam width of 4.5 fits within the diameter of the cluster core (typically 8'). It is natural that galaxies other than the cD will fall in the beam, although most of these are far less luminous and of early type. Should a spiral galaxy fall in the beam at close to the location of the cD, the cluster is rejected for possible contamination, since optically bright gas-rich spiral galaxies would be detectable at these distances. After discarding clusters where the cD might be contaminated by a spiral galaxy, where the baseline fit was suspect, or where IRAS did not obtain data, 148 objects remain in the sample.

The uncertainty in a measurement provided by either the ADDSCAN software or our analysis seemed optimistic compared to a subjective examination of the scan. Rather than use the quoted errors, the results were divided up into four categories (after Haynes et al. 1990): no detection (114 objects; 77%), possible (17; 11.5%), probable (13; 8.8%), and very likely (4; 2.7%). For comparison, a comparably sized control sample was obtained of random blank field regions around the sky with the same latitude/longitude distribution as the source sample. This control sample was identically processed, leading to 112 objects with the distribution: no detection (100; 89%), possible (10; 8.9%), probable (2; 1.8%), and very likely (0; 0%). This indicates that most of the objects in the possible category result from chance and are not true detections while most of the objects in the probable and very likely categories are secure detections. *Approximately 12% of the object sample are true detections*; similar results occur when binned by the actual S/N. A disturbing aspect of this analysis is that the uncertainty given by the ADDSCAN software, which is widely used, is lower than the rms deduced from the distribution of detections given by the control sample (to be discussed more thoroughly in Cox, Bregman, and Schombert 1992). A significantly smaller fraction of this sample was detected compared to previous samples (Bregman, McNamara, and O'Connell 1990; Grabelsky and Ulmer 1990), and the presence of false (weak, random) detections can account only for part of the difference. Sample criteria must be an important reason for the detection rate difference since these earlier samples have a lower mean distance. For all objects listed as probable or likely detections in the new sample, IRAS images were obtained and inspected in order to check for anomalous contaminating cirrus emission.

The typical detection characteristics of a cD are $<z> = 0.06$, $F_{100\mu m} = 1$ Jy, $F_{60\mu m} = 0.3$ Jy, and the ratio of these two flux densities implies $T_{dust} = 24\text{-}33$ K, depending upon the emissivity law used. From these flux densities, one can estimate the luminosity in a band of width 80 μm and centered at 82.5 μm:

$$L_{FIR} = 4\pi D^2 (3.25 \times 10^{-14} F_{60\mu m} + 1.26 \times 10^{-14} F_{100\mu m}) \text{ erg/sec}$$

where $F_{60\mu m}$ and $F_{100\mu m}$ are expressed in Jy. This is a lower limit on the total far infrared luminosity, which can be 1-3 times greater due to emission at wavelengths longer than 120 μm or shorter than 40 μm. According to the above equation, the resulting value for the luminosity of a typical detection is 3×10^{44} erg/sec. If this is interpreted as thermal dust emission, then the inferred dust mass is $10^7\text{-}10^8$ $M_{\odot}$, depending on whether the dust temperature is 30 K or 20 K, respectively.

This large far infrared flux places significant constraints on the origin of possible heating mechanisms. For comparison, the blue luminosity of the cD galaxy is 3×10^{44} erg/sec, and the bolometric luminosity is 4-5 times greater. If starlight heats the dust, there is not sufficient energy in the ultraviolet bands, so absorption of blue and visual light is necessary. This implies that a significant amount of the blue luminosity must be absorbed and an extinction of $A_B = 0.5\text{-}1$ mag is estimated, with the associated $E(B-V) = 0.2\text{-}0.3$ mag. Optical imaging sometimes shows extinction in these systems, although it is either rare or unprecedented to find significant extinction across the entire galaxy extending tens of kiloparsecs from the center (Romanishin and Hintzen 1988; Sparks, Macchetto, and Golombek 1989; McNamara and O'Connell 1991). Furthermore, only in the central few kiloparsecs of the galaxy would the photon energy density be sufficient to heat the dust so that it radiates in the IRAS band. We conclude that it is unlikely for the dust to be heated by starlight.

Another potential heating agent is the hot gas that pervades the cluster. Direct heating of the dust by the absorption of X-ray photons can be ruled out from an energy budget argument since the X-ray luminosity within the cooling radius (about the size of the core radius of the cluster, 100-200 kpc) is 3×10^{43} ergs/sec, one-tenth that of the infrared luminosity. Even summed over the entire cluster, the X-ray luminosity, typically 3×10^{44} erg/sec, is only comparable to the far infrared luminosity. A more promising heating mechanism is collisional electron heating between the dust particles and the hot electrons (Dwek, Raphaeli, and Mather 1990; Dwek 1992), a process that also destroys the dust through sputtering. For hot gas temperatures of $3-10 \times 10^7$ K, this heating mechanism is proportional to the electron density, and since it decreases on lengthscales of tens of kiloparsecs, it should be an effective heating mechanism in the region around the cD galaxy. Furthermore, the temperature to which the dust would be heated is similar to the observed value, although possibly on the low side. Heating by collisions of hot electrons appears to be a promising mechanism.

An implication of this heating process is that the destruction of dust by sputtering requires a fresh supply if steady-state conditions are to be maintained. A natural source of new dust is the material shed by stars during the normal course of stellar evolution. For the old stellar populations generally present in clusters, the total mass loss rate of gas from stars is approximately 10 $M_\odot$/yr from the cD and 100 $M_\odot$/yr from the sum of all galaxies in the cluster (Soker, Bregman, and Sarazin 1991). For a gas to dust ratio of 100, this implies dust supply rates of 0.1 $M_\odot$/yr and 1 $M_\odot$/yr for the cD and the cluster. The sputtering time for dust is about 3×10^7 yr, so the required dust supply rate is about 0.3 $(T_{dust}/30 \text{ K})^{-5.9}$ $M_\odot$/yr, which is a bit too large to be supplied entirely by the cD (see Bregman, McNamara, and O'Connell 1990 for more details). However, it is unlikely that a steady-state exists between dust creation and destruction, otherwise the far infrared to optical luminosity ratio from all clusters would be similar. Dust and cold gas may enter the cluster in discrete units through the stripping of a galaxy as it passes within the cluster core. The amount of dust produced in this fashion is adequate and the sporadic nature of the process naturally introduces cluster-to-cluster variation in the infrared emission.

The production of fresh dust from stellar mass loss and the heating by hot electrons appears to produce a consistent picture for the observed infrared luminosity. However, the implications of this model for the cooling of the hot gas are dramatic. The amount of energy radiated by the dust, 3×10^{44} erg/sec, can be expressed in terms of the rate at which energy is being transferred from the hot gas to the dust. Taking the hot gas temperature as the typical temperature seen in X-ray observations (scaled to 5×10^7 K here), the implied cooling rate is 500 $(T/5 \times 10^7)^{-1}$ $M_\odot$/yr. That is, if electrons heat the dust, one must transfer all of the energy in 500 $M_\odot$ of gas every year. For comparison, the rate at which the hot gas is losing energy through X-ray emission is roughly 10-300 $M_\odot$/yr, with 50 $M_\odot$/yr being a typical value. Therefore, *for the clusters with infrared detections, the cooling rate for the hot gas due to the far infrared emission is ten times greater than from its X-ray emission.* Even for clusters not positively detected by IRAS, the upper limits to the infrared luminosities are large enough (1×10^{44} erg/sec from the central region) that far infrared emission can dominate X-ray emission as the primary coolant.

We are in the process of obtaining additional observations at other wavebands in order to understand this emission phenomenon more deeply. Optical observations of the clusters with the best far infrared detections are being obtained to search for the presence of extinction and any anomalous properties of the cD galaxy. We have proposed to obtain X-ray images of these clusters, most of which have either never been observed, or the X-ray observation was too short to provide quantitative morphological information. Finally, future observations with ISO will prove invaluable in obtaining the high-quality far infrared data so sorely needed.

3. Diffuse Far Infrared Emission From Clusters

In addition to the compact far-infrared emission from the vicinity of the cD galaxy, Wise et al. (1992) have begun a search for extended low-surface brightness emission associated with the cluster as a whole. The dimension of a typical cluster of galaxies at z = 0.06 is tens of arcminutes, and unfortunately, diffuse emission from galactic cirrus is present on the same scale. Consequently, finding an effective method of separating diffuse galactic emission from diffuse cluster emission is the central issue of this study. This separation can be performed statistically by determining the distribution of diffuse emission in a control sample and comparing it to the object sample.

The object sample is comprised of 56 clusters of galaxies with a variety of richness properties as well as a range of X-ray emission properties. The five categories are: (1) rich, regular X-ray clusters; (2) poor X-ray clusters; (3) nXD clusters (X-ray clusters without a central dominant galaxy); (4) clusters where the X-ray cooling time is longer than a Hubble time; (5) low X-ray luminosity clusters. For each cluster, a 6°x6° image was obtained at 60 μm and 100 μm with 5' angular resolution. These images were flattened and the total flux was measured in five apertures centered on the cluster; the aperture diameters were 4', 10', 20', 30', and twice the Abell radius. To form the control sample, the flux was measured in each 6°x6° field for these same size apertures in 100 non-overlapping and randomly chosen locations outside the cluster region. The difference between the cluster location and the background fields yields the net flux, with the uncertainty being determined by the distribution of control fields. Although there are no individual clusters with strong diffuse emission ($>3\sigma$), there are more positive 2σ detections than would be expected by chance; examples of such clusters are A262 and A2670. Although Hu (1988) claims to have detected diffuse emission from the Perseus cluster, we do not confirm this result.

Another approach to testing the presence of diffuse emission is to compare the distribution of extracted fluxes for the entire sample (56 objects) to the entire control sample (5600 measurements). The non-parametric Kolmogorov-Smirnov test was used to make this comparison and it showed that the two samples are not drawn from the same population ($>>99\%$ confidence; 60μm and 100μm, 4-30' apertures) in the sense that the cluster fluxes are greater than the control fluxes. When examined by cluster type, we find that all groups are consistent with zero diffuse emission except for the category 2 objects (poor X-ray clusters). The most reliable measurements of the diffuse emission are found with the 10' aperture, and the typical flux of group 2 detections is a few tens of Jy at 100 μm with a 100 μm to 60 μm flux density ratio of ≈ 3, which indicates a temperature (≈ 25 K) near the limit of detectability with IRAS. Measurements with the 30' aperture suggest a diffuse flux that is nearly an order of magnitude larger, but the confidence in that result is lower. For the 10' aperture, a detection of 30 Jy corresponds to a total far infrared luminosity of $L_{FIR} = 10^{46}$ erg/sec and a dust mass of about 10^9 M$_\odot$.

In trying to understand the cause of the emission, the contribution by point sources was estimated separately since these were not removed from the images. For fields with the greatest diffuse emission, point sources were extracted from the entire field. The contribution from point sources in the cluster that are below the detection threshold was estimated (after Corbelli, Salpeter, and Dickey 1991) and the total contribution from point sources was found to be less than 30%. Although work on this is continuing, the analysis shows that diffuse emission is present in a small fraction of clusters, with poor clusters such as A262 being the most common sites of emission.

We speculate that the cause of this diffuse emission might be related to a merger event between two clusters. Prior to the merger the galaxies might be able to retain their gas if stripping is ineffective, but after the merger, the increased velocity dispersion of the galaxies might make stripping significantly more effective, rendering large quantities of cold gas (10^{11} M$_\odot$) and dust (10^9 M$_\odot$) unbound from their host galaxies. Once immersed in the hot cluster medium, thermal evaporation of the dusty gas clouds could be a potential heat source for the dust. If such a collision and the subsequent stripping lasts for $10^{8.5}$ yr, $10^{62.5}$ ergs of energy would be emitted in the form of far infrared radiation, which is about half an order of magnitude less than the energy involved in the collision between two clusters and at least an order of magnitude less than the thermal energy stored in the hot cluster medium.

4. Warming Flows

In addition to cluster cooling flows, several mechanisms can lead to the production of cold gas within clusters of galaxies at rates comparable to or exceeding 100 M$_\odot$/yr. Mass is shed by stars within the ensemble of galaxies at a rate of about 100 M$_\odot$/yr, and this cold material is likely to be stripped from the galaxies and enter the intracluster medium (Soker, Bregman, and Sarazin 1991). The other mechanism, discussed above (sec. 2), is that radiative cooling by dust in the central region would rob thermal energy from hot gas at a rate of ~500 $M_\odot$/yr, leading to the production of cool gas. These rates for the production of cool gas are in addition to the amount inferred from the cooling by X-ray emitting gas (i.e., cooling flows), typically 10-300 $M_\odot$/yr. When these three production mechanisms are taken together, cool gas is being produced at a rate approaching 1000 $M_\odot$/yr, unless reheating processes are important. If this gas is to be hidden in the form of stars, then one might expect to find evidence of star formation in nearly all rich clusters, not just those with high inferred X-ray cooling rates.

Here we consider an alternative model in which gas is not cooling, but is being warmed to the ambient temperature of the cluster (henceforth, "warming flows"). A variety of models involving heating mechanisms have been suggested previously, with conduction being the most commonly discussed process (Takahara and Takahara 1979; Bertschinger and Meiksin 1986; Bregman and David 1988; Rosner and Tucker 1989; Sparks, Macchetto, and Golombek 1989). Conductive heating carried by electrons is an attractive process in the cluster environment. The radius at which the instantaneous radiative cooling time equals the Hubble time is typically 100-200 kpc, and only a fraction of the hot gas mass is contained within this region. Surrounding this cooling region is a dilute gas with a thermal energy content so immense (10^{63}-10^{64} ergs) that is one or two orders of magnitude greater than the total inferred radiative losses over a Hubble time. In several models, conduction transfers energy from the outer dilute region to the more rapidly cooling inner region. Most of these models seek a balance between radiative cooling and conductive heating, although detailed calculations suggest that such a balance would not naturally occur, leading to dismissal of these models. One of the non-equilibrium models suggests that cool gas is being evaporated, although only near the center of a giant elliptical galaxy and at a rate of ~10 $M_\odot$/yr (de Jong et al. 1990).

Conductive evaporation can be extended to encompass all cool gas and dust that enters the cluster environment. Stellar mass loss within a cD galaxy is about 10 $M_\odot$/yr, about one-tenth of the total cluster value and a contribution considerably greater than any other single galaxy. Individual galaxies within the cluster have their gas stripped most easily at high orbital velocity and at high ambient cluster gas density, both of which occur toward the core region. As the stripped gas begins to fall inward in the cluster, it is likely to become shredded into small gas parcels that may undergo magnetic reconnection to the hot surroundings; conductive evaporation ensues. Evaporation consists of two phases, the first being the heating of the cool gas, and the second being the heating and destruction of the dust. The second stage consumes about five times as much energy as the first stage.

The conductive heat flux is adequate to heat the gas and dust, provided that the mean free path of the electrons is not too small. Conduction is most efficient when the electron mean free path is determined by electron-electron collisions. This mean free path will be reduced if the magnetic field is kinked or rippled on scales comparable to the gyroradius, leading to a reduction in the conductive heat flux (to represent this reduction, we adopt a conductive efficiency parameter μ with has a range of 0-1). Gas deposition per unit volume is largest in the central parts of the cluster, and certainly the X-ray and far infrared cooling rates are greatest toward the center, so the conductive heat flux vector must point from the outer to the inner regions. If the temperature scale length for the hot gas (dr/dlnT) is comparable to the size of the cluster (1 Mpc), then the temperature gradient required to evaporate continuously the incoming cool gas is

$$\mathrm{dlnT/dlnr} = 0.2 \; (\dot{\mathrm{M}}/100 \; \mathrm{M}_\odot/\mathrm{yr}) \; (\mathrm{T}/5\mathrm{x}10^7)^{-5/2} \; \mu^{-1}$$

At high conductive efficiency ($\mu = 1$), the required temperature gradient is comparable to the temperature gradient observed in clusters (see review by Fabian, this meeting).

This warming flow model has a variety of virtues relative to cooling flows. In contrast to the cooling flow prediction that the cold gas is consumed in nearly invisible star formation at a high rate, most of the cool gas is evaporated so that little, if any material engages in star formation. Some cool gas may not necessarily be evaporated, such as in the central parts of a cD galaxy where the mass loss rate is relatively high. Should some of the stellar mass loss from the central dominant galaxy ($\leq 10 \; \mathrm{M}_\odot/\mathrm{yr}$) form into stars with a Salpeter initial mass function, it would account for the small amount of young stars found toward some of these systems. Cooling flows face at least two other potential problems that are avoided with warming flows. Thermal instabilities are needed in cooling flows for cool gas to condense in an extended region (100-200 pc) but only a limited range of modes will grow (Balbus 1991). Also, cooling flows predict that cool gas should be present throughout the region defined by the cooling radius (100-200 kpc) yet the tracers of cooled gas, Hα emission line filaments or young stars, only exist in the inner 20 kpc.

Another observation explained by warming flows is that the central dominant galaxy lies in the center of the cluster X-ray emission, which is not always the optical center of the cluster. In the warming flow picture the greatest temperature gradient is toward the cD galaxy, where the X-ray temperature should reach a minimum and the X-ray flux a maximum. This type of "conductive focusing" will align the X-ray emission with the cD position, even if it does not lie exactly at the center of the cluster potential. For cooling flows, the cooling time of the X-ray gas should depend on the potential well and the total gas content, but not necessarily on a single large galaxy. However, only those clusters with central dominant galaxies have an inferred X-ray cooling time less than a Hubble time. This observation is easily understood within warming flows, where the apparent cooling time ($\propto \mathrm{L}_x/\mathrm{T}_x$) is shortest where the X-ray emission is most intense, which is necessarily the regions of greatest conductive evaporation. Furthermore, the presence of dust in the optical emission line gas (Sparks et al. 1990) is not expected if the gas cooled from 10^8 K (i.e., cooling flows), but in the warming flow picture is naturally identified as stellar mass loss with a normal dust to gas ratio. Finally, warming flows would predict a higher inferred $\dot{\mathrm{M}}$ at higher redshift because stellar mass loss is greater at earlier times, and this is consistent with the high cooling rate values suggested for redshifted clusters (Crawford and Fabian 1989; also, Fabian, this meeting).

Although both the cooling and warming flow models have substantial amounts of gas between 10^8 K and a much cooler phase (10-10^4 K), it is possible to discriminate between cooling and warming gas, and therefore eliminate one of the models from consideration. The temperature distribution of cooling gas is determined by the rate at which thermal energy is radiated, while for evaporating gas, the temperature distribution is determined by the heat flux into the system. In both cases, the gas spends most of its time near the temperature of the hot ambient gas, either because the cooling time is at a maximum or because the temperature gradient (and the heat flux vector) is at a minimum. However, the amounts of gas at intermediate temperatures (10^5-10^7 K) in the two models should be dramatically different.

For distinguishing between evaporating and cooling gas, one of the most important diagnostics is the Fe XVII emission complex, which has a series of lines at 12 - 17 A. For cooling material, these lines dominate the radiative losses in the temperature region 10^6-10^7 K, even if the metal abundance is one-tenth of the solar value (Sarazin and Wise 1991). Consequently, approximately one Fe XVII photon is emitted per hydrogen atom in order to cool the gas significantly (more detailed calculations indicate that the number is closer to 0.6 photons/H atom; Sarazin and Wise 1991). For material warming through conductive heating, the time that an iron atom spends as Fe XVII depends on the rate of collisional ionization. Because the rate of line emission also depends upon the collisional rate, the number of Fe XVII photons per Fe atom is approximately the ratio of the sum of the line cross sections to the ionization cross section. This ratio (approximately 3-10) times the Fe/H abundance ratio gives about 10^{-3} Fe XVII photons per hydrogen atom, or about 10^3 less than for an equivalent mass flux of cooling gas. Presently, there are few published Fe XVII observations, although the observation of M87 and Perseus (Canizares, Markert, and Donahue 1988; Mushotzky and Szymkowiak 1988) is consistent with cooling material in the cooling flow picture. However, these observations do not rule out the presence of evaporating material since the line emission from such material would be below the detection threshold of instruments in the near future.

I would like to thank Craig Sarazin and John Raymond for their valuable comments on the subject of warming flows and to Andy Fabian and the organizing committee for their hospitality and efforts in arranging an interesting and productive meeting.

5. References

Balbus, S.A. 1991, ApJ, 372, 25.

Bertschinger, E, and Meiksin, 1986, ApJ, 306, L1.

Bregman, J.N., and David, L.P. 1988, ApJ, 326, 639.

Bregman, J.N., McNamara, B.R., and O'Connell, R.W. 1990, ApJ, 351, 406.

Canizares, C.R., Markert, T.H., and Donahue, M.E. 1988, in "Cooling Flows in Clusters and Galaxies", ed. A.C. Fabian, (Kluwer: Dordrecht), p. 63.

Corbelli, E., Salpeter, E.E., and Dickey, J.M. 1991, ApJ, 370, 49.

Cox, C., Bregman, J.N., and Schombert, J.S. 1992, in preparation.

Crawford, C.S., and Fabian, A.C. 1989, MNRAS, 238, 41P.

de Jong, T., Norgaard-Nielsen, H.U., Jorgensen, H.E., and Hansen, L. 1990, A&Ap, 232, 317.

Draine, B.T. 1990, in "The Interstellar Medium in Galaxies", ed H.A. Thronson, Jr., and J.M. Shull (Dordrecht: Kluwer), p. 483.

Draine, B.T., and Lee, H.M. 1984, ApJ, 285, 89.

Dwek, E., Rephaeli, Y., and Mather, J.C. 1990, ApJ, 350, 104.

Fabian, A.C., Nulsen, P.E.J., and Canizares, C.R. 1991, A&ApRev, 2, 191.

Grabelsky, D.A., and Ulmer, M.P. 1990, ApJ, 355, 401.

Haynes, M.P., Herter, T., Barton, A.S., and Benesohn, J.S. 1990, AJ, 99, 1740.

Hildebrand, R.H. 1983, QJRAS, 24, 267.

Mathis, J.S., Rumpl, W., and Nordsieck, K.H. 1977, ApJ, 217, 425.

Mathis, J.S., and Whiffen, G. 1989, ApJ, 341, 808.

McNamara, B.R., and O'Connell, R.W. 1991, ApJ, in press.

Mushotzky, R., and Szymkowiak, A.E. 1988, in "Cooling Flows in Clusters and Galaxies", ed. A.C. Fabian, (Kluwer: Dordrecht), p. 47.

Neugebauer, G., et al. 1984, ApJ, 278, L1.

Romanishin, W., and Hintzen, P. 1989, ApJ, 341, 41

Rosner, R., and Tucker, W.H. 1989, ApJ, 338, 761.

Rowan-Robinson, M. 1986, MNRAS, 219, 737.

Sarazin, C.L., and Wise, M.W. 1991, in "Iron Line Diagnostics in X-Ray Sources", ed. A. Treves (Berlin: Springer-Verlag), in press.

Soker, N., Bregman, J.N., and Sarazin, C.L. 1991, ApJ, 368, 341.

Sparks, W.B., Macchetto, F.D., and Golombek, D. 1989, ApJ, 345, 153.

Takahara, M., and Takahara, F. 1979, Prog. Theor. Phys., 62, 1253.

Wise, M.W., O'Connell, R.W., Bregman, J.N., and Roberts, M.S. 1992, in preparation.

Young, J. 1990, in "The Interstellar Medium in Galaxies", ed H.A. Thronson, Jr., and J.M. Shull (Dordrecht: Kluwer), p. 67.

THE INTRACLUSTER MEDIUM

C.L. Sarazin
Department of Astronomy
University of Virginia
P. O. Box 3818
Charlottesville VA 22903-0818
U.S.A.

ABSTRACT. Clusters of galaxies are full of hot, diffuse plasma, with typical temperatures and electron densities of $T \sim 7 \times 10^7$ K and $n_e \sim 10^{-3}$ cm^{-3}. This intracluster medium (ICM) produces copious X-ray emission, and typical cluster X-ray luminosities are $L_X \sim 10^{43} - 10^{45}$ ergs s^{-1}. The total mass of intracluster gas generally exceeds the mass of all of the galaxies in a rich cluster. The gas is probably nearly hydrostatic. From the distribution of the intracluster plasma, one can determine the total gravitational mass of the cluster and its distribution. These measurements indicate that much of the cluster mass is in the form of "dark matter." However, the X-ray observations also favor a distribution in which the dark matter is more centrally condensed than the galaxies, which are in turn more centrally condensed than the gas. This suggests that at least part of the dark matter is baryonic. The X-ray emission processes and physical properties of the ICM are reviewed. The ICM is probably inhomogeneous, with blobs of denser and cooler gas intermixed with hotter, more diffuse gas. If this is correct, the density fluctuations may complicate efforts to use the ICM to determine cluster masses and distances. The intracluster medium also contains a significant magnetic field. Tangled magnetic field lines probably inhibit thermal conduction, and allow multiphase gas and cooling to occur. The magnetic fields are dramatically amplified by inflow and compression in cooling flow regions. These strong magnetic fields produce large Faraday rotations in many nearby radio sources associated with the central galaxies in cooling flows. I argue that similar cooling flow magnetic fields depolarize distant radio galaxies and quasars. In cooling flows, the magnetic fields reach equipartition with the thermal gas, and are probably quite important dynamically.

1. Introduction

One of the more surprising results from X-ray astronomy is that the great volumes of space between galaxies in clusters of galaxies are not empty, as they appear in optical images. Instead, they are filled with a diffuse, hot plasma, with typical temperatures of $T \sim 10^7 - 10^8$ K. At this temperature, the sound speed in the gas is comparable to the orbit velocities of the galaxies in the cluster, which is consistent with the gas being in hydrostatic equilibrium with the same gravitational potential as binds the galaxies. This intracluster medium (ICM) is highly rarefied, with proton number densities of $n_p \sim 10^{-4} - 10^{-2}$ cm^{-3}. At least on large scales, the gas is stably stratified, with the density decreasing with increasing radius r. The gas extends out to distances of $r \gtrsim$ Mpc from the cluster center. The total

A. C. Fabian (ed.), Clusters and Superclusters of Galaxies, 131–150.

mass of hot gas is typically $M_{gas} \sim 10^{14} M_{\odot}$; this mass exceeds the total mass of all the galaxies in a typical rich clusters, although even more of the mass is in the form of unseen "dark matter." (All comparisons to observations in this paper assume a Hubble constant of $H_o = 50$ km s^{-1} Mpc^{-1}.)

At temperatures of $10^6 - 10^8$ K, the dominant radiation mechanism of a plasma is X-ray emission. As a result, clusters of galaxies are generally very luminous X-ray emitters, with luminosities of $L_X \sim 10^{43} - 10^{45}$ ergs s^{-1}. Clusters are second only to quasars as the most luminous X-ray sources in the Universe. While X-ray emission is the primary observational diagnostic for the intracluster medium, the ICM has a number of other important physical effects. It confines and distorts radio galaxies within the cluster. The cosmic ray and magnetic field components of the intracluster medium can also produce diffuse radio emission (e.g., Jaffe 1992) The ICM can strip interstellar gas from galaxies as they move through the cluster. Intracluster gas cools at the centers of many clusters, producing lower temperature gas. If the ICM contains dust, the dust will be strongly heated by the plasma, and will emit strongly in the infrared (Bregman 1992). The ICM also has a number of opacity effects; for example, it scatters and heats the cosmic background radiation which passes through it (e.g., Lasenby 1992). The magnetic field in the ICM leads to Faraday rotation and depolarization (section 8).

In this paper, I will review the X-ray emission processes and physical state of the ICM (sections 2 and 3). Models for the distribution of the gas, and the use of the gas to determine the total mass distributions of clusters are described in sections (4) and (5). The effects of thermal conduction and its possible suppression by tangled magnetic fields are discussed in section (6). If conduction is suppressed, the intracluster gas is probably fairly inhomogeneous (section 7). The role of the magnetic field in the ICM is reviewed in section (8), where special emphasis is placed on the amplification of the field which occurs in cluster cooling flow regions. Models for cooling flow magnetic fields are compared to observations of Faraday rotation in nearby radio galaxies in section (8.3) and observations of the depolarization of distant radio sources is section (8.4). Section (9) gives some concluding remarks. Many important topics have been omitted in the interest of brevity. For example, in my talk I discussed models for the chemical and thermal evolution of the ICM (see also Sarazin 1988, Jones 1992, and Henry 1992).

2. X-ray Emission

The X-ray emission of the intracluster gas is mainly due to thermal bremsstrahlung and line emission. At the low gas densities in clusters, the X-ray emissivity is just proportional to the square of the gas density. The emissivity due to thermal bremsstrahlung (free–free emission) is given by

$$\epsilon_{\nu}^{ff} = \frac{2^5 \pi e^6}{3 m_e c^3} \left(\frac{2\pi}{3 m_e k} \right)^{1/2} n_e T^{-1/2} \exp(-h\nu/kT) \sum_i Z_i^2 n_i g_{ff}(Z_i, T, \nu), \qquad (1)$$

where the emissivity ϵ_{ν} is defined as the emitted energy per unit time, frequency, and volume. The sum is over the various ions in the plasma, but is dominated by hydrogen and helium for solar abundances. The Gaunt factor $g_{ff}(Z_i, T, \nu)$ corrects

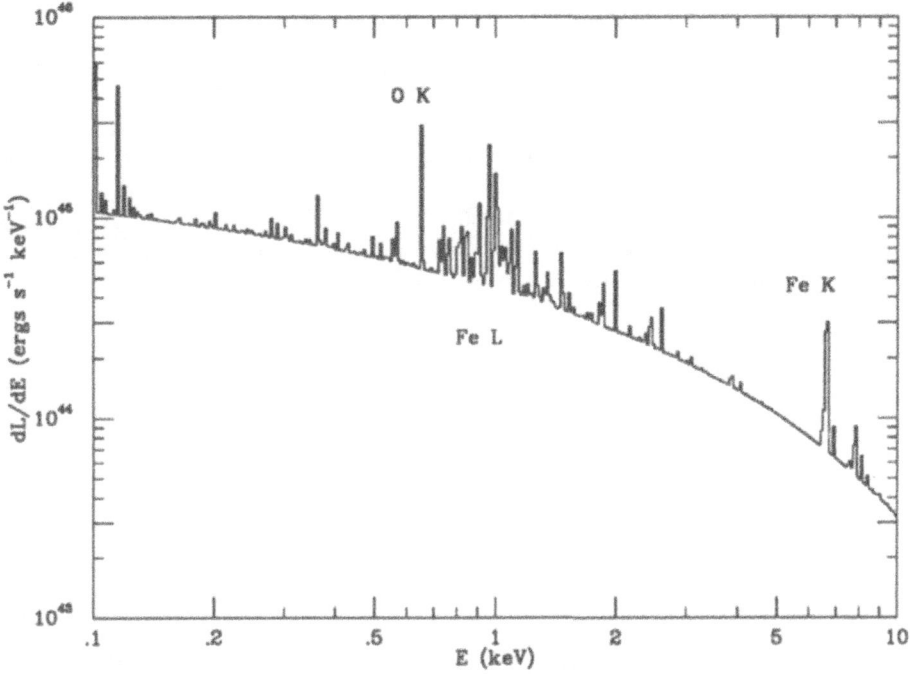

Fig. 1. The predicted X-ray spectrum of a cluster of galaxies. The model cluster is isothermal at $T = 8 \times 10^7$ K in its outer regions, and has a cooling flow with $\dot{M} = 300\,M_\odot\,\mathrm{yr}^{-1}$ in its inner regions.

for quantum mechanical effects and for the effect of distant collisions, and is a slowly varying function of frequency and temperature. As a result, the dominant dependence of the free–free emissivity on frequency is the Boltzmann exponential factor, and the main dependences on temperature are this factor and the square–root factor $T^{-1/2}$. Thermal bremsstrahlung produces a roughly exponential continuum component in the X-ray spectrum. At high temperatures $T \gtrsim 3 \times 10^7$ K, thermal bremsstrahlung is the dominant emission mechanism.

At lower temperatures, the main X-ray radiation is from lines. The strongest line feature observed from most clusters of galaxies is the complex of iron Fe Kα lines at about 6.7 keV. This line feature is actually a blend of lines from iron ions (mainly Fe^{+24} and Fe^{+25}) and weaker lines from nickel ions. The notation "Kα" gives the principal quantum number n of the lower level of the transition and the change in the principal quantum number $\Delta n \equiv n' - n$, where n' is the principal quantum number of the upper level of the transition. K indicates that the lower level is in the K-shell ($n = 1$), L indicates the lower level is in the L-shell ($n = 2$), and so on, while α indicates that $\Delta n = 1$, β indicates that $\Delta n = 2$, etc.

In addition to the Fe K line complex, the X-ray spectra of clusters of galaxies contain a large number of lower energy lines. These include the K lines of the common elements lighter than iron, such as C, N, O, Ne, Mg, Si, S, Ar, and Ca, as well as the L lines of Fe and Ni. These lines become very strong at lower temperatures

($T \lesssim 10^7$ K), and the presence of these lines is the most direct evidence for the presence of cooling gas in clusters (Fabian et al. 1991; Fabian 1992; Mushotzky 1992).

As a illustration, Figure (1) shows the predicted X-ray spectrum of an X-ray cluster (Wise and Sarazin 1992). The model cluster is isothermal in its outer regions (with a temperature of 8×10^7 K), and has a cooling flow in its inner regions with a cooling rate of $\dot{M} = 300 \, M_\odot \, \mathrm{yr}^{-1}$. The figure shows the overall exponential continuum from thermal bremsstrahlung, the Fe K lines at about 7 keV (which come mainly from the region of the cluster outside of the cooling flow), and the lower energy lines from the cooling flow. The latter include the strong O K line at 0.65 keV, and the complex of Fe L lines around 1 keV.

Most X-ray lines are excited by collisional excitation by electrons, although radiative and dielectronic recombination and inner shell collisional ionization also play a role. The emissivity due to a collisionally excited line is usually written (e.g., Osterbrock 1974)

$$\int \epsilon_\nu^{line} d\nu = n(X^i) n_e \frac{h^3 \nu \Omega(T) B}{4 \omega_{gs}(X^i)} \left[\frac{2}{\pi^3 m_e^3 kT} \right]^{1/2} e^{-\Delta E / kT} , \qquad (2)$$

where $h\nu$ is the energy of the transition, ΔE is the excitation energy above the ground state of the excited level, B is the branching ratio for the line (the probability that the upper state decays through this transition), and Ω is the 'collision strength', which is often a slowly varying function of temperature. The intracluster gas is almost certainly in collisional ionization equilibrium; under these circumstances, the ionization fractions depend only on the electron temperature T, and are independent of the density of the gas. The emissivity of a line is then proportional to the square of the density and to the abundance of the relevant element, and depends significantly on the electron temperature. Because the thermal bremsstrahlung emissivity also is proportional to the square of the density (equation 1), the ratio of line emission to thermal bremsstrahlung continuum emission is independent of density. Line ratios or the shape of the X-ray continuum spectrum can be used to derive a temperature for the gas in a cluster. Then, the ratio of line emission to thermal bremsstrahlung continuum emission can be used to determine the abundance of the heavy element responsible for the line.

When this technique is applied to the Fe K lines from clusters, the iron abundances (by number of atoms) are all roughly Fe/H $\approx 1 \times 10^{-5}$, which is about one-third of the solar value (e.g., Mushotzky 1992). FPCS observations of the Fe L and O K lines in the Perseus cluster indicate that the relative abundance of oxygen and iron is about three times the solar value, O/Fe ≈ 3(O/Fe)$_\odot$ (Canizares et al. 1988).

3. Physical Properties of the ICM

The mean free paths of electrons and ions in a plasma without a magnetic field are determined by Coulomb collisions. The mean free path λ_e for an electron to suffer an energy exchanging collision with another electron is given by (Spitzer 1956)

$$\lambda_e = \frac{3^{3/2}(kT)^2}{4\pi^{1/2}n_e e^4 \ln \Lambda} \approx 23\,\text{kpc} \left(\frac{T}{10^8\,\text{K}}\right)^2 \left(\frac{n_e}{10^{-3}\,\text{cm}^{-3}}\right)^{-1}, \tag{3}$$

where n_e is the electron number density, and $\ln \Lambda \approx 38$ is nearly independent of density or temperature. (Λ is the ratio of largest to smallest impact parameters for the collisions.) The mean free path for ions (protons) is essentially equal to that for electrons, $\lambda_i = \lambda_e$.

These mean free paths are smaller than most scales of interest in clusters. Thus, it is reasonable to treat the ICM as a fluid under most circumstances. The fluid approximation might breakdown in the outer parts of a cluster (where the density is low), in interactions with galaxies (whose sizes are comparable to λ_e), or if the ICM is very inhomogeneous. In any case, the ICM apparently contains a significant magnetic field $B \sim 1\mu G$ (section 8). The gyroradii of electrons and ions in such a field are very small. For example, the gyroradius of a typical electron is

$$r_g = 3.1 \times 10^8\,\text{cm} \left(\frac{T}{10^8\,\text{K}}\right)^{1/2} \left(\frac{B}{1\,\mu G}\right)^{-1}. \tag{4}$$

These small gyroradii probably insure that the ICM acts as a fluid even when the Coulomb mean free paths are long.

Collisions between electrons will bring the electrons into equilibration (an isotropic Maxwellian velocity distribution) on a time scale of roughly $t_{eq}(e,e) \equiv \lambda_e/\langle v_e \rangle_{rms}$, where the denominator is the rms electron velocity. Numerically, this gives

$$t_{eq}(e,e) \approx 3.3 \times 10^5\,\text{yr} \left(\frac{T}{10^8\,\text{K}}\right)^{3/2} \left(\frac{n_e}{10^{-3}\,\text{cm}^{-3}}\right)^{-1}. \tag{5}$$

The time scale for protons to equilibrate among themselves is $t_{eq}(p,p) \approx (m_p/m_e)^{1/2} t_{eq}(e,e)$, or roughly 43 times longer than the value in equation (5). Following this time, the protons and ions would each have Maxwellian distributions, but generally at different temperatures. The time scale for the electrons and ions to reach equipartition (equal temperatures) is $t_{eq}(p,e) \approx (m_p/m_e)t_{eq}(e,e)$, or roughly 1870 times the value in equation (5). Under the conditions in clusters, this longest equilibration time is $t_{eq}(p,e) \lesssim 10^9$ years. Since this is shorter than the age of the cluster or the cooling time, the intracluster plasma can generally be characterized by a Maxwellian distribution at a single kinetic temperature T.

The dynamical equation for a single component fluid is

$$\rho\frac{D\mathbf{v}}{Dt} + \nabla P + \rho\nabla\Phi = 0, \tag{6}$$

where ρ is the mass density of the gas, P is the pressure, Φ is the gravitational potential, and D/Dt is the Lagrangian derivative with respect to time. Equation (6) ignores nongravitational forces, such as magnetic stresses. In section (8), we show that magnetic fields may be important in clusters. Many clusters may currently be in the processes of collapse and virialization, and the the gravitational potential may vary with time. Consider a cluster which is relaxed (so that $\partial\Phi/\partial t = 0$), and which is not subject to very rapid cooling or heating of the gas. Then, the gas will relax into hydrostatic equilibrium (with pressure forces balancing gravitational forces) after a period comparable to the time required for a sound wave to cross the cluster,

$$t_s \equiv \frac{D}{c_s} \approx 6.6 \times 10^8 \, \text{yr} \left(\frac{T}{10^8 \, \text{K}}\right)^{-1/2} \left(\frac{D}{\text{Mpc}}\right) . \tag{7}$$

Here, D is the diameter of the cluster, and c_s is the sound speed.

4. Gas Distributions

Following the arguments given above, simple models for the distribution of the intracluster gas generally assume that the gas is hydrostatic. If, in addition, the cluster is assumed to be spherically symmetric, equation (6) reduces to

$$\frac{1}{\rho}\frac{dP}{dr} = -\frac{GM(r)}{r^2} , \tag{8}$$

where $M(r)$ is the total cluster mass within r. The galaxies in a cluster respond to the same gravitational field. If the galaxies are assumed to move on isotropic orbits, they also satisfy a hydrostatic equation,

$$\frac{1}{n_{gal}}\frac{d(n_{gal}\sigma^2)}{dr} = -\frac{GM(r)}{r^2} , \tag{9}$$

where n_{gal} is the number density of galaxies, and σ is their one-dimensional velocity dispersion (e.g., White 1992). Equations (8) and (9) can be combined to eliminate the total mass and relate the gas and galaxy distributions:

$$\frac{1}{\rho}\frac{dP}{dr} = \frac{1}{n_{gal}}\frac{d(n_{gal}\sigma^2)}{dr} . \tag{10}$$

Because equations (8) or (10) give a single relation for two gas properties (density and pressure), one must also specify the entropy distribution of the gas to determine its distribution. A very simple model follows if the gas is assumed to be isothermal ($T = $ constant); isothermality might result is thermal conduction were efficient in the cluster (see section 6). If the galaxies are also assumed to be "isothermal" ($\sigma = $ constant), equation (10) is easily solved to yield

$$\rho \propto n_{gal}^\beta \tag{11}$$

(Cavaliere & Fusco-Femiano 1976; Sarazin & Bahcall 1977). Here,

$$\beta \equiv \frac{\mu m_p \sigma^2}{kT} = 0.76 \left(\frac{\sigma}{10^3 \, \text{km/s}}\right)^2 \left(\frac{T}{10^8 K}\right)^{-1} , \tag{12}$$

where $\mu = 0.63$ is the mean mass per particle in the gas in units of the proton mass. The quantity β is the square of the ratio of the average galaxy and gas particle speeds. If $\beta > 1$, the galaxies are "hotter" than the gas (the galaxies move faster). If $\beta < 1$, the gas is "hotter" than the galaxies.

Often, the galaxy distributions in clusters are fit to the "King analytic" form

$$n_{gal} = n_{gal}^o \left[1 + \left(\frac{r}{r_c} \right)^2 \right]^{-3/2} , \tag{13}$$

where n_{gal}^o is the central value, and r_c is defined as the "core radius" of the cluster. If this form is substituted into equation (11), the gas distribution is found to be

$$\rho(r) = \rho_o \left[1 + \left(\frac{r}{r_c} \right)^2 \right]^{-3\beta/2} . \tag{14}$$

Again, ρ_o is the central value of the gas density.

Equation (14) has been very widely used to fit or model gas distributions in clusters. It has several attractive properties; one is that essentially all necessary integrals of this function can easily be done analytically. For example, the resulting X-ray surface brightness in any spectral band can be shown to be

$$I_X(r) = I_X^o \left[1 + \left(\frac{r}{r_c} \right)^2 \right]^{-3\beta+1/2} . \tag{15}$$

Jones and Forman (1984) found that equation (15) provides an adequate fit to the radial surface brightness profiles, except for the innermost regions of cooling flow clusters (e.g., Fabian 1992). The average value of β determined by these fits to the X-ray surface brightness of a large number of clusters was found to be (Jones & Forman 1984)

$$\langle \beta_{image} \rangle \approx \frac{2}{3} . \tag{16}$$

Thus, the X-ray surface brightness and implied gas density vary on average as

$$I_x(r) \propto \left[1 + \left(\frac{r}{r_c} \right)^2 \right]^{-3/2} , \tag{17}$$

and

$$\rho(r) \propto \left[1 + \left(\frac{r}{r_c} \right)^2 \right]^{-1} . \tag{18}$$

This suggests that the gas density falls off less rapidly with radius than the galaxy density, and that the gas is "hotter" than the galaxies.

Unfortunately, equation (16) does not agree very well with determinations of the X-ray spectral temperatures and the galaxy velocity dispersions of clusters (Mushotzky 1984; Edge & Stewart 1991). When observed gas temperatures and galaxy velocity dispersions are substituted directly into equation (12), the average value of β is found to be

$$\langle \beta_{spect} \rangle \approx 1 . \tag{19}$$

What is the cause of this discrepancy? First, the gas may very well not be isothermal, although Mushotzky (1984) has argued that the same problem occurs for other global thermal distributions in the gas. The gas may have small scale inhomogeneities, which would cause the X-ray spectra to underestimate the temperature of the most diffuse gas (see section 7). Second, it may be that the velocity dispersion does not accurately represent the energy per unit mass of the galaxies. The galaxy orbits may be anisotropic, or nonisothermal. It may be that many cluster velocity dispersions are contaminated by foreground or background galaxies or by the effects of subclustering (Geller & Beers 1982). Any of these effects could cause the data to overestimate the actual cluster potential and β_{spect}. Finally, it may be that the galaxy distributions are not well-represented by the analytic King form (equation 13). At large radii, the analytic King model varies as $n_{gal} \propto r^{-3}$, while a self-gravitating isothermal sphere varies as $n_{gal} \propto r^{-2}$. If the latter were the correct representation of the galaxy distribution, the values of β implied by the X-ray images would be $\langle \beta_{image} \rangle \approx 1$, which is consistent with the spectral observations.

5. Masses of Clusters

For the average value of β_{image} derived from X-ray images, the gas density varies according to equation (18), which implies that $\rho \propto r^{-2}$ at large radii. Thus, the total mass of gas diverges with radius as $M_{gas} \propto r$. At the same time, the X-ray surface brightness declines very rapidly, with $I_X \propto r^{-3}$ (equation 17). Thus, it is very difficult to determine the total mass of intracluster gas; most of the gas is at large radii where it is very difficult to observe. Within about 1 Mpc of the cluster center for the best-studied clusters, the total mass of hot gas is $M_{gas} \sim 10^{14} M_\odot$; In rich clusters, this is typically five times the mass of all of the optical galaxies in the cluster.

The condition of hydrostatic equilibrium can be used to determine the total mass and density distribution of a cluster. Equation (8) can be solved for the total mass of a cluster:

$$M(r) = -\frac{kT(r)r}{\mu m_p G} \left(\frac{d \log \rho}{d \log r} + \frac{d \log T}{d \log r} \right) \tag{20}$$

(Bahcall & Sarazin 1977; Mathews 1978; Fabricant et al. 1980, 1984; Fabricant & Gorenstein 1983). This method has many advantages over optical determinations of the cluster mass based on galaxy orbits (e.g., White 1992). The cluster mass is not very dependent on the gas density profile in the cluster, since only the logarithm of the density enters equation (20). However, it is very sensitive the distribution of gas temperatures. This is unfortunate, because the most of the existing X-ray observations of clusters do not have very good spectral resolution, and give more accurate profiles of the density than the temperature.

The most successful application of the hydrostatic method to determine cluster masses has been with M87 in the Virgo cluster (Fabricant & Gorenstein 1983; Stewart et al. 1984). The X-ray observations are consistent with a total mass of

$$M(r) = (3 - 6) \times 10^{13} \, M_\odot \left(\frac{r}{300 \, \text{kpc}} \right) . \tag{21}$$

At large radii $r > 100$ kpc, the implied total mass–to–light ratio is $(M/L_B)_{tot} > 150 \, M_\odot/L_\odot$, and the local value is even higher, $(M/L_B)_{local} > 500 \, M_\odot/L_\odot$. This implies that, in the region around M87, most of the mass is in the form of "dark matter." Unfortunately, because this mass refers to a small region around M87 in a cluster which is not terribly rich (the Virgo cluster), it is unclear whether this dark matter is really a halo around the galaxy M87 or a distributed element of the entire cluster.

The hydrostatic method has been applied to the mass distributions in several other clusters, particularly Coma and Perseus (Cowie et al. 1987; The and White 1986, 1988; Hughes 1989; Eyles et al. 1991). Because of the lack of observations combining good spectral and spatial resolution from which to derive the density and temperature as a function of radius, the results for these clusters are somewhat uncertain. In the Coma cluster, Hughes (1989) derives a total mass within $r < 5$ Mpc of about $2 \times 10^{15} \, M_\odot$, and a total mass-to-light ratio of $(M/L_B)_{tot} \approx 165 \, M_\odot/L_\odot$. This is somewhat lower than many optical determinations based on galaxy orbits. When the total mass is compared to the mass in intracluster gas M_{gas} and the mass in optical galaxies M_{gal}, a significant residual of dark matter is required. In most rich clusters, the different components of the mass satisfy the inequality

$$M_{gal} < M_{gas} < M_{dark \ matter} . \tag{22}$$

However, the fraction of the total mass which is due to hot intracluster gas is not insignificant, with $M_{gas} \sim 30\% \, M_{tot}$.

The hydrostatic method allows one to derive the distribution of the total mass (equation 20). When this is compared to the distributions of the gas and galaxy densities, the density distribution of dark matter can be derived. Figure (2) shows the derived distributions of the mass density in dark matter, gas, and galaxies from the Perseus cluster from Eyles et al. (1991). Similar results have been found for Coma and several other clusters by Cowie et al. (1987) and Hughes (1989). Of course, these distributions are rather uncertain with the present data.

These X-ray observations suggest that the dark matter is *more* centrally condensed than the galaxies, which are themselves *more* centrally condensed than the gas. Now, when extended dark matter halos were inferred to exist around individual spiral galaxies, their extent relative to the normal material was taken as evidence that the dark matter was not dissipational. If the dark matter in clusters is really *less* extended than the dissipational gas and galactic matter, this may indicate that the cluster dark matter is actually dissipational. This argument might eliminate weakly interacting particles as a candidate for the cluster dark matter. The X-ray observations may indicate that the cluster dark matter (or, at least the

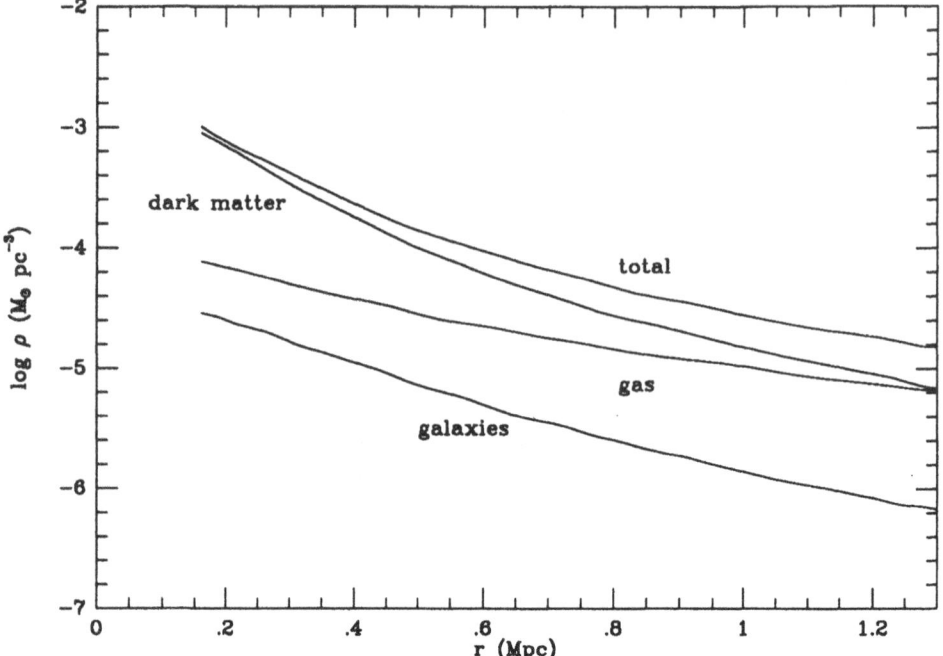

Fig. 2. The density of mass in galaxies, in hot gas, and in dark matter, and in total, as derived by Eyles et al. (1991) from X-ray observations of the Perseus cluster.

portion near the cluster center) is baryonic. This is particularly interesting given the evidence that cluster cooling flows may be generating baryonic dark matter at the present time (Thomas and Fabian 1990; Fabian 1992).

6. Thermal Conduction

Unless suppressed in some manner, thermal conduction should be extremely effective in the hot gas in clusters. In a plasma with a gradient in the electron temperature, heat is conducted down the temperature gradient. If the scale length of the temperature gradient is much longer than the mean free path of electrons λ_e (equation 3), then the heat flux is given by

$$\mathbf{Q} = -\kappa \, \nabla T,$$ (23)

where the thermal conductivity κ for a hydrogen plasma is (Spitzer 1956)

$$\kappa = 1.31 n_e \lambda_e k \left(\frac{kT}{m_e}\right)^{1/2} \approx 4.6 \times 10^{13} \left(\frac{T}{10^8 \, \mathrm{K}}\right)^{5/2} \mathrm{ergs \, s^{-1} \, cm^{-1} \, K^{-1}}.$$ (24)

If the very weak dependence of $\ln \Lambda$ on density is ignored, then κ is independent of density, but increases very rapidly with increasing temperature.

If a cluster initially had a significant radial temperature gradient, thermal conduction would tend to make the central portions of the cluster isothermal. Figure

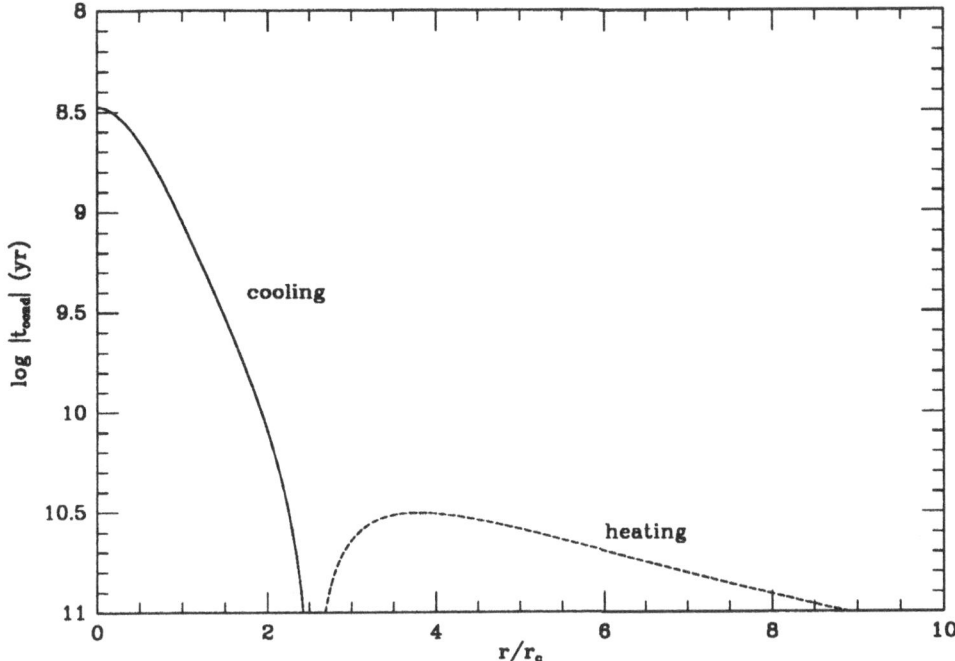

Fig. 3. The logarithm of the time scale (in years) for thermal conduction to modify the temperature in a cluster, as a function of the radius r in units of the cluster core radius r_c. Note that the vertical scale is inverted and that shorter time scales correspond to more effective conduction. The cluster gas in this figure is initially adiabatic, with a central temperature of $T = 10^8$ K, a central electron density of $n_e = 10^{-3}\,cm^{-3}$, and a core radius of $r_c = 0.25$ Mpc. The solid curve indicates regions where the gas is cooled by conduction, while the dashed curve indicates where the gas is heated.

(3) illustrates this point. The quantity $t_{cond} \equiv -(d\ln T_e/dt)^{-1}$ is the time required for thermal conduction to significantly change the temperature. The gas in the cluster in this figure initially was adiabatic, with a radially decreasing temperature. Initially, the central temperature and electron density are $T = 10^8$ K and $n_e = 10^{-3}\,cm^{-3}$, and the cluster core radius (equation 13) is $r_c = 0.25$ Mpc. Because conduction only transports heat, the average temperature of the gas is not changed; in the inner parts the gas is cooled and in the outer parts the gas is heated. At radii $r \gtrsim 2r_c$, the conduction time is typically longer than the Hubble time. Heat conduction is most effective within the cluster core, and $|t_{cond}|$ increases very rapidly with radius.

Spectral observations of the Coma cluster suggest that the gas is nearly isothermal in the inner regions, but the temperature drops in the outer parts of the cluster (Hughes et al. 1988). It is possible that this is due to conduction; alternatively, it may reflect the dynamical history of the cluster. Because it contains two large D galaxies, it is likely that Coma recently had a subcluster merger.

Thermal conduction can be reduced by the presence of a tangled magnetic field in the ICM (section 8). Electrons gyrate around magnetic field lines on orbits with a radius of r_g (equation 4), which is much smaller than the mean free path or any length scale of interest in clusters. Gyroradii are so small that heat conduction perpendicular to the magnetic field is greatly reduced in the intracluster gas, even if the field is very weak. If the magnetic field lines are highly tangled, it seems plausible that the conduction would be very highly suppressed (David and Bregman 1989; Tribble 1989). Because conduction occurs mainly along field lines, the lengths over which the heat must diffuse are greatly increased by a tangled magnetic field. Further, if the topology of the field is such that hot and cold regions are not connected by the same field lines, heat conduction would require diffusion across field lines.

The presence of cooling flows provides evidence that thermal conductivity of the intracluster gas is suppressed. On a global scale, cooling flows contain cooler gas near the cluster center and hotter gas further out. Models show that heat conduction would erase the observed temperature and density gradients in cooling flows, unless the thermal conduction is suppressed (David and Bregman 1989). On smaller scales, the observations of cooling flows indicate that cooler blobs of gas are immersed in the hotter, more diffuse gas (e.g., Fabian 1992). Again, these blobs would evaporate if thermal conduction were unimpeded. The overall conclusion is that the effective heat conductivity of the intracluster gas must be much lower than the 'Spitzer' value (equation 24). While it is likely that this results from tangled magnetic fields, I am not certain that this problem has yet been treated in a completely self-consistent manner.

7. Is the ICM Inhomogeneous?

Most of the discussion of the ICM assumes that it is a homogeneous plasma, with all of the gas at a given radius having the same density and temperature. Observations of cooling flows in clusters indicate that the gas there is quite inhomogeneous (e.g., Fabian 1992). The inhomogeneity of the cooling flow gas is probably enhanced by thermal instabilities and other processes (like magnetic effects; section 8), but calculations indicate that the gas entering the central regions must already have very significant density fluctuations. All of this suggests that the ICM may not be homogeneous.

What is the nature of the density variations? The sound crossing time for the intracluster gas is quite short (equation 7). This indicates that non-hydrostatic pressure variations in the gas damp rather quickly, except during dynamical phases of the cluster evolution (such as subcluster mergers). On the other hand, isobaric entropy variations may survive much longer if thermal conductivity is suppressed (section 6). In such fluctuations, higher density is accompanied by lower temperature, keeping the pressure nearly constant. Unless the denser regions are supported in some way, they will tend to fall through the lower density ambient gas. Ram pressure and Kelvin–Helmholtz instabilities will break up the large blobs into smaller blobs. The smaller blobs have more surface area per mass and drag, and will fall less rapidly. Thus, one might expect that the higher density material will mainly consist of a large number of rather small blobs. If the contrast in density of the blobs is sufficiently large, they may be able to cool (which further increases their density).

What are the possible sources for inhomogeneities in the ICM? First, mergers of subclusters may intermix regions of plasma with different entropies. A second mechanism is the stripping of interstellar gas from cluster galaxies. As noted by Soker et al. (1990), this should introduce about 100 $M_\odot$ per year of low entropy gas into a typical rich cluster; this is comparable to the amount of gas which cools in a cooling flow. Inhomogeneities may also be caused and supported by other physical effects, such as magnetic fields, turbulence, and galactic wakes.

There have been several attempts to search for inhomogeneities in the ICM. Soltan and Fabricant (1990) found fluctuations in the X-ray surface brightness of a sample of clusters. The fluctuations had individual X-ray luminosities of $L_X \approx 4 \times 10^{41}$ ergs s^{-1} on angular scales of $\lesssim 1$ arcmin. One problem with attempts to detect clumping with X-ray images is that the clumps may be too small to be resolved. However, as long as they are nearly hydrostatic, denser clumps will have cooler temperatures and may be detectable spectroscopically. Allen et al. (1991) find that the spectrum of the Perseus cluster is consistent with a a temperature spread of $\Delta(kT) \lesssim 3.5$ keV. Canizares et al. (1988) gave an upper limit of the rate of cooling of diffuse gas in the Coma cluster; this could occur if density fluctuations were large enough that the denser gas cooled rapidly. Their upper limit was $\dot{M}_{cool} \lesssim 125\, M_\odot$ yr^{-1}.

Given the degree of clumping of the gas entering cooling flows, it seems likely that the general intracluster medium is also quite inhomogeneous. Unfortunately, clumping of the ICM could confuse cluster mass determinations based on the hydrostatic method (section 5), and attempts to use the Sunyaev–Zeldovich effect to determine cluster distances (e.g., Lasenby 1992). Spectral observations of clusters with Astro–D should provide a definitive test of the clumping of the ICM.

8. Cluster Magnetic Fields

This review has concentrated on the non-relativistic, thermal plasma component of the intracluster medium. The ICM also contains relativistic components, including cosmic ray particles and magnetic fields (e.g., Jaffe 1992). The intracluster magnetic field can produce a number of important physical effects. If the pressure in the magnetic field $P_B = B^2/8\pi$ is greater than or comparable to the thermal gas pressure, the magnetic field can strongly affect the dynamics of the intracluster gas. At other extreme, relatively weak but tangled magnetic fields may effectively suppress thermal conduction (section 6), as seems to be required by observations of cluster cooling flows.

8.1. FARADAY ROTATION

The most direct information with have about the intracluster magnetic field comes from its effect on polarized radio emission from cluster or background radio sources. A plasma containing a magnetic field is birefringent; the speed of propagation of an electromagnetic wave depends on its circular polarization. While natural sources of circularly polarized radiation are rare, synchrotron emission in an ordered magnetic field produces linearly polarized radiation, and many radio galaxies and quasars produce radio emission that is somewhat linearly polarized. One can consider linearly polarized radiation to be a superposition of equal amounts of two circularly polarized beams. Because of the difference in the propagation speeds of the two circular polarizations, the plane of polarization of linearly polarized radiation is rotated during passage through a magnetized plasma. This effect is referred to a "Faraday rotation." The angle of rotation of the plane of polarization is

$$\phi = (RM)\lambda^2 \,, \tag{25}$$

where λ is the wavelength of the radiation. If the wavelength is given in meters, the "rotation measure" RM has units of radians m^{-2}. The rotation measure is given by

$$RM = \frac{e^3}{2\pi m_e^2 c^4} \int n_e B_\parallel dl = 8.12 \times 10^5 \,\text{rad m}^{-2} \int n_e B_\parallel dl_{pc} \,, \tag{26}$$

where l is the path length through the medium, and $B_\parallel$ is the component of the magnetic field parallel to the direction of propagation of the radiation. On the far right of equation (26), l is given in pc, n_e in cm^{-3}, and $B_\parallel$ in gauss (G). The wavelength dependence of the rotation is the feature that allows the observational separation of the initial polarization angle and the amount of rotation. Note that Faraday rotation only measures the average value of a single component of the field. If the magnetic field has components which are very tangled, the integral in equation (26) mainly cancels, and the Faraday rotation may be small even if the field is strong. Crudely speaking, Faraday rotation is only sensitive to the largest scale components of the magnetic field.

Under a number of circumstances, Faraday rotation can have the effect of depolarizing a radio source, rather than merely rotating the direction of polarization. First, consider a source of linearly polarized radio emission in which the region of radio emission is intermixed with thermal plasma. Assume that the total Faraday rotation through the source is large, $\phi \gg 1$ (equation 25). Then, radio emission at differing distances l along the same line-of-sight will be subject to widely varying Faraday rotations, and the result will be an admixture of all directions of polarization. This results in unpolarized radiation. This argument indicates that if one detects very large Faraday rotations from a radio source, the thermal plasma producing the Faraday rotation must lie in front of (and not be mixed with) the nonthermal plasma producing the radio emission.

Observations of a polarized source with a large Faraday rotation can also result in the detection of very little polarized flux if the observations have insufficient spectral or spatial resolution. If a source is observed with an instrument with a wide bandwidth $\Delta\lambda$, such that the variation in Faraday rotation across the bandwidth is large $\Delta\phi \gg 1$, the source will appear unpolarized. Alternatively, if the Faraday rotation in front of a radio source is large and varies rapidly across the source, the source will appear unpolarized if viewed with a large beam with mixes regions with widely varying polarization directions. Because of these observational depolarization effects, radio observations of sources with large Faraday rotations require an instrument with very good spectral and spatial resolution, such as the VLA.

8.2. GENERAL CLUSTER FIELDS

In clusters with diffuse radio emission (e.g., Jaffe 1992), X-ray observations can give a lower limit to the strength of the magnetic field. The radio emission is proportional to the product of the energy density in cosmic ray electrons and in the magnetic field. The cosmic ray electrons can also scatter photons from the cosmic microwave background to produce X-ray photons. This "inverse Compton" X-ray emission also depends on the energy density in the cosmic ray electrons. Hard X-ray observations of clusters yield upper limits on the amount of inverse Compton emission, which correspond to upper limits on the energy density of cosmic ray electrons. Because the amount of diffuse radio emission is known, an upper limit on the density of relativistic electrons leads to a lower limit on the strength of the magnetic field. Typically, these limits are $B \gtrsim 0.1 \, \mu\text{G}$ (Rephaeli et al. 1987; Bazzano et al. 1990). One attractive feature these limits is that they apply to the total energy density in magnetic field, whatever its spatial scale. The limiting fields are typically an order of magnitude smaller than the magnetic field which would be in equipartition with the cosmic ray particles.

Faraday rotation measurements towards background or cluster radio sources have also been used to determine the intracluster magnetic field (Jaffe 1980; Lawler and Dennison 1982; Hennessy et al. 1989; Kim et al. 1990). The Faraday rotation measurements have been variously interpreted as giving actual detections or only upper limits on the field. The measured values of the Faraday rotation are $RM \lesssim 100 \, \text{rad m}^{-1}$. These measurements (or limits) are generally inconsistent with the lower limits on the field required by the inverse Compton X-ray observations unless the field is significantly tangled. These observations are all more or less consistent with an intracluster field strength and coherence length of roughly

$$B \sim 1 \, \mu\text{G} \qquad l_B \lesssim 10 \, \text{kpc} \,. \tag{27}$$

Here, the magnetic field coherence length l_B is the average distance over which the field changes its direction. With this value for the field strength, the ratio of magnetic to gas pressure is roughly $(P_B/P) \lesssim 10^{-3}$, implying that the field is too weak to have important dynamical effects. However, the Faraday measurement can only detect large-scale magnetic fields; a dynamically important but highly tangled field cannot be excluded. The source of the intracluster magnetic field is uncertain, although ejection from galaxies and turbulent amplification have been suggested (Jaffe 1980, 1992; Eilek 1991; De Young 1991).

8.3. MAGNETIC FIELDS IN CLUSTER COOLING FLOWS

Although general cluster magnetic fields may be weak, magnetic fields are enormously amplified by the compression and inflow in cluster cooling flows (Soker & Sarazin 1990). Assuming only frozen–in fields, the magnetic field in a homogeneous cooling flow with spherical inflow varies as $B \propto r^{-2}$, and the Faraday rotation

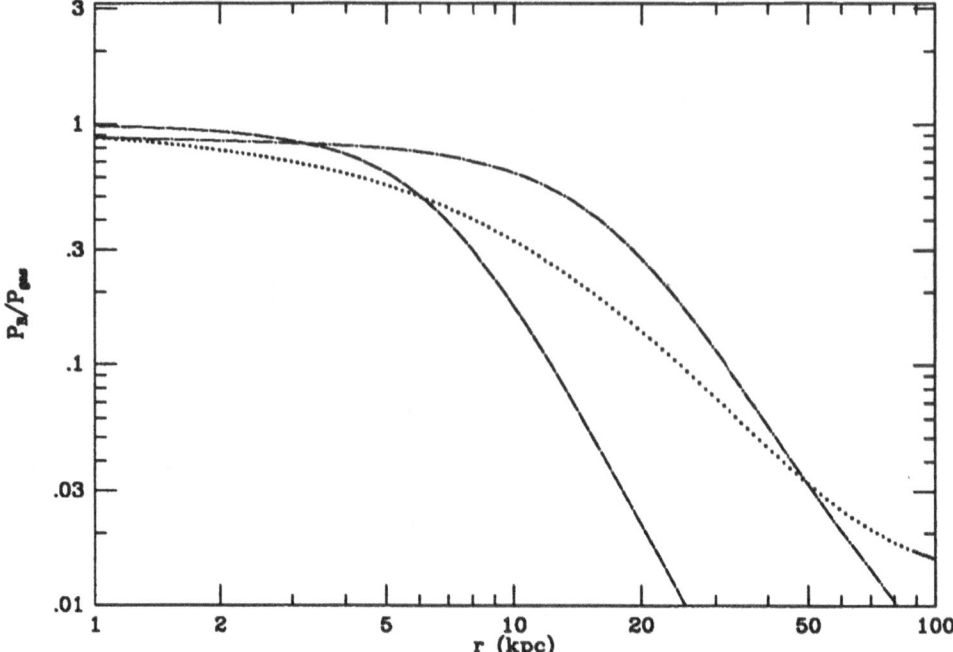

Fig. 4. The ratio of the magnetic pressure to the gas pressure is plotted versus radius for three cooling flow models from Soker and Sarazin (1990).

increases in the same way. For purely spherical inflow, the magnetic field would become more radial as the radius decreased. Real cooling flows are apparently inhomogeneous and turbulent; in this case, the magnetic field is likely to be more isotropic, and the increase of B and RM with decreasing radius may be slower ($B \propto r^{-1}$). The pressure associated with the magnetic field increases very dramatically, with $P_B \propto r^{-4}$ for homogeneous inflow. The gas pressure increases only slightly, and the magnetic pressure comes into equipartition with the gas pressure in the inner regions of the cooling flow. The magnetic field reaches equipartition at a radius r_B of roughly

$$r_B \sim 10 \, \text{kpc} \left(\frac{B}{1 \, \mu\text{G}} \right)^{1/2} \left(\frac{\dot{M}}{100 \, M_\odot \, \text{yr}^{-1}} \right)^{1/3} , \tag{28}$$

where B is the field in the cluster outside of the cooling flow, and $\dot{M}$ is the total cooling rate. Figure (4) illustrates this increase in the ratio of the magnetic pressure to the gas pressure for several detailed models for cooling flows from Soker & Sarazin (1990). In the inner regions of cooling flows, the magnetic field should be very important dynamically.

This rapid increase of the magnetic field in cooling flows also implies a large increase in the rotation measure. In the inner regions ($r \lesssim r_B$), the rotation measure is expected to reach values of

$$RM \sim 4000\,\mathrm{rad\,m^{-2}} \left(\frac{l_B}{10\,\mathrm{kpc}}\right)^{1/2} \left(\frac{\dot{M}}{100\,M_\odot\,\mathrm{yr^{-1}}}\right)^{1/2} \qquad (29)$$

where l_B is the coherence length of the field outside of the cooling flow region (Soker and Sarazin 1990). It is clear that an important observable consequence of cooling flows should be a rapid increase in the rotation measure with decreasing radii, with RM achieving rather large values within the cooling flow region.

Recently, radio astronomers have found a number of radio source which show very strong Faraday rotation. In every case found so far, these radio sources are associated with the central galaxies in a cluster with a very strong cooling flow, and the radio source is confined to the region of the cooling flow. Examples include M87/Virgo (Owen et al. 1990), Cygnus A (Dreher et al. 1987), Hydra A (Taylor et al. 1990), 3C295 (Perley & Taylor 1991), and A1795, A2199, A2052 (Ge 1991). These radio sources have rotation measures of $RM \approx 10^3 - 2 \times 10^4$ rad m^{-2}, and imply magnetic fields with

$$B \gtrsim 10\,\mu\mathrm{G} \qquad l_B \sim 1\,\mathrm{kpc}\,. \qquad (30)$$

Of particular interest is the recent Faraday rotation survey of cluster radio sources by Ge (1991). He concludes that "all sources in the centers of strong cooling flows have high RM ($\gtrsim 1000$ rad m^{-2})," and that all other sources (in the centers of noncooling flow clusters, or in the outer parts of cooling flow clusters) have much smaller rotation measures ($RM \lesssim 100$ rad m^{-2}). The rotation measures do not correlate with any of the radio properties of the sources (radio power, structure, or size). As noted above, the material providing the large Faraday rotation must lie in front of (and not within) the radio source itself. All of this strongly suggests that the Faraday rotation is created by the cluster cooling flow.

These observations agree with the simple prediction that cooling flows should have very strong magnetic fields and Faraday rotation, based only on inflow and frozen–in fields. It is also probable that more complicated physical processes play a role, such as dynamo field amplification (Eilek 1991). It may be difficult to interpret the observations of cluster rotation measure in detail, because of the dependence of the rotation measure on the unknown geometry of the magnetic field (equation 26). The magnetic field strengths implied by the Faraday measurement are large enough (in some cases) that the magnetic fields must be nearly in equipartition in the inner regions of the cooling flow, as predicted by the models (figure 4).

It is interesting that most of the observational signatures of cooling flows show the presence of *cooling* gas, rather than *flowing* gas. (Flow is required by mass conservation, however.) X-ray observations have too little spectral resolution of measure the small Doppler shifts associated with the slow inflow velocities ($\lesssim 30$ km s^{-1}). However, the increase in the magnetic fields in cooling flows is due directly to inflow rather than cooling. The detection of large rotation measures in cooling flows may be the first direct evidence we have that inflow is really occurring.

8.4. DEPOLARIZATION OF DISTANT RADIO GALAXIES AND COOLING FLOWS

Distant, powerful radio sources show a remarkable asymmetry in their polarization properties (Garrington et al. 1988, 1991; Laing 1988; Garrington & Conway 1991). These sources have extended, radio emitting lobes on opposite sides of the nucleus of the central galaxy. A single radio jet connects the nucleus with one of the lobes. One interpretation of this double–lobe/single–jet structure is that both lobes are fueled by relativistic jets emerging from the nucleus. Because of relativistic beaming and Doppler boosting, the jet which is pointed towards us is much brighter than the jet pointed away from us. Thus, the jet serves to "mark" the radio lobe which is closer to us. Now, the remarkable asymmetry is that the lobe without the jet is strongly depolarized relative to the lobe with the jet. The simplest interpretation of this result is that the region around the radio source contains thermal plasma with a magnetic field, and the depolarization is due to strongly varying Faraday rotation (Laing 1988). In the relativistic beaming model for the radio source, the lobe with the jet is closer to us; thus, our line of sight to this source passes through less of the Faraday medium. This results in less depolarization.

What is the nature of the depolarizing plasma around these radio sources? I would like to suggest that these sources are located in cluster cooling flows. The amount of Faraday rotation required, the size of the sources, and the needed variations in the Faraday rotation are all very similar to the values of the Faraday rotation observed in nearby cooling flows. At the greater distances to these sources, similar gradients in the Faraday rotation would result in depolarization.

It is sometimes argued that the depolarizing medium around these radio galaxies must be generated by the radio source, since it appears to "know" how big the radio source is. If the depolarizing medium is relatively uniform, then it must have about the same spatial extent as the radio source. If the depolarizer were much larger than the radio source, then most of the depolarization would occur outside of the radio source, and it would depolarize both lobes equally. If the depolarizer were much smaller than the radio source, it wouldn't depolarize either lobe. If the size of the depolarizer is always the same as the radio source, this would suggest that the two are related. This would be hard to understand if the depolarizer was a feature of the general environment, as would be the case for a cooling flow.

A possible flaw in this argument is the implicit assumption that depolarizer is uniform. As noted above, the Faraday rotation in a cooling flow increases rapidly towards the center of the flow. In this situation, most of the Faraday rotation and polarization is produced at a radius comparable to the size of the radio source (Fabian, private communication). If the rotation measure increases rapidly with decreasing radii, the effective size of the depolarizer is always the size of the radio source! Thus, a cooling flow origin for the depolarization naturally explains this feature of the observations.

There are other indications that the radio sources with the strongest depolarization asymmetries are located at the centers of clusters of galaxies with cooling flows. The radio quasar 3C 275.1 is a good example. First, many of these sources have distorted, Wide–Angle–Tail structures which are characteristic of radio sources associated with the central galaxies in clusters (Hintzen et al. 1983). Many of these galaxies have extended emission line systems (e.g., Hintzen & Stocke 1986), similar to those in nearby cooling flow clusters (e.g., Fabian 1992). Several of these sources are actually seen to lie at the centers of clusters of galaxies (e.g., Hintzen et al. 1981). This suggests that depolarization observations may be useful to detect distant clusters and cooling flows.

9. Concluding Remarks

The dominant observed form of matter in clusters of galaxies is hot gas. Gas at $T \sim 10^8$ K and $n_e \sim 10^{-3}$ cm^{-3} occupies the great volumes of space between the galaxies. The distribution of the total gravitational mass in a cluster can be derived from the distribution of the ICM and the assumption of hydrostatic equilibrium. The present X-ray observations suggest that the dark matter in clusters is concentrated towards the cluster center; this favors the hypothesis that the dark matter is baryonic.

Observations of the distribution and X-ray spectrum of the intracluster gas indicate that it has be enriched with heavy elements and heated, perhaps by supernova driven galactic winds. Observations of the chemical and thermal structure of the ICM provide important constraints on the formation and evolution of galaxies.

Although the ICM has mainly been treated as a single component hydrostatic fluid, it is probably much more complex. The intracluster gas is likely to be inhomogeneous, with denser, cooler lumps immersed in hotter, more diffuse gas. The plasma probably undergoes important dynamical processes, particularly during the subcluster mergers. The ICM is likely to be turbulent. Magnetic fields may play an important role. It seems likely that magnetic fields inhibit transport processes like thermal conduction, although this is poorly understood at present.

Magnetic fields are dramatically amplified in the cooling flow regions at the centers of many clusters. It is expected that the field achieve equipartition within the inner regions of cooling flows, and have important dynamical consequences. This result has been confirmed by recent observations of very large Faraday rotations in the radio sources associated with the central galaxies in many nearly cluster cooling flows. Cooling flows could also explain the depolarization asymmetry seen in more distant radio sources with one-sided jets.

Acknowledgements

I would like to thank Andy Fabian for organizing such a delightful and useful meeting, and for several very useful discussions. I am also very grateful to Judith Moss and Michael Ingham for their help during the meeting. This work was supported by NASA Astrophysical Theory Program Grant NAGW-2376.

References

Allen, S.W., Fabian, A.C., Johnstone, R.M., Nulsen, P.E., & Edge, A.C. 1991, preprint
Bahcall, J.N. & Sarazin, C.L. 1977, ApJ, 213, L99
Bazzano, A., et al. 1990, ApJ, 362, L51

150

Bregman, J.N. 1992, this volume
Canizares, C.R., Markert, T.H., & Donahue, M.E. 1988, in Cooling Flows in Clusters and Galaxies,
    ed. A.C. Fabian, (Dordrecht: Kluwer), 63
Cavaliere, A. & Fusco-Femiano, R. 1976, A&A, 49, 137
Cowie, L.L., Henriksen, M.J., & Mushotzky, R. 1987, ApJ, 317, 593
David, L.P. & Bregman, J.N. 1989, ApJ, 337, 97
De Young, D.S. 1991, preprint
Dreher, J.W., Carilli, C.L., & Perley, R.A. 1987, ApJ, 316, 611
Edge, A.C. & Stewart, G.C. 1991, preprint
Eilek, J.A. 1991, preprint
Eyles, C.J., et al. 1991, ApJ, 376, 23
Fabian, A.C. 1992, this volume
Fabian, A.C., Nulsen, P.E., & Canizares, C.R. 1991, A&ARev, 2, 191
Fabricant, D. & Gorenstein, P. 1983, ApJ, 267, 535
Fabricant, D., Lecar, M., & Gorenstein, P. 1980, ApJ, 241, 552
Fabricant, D., Rybicki, G., & Gorenstein, P. 1984, ApJ, 286, 186
Garrington, S.T. & Conway, R.G. 1991, MNRAS, 250, 198
Garrington, S.T., Conway, R.G., & Leahy, J.P. 1991, MNRAS, 250, 171
Garrington, S.T., Leahy, J.P., Conway, R.G., & Laing, R.A. 1988, Nature, 331, 147
Ge, J.-P. 1991, Ph.D. thesis, New Mexico Institute of Mining and Technology
Geller, M.J. & Beers, T.C. 1982, PASP, 94, 421
Hennessy, G., Owen, F.N., & Eilek, J.A. 1989, ApJ, 347, 144
Henry, J.P. 1992, this volume
Hintzen, P., Boeshaar, G.O., & Scott, J.S. 1981, ApJ, 246, 1
Hintzen, P. & Stocke, J. 1986, ApJ, 308, 540
Hintzen, P., Ulvestad, J., & Owen, F. 1983, AJ, 88, 709
Hughes, J.P. 1989, ApJ, 337, 21
Hughes, J.P., Gorenstein, P., & Fabricant, D. 1988, ApJ, 329, 82
Jaffe, W. 1980, ApJ, 241, 924
Jaffe, W. 1992, this volume
Jones, C. 1992, this volume
Jones, C. & Forman, W. 1984, ApJ, 276, 38
Kim, K.-T., Kronberg, P.P., Dewdney, P.E., & Landecker, T.L. 1990, ApJ, 335, 29
Laing, R.A. 1988, Nature, 331, 149
Lasenby, T. 1992, this volume
Lawler, J.M. & Dennison, B. 1982, ApJ, 252, 81
Mathews, W.G. 1978, ApJ, 219, 408
Mushotzky, R.F. 1984, PhyScr, T7, 157
Mushotzky, R.F. 1992, this volume
Osterbrock, D.E. 1974, Astrophysics of Gaseous Nebulae, (San Francisco: Freeman), 46
Owen, F.N., Eilek, J.A., & Keel, W.C. 1990, ApJ, 362, 449
Perley, R.A. & Taylor, G.B. 1991, AJ, 101, 1623
Rephaeli, Y., Gruber, D., & Rothschild, R. 1987, ApJ, 320, 139
Sarazin, C.L. 1988, X-ray Emission from Clusters of Galaxies, (Cambridge: Cambridge Univ.
    Press)
Sarazin, C.L. & Bahcall, J.N. 1977, ApJS, 34, 451
Soker, N., Bregman, J.N., & Sarazin, C.L. 1991, ApJ, 368, 341
Soker, N. & Sarazin, C.L. 1990, ApJ, 348, 73
Soltan, A. & Fabricant, D.G. 1990, ApJ, 364, 433
Spitzer, L.Jr. 1956, Physics of Fully Ionized Gases, (New York; Interscience), 120
Stewart, G.C., Canizares, C.R., Fabian, A.C., & Nulsen, P.E. 1984, ApJ, 278, 536
Taylor, G.B., Perley, R.A., Inoue, M., Kato, T., Tabara, H., Aizu, K. 1990, ApJ, 360, 41
The, L.S. & White, S.D. 1986, AJ, 92, 1248
The, L.S. & White, S.D. 1988, AJ, 95, 15
Thomas, P.A. & Fabian, A.C. 1990, MNRAS, 246, 156
White, S.D. 1992, this volume
Wise, M.W. & Sarazin, C.L. 1992, preprint

COOLING FLOWS IN CLUSTERS OF GALAXIES

A.C. Fabian
Institute of Astronomy
Madingley Road
Cambridge CB3 0HA
U.K.

ABSTRACT. The cooling time in the dense gas within 50 – 300 kpc of the central galaxy in most clusters is found from X-ray images to be less than about 10^{10} yr. The weight of the overlying gas then causes a net inflow which is called a cooling flow. X-ray spectra confirm that the gas is cooling and loses at least 90 per cent of its thermal energy. The rate at which the gas cools ranges from $\sim 10 - 500 \, M_\odot \, yr^{-1}$. The soft X-ray absorption now discovered in cooling flows suggests that the cooled gas accumulates as very cold, small, gas clouds. Any large–scale star formation must be biased to low mass objects, except in the centres of some flows where some massive star may form, possibly from larger clouds assembled from cloud collisions and aggregation.

A possible evolutionary scenario for cooling flows is outlined, in which cluster–cluster mergers is the driving force. At redshifts of about 1, the observations of the ionization state of the optical nebulosity and measurements of Faraday rotation and depolarization of radio lobes show that many luminous radio–loud objects are surrounded by massive cooling flows. They may therefore be an important phase in the formation of the most massive galaxies.

1. Introduction

Cooling flows occur in the densest gas in the cores of most clusters of galaxies where the radiative cooling time t_{cool} is a Hubble time H^{-1} or less. The intracluster gas loses energy by the radiation of bremsstrahlung and line emission, both of which are 2-body processes so that the cooling time is shortest in the densest gas. Cooling causes the gas temperature to drop which would, in the absence of gravity, cause the pressure to drop. This cannot however occur in the core of a cluster, for the pressure has to support, and is determined by, the weight of the overlying gas. The only means for the pressure to be maintained is for the gas density to rise, which means that gas must become more concentrated, or in other words it must flow inwards. This is the cooling flow.

In most nearby clusters, the inward cooling flow velocity, v, at most radii R is highly subsonic, $v \sim R/t_{cool}$. In extreme situations where the gas is very dense and the cooling time very short, the flow is limited by gravitational free-fall to $v \leq R/t_{grav}$. The thermal pressure then drops in the freely-falling gas. Cooling flows take place where

$$H^{-1} > t_{cool} > t_{grav},$$

A. C. Fabian (ed.), Clusters and Superclusters of Galaxies, 151–169.

which applies to the regions within the central 100–250 kpc of most clusters of galaxies, and to the hot interstellar medium of many massive elliptical galaxies outside clusters.

The hot intracluster medium is at temperatures of about $10^7 - 10^8$ K and so is only directly observed through its X-ray emission, the radiation of which is the cause of the cooling. The rate at which gas cools can be inferred from these observations and ranges from less than 10 to more than $300 \, \mathrm{M_\odot \, yr^{-1}}$. These rates are often so large that if the cooled gas forms stars with an intial-mass-function (IMF) similar to that in the Solar Neighbourhood, the central galaxy would be very bright and blue at optical wavelengths. Since they are clearly not so bright and blue, nor is there much sign of the cooled gas at optical (or any other) wavelengths, it is often assumed that the gas does not cool at the X-ray inferred rates. Here, it is emphasized that the X-ray observations require that the gas cools, losing at least 90 per cent of its thermal energy in that band alone. The cooled gas must reside in some unobservable form such as low-mass stars or very cold gas clouds. Any star formation must take place with an IMF that is very different from that in the Solar Neighbourhood. This is not surprising since the physical conditions in a cooling flow are also very different.

Cooling flows have been previously reviewed at length by Fabian, Nulsen & Canizares (1984, 1991) and by Sarazin (1986, 1988). Much of the present review is an update of our most recent one. In particular the evidence for large quantities of cold gas is discussed and a possible evolutionary scenario for cooling flows is outlined.

2. The X-ray Observations

Cooling flows are observable both in X-ray images and in X-ray spectra. The peaked X-ray surface brightness distribution in many clusters, as imaged by the *Einstein Observatory* Imaging Proportional Counter (IPC) and High Resolution Imager (HRI), the EXOSAT Channel Multiplier Array (CMA) and the ROSAT Position Sensitive Proportional Counter (PSPC), shows that the density of the intracluster gas rises steeply into the centre of most clusters (*e.g.* Fig. 1). (The soft X-ray surface brightness of a cluster is principally sensitive to the gas density and only weakly dependent on the temperature.) If the pressure is roughly constant there (*i.e.* the gravitational potential is flat) then this means that the temperature falls into the centre.

The shortest radiative cooling time inferred from the X-ray images depends on the instrumental resolution, but is less than a Hubble time (roughly the age of the cluster $\sim 2 \times 10^{10}$ yr) in the centres of more than 70 per cent of all clusters studied in detail (Arnaud 1988; Edge, Stewart & Fabian 1991). This means that cooling flows are common and long-lived. They are not a transient phase in the lives of a few clusters.

The rate at which gas is cooling out of the hot medium (usually called the mass deposition rate $\dot{M}$ in the sense that the gas is deposited cold from the flow) is determined by the luminosity of the cooling flow L_{cool} within the cooling radius

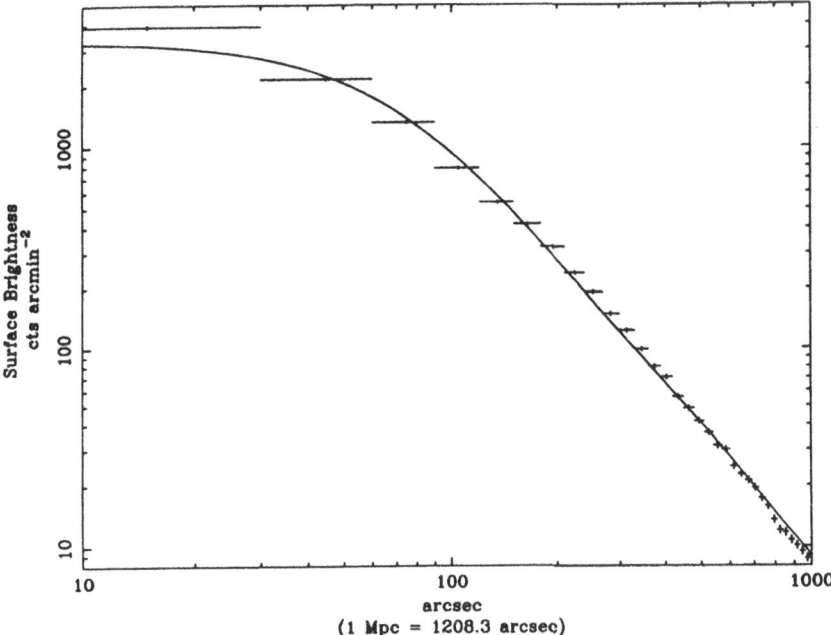

Fig. 1. PSPC surface brightness profile of the rich cluster A2199, courtesy of G.C. Stewart and R.M. Johnstone.

r_{cool} where $t_{cool} = H^{-1}$.

$$L_{cool} \sim \frac{5}{2}\frac{\dot{M}}{\mu}kT,$$

where μ is the mean molecular weight and T is the initial (hot) gas temperature. This just equates the loss of thermal energy to the observed radiation, with a factor of 5/2 used to include the PdV work (i.e. the enthalpy is used). This is a robust result since X-rays such as those detected are the radiation which causes the gas to cool. Typical values for $\dot{M}$ within r_{cool} are $50-200\,M_\odot\,\mathrm{yr}^{-1}$ with some estimated at $\sim 500\,M_\odot\,\mathrm{yr}^{-1}$ (e.g. A2597, A478, A1795, PKS 0745-191, Hydra A). Gravitational work done in the flow if the cluster potential is peaked, or reducing the estimated age of the cluster by a factor of 10, both can reduce these rates by a factor of about 2 (Arnaud 1988). Values of $\dot{M}$ for many clusters can be found from a list circulated by K. Arnaud and from Stewart et al. (1984), White et al. (1991a,b), Edge & Stewart (1991) and from Edge, Stewart & Fabian (1991).

In the case of an image, we measure the surface brightness profile $L(< R)$ and so can determine the mass deposition profile $\dot{M}(< R)$, the mass deposited within radius R. This is generally found to be approximately

$$\dot{M}(< R) \propto R$$

(Fabian, Nulsen & Canizares 1984; Thomas, Fabian & Nulsen 1987; White & Sarazin 1987abc). The gas does not all flow into the very centre of the cluster but is deposited in a distributed manner, roughly so that its density varies as R^{-2}.

The other way in which a cooling flow is observed is by the detection of low X-ray temperature components in the cluster core. As the intracluster medium cools from a typical cluster temperature of 5 keV it emits much soft X-radiation and particularly much line emission which would not otherwise be detected. Such emission has been observed with the *Einstein Observatory* Focal Plane Spectrometer (FPCS; see *e.g.* Canizares, Markert & Donahue 1988) and Solid State Spectrometer (SSS; see *e.g.* Mushotzky & Szymkowiak 1988).

In particular, the FeXVII line detected with the FPCS from the Perseus cluster shows gas at $T \lesssim 5 \times 10^6$ K with an emission measure consistent with that inferred from the cooling flow indicated by the X-ray images. The cooling time of this gas (assuming the pressure is that implied by the imaging surface brightness) is less than about 30 million years, whereas the cooling times implied by the images range from 100 million to more than a billion years. The good agreement in the value of $\dot{M}$ at all these temperatures and cooling times means that the cooling flow in the Perseus cluster, at least, is steady. The cooling of gas from a temperature exceeding 5 keV to less than 0.5 keV, as implied by these spectra, shows that the gas really does cool, losing more than 90 per cent of its thermal energy in the observed X-ray band.

The SSS data cover a wider energy band (about 0.6 – 4 keV) and show gas at a range of temperatures. The spectra for many clusters are well fit by a model including a cooling flow spectrum in which gas is assumed to be steadily cooling from some upper temperature, usually equal to that of the rest of the intracluster gas (Mushotzky & Szymkowiak 1988; White *et al.* 1991a). Also, the total broad band (2–20 keV) X-ray spectrum of the Perseus cluster has been shown to require a cooling flow spectrum (Allen *et al.* 1991). In all cases the value of $\dot{M}$ inferred from the spectral analysis is consistent with that deduced from the images (Fig. 2).

It is worth noting that simple recombination radiation from the gas cooling below the X-ray band (*i.e.* $T \lesssim 10^6$ K), spread out to the cooling radius of 100 – 200 kpc, should not be detectable in the optical or UV bands with present techniques.

It should be emphasized that the detailed analyses carried out so far ignore any significant heat sources (apart from gravitational work done in the flow). This is justified by the good spectral results, which show that the gas is cooling and that the rate is steady. Despite this strong evidence, there have been several attempts to invoke various heat sources such as cosmic rays (*e.g.* Tucker & Rosner 1982), thermal conduction (Bertschinger & Meiksin 1986; Friaca 1988), galaxy motions and other kinetic energy (Miller 1986; Pringle 1989), supernovae (Silk *et al.* 1986) and gravitational heating due to cluster collapse (Meiksin 1991). None of these attempts confront the spectral data, which clearly shows that the gas is cooling at a rate close to that inferred by the images. There are also several discussions which show that there are major problems with strong heat sources in the from of cosmic rays (Schwarz Bohringer & Morfill 1991; Loewenstein, Zweibel & Begelman 1991), conduction (Nulsen *et al.* 1982; Bregman & David 1989), galaxy motions and noise (Balbus & Soker 1989) and supernovae (Canizares *et al.* 1981; Arnaud 1988). Any heat source that balances the cooling within a flow has to be a major energy flow in the Universe, since it requires more than about 10^{62} erg for a typical flow over

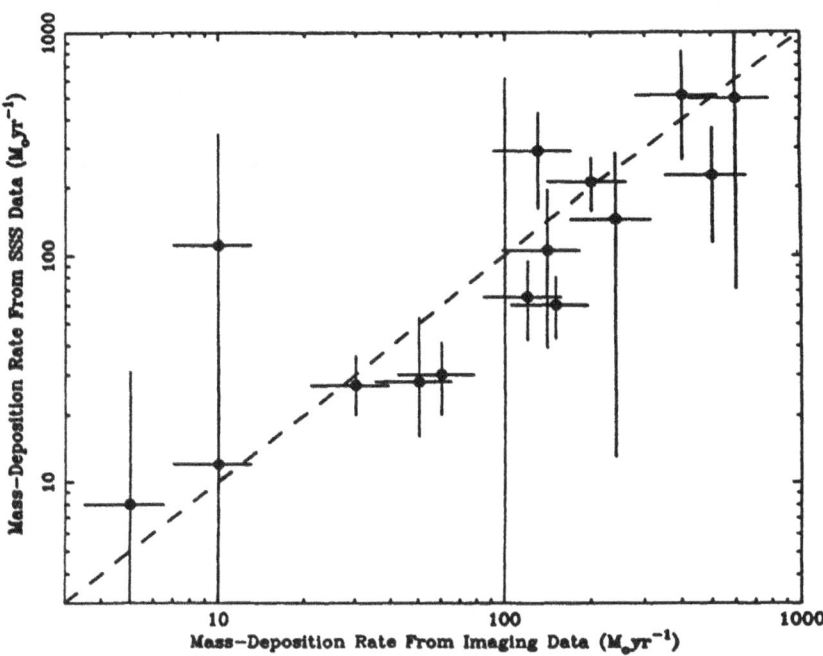

Fig. 2. Mass deposition rate estimates from SSS spectra $\dot{M}_{SSS}$ compared with those from imaging analyses $\dot{M}_{image}$, from White *et al.* (1991a).

a Hubble time. This is about 100 times the energy in the luminous radio source Cygnus A (which is surrounded by a $100\,\mathrm{M}_\odot\,\mathrm{yr}^{-1}$ cooling flow).

This does not, of course, mean that there is no heating at all within a flow. The central radio source (most central cluster galaxies have a detectable radio source, Zhao, Burns & Owen 1989) and the dissipation of turbulence within the intracluster gas must inject some heat. What the X-ray data show is that the heating is much less than the cooling that takes place. The X-ray spectra show that the gas loses at least 90 per cent of its initial energy.

3. The Fate of the Cooled Gas

The total mass of cooled gas accumulated from a cooling flow is

$$\dot{M} \approx 2 \times 10^{12} \left(\frac{\dot{M}}{100\,\mathrm{M}_\odot\,\mathrm{yr}^{-1}} \right) \left(\frac{H_0}{50\,\mathrm{km\,s^{-1}\,Mpc^{-1}}} \right)\,\mathrm{M}_\odot.$$

This is a significant fraction of the mass of the central galaxy. It could account for most of that mass if $\dot{M} \sim 1000\,\mathrm{M}_\odot\,\mathrm{yr}^{-1}$ for most of the time.

The total colours and magnitudes of central cluster galaxies show that the cooled gas is not forming stars with a solar-neighbourhood IMF. If it were, the galaxies would be much brighter and bluer than they appear. The mean mass of any stars forming from a cooling flow must be approximately less than $1\,\mathrm{M}_\odot$. There are some stars formed in some objects such as NGC1275 in the Perseus cluster, which shows an A spectrum, and a few other objects which have 'excess blue light' (see *e.g.* Johnstone, Fabian & Nulsen 1987; O'Connell & McNamara 1989). Also there are objects which show no evidence for any hot blue stars at the present level of detectability.

Alternatively, star formation may be inefficient and most of the cooled gas may just accumulate as small cold clouds. Recent studies at 21 cm of HI and the presence of X-ray absorption in many flows show that this does occur. The X-ray absorption show that the mass of cold gas,

$$M_{cold} \approx 10^{11} - 10^{12} \, M_\odot.$$

This means that star formation need not be efficient (and need not occur at all).

Early HI radio studies of clusters, both searching for 21 cm emission or absorption, suggested that the mass of HI is less than $\sim 10^9 \, M_\odot$. Recently, however, Jaffe (1990, 1991) has found HI absorption in the Perseus and Virgo clusters, distributed over ten kpc or so. The mass of HI inferred from these studies is about $10^{10} \, M_\odot$, depending on the (spin) temperature of the gas. One large difference between the detection of HI from cold clouds in our galaxy and in a cluster is that even small clouds at the pressure of a cooling flow are optically thick to 21 cm radiation. The strongest limits are perhaps from 21 cm emission, but then its detectability depends strongly on the gas temperature. Extended CO emission has also been detected from the inner few kpc of NGC1275 in the Perseus cluster (Lazareff *et al.* 1989; Mirabel, Sanders & Kazes 1989).

The X-ray absorption is apparent in SSS spectra of many cooling flows (White *et al.* 1991a; Johnstone *et al.* 1991). When compared with a simple bremsstrahlung spectrum at the cluster temperature, absorbed by the Galactic interstellar medium, the data require two additional components; an emission one around 1 keV, which is due to the line emission from the cooling flow, and an absorption component below about 1.5 keV, which implies an excess column density of absorbing gas along the line of sight (Fig. 3). The excess column density is typically about $10^{21} \, cm^{-2}$ which, since the 3 arcmin radius of the SSS beam corresponds to about 250 kpc at the distance of the average cluster studied, means that the absorbing mass is about $5 \times 10^{11} \, M_\odot$. The main absorber is oxygen in the SSS band and this could be overabundant in cooling flows (Canizares *et al.* 1988). Consequently this mass (which assumes a cosmic abundance ratio with respect to hydrogen) could be an overestimate by a factor of maybe up to 3. It could also be an underestimate if the clouds are clumped. Simple arguments on the easy detection of such large masses of gas at intermediate temperatures where the cooling curve is high strongly indicates that the absorbing gas must be cold, probably much colder than 100 K (see Fig. 4). X-ray absorption is also apparent in IPC spectra (White *et al.* 1991c) of the inner parts of several clusters and in BBXRT data (Arnaud *et al.* 1991) of the Perseus cluster.

The pressure of the cooling flow is so high ($nT \gtrsim 10^5 P_5 \, cm^{-3} \, K$, where n is the gas density) that a column density of $\sim 10^{21} \, cm^{-2}$ has a total thickness of only $10^{18} P_5^{-1} T_2 \, cm$, where the gas temperature is $100 T_2 \, K$. Since this is spread over a radius of at least several hundred kpc the cold gas is very likely to be in the form

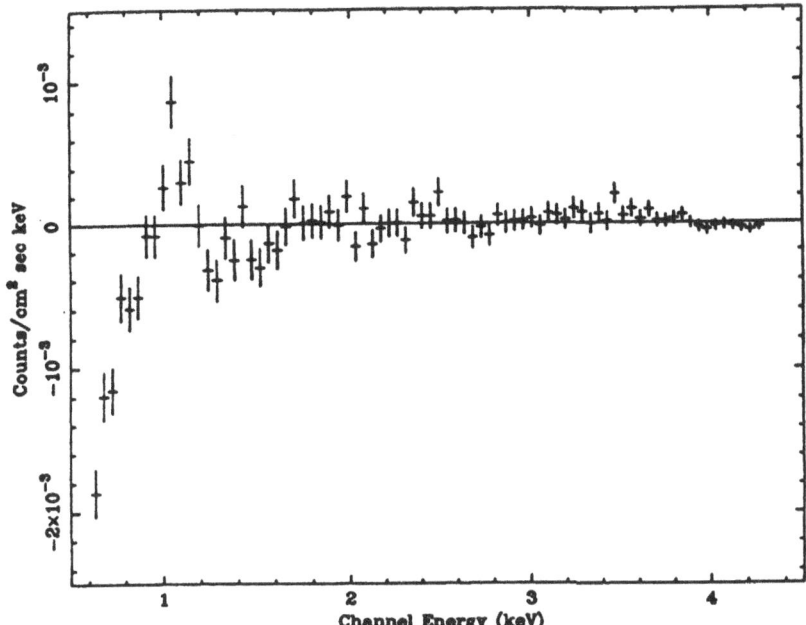

Fig. 3. Spectral residuals from a simple bremsstrahlung model fitted above 1.6 keV to the SSS data of A2199. Note the general drop below about 1.5 keV which is due to the excess absorption and the peak around 1 keV which is due to line emission from the cooling gas.

of clouds or sheets. For a covering fraction of unity, the radius of a cloud is similar to the above estimated thickness, or about one third of a parsec. If the gas is colder still, then the cloud size could be even smaller. Of course, in practice there is likely to be a cloud distribution with a mean covering fraction of about 1, but in some places there could be large holes and in others it could be quite thick. We do not yet know the cloud distribution or even the origin of the cold gas. Deposition by the cooling flow is likely, although it is difficult to explain the large column density in a low $\dot{M}$ cluster such as M87 in this way unless $\dot{M}$ was much higher in the past. The cooling flow may just serve to concentrate cold gas clouds that exist throughout the core of a cluster (from stripped galaxies and past cooling flow activity) so that the covering fraction rises from say 30 per cent at 300 kpc to 1 within 100 kpc. The SSS spectra require that the mean covering fraction within the inner 250 kpc of A2199 exceeds $\sim$ 70 per cent.

Clumps of cold clouds in the centres of cooling flows should cause soft X-ray images to appear patchy. It may explain the 'hole' in the soft X-ray image of the Perseus cluster (Fabian et al. 1981).

A detailed calculation of cold gas irradiated by a cooling flow carried out with Ferland's photoionization program CLOUDY shows (Ferland, Fabian & Johnstone 1991) that within a small fraction of a parsec the high pressure gas becomes molecular, even if dust is not present since H_2 forms via H^-, and cools (through C^+ and CO etc) down to the temperature of the microwave background (now 3 K). The

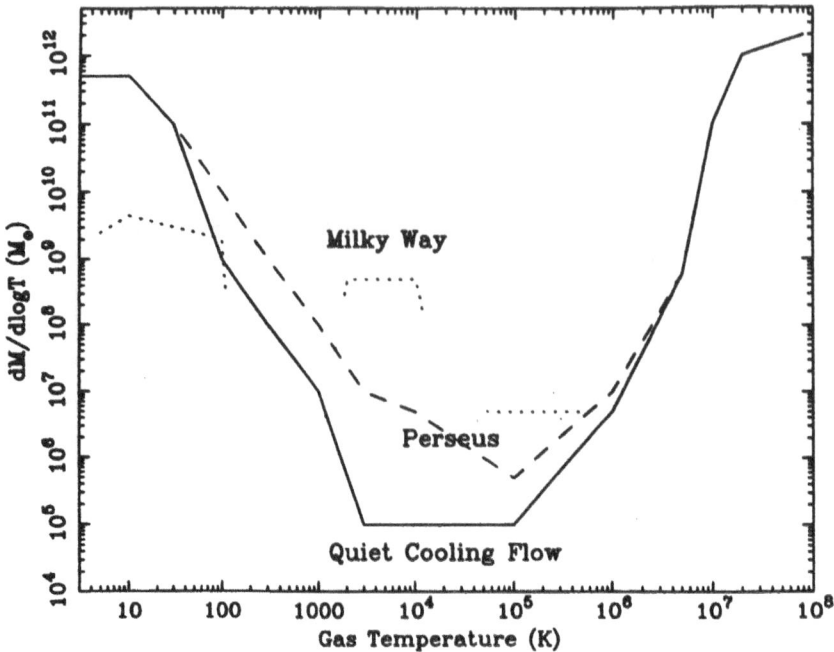

Fig. 4. Schematic representation of the mass distribution of gas in a $\sim 200\,\mathrm{M_\odot\,yr^{-1}}$ cooling flow and in our Galaxy. A 'quiet cooling flow' in which there is little line emission has a distribution approximately following the solid line. It is bimodal, with most of the gas (within 200 kpc) either very hot or very cold. The lack of gas around $T \approx 10^4$ K is due to the rapid cooling there. The cooling flow around NGC1275 in the Perseus cluster has an optical nebulosity of galaxy–size, ~ 10 kpc, in which there is more gas at $T \approx 10^4$ K. The 3–phase estimates for our Galaxy are taken from Knapp (1990). Note that the pressure in a cooling flow is $10^2 - 10^3$ times higher than that in the interstellar medium of the Milky Way. This figure emphasizes the difficulty in detecting gas in a cooling flow at optical wavelengths.

Jeans mass of such cold gas is then about one tenth of a solar mass.

In order that a cooling flow can occur at all, thermal conduction must be suppressed below the Spitzer value. For very cold clouds to form on small length scales of about one pc, the conduction must be suppressed by a factor of about 10^4. This presumably means that tangled magnetic fields are present which prevent free thermal conduction occurring. How that happens is not clear. Tribble (1989) has presented an interesting model for conduction in a tangled field which shows that the conductivity must include the length scale of the whole system. Jafelice (1991) has looked into the effects of plasma instabilities. Perhaps particle mirroring in tangled magnetic field lines is very important. Some further effects of magnetic fields are discussed by Soker & Sarazin (1990).

Magnetic fields are also need to allow cooling clouds to form yet not be destroyed by their motions relative to the surrounding hot gas. Free gas clouds which are not self-gravitating would otherwise be rapidly destroyed. Magnetic fields are inferred to occur throughout the hot gas in cooling flows through observations of Faraday rotation and depolarization (Dreher *et al.* 1987; Owen, Eilek & Keel 1990). The cold gas presumably expels the magnetic field through the process of ambipolar diffusion. Only then can stars form, in a manner similar to that of star formation in our own Galaxy (see *e.g.* Lada & Shu 1990). There, magnetic fields delay the collapse of Giant Molecular Clouds, yet must be largely expelled when a region collapses.

Whether some star formation occurs in present day cooling flows or not, the conditions are much different from those in our Galaxy. A Giant Molecular Cloud placed at say 100 kpc radius in a cooling flow would fall inward and be shredded into many smaller clouds by its motion relative to the hot gas. Star formation must therefore be different from that in our Galaxy. Only small cold clouds can exist within much of a cooling flow and this probably implies that star formation is inefficient (as when averaged over a Giant Molecular Cloud) and when it does occur it must take place in small clouds, leading to the formation of low-mass stars. As mentioned above, the high pressure of a flow means that the Jeans mass of a very cold cloud can be $0.1\,M_\odot$. Whether that determines the final mass of the star or not is unclear (and unlikely) but it is consistent with star formation being biased to low masses.

Much of the cold absorbing gas may be in the form of small shreds of gas below the Jeans mass. Some of this may mix in with hotter gas or radiate energy conducted from the surrounding gas and thereby speed up the overall cooling of the region (then the X-ray estimates of $\dot{M}$ are lower limits). The covering fraction is probably limited by the total weight of cold gas that can be supported by the hot gas. The mass of cold gas cannot much exceed that of the hot phase, and if more is deposited then it probably falls to the inner regions of the flow where clouds may collide and possibly coagulate into larger clouds. Cloud-cloud interactions may there heat the cold gas, raising the Jeans mass and making large clouds in which more massive stars can form.

Whether or not there is dust present in the cold clouds is not clear. If they are 'fossils' from gas earlier stripped from cluster galaxies, then they may be dust rich. Radiation from hot dust in the clouds may partly explain the IRAS $100\mu m$ emission found by Bregman *et al.* (1990) and discussed by Bregman (1992).

4. The Inhomogeneous Intracluster Medium

If the intracluster medium were initially homogeneous and unperturbed, then it would cool into a central singularity. This is due to the action of gravity on gas which would otherwise be thermally unstable (see *e.g.* Malagoli, Rosner & Bodo *et al.* 1987; Balbus & Soker 1989). Denser, cooler gas falls ahead of the hotter gas which initially surrounded it and joins gas with similar properties to itself at smaller radii. Drag and buoyancy forces make this process very complicated in that dense gas blobs are torn apart by their motions relative to the hotter gas until (Nulsen 1986). Magnetic fields are presumably responsible for limiting the maximum size of dense blobs and they may also pin such blobs into the flow so that they do cool out at large radii. The observed mass deposition profiles do however require that there is an initial wide range of densities at all radii at least out to r_{cool}. The origin of this spread of densities has become an important problem.

Daines (private communication) has pointed out that mixing alone can make a cooling flow inhomogeneous by causing cooler gas from smaller radii to move to outer regions, provided that the cooler gas remains cooler and does not completely mix into the hotter gas. Probably, chaotic motions exceeding the small inflow velocities of denser, cooler blobs forming from small perturbations in an initially homogeneous flow are sufficient to allow the flow to become thermally unstable.

There are many sources of kinetic energy and of cooler gas in a cluster. The intracluster medium is enriched in metals by gas ejection from galaxies and particularly by cluster–(sub)cluster collisions, which provide injections of both cooler gas and of kinetic energy. The recent evidence that clusters are evolving through subcluster mergers (Edge *et al.* 1990; Gioia *et al.* 1990) shows that this must occur. The intracluster gas must be turbulent and inhomogeneous. The recent ROSAT observations of A2256 by Briel *et al.* (1991) show a denser infalling subcluster close to the core of the main cluster. Daines & I (Fabian & Daines 1991) have shown that the lower entropy gas in the core of an infalling subcluster need not be stripped from the potential well of that cluster until it reaches the core of the main cluster. The gas need not be shocked (the air in the cabin of the Space Shuttle is not shocked on re-entry) and is only adiabatically compressed. This dense gas is then mixed into the cluster core, breaking into many small blobs the size of which is limited by magnetic fields. This will promote a cooling flow to form in the core of the main cluster. There will of course also be a considerable amount of kinetic energy in the core.

This suggests the following scenario for the present-day evolution of cooling flows in clusters. Cold clouds occur throughout the cores of many clusters, both as fossils of the formation of the intracluster medium and from early cooling processes. These clouds are focussed and added to by the central cooling flow in the cluster, possibly accounting for a significant fraction of the cooled mass of the cluster. Some star formation may be continuously occurring.

Most of the clouds at large radii are very cold, dynamically quiet and very optically thick at 21 cm. The mean total covering fraction is probably less than about 30 per cent. Chaotic motions and turbulence of the gas in the central few kpc of some clusters cause turbulent mixing layers on the surfaces of some clouds (Begelman & Fabian 1990) which produce the low ionization lines seen in some optical spectra (Baum 1991; Johnstone *et al.* 1987). Cloud–cloud collisions lead to shocks. The post–shock gas is then at very high pressures (the gas densities exceed $\sim 10^4\,\mathrm{cm}^{-3}$) where most forbidden lines (characteristic of the mixing layer spectra) are de-excited. The shocks are therefore rich in resonance lines and molecular emission. Where the collisions lead to clumping of gas, some massive star formation can occur.

A major driving force for the chaotic motions is the effect of cluster–cluster collisions and mergers. A lesser process is the effect of the central radio source. Cluster collisions shake up the cluster potential, introduce (moving) dense gas into the cluster core and probably also introduces another central cluster galaxy into the core.

If the two (ex)central cluster galaxies merge together rapidly then the cooling flow need not be much disrupted. The increased amount of dense gas in the core and the increased depth of the central potential may just increase the value of $\dot{M}$. Since the cooling processes are 2-body ones, the cooling rate is not linear. If 2 equal clusters merge such that their potentials add then the X-ray luminosity increases as $2^{2.5} = 5.6$. The cooling flow will however have much chaotic energy from the injection of the core gas from the other cluster and from the motions induced by the displacement of the central cluster galaxies as they merge. This may cause the cold clouds throughout the inner region of the flow to have extensive mixing layers and collide more frequently. It may therefore become bright in optical and UV line emission and massive star formation. If no further merger occurs in the time that it takes this energy to dissipate, the flow may become quiet in the centre, with little optical line emission detectable.

If the merger occurs so that the central cluster galaxies have much angular momentum, then they may orbit each other and the flow widely disrupted. There will be no focus for the flow and, if the cold and cooling clouds from the original clusters are not mixed into the hot gas, more cold clouds may accumulate throughout the new cluster core. This may be the situation in the Coma cluster (see also discussion in Fabian, Nulsen & Canizares 1984). This situation probably requires that the original clusters are relatively equal.

I suspect that mergers and chaotic motions in the dense intracluster gas somehow stimulates radio emission from the central cluster galaxy. This is strongly suggested by the large number of bright head–tail radio sources in the infalling subcluster of A2256 (Bridle & Fomalont 1976; Fabian & Daines 1991). Why does it contain so many sources if the interaction is not causing them to be luminous? Perhaps the chaotic motions cause dense gas to reach the central engine of the radio source and thereby power it? Consequently, the observation that many central galaxies of cooling flows are radio bright when they are also optically–line bright may be due to both emisions being secondary to the kinetic energy and stirring of the cluster core gas. It does also allow for the observation that some strong cooling flows are luminous radio sources and some are not.

The picture put forward here then is that cooling flows are common in subclusters which merge to form richer clusters. The merger causes most cooling flows to become stronger, due to the increased amount of dense core gas and deeper potential. The kinetic energy of the merger of the cores and central galaxies may help to power the optical and UV line emission and perhaps stimulate the central radio source, creating a region like that around NGC1275 for a billion years or so. If the cooling flow is not further triggered by another merger, it settles down to become an optically– and radio–quiet cluster like A478. If the merger caused the central galaxies to orbit each other then the flow is disrupted and the result is something like the Coma cluster.

Of course, much of the above scenario is highly speculative and needs a considerable amount of work to justify. In many places it may be difficult to make sensible estimates without recourse to combined N-body and hydro codes covering a wide range of scales. At some level it involves unknowns like the suppression of conduction and dissipation of turbulence, and the rôle of magnetic fields. The initial conditions for such studies are also highly uncertain. The scenario is put forward here as one possible way in which the observations can be reconciled.

5. The Early Evolution of Cooling Flows

Present–day clusters are a poor guide to the conditions in clusters at redshifts $z \sim 1$ owing to the effects of cluster mergers discussed above. Although there may now be a tendency for cooling flows to become stronger with time, that does not rule out an early phase of strong flows in young clusters when the intracluster gas may have been cooler and denser and the core smaller in radius.

Observationally, there are several clues that cooling flows were important in the past. They are observed in X-rays in the *Einstein Observatory Medium Sensitivity Survey* (Gioia *et al.* 1991; Donahue *et al.* 1991) out to $z = 37$ and in the 3C295 cluster at $z = 0.5$? (Henry & Henriksen 1987). Extended optical nebulosity reminiscent of nearby cooling flows is also observed around the radio-loud quasar 3C275.1 (Hintzen & Stocke 1985) and many distant radio galaxies (Fabian *et al.* 1985). Both radio-loud quasars and luminous radio galaxies are found in elliptical galaxies (note that radio-loud quasars are often thought to be radio galaxies observed within about 45° of the radio axis, see *e.g.* Barthel 1989) which are embedded in clusters at least out to $z \sim 0.6$ (Yee & Green 1987). The optical magnitude and continuum profile of the host galaxy resembles a central cooling flow galaxy (Romanishin & Hintzen 1989).

We proposed (Fabian *et al.* 1986) that radio-loud quasars and radio galaxies are in cooling flows and that optical spectroscopy of the optical oxygen lines can be used to estimate the pressure of the confining gas and thereby test the hypothesis. We have followed that by observations of many radio-loud quasars in the redshift interval $0.3 < z < 1.1$ (Crawford, Fabian & Johnstone 1988; Crawford & Fabian 1989) and find that the pressure of the nebulosity rises steeply above $z \sim 0.6$ (Forbes *et al.* 1990, Fig. 5; Bremer *et al.* 1991). Heckman *et al.* (1991) have shown that the pressure is still high at $z > 2$.

The pressure is inferred by comparing the observed and predicted ionization state of the emission-line nebulosity, assuming that it is photoionized by the quasar radiation of ionizing luminosity L_{ion}. Other sources of ionization would increase ionization state so the estimate obtained is a lower limit. We use the [OIII] and [OII] lines intensities measure at distance R in the nebula to estimate the estimate the gas pressure P from

$$\frac{[OIII]}{[OII]} = f\left(\frac{L_{ion}}{PR^2}\right),$$

where the function f is computed with Ferland's code CLOUDY. Basically, the gas temperature where the oxygen lines is emitted is about 10^4 K so P is mainly determined by the gas density. If it is high then recombinations are better able to balance photoionizations than when it is low. We typically find that $P = nT \sim 10^5 - 10^7$ cm^{-3} K around radio-loud quasars (Fig. 6). At a fixed radius it rises with z (and also with L_{ion} which also rises with z).

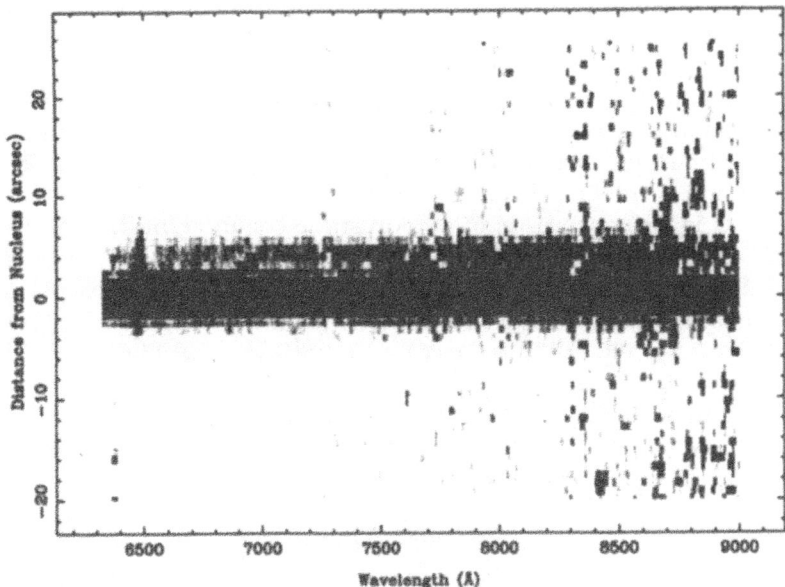

Fig. 5. Long slit spectrum of the radio-loud quasar 3C254, from Forbes *et al.* (1990). Note the highly extended emission lines of [OII] and [OIII].

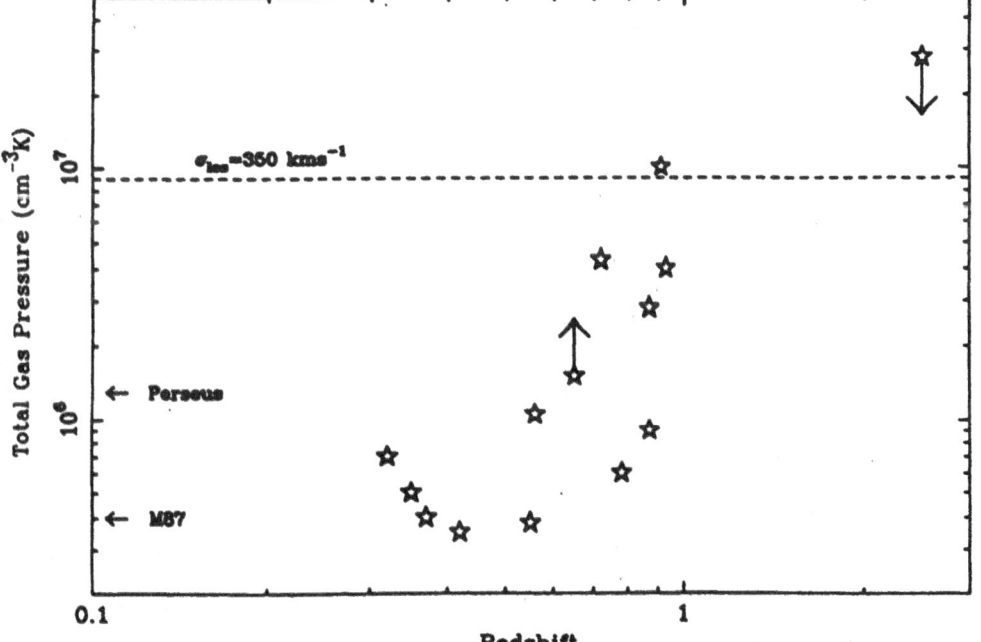

Fig. 6. Pressure inferred from the ionization state of oxygen in nebulosities at 20 kpc radius around radio–loud quasars, from Bremer *et al.* (1991).

Now the above simply gives an estimate of the pressure of the optical nebulosity. This is usually so high, and the ionized layer so thin, that it would disperse at its own sound speed in $10^5 - 10^6$ yr if not confined in some way. The simplest way to confine the gas is to assume that it is surrounded by a hot intracluster medium at the same pressure. (It is most unlikely to be gravitationally bound, and if unbound, some $10^{11} - 10^{12}$ M$_\odot$ of unexposed cold gas is required to make the nebulosity last for say a dynamical crossing time.) The pressure profiles are found to increase to smaller radii as is a characteristic of a hydrostatic confining medium at the virial temperature of the galactic potential ($\sim 10^7$ K if typical of a central cluster galaxy).

The pressure of the confining gas is often so high, especially above $z \sim 0.6$ that its cooling time, if comparable to the virial temperature of a galaxy is very short and the implied $\dot{M} > 100$ and in some cases 1000 M$_\odot$ yr^{-1} within 100 kpc. We find that the highest inferred pressures are close to the maximum that can be obtained by gas cooling in a potential well of a galaxy (*i.e.* it is close to the limit where $t_{cool} \sim t_{grav}$). We dub these 'maximal cooling flows', where $\dot{M} \sim 2000$ M$_\odot$ yr^{-1} and find them to be common around radio-loud quasars at $z \sim 1$.

Corroborative evidence that radio–loud quasars are surrounded by hot dense gas is provided by observations of strong differential Faraday rotation and depolarization between the radio lobes of radio–loud quasars, as reported by Laing (1988) and by Garrington *et al.* (1988; 1991). The geometry of the radio source is considered to be such that the radio axis is relatively close to the line of sight and the nearer lobe is on the side of the radio jet, which is boosted into visibility by relativistic beaming. This lobe always shows much less depolarization than the further lobe. This is explained by the difference in path length of our line-of-sight to the further lobe if the quasar and the lobes are surrounded by a dense magnetized plasma. The explanation only works well if the scale size of the dense gas is similar to the scale size of the radio lobes, which is the case if the gas density is strongly peaked around the quasar, as in a cooling flow.

A further observational effect that may also occur is that the light from the quasar nucleus is scattered by the electrons in the flow (Fabian 1989), so the quasar continuum appears extended along the radio axis. This may have been observed in distant radio galaxies (Chambers, Miley & van Breugel 1987; McCarthy *et al.* 1987) quasars (Heckman *et al.* 1991).

The clearest evidence that this picture is correct will be the discovery of extensive X-ray emitting haloes around distant radio-loud quasars and radio galaxies. They should in particular be luminous *soft* X-ray sources with observed temperatures of about 0.5 keV or so, since very rich massive clusters should not then be common, and much of the cooling in the core takes place around the virial temperature of the central galaxy, which must be continuing to form in this way. Indeed, this must be one of the most rapid phases in the formation of a central cluster galaxy (and its dark halo). X-ray absorption may take place in the cooled gas if star formation is not rapid and efficient.

A distant cooling flow may have already been directly observed around the double (lensed) quasar 0957+56 ($z = 1.4$), as imaged by the *Einstein Observatory* HRI (Jones *et al.* 1991). This X-ray image shows a partial Einstein ring around the two quasar images, particularly in the form of a bright arc around the Northern component. Ray tracing assuming a symmetric lens shows that the quasar must be surrounded by a 10 kpc region of diffuse soft X-ray emission. A maximal cooling flow is the simplest interpretation.

A luminous quasar surrounded by a cooling flow may be fuelled by gas from the flow, if the Compton temperature of the quasar radiation is less than the virial temperature of the host galaxy. Crawford and I have investigated such Compton–cooled accretion flows (Fabian & Crawford 1990) and note that gas within the radius at which the Compton cooling rate exceeds the bremsstrahlung cooling rate (about 250 pc for a luminous quasar in a maximal flow) will be thermally stable and all accrete onto the central object. This leads to an accretion luminosity $L_C \sim 6 \times 10^{46} T_7^{1/2} P_8^{-1} \, \mathrm{erg\,s^{-1}}$, which is much larger than the Bondi accretion luminosity $L_B \sim 3 \times 10^{45} M_9^2 T_7^{-5/2} P_8 \, \mathrm{erg\,s^{-1}}$, where the mass of the central object is $10^9 M_9 \, \mathrm{M_\odot}$, the gas temperature is $10^7 T_7 \, \mathrm{K}$ and its pressure $10^8 P_8 \, \mathrm{cm^{-3}\,K}$.

If the quasar luminosity exceeds L_C, perhaps due to a plentiful supply of fuel when it is young, then the Compton feedback will cause to it to increase to the Eddington luminosity. However, when the parent (sub)cluster merges with another one, the cooling flow may be temporarily disrupted and P drop (the high value assumed above assumes that the gas is cooling in the host galaxy potential). Compton feedback is lost and the luminosity drops to L_B for the new, lower pressure. Provided that $L_B < L_C$, then the quasar cannot restart by accretion of matter from the cooling flow.

This picture gives one explanation for why the richness of the host cluster of radio-loud quasars increases with z. The most luminous quasars at each epoch are in the richest (sub)clusters that have avoided disrupting collisions. Only poor clusters have survived to the present without an 'equal' merger of the kind that disrupted the flow. The evolution of the luminous radio-loud quasars is thereby seen as due to the evolution of the host clusters. Since radio-quiet quasars undergo similar evolution, clusters might also be responsible there too, but in a less direct manner (perhaps due to the stripping or compression of gas from their host spiral galaxies). Quasars do have a correlation length indicative of at least groups of galaxies (Bahcall & Choksi 1991) and these groups might also evolve by group–group mergers.

If massive, maximal cooling flows were common in the past then they deposit cooled gas at rates of several thousands of solar masses per year. Unless they are very transient, they thereby cause the formation of most of the central galaxy and must presumably form low mass stars with some reasonable efficiency. Peter Thomas and I have investigated whether much of the inferred dark matter in the Universe (apart from that required if the closure density is unity) could be formed from cooling flows (Thomas & Fabian 1990). We propose that systems in which $t_{grav} < t_{cool}$ allow large clouds of cold gas to exist and form stars and so make 'normal' galaxies. If $t_{grav} > t_{cool}$, however, only low mass stars form from cooled gas since any large clouds are shredded by infall through the surrounding hot gas.

In a hierarchical picture, galaxy formation then proceeds from smaller ($10^{10} - 10^{11} \, \mathrm{M_\odot}$) objects to higher mass objects with deeper potentials. The metallicity also rises as the gas from the smaller objects is trapped by, and forms into, the larger objects. The situation then passes from one in which normal star formation

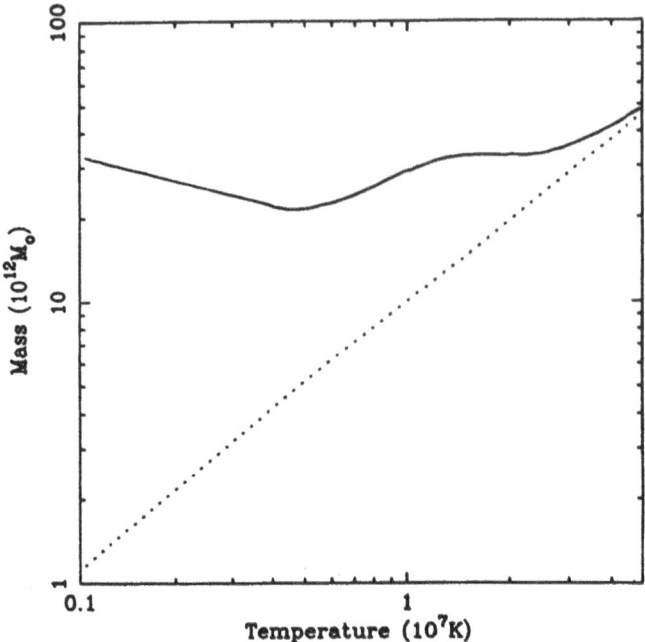

Fig. 7. $t_{cool} = t_{grav}$ lines for self–gravitating gas clouds of 0.4 solar (solid line) and zero metallicity (dashed line) in the gas mass – temperature plane, from Thomas & Fabian (1990). Cooling flows occur above these lines according to the metallicity of the gas. In our hierarchical picture for structure formation, objects begin at the lower left with low metallicity, form a first generation of stars which enriches the gas and are later assembled into objects of higher mass and (virial) temperature. The largest structures at a given epoch therefore move further up the figure and to the right. Maximal cooling flows occur as the objects cross the $t_{cool} = t_{grav}$ line and 'normal' cooling flows above that. Most present-day cluster cores are in this regime.

takes place to one where only low-mass stars are formed as the mass of the objects rises above a few $10^{12} M_\odot$ for a low metallicity to $\sim 3 \times 10^{13} M_\odot$ for a metallicity appropriate to clusters (see Fig. 7). As the critical $t_{grav} = t_{cool}$ line is passed, the object becomes a 'maximal cooling flow' until the next stage of the hierarchy is reached. Therefore all subclusters pass through this stage and much of the dark matter in clusters and around early type galaxies can be produced as low-mass stars. The soft X-ray emission from the cooling exceeds the present limits on the soft X-ray Background unless the phase when most of the dark matter is formed occurs beyond $z > 2$ or X-ray absorption by the cold gas reprocesses most of the cooling X-ray luminosity to lower photon energies.

The (distant) future of cooling flows is that they must become more common. As clusters evolve by mergers, the deepening of the cluster potential and increased gas density means that t_{cool} decreases (roughly as $L_X^{-1/5}$). Those clusters which are not yet cooling will do so and those that are, the majority, must do so at a higher rate. Both L_X and $\dot{M}$ will increase over the next few Hubble times. The X-ray emission from some clusters may eventually rise to $10^{46}\,\mathrm{erg\,s^{-1}}$ and so rival the luminosity of past quasars. The thermal energy of $10^{14}\,\mathrm{M_\odot}$ of gas at $10^8\,\mathrm{K}$ is comparable to the rest mass energy of $10^{10}\,\mathrm{M_\odot}$! Note also that the increased pressure might also resurrect some of the 'dead' central engines in the cores of clusters.

6. Summary

• Cooling flows are very common at the present epoch in clusters of galaxies. The X-ray evidence that the gas is cooling below $5 \times 10^6\,\mathrm{K}$ is very strong.

• $\dot{M} \sim 10 - 500\,\mathrm{M_\odot\,yr^{-1}}$ in these clusters. Much of the cooled gas now resides as cold gas clouds. Any significant amount of star formation must be in low-mass stars.

• There is growing evidence for high pressure, cooling gas around radio-loud quasars (and by analogy around distant radio galaxies), where $\dot{M} \gtrsim 1000\,\mathrm{M_\odot\,yr^{-1}}$.

• If cooling flows form low-mass stars with reasonable efficiency, then much dark matter may be baryonic, formed in this way.

Acknowledgements

I thank Malcolm Bremer, Carolin Crawford, Stuart Daines, Alastair Edge, Roderick Johnstone, Peter Thomas and Dave White for help and discussions on the topics presented here.

References

Allen, S.W., Fabian, A.C., Johnstone, R.M., Nulsen, P.E.J. & Edge, A.C., 1991. *Mon. Not. R. astr. Soc.* in press.

Arnaud, K.A., 1988. In *Cooling Flows in Clusters and Galaxies*, ed. A.C.Fabian, Kluwer, 31.

Arnaud, K.A. *et al.*, 1991. In *Proc. 28th Yamada Conference*, eds K. Koyama & Y. Tanaka, in press.

Bahcall, N.A. & Choksi, A., 1991. *Astrophys. J.*, **380**, L9.

Balbus, S., 1988. *Astrophys. J.*, **328**, 395.

Balbus, S. & Soker, N., 1989. *Astrophys. J.*, **341**, 611.

Barthel, P.D., 1989. *Astrophys. J.*, **336**, 606.

Baum, S., 1991. These Proceedings.

Begelman, M.C. & Fabian, A.C., 1990. *Mon. Not. R. astr. Soc.*, **244**, 26P.

Bertschinger, E. & Meiksin, A., 1986. *Astrophys.J*, **306**, L1.

Bregman, J.N., 1991. These Proceedings.

Bregman, J.N. & David, L.P., 1988. *Astrophys. J.*, **326**, 639.

Bregman, J.N., McNamara, B.R. & O'Connell, R.W., 1990. *Astrophys. J.*, **351**, 406.

Bremer, M., Crawford, C.S., Fabian, A.C. & Johnstone, R.M., 1991. *Mon. Not. R. astr. Soc.*, in press.

Bridle, A.H. & Fomalont, E.B., 1976. *Astr. Astrophys.*, **52**, 107.

Briel, U.G., Henry, J.P., Schwarz, R.A., Böhringer, H., Ebeling, H., Edge, A.C., Hartner, G.D., Schindler, S., Trümper, J. & Voges, W., 1991. *Astr. Astrophys.*, **246**, L10.

Canizares, C.R., 1981. In *X-ray Astronomy with the Einstein Satellite* ed. R. Giacconi, Reidel, 215.

Canizares, C.R., Markert, T.H. & Donahue, M.E., 1988. In *Cooling Flows in Clusters and Galaxies*, ed. A.C.Fabian, Kluwer, 63.

Chambers, K.C., Miley, G.K. & van Breugel, W., 1987. *Nature*, **329**, 604.

Crawford, C.S., Fabian, A.C. & Johnstone, R.M., 1988. *Mon. Not. R. astr. Soc.*, **235**, 183.

Crawford, C.S. & Fabian, A.C., 1989. *Mon. Not. R. astr. Soc.*, **239**, 219.

168

Daines, S.J., Fabian, A.C., Thomas, P.A., Johnstone, R.M., 1990. *Mon. Not. R. astr. Soc.*, submitted.
Donahue, M., Stocke, J.T. & Gioia, I.M., 1991. Preprint.
Dreher, J.W., Carilli, C.L. & Perley, R.A., 1987. *Astrophys. J.*, **316**, 611.
Edge, A.C. & Stewart, G.C., 1991. *Mon. Not. R. astr. Soc.*, **252**, 414.
Edge, A.C., Stewart, G.C., Fabian, A.C. & Arnaud, K.A., 1989. *Mon. Not. R. astr. Soc.*, **245**, 559.
Edge, A.C., Stewart, G.C. & Fabian, A.C., 1990. *Mon. Not. R. astr. Soc.*, submitted.
Fabian, A.C., 1989. *Mon. Not. R. astr. Soc.*, **238**, 41P.
Fabian, A.C. & Crawford, C.S., 1990. *Mon. Not. R. astr. Soc.*, **247**, 439.
Fabian, A.C. & Daines, S.J., 1991. *Mon. Not. R. astr. Soc.*, **252**, 17P.
Fabian, A.C., Nulsen, P.E.J. & Canizares, C.R., 1984. *Nature*, **311**, 733.
Fabian, A.C., Nulsen, P.E.J. & Canizares, C.R., 1991. *Astr. Astrophys. Rev.*, **2**, 191.
Fabian, A.C., Arnaud, K.A., Nulsen, P.E.J. & Mushotzky, R.F., 1986. *Astrophys. J.*, **305**, 9.
Fabian, A.C., Hu, E.M., Cowie, L.L. & Grindlay, J.E., 1981. *Astrophys. J.*,
Fabian, A.C. *et al.* 1985. *Mon. Not. R. astr. Soc.*, **216**, 923.
Ferland, G., Fabian, A.C. & Johnstone, R.M., 1991. Preprint.
Forbes, D.A., Fabian, A.C., Crawford, C.S. & Johnstone, R.M., 1989. *Mon. Not. R. astr. Soc.*, **244**, 680.
Garrington, S. T., Leahy, J. P., Conway, R. G., Laing, R. A., 1988. *Nature*, **331**, 147.
Garrington, S. T., Leahy, J. P. & Conway, R. G., 1991. *Mon. Not. R. astr. Soc.*, **250**, 171.
Gioia, I.M., Henry, J.P., Maccacaro, T., Morris, S.L. & Wolter, A., 1990. *Astrophys. J.*, **356**, L35.
Heckman, T.M., Baum, S.A., van Breugel, W.J.M. & McCarthy, P.,1989. *Astrophys. J.*, **338**, 48.
Heckman, T.M., Lehnert, M., van Breugel, W. & Miley, G., 1991. *Astrophys. J.*, **370**, 78.
Hintzen, P. & Stocke, J., 1986. *Astrophys. J.*, **308**, 540.
Henry, J.P & Henriksen, M.J., 1986. *Astrophys. J.*, **301**, 689.
Jafelice, L.C., 1991. In *Clusters & Superclusters of Galaxies*, eds. Colless, M. *et al.*, Institute of Astronomy, Cambridge.
Jaffe, W., 1990. *Astr. Astrophys.*, **240**, 254.
Jaffe, W., 1991. These Proceedings.
Johnstone, R.M., Fabian, A.C. & Nulsen, P.E.J., 1987. *Mon. Not. R. astr. Soc.*, **224**, 75.
Johnstone, R.M., Fabian, A.C., Edge, A.C. & Thomas, P.A., 1990. *Mon. Not. R. astr. Soc.*, in press.
Jones, C., Stern, C., Falco, E., Forman, W., David, L., Shapiro, I. & Fabian, A.C., 1991. *Astrophys. J.*, in press.
Knapp, G., 1990. In *The Interstellar Medium in Galaxies*, eds Thronson, H.A. & Shull, J.M., Kluwer, p3.
Lada, C.J. & Shu, F.H, 1990. *Science*, **248**, 564.
Laing, R.A., 1988. *Nature*, **331**, 145.
Lazareff, B., Castets, A., Kim, D-W. & Jura, M., 1989, **336**, L13.
Loewenstein, M., Zweibel, E.G. & Begelman, M.C., 1991. *Astrophys. J.*, **377**, 392.
Malagoli, A., Rosner, R. & Bodo, G., 1987. *Astrophys. J.*, **319**, 632.
McCarthy, P.J., van Breugel, W., Spinrad, H. & Djorgovski, S., 1987. *Astrophys. J.*, **321**, L29.
Meiksin, A., 1990. *Astrophys. J.*, **352**, 466.
Miller, L., 1986. *Mon. Not. R. astr. Soc.*, **220**, 713.
Mirabel, I.F., Sanders, D.B. & Kazes, I., 1989. *Astrophys. J.*, **340**, L9.
Mushotzky, R.F. & Szymkowiak, A.E., 1987. In *Cooling Flows in Clusters and Galaxies*, ed. A.C.Fabian, Kluwer, 47.
Nulsen, P.E.J., Stewart, G.C., Fabian, A.C., Mushotzky, R.F., Holt, S.S., Ku, W.H.M. & Malin, D.F., 1982. *Mon. Not. R. astr. Soc.*, **199**, 1089.
Nulsen, P.E.J., 1986. *Mon. Not. R. astr. Soc.*, **221**, 377.
O'Connell, R. & McNamara, B., 1989. *Astrophys. J.*, **98**, 180.
Owen, F., Eilek, J.A. & Keel, W.C., 1990. *Astrophys. J.*, **346**, 449.
Pringle, J.E., 1989. *Mon. Not. R. astr. Soc.*, **239**, 479.
Romanishin, W. & Hintzen, P., 1989. *Astrophys. J.*, **341**, 41.
Sarazin, C.L., 1986. *Rev. Mod. Phys.*, **58**, 1.
Sarazin, C.L., 1988. *X-ray Emission from Clusters of Galaxies*, C.U.P.
Schwarz, R., Bohringer, H. & Morfill, G. 1990. Preprint.

Silk, J., Djorgovski, G., Wyse, R.F.G. & Bruzual, G.A., 1986. *Astrophys. J.*, **307**, 415.

Soker, N. & Sarazin, C.L., 1990. *Astrophys. J.*, **348**, 73.

Stewart, G.C., Fabian, A.C., Jones, C. & Forman, W., 1984. *Astrophys. J.*, **285**, 1.

Thomas, P.A., Fabian, A.C. & Nulsen, P.E.J., 1987. *Mon. Not. R. astr. Soc.*, **228**, 973.

Thomas, P.A. & Fabian, A.C., 1990. *Mon. Not. R. astr. Soc.*, **246**, 156.

Tribble, P. 1989. *Mon. Not. R. astr. Soc.*, **238**, 1247.

Tucker, W.H. & Rosner, R., 1982. *Astrophys. J.*, **267**, 547.

White, R.E. & Sarazin, C.L., 1987a. *Astrophys. J.*, **318**, 612.

White, R.E. & Sarazin, C.L., 1987b. *Astrophys. J.*, **318**, 621.

White, R.E. & Sarazin, C.L., 1987c. *Astrophys. J.*, **318**, 629.

White, D.A., Fabian, A.C., Johnstone, R.M., Arnaud, K.A. & Mushotzky, R.F., 1991a. *Mon. Not. R. astr. Soc.*, **252**, 72.

White, D.A. *et al.* 1991b. In preparation.

White, D.A., Jones, C., Forman, W. & Fabian, A.C., 1991c. Preprint.

Yee, H.K.C. & Green, R.F., 1987. *Astrophys. J.*, **319**, 28.

Zhao, Burns, J. & Owen, F., 1989. *Astr. J.*, **98**, 64.

EMISSION LINE NEBULAE IN CLUSTERS OF GALAXIES

Stefi A. Baum
Space Telescope Science Institute
3700 San Martin Dr.
Baltimore
MD 21218
USA

ABSTRACT. The properties of emission line nebulae in the centers of clusters of galaxies are reviewed. Though the emission line gas (i.e., the gas at 10^4K) composes only a very small fraction of the mass and energy content of the ICM as a whole, the properties of this gas provide a sensitive probe of the conditions in the ICM. The traditional view holds that the emission line filaments have formed from thermal instabilities which condense out of the hot ICM. Emission line nebulae are found to be associated only with central dominant galaxies in clusters for which there is x-ray evidence of a cooling accretion flow, consistent with this interpretation. However, detailed studies of the morphology, kinematic properties, line ratios, dust content, and the mass and energy requirements of the 10^4K gas in cooling flow clusters raises serious questions about both the origin of the filaments and the energy source required to maintain the ionization of the gas. The physical properties of the nebulae are presented, the difficulties posed by the nebulae for the standard cooling flow scenario are discussed, and possible sources for the origin and ionization of the gas are explored. The evidence points to the necessity for either alternate theories for the origin of the optical emission line gas, or for a revised picture of the ICM, which allows for spatial and kinematic inhomogenities in the 10^{7-8}K gas.

1. Introduction

An optical line emitting nebula in the center of a cluster of galaxies comprises a very small fraction, both energetically and in mass fraction, of the ICM as a whole. As such, the nebula is an insignificant component of the ICM in physical terms (the mass of the gas at 10^4K is $\sim 0.0001\%$ of the mass of the hot component of the ICM, and the luminosity is $\sim 1\%$ of that of the hot component). However, observationally the emission line nebulae are a very important component. The nebulae are bright emitters of recombination and forbidden lines which can be studied at high spectral and spatial resolution from the ground through the combined use of narrow band imaging and long slit spectroscopy in the optical part of the spectrum. Through such studies a wealth of physical information has been gleaned directly about the properties of the optical emission line nebulae, and indirectly about the hot component of the ICM in which the 10^4K filaments are bathed. As will become evident to the reader of this review, the more we have come to learn about the physical parameters of the optical line emitting gas in clusters, the more difficult

A. C. Fabian (ed.), Clusters and Superclusters of Galaxies, 171–198.

it has become to explain the nature of these nebulae within the context of standard models for the hot ICM. The kinematics, line luminosities, physical extents, morphological appearances, and energy requirements of the optical emission line nebulae all put severe constraints on both theories which seek to explain the origin and ionization of these filaments and models of the ICM itself, which must explain the coexistence of the 10^4K filaments with the pervading, 'homogeneous' 10^{7-8}K hot intracluster gas. The resolution of these dilemmas may lie either in alternate theories for the origin of the optical emission line gas, or in a revised picture of the ICM, which allows for spatial and kinematic inhomogenities in the 10^{7-8}K gas.

2. The Good Old Days

The complexity of the emission line nebulae in clusters has only recently become apparent. The situation did not always appear so complicated! As early as 1965, Field recognized that gas at 10^{7-8}K would be thermally unstable; that is it would tend to cool and condense into filaments if perturbed. Applications of linear theories of thermal instabilities to the study of the hot ICM, led to the prediction that gas should be condensing out of the hot ICM in filaments (Fabian and Nulsen 1977; Mathews and Bregman 1978). Further, these filaments might be easily detectable in the Hα or Hβ lines in the optical part of the spectrum, as the gas cooled through 10^4K and hydrogen recombined. With the theory predicting success in hand, observers went out and detected optical emission line nebulae in the centers of clusters of galaxies (e.g., Ford and Butcher 1979; Heckman 1981; Cowie *et al.* 1983).

The nebulae were detected, in fact (see section 6) in a very specific subset of clusters - those with x–ray inferred cooling accretion flows. These clusters have peaked x–ray surface brightness profiles and central gas densities in the hot ICM such that the cooling time of this hot gas is shorter than the lifetime of the cluster (taken to be some fraction of the Hubble time). In these 'cooling flow' clusters, the cooling of the gas in the center of the cluster can lead to a loss of pressure support in the ICM; the cooling gas in the center of the ICM can no longer support the weight of the overlying ICM, leading to a subsequent (subsonic) inflow of the ICM and an increase in the central gas density (i.e., a cooling accretion flow) (Silk 1976; and review by Fabian, this volume). The detection of optical emission line nebulae in the centers of only the cooling flow clusters provided, apparently, both striking confirmation of the theory of cooling accretion flows and striking confirmation of the theory of thermal instabilities.

3. The times they are achanging

As we have all been told a little knowledge (data/theory) is a dangerous thing! What once looked so consistent and understandable has now, with the accumulation of detailed observations and more careful theoretical investigations, become increasingly complex and poorly understood. Accordingly, I will use the remainder of this review to detail the physical properties of the emission line nebulae and the problems those properties pose for our understanding of the ICM, the evolution of the ICM, and the past histories of the galaxies in the centers of clusters. There are three points to keep in mind when reading what follows. First, the observed gas at 10^4K comprises an insignificant fraction of the mass and energy in the ICM as a whole, i.e., when we look at the optical line emitting gas, we are not looking at the dominant component of the ICM. Second, the x–ray observations of clusters paint a (mostly) smooth picture of hot gas in hydrostatic equilibrium, however x–ray telescopes are only just beginning to have the spectral and spatial resolution needed to see deviations from such an equilibrium, and what looks smooth and easily modeled from 'far away' may well show increased complexity when we look close up.

Finally, it is important to point out that most (if not all) of the central galaxies in clusters which have emission line nebulae are active radio galaxies (see e.g., W. Jaffe's review in this volume, O'Dea and Baum 1987a, Burns 1990; Zhao, Burns, and Owen 1989). In many cases, the radio emission can be resolved into two jet like components straddling the central nucleus and ending in diffuse plumes or lobes. These 'jets' are presumed to be the channels along which cosmic rays, magnetic fields, and momentum flux flow from the active nucleus and feed into the radio lobes. It is now well established that radio galaxies, both in and out of clusters, have emission line nebulae and that the properties of these nebulae correlate with the radio activity (e.g., Baum and Heckman 1990; Rawlings et al. 1989; Morganti et al. 1991). Thus, if we are to understand the connection between the ICM and the emission line nebulae at the centers of clusters, we must first disentangle the causal effects between the radio activity and the nebulae and isolate those properties of the nebulae which can be attributed to the unique location of these nebulae in the centers of clusters.

4. Physical Properties of the Nebulae

In this section, the physical properties of the emission line gas are detailed in as much as we are able to determine them from spectroscopic and imaging observations. The observations that allow us to determine these parameters are described as are the implications of the measurements. The results presented in this section are taken mainly from the four definitive papers on emission line gas in clusters, Cowie et al. 1983; Hu et al. 1985; Johnstone et al. 1987 and Heckman, Baum, van Breugel, McCarthy 1989 (hereafter HBvBM). CGS units are used throughout, unless otherwise noted.

4.1. OBSERVATIONAL CONSTRAINTS

When an observer sets about to observe an emission line nebulae in a cluster what does s/he do? Detection experiments for such nebulae fall into two classes (1) long slit spectroscopic observations and (2) narrow band imaging observations. Spectroscopic experiments allow determination of line profiles and centers which give information on the kinematic properties of the gas. They also allow measurement of the ratios of lines of different ionization levels and different atoms, which in turn are diagnostics for the temperature and density in the gas, the ionization/heating mechanism for the gas, and the chemical abundance in the gas. The narrow band imaging observations allow a two dimensional surface brightness image to be made in the light of a given emission line, thereby providing the observer with the spatial distribution and brightness profile of the emission.

What is required, in physical terms, for the observer to make a detection of line emission in a recombination line such as Hα? The Hα surface brightness is proportional to the mildy temperature sensitive recombination coefficient, times the emission measure of the gas (i.e., the electron density $[n_e]$ times the hydrogen ion density $[n_{H+}]$ times the pathlength through the cloud $[dL]$), averaged over the area of the resolution element. Thus, for an unresolved cloud, the surface brightness is proportional to the density squared times the size of the cloud.

4.2. MORPHOLOGY

Figure 1 shows four examples of emission line images (i.e., images in the light of e.g., Hα alone) of nebulae at the centers of clusters overlayed on VLA continuum images of the associated radio source. They are several immediate points to note. The line emission is often asymmetrically distributed with respect to the nucleus of the central galaxy. It is typically filamentary, and the filaments can be quite elongated with apparent widths of < 1 to a few arcseconds and lengths of 10 to 20 arcseconds. The bright line emission is confined to the inner parts of the central galaxy and is observed on the same spatial scales as the radio source, though there is not, typically, a detailed correspondence of radio and emission line features. Faint line emission is detected down to the limiting detectable surface brightness of the observations.

4.3. DENSITY

Estimates of the density in the emission line gas come from two approaches. First, if we assume that the emission line gas is in pressure equilibrium with the hot, x–ray emitting gas which dominates the ICM, then, assuming that the temperature in the nebula is 10^4K (which should be accurate to a factor of two), and using the x–ray derived pressures for the hot ICM, we can estimate the density of the emission line nebula. Typical central temperature and density for the hot ICM are 10^7K and 0.1cm^{-3}, respectively (e.g., Jones and Forman 1984; Sarazin 1986), leading to a generic estimate for the density in the emission line gas (on the same physical size scales, i.e., at ~ 3kpc), of 100 cm^{-3}.

The second method that has been used to estimate the density in the emission line gas is to observe the relative intensities of two lines of the same ion emitted by different levels which have nearly the same excitation energy, such as the [SII] lines at 6717Å and 6731Å. In the regime between the low density limit ($n_e \sim 80$ cm^{-3}; in which the relative strengths are set by the statistical weights) and the high density limit ($n_e \sim 10^4$; in which the relative strengths are set by the Boltzmann population ratio), the ratio of the two lines is dependent primarily on the electron density and only secondarily on the temperature.

Thus, the observed ratio of the two [SII] lines can be used to determine an (emission-weighted) average electron density in the region of the emission line nebula from which the [SII] lines arise. This can be corrected to an average gas density

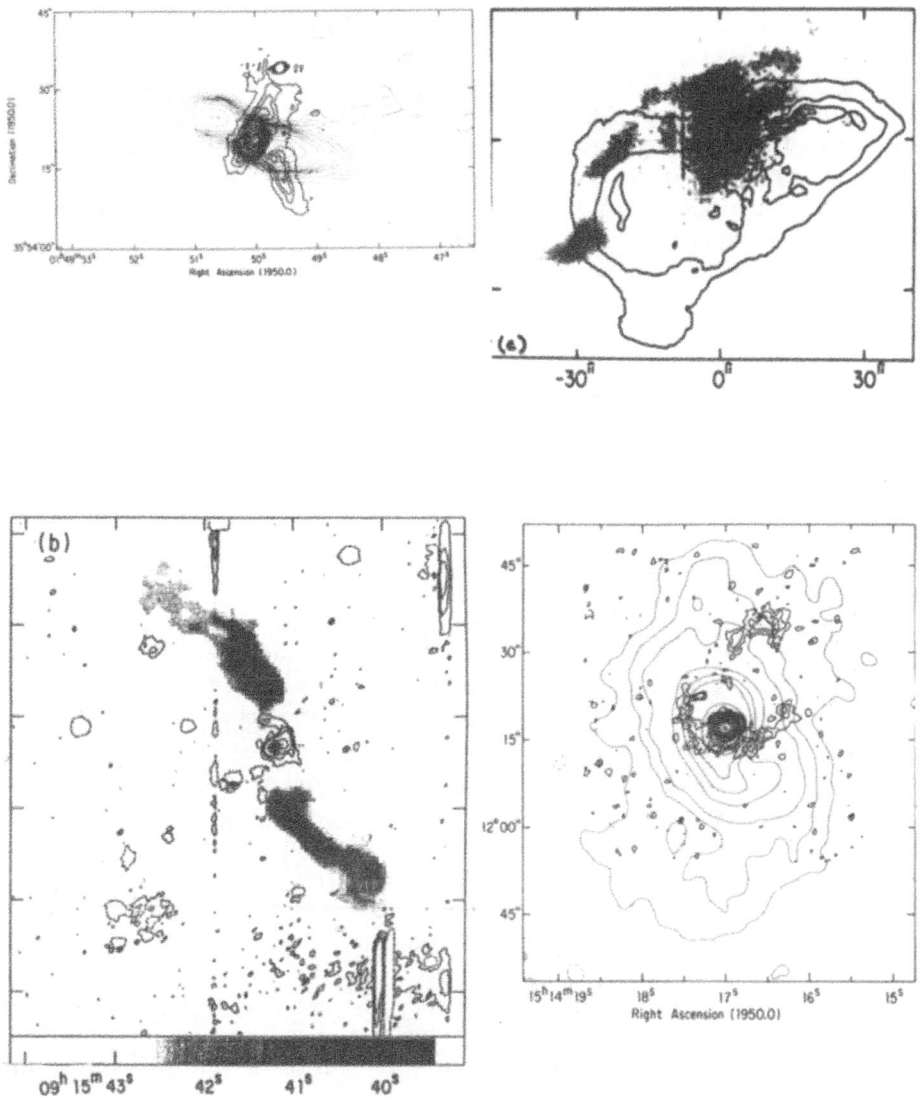

Fig. 1. Overlays showing line emission on radio emission. Clockwise from upper right; Abell 262, dark Hα contours on grey radio contours; Virgo, greyscale Hα on radio contours; Abell 2052, dark Hα contours on grey radio contours; Hydra, Hα contours on greyscale radio.

in this region through the assumption of a electron to atom and ion ratio (dependent primarily on the fractional ionization). The [SII] line arises partly in the warm, semi–ionized region and partly in the ionized zone. Thus, the 'correction factor' needed to translate from an electron to a gas density is sensitive to the extent of the semi–ionized region which in turn reflects the way in which the gas is ionized (shock heated or photo–ionized, as well as the spectrum of any ionizing continuum), and the geometry of the ionized cloud (photon–bounded or matter–bounded).

HBvBM, have 'measured' densities from the ratio of the two [SII] lines for seven nebulae in cooling flows. The basic results of this investigation are (1) typical densities are a few hundred within a few kpc, (2) densities rise in the nebulae by a factor of ~ 2 over the inner few kpc, (3) the densities are roughly consistent with x-ray estimated densities. That is, there is no evidence for overpressure or underpressure in the emission line gas relative to the hot ICM, and there is evidence for an inferred rise in pressure in the ICM by a factor of $\sim$ two within the inner few kpc.

4.4. RELEVANT TIMESCALES

We can next ask what the recombination and cooling time for such a nebula is. The recombination time, $t_{rec} \sim 10^5 n_e^{-1}$ years, while the cooling time is $t_{cool} \sim 10^4 n_e^{-1}$ years (with n_e is cm^{-3}). Comparing these times with the other pertinent timescales, e.g., the lifetime of the cluster, the inferred lifetime of the cooling flow, and the guestimated lifetime of the central radio sources in clusters, we immediately conclude that energy must be continually supplied to the nebulae to maintain their temperature and ionization state. Thus, detectable Hα emission implies the presence of cold gas of high enough density and pathlength, as well as the presence of a source of (re–)ionizing energy for that gas.

In addition, the timescales for evaporation of the cold gas in the hot ICM, cloud destruction to due Rayleigh–Taylor instabilities if the clouds are not co–moving with the hot ICM, and the free fall collapse times (all of which are dependent on the unknown magnetic field strength and configuration in the clouds) are likely to be quite short ($< 10^6$ years, e.g., Loewenstein and Fabian 1988). Thus not only energy, but matter may need be resupplied to the nebulae, if the nebulae are to be visible over a subtantial fraction of the cluster and radio source lifetimes. What we will see later (see section 7.7) is that the *amount* of re–ionizing energy which needs to be supplied to the nebulae is so great that it seriously strains all the viable energy sources. Further, since there may be as much as $10^8 M_\odot$ of 10^4K gas in some clusters (see section 4.5), and since this gas may need to be replenished every $\sim 10^6$ years, the total matter requirements may approch $10^{12} M_\odot$ over the cluster lifetime. Interestingly, this is roughly the amount of matter thought to be cooling out of the accretion flow.

4.5. MASS OF GAS AT 10^4K

The mass of the detected emission line gas can be derived from the observed Hα luminosity, assuming case B recombination (e.g., Osterbrock 1989) and assuming that hydrogen is totally ionized as

$$M_{\mathrm{gas}} = L_{\mathrm{H\alpha}} m_{\mathrm{p}}/(\alpha_{\mathrm{H\alpha}}^{\mathrm{eff}} h\nu_{\mathrm{H\alpha}} n_e) = 4.6 \times L_{\mathrm{h\alpha}} \times n_e^{-1} \text{ grams},$$

using cgs units throughout, and where m_{p} is the mass of the proton, $\alpha_{\mathrm{H\alpha}}^{\mathrm{eff}}$ is the effective recombination coefficient for Hα emission, and n_e is the electron density (e.g., Baum and Heckman 1990a, Osterbrock 1989). Thus, using the [SII] measured densities in the gas, we can estimate the gas mass from the observed Hα luminosity.

In this way HBvBM have estimated that the gas in 10^4K gas is $\sim 10^5 - 10^7 M_\odot$ within the inner few kpc in seven cooling flow clusters. The total volume filling factors they find are small, $\sim 10^{-5}$ and less, where the volume filling factor is defined here to be the fraction of the spherical volume within a few kpc of the nucleus which is occupied by emission line gas.

Beyond a few kpc from the galaxy nucleus, the surface brightness of the line emission is low, and we typically either have no reliable estimate of the gas density or the [SII] line ratios are within the low density limit. We can calculate a rough estimate of mass of the detected 10^4K gas beyond a few kpc, by assuming that the density of the gas is 100 cm^{-3}. When this is done for the seven sources from HBvBM, masses in ionized gas as high as a $5 \times 10^7 M_\odot$ are found for A2597, A1795, PKS 0745–191, and 3C84. These are in some sense lower limits to the mass in ionized gas in the low surface brightness component of the nebula, since the flux weighted average density in the extended emission line gas may be considerably below 100 cm^{-3}, and $M_{10^4} \propto n_{\mathrm{H^+}}^{-1}$. Thus in at least some of the high mass accretion rate clusters, the mass in 10^4K gas may well approach $10^8 M_\odot$. As mentioned in section 4.4 above, if this mass in ionized gas must be replenished on a timescale given by the cloud destruction time ($\sim 10^6$ years), and if this resupply is continuous over the lifetime of the cluster, then $\sim 10^{12} M_\odot$ of matter must pass through a 10^4K stage, over the lifetime of the cluster.

4.6. COVERING FACTOR

An interesting parameter for many models of the energization of the emission line gas is the covering factor of the gas. The higher the covering factor the less extreme are the total energy requirments, since the nebulae then intercept a larger fraction of the available energy. There are two important points to make here, a theoretical one, and an observational one. First, though the volume filling factors are very small, it is still possible to have high covering factors if one of two situations holds. If the emission line filaments consist of a fine mist of very small ionized clouds, then since the covering factor $cf \propto N_{\mathrm{clouds}} \times (r_{\mathrm{cloud}}/r_{\mathrm{nebula}})^2$, while the volume filling factor f is $\propto N_{\mathrm{clouds}} \times (r_{\mathrm{cloud}}/r_{\mathrm{nebula}})^3$ then cf can still approach unity with $f = 10^{-5}$, if the cloud radius is a very small fraction (10^{-5}) of the radius of the nebula, and there are a very large number of clouds. For example, for $r_{\mathrm{nebula}} = 10$kpc, $r_{\mathrm{cloud}} = 3 \times 10^{17}$ cm, and $N_{\mathrm{clouds}} = 10^{10}$.

An alternate way in which the covering factor can be high, despite the very small observed filling factor of the gas, is if the emission lines arise from a very thin ionized skin covering thick, cold, neutral clouds. HBvBM present arguments that it is not unreasonable to think that the mass in cold, neutral matter could be roughly 100 times greater than the mass in ionized gas (see also section 4.8). The tightest constraints on the mass of cold matter must come from HI and CO observations of clusters looking for both atomic and molecular gas (see review by Jaffe, this volume, and references therein).

Observationally, we can attempt to measure a covering factor by considering what fraction of the sky is covered by line emission within a given radius. In actuality, we can only obtain an upper limit to the covering factor in this way because individual clouds may be unresolved by our observations. While the nebulae appear in some cases to be quite diffuse and therefore to have a high covering factor (e.g., PKS 0745–191), in many cases, particularly those which are studied at higher physical resolution, (e.g., M87, Hydra A, Abell 2052) the upper limit to the covering factor is very low indeed (see Figure 1).

4.7. KINEMATICS

We consider next the kinematic properties of the gas. Typically, the emission line nebulae have the following properties. The emission line nebulae exhibit relatively large line widths in the inner kpc of $\sim 500 - 1000$ km sec^{-1} FWHM followed by a drop with radial distance to ~ 200 km sec^{-1} or less at radii of 3–15 kpc. Since the typical size scale of bright radio emission is also several to ten kpc, the rapid drop of line width with radial distance can also be interpreted as a rapid drop in the line width outside of regions of bright radio emission.

The nebulae exhibit relative velocity fluctuations of $\sim 100 - 200$ km sec^{-1} on scales of a few kpc, suggesting that this is roughly the size scale of an individual cloud. The relative motions appear predominantly turbulent; there is typically no clear evidence for ordered motion (e.g., rotation or infall or outflow) in the nebulae (see Section 5.3). A small subset of cooling flow nebulae do show apparent rotation with $v_{rot} \sim 300$, within a few kpc. However, these same nebulae can show disordered velocity patterns in the gas on larger scales (e.g., Hu et al. 1985).

4.8. LINE RATIOS

Finally, we briefly discuss the line ratios exhibited by the nebulae. Figure 2 shows characteristic examples of a cooling flow nebula spectra, both on and off the nucleus from M87.

The spectra can be classified as LINER type spectra. That is, the lines from singly ionized and neutral species (e.g., [OI] 6300Å, [OII] 3727Å, [SII] 6716,6731Å) are strong relative to HII regions, and the lines of high ionization species (e.g., [OIII] 5007Å, [NeV] 3426Å) are weak relative to Seyferts and powerful radio galaxies (e.g., Heckman 1987). Most of the discussions of LINERs concern low level activity in spiral and irregular galaxies. For years, there have been discussions of the ionization source for LINERS (e.g, Heckman 1987 and references therein), and most if not all of these discussion apply to the nebulae in cooling flows as well. The spectra are typically reproducible either by low velocity shocks ($v_{shock} < 100$ km sec^{-1}) or by photo–ionization by a relatively dilute power law continuum such that the ratio of the flux of ionizing photons to the electron density is small, (i.e., the ionization

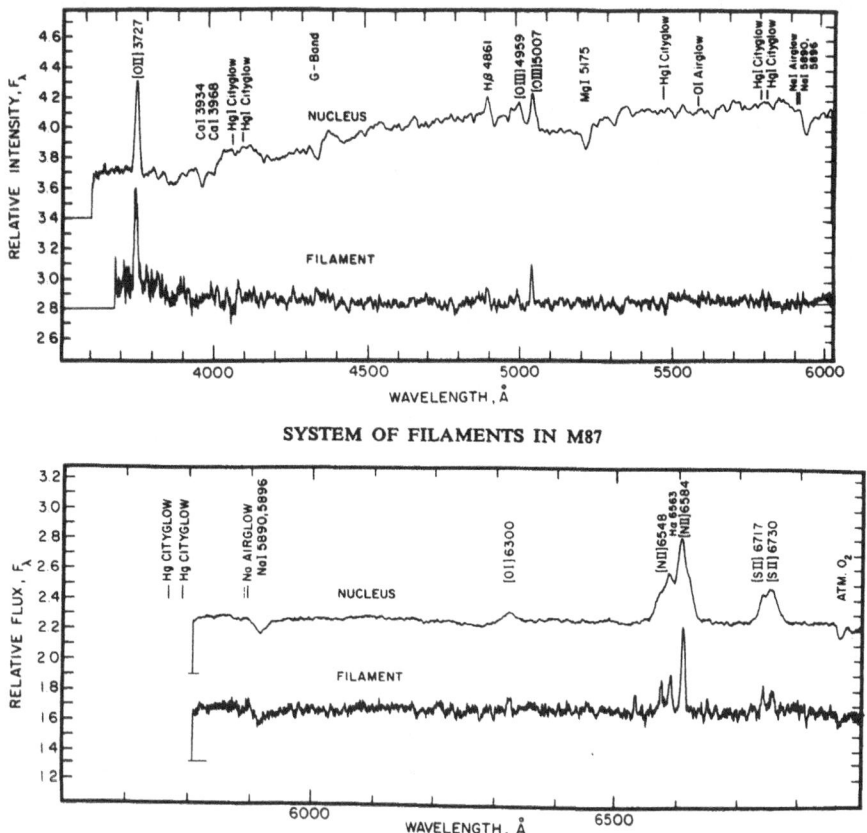

Fig. 2. Spectra of filaments in M87 reproduced from Ford and Butcher 1979.

parameter, defined to be the ionizing flux in photon sec^{-1} cm^{-2} divided by the speed of light times the electron density, is $\sim 10^{-3}$ to 10^{-4}).

From inspection of the spectra of seven cluster nebulae HBvBM classified the nebulae into two spectral types, those with relatively strong [NII]/Hα and [SII]/Hα and relatively weak [OI]/Hα (Class I), and those with weaker [NII] and [SII] to Hα ratios and stronger [OI]. (Class II). The high [NII] to Hα ratios exhibited by the Class I sources are not reproduced by standard photoionzation or shock models with solar abundances. This high ratio indicates a high rate of heating per ionization (e.g., Donahue and Voit 1991).

While the apparent bimodality of the sample of HBvBM was probably based on the small number of sources, the basic points noted by HBvBM are valid. Most importantly, HBvBM noted that, relative to the Class II clusters, the Class I clusters have low X-ray and Hα luminosities, small optical nebulae and low mass accretion rates. Thus, it appears that the line ratios exhibited by the nebulae correlate, in some sense, with the depth of the potential in the cluster center and the inferred magnitude of the cooling accretion flow. The origin of these line ratio differences may be the metallicity of the ICM (e.g., HBvBM) or the size scales (and hence optical depths) of filaments which condense out of the ICM (Voit and Donahue 1991) in these different types of clusters.

Finally, the line ratios can be used to constrain models for the geometry of the emission line filaments. The [OI] 6300 Å line is produced (almost exclusively) in the semi–ionized region. Thus, the strength of this line is a diagnostic for the presence of such a region. In photo–ionized nebulae, there will only be a semi–ionized region if (1) the ionizing continuum contains relatively high energy photons for which the absorption cross section is large (i.e., the continuum extends into the x-ray) and (2) the nebulae are matter bounded (i.e., there are more than enough cold atoms to absorb all of the ionizing photons emitted by the continuum source). In fact, the [OI] 6300 line is relatively strong in cluster nebulae (see Figure 2), thus there must be a semi–ionized zone, and the filaments cannot all be optically thin (i.e., photon bounded). This allows the possibility that there is substantially more un–ionized, cold, material, than there is 10^4K gas (i.e., that the emission lines come from an ionized skin of a cold cloud).

5. Separating Out the Radio Connection

As mentioned in the introduction, most, but interestingly, perhaps not all central galaxies in clusters with emission line nebulae are active radio sources (e.g., HBvBM). It is also known that most, if not all, active radio sources have emission line nebulae, and that the properties of those nebulae correlate, in a global sense, with the radio properties (Baum and Heckman 1989; Rawlings et al. 1989). Thus, before we invest too much time and effort associating properties of the emission line nebulae in clusters with the cluster environment, we must first determine whether the cluster emission line nebulae are, indeed, 'special' in some way. We consider this question by comparing the properties of the emission line nebulae in sources chosen because they are the central dominant galaxies in clusters with cooling accretion flows (hereafter the 'cooling flow sample', HBvBM) with the properties of a radio flux density selected sample of radio galaxies (hereafter the 'radio selected sample' (Baum et al. 1989; Baum, Heckman and van Breugel, 1990, 1991). Below, we discuss three distinguishing properties of the emission line nebulae in clusters; (1) the strength of their line luminosity, (2) their characteristic line ratios, and (3) the lack of organized kinematic motions in the gas.

5.1. CORRELATION OF RADIO AND LINE LUMINOSITIES

As state above, it has now been shown by a host of authors that radio luminosity and line luminosity (as measured by a narrow emission line such as [OII] 3727, [OIII] 5007, or Hα) correlate in radio galaxies, and that this correlation extends over at least four orders of magnitude in both line and radio luminosity. There is however much scatter in this relation (about one order of magnitude in each parameter). In Figure 3, a plot from HBvBM is reproduced showing line luminosity versus radio

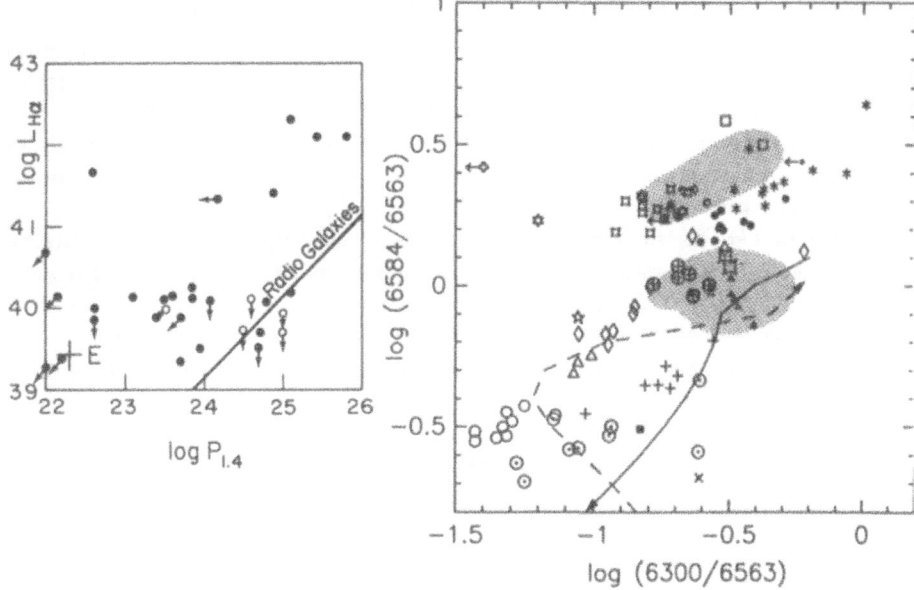

Fig. 3. Left, log of Hα luminosity versus log of radio power measured at 1.4 GHz for brightest cluster galaxies, reproduced from HBvBM. The line showing the mean relation for radio galaxies is superposed. Filled circles are known cooling flows.

Fig. 4. Right. Plot of the [NII]/Hα versus [OI]/Hα ratio as determined from long slit spectroscopy for the radio selected sampled of BHvB, reproduced from BHvB. Stippled areas show regions occupied by cooling flow sources from HBvBM.

luminosity for brightest cluster galaxies. That subset of radio sources which are at the centers of 'cooling flow clusters' are shown as filled circles. Superposed is a diagonal line which indicates the mean relation for the radio flux density selected sample of radio galaxies As with radio sources as a whole, there is a correlation between radio and line luminosity for the cluster center sources, but what is most interesting is that the radio sources which are at the centers of cooling flow clusters have emission line nebulae which are, in the mean, about a factor of 10 brighter in emission lines, relative to their radio luminosity, than do radio sources not chosed for membership at the centers of clusters. Thus *cooling flow nebulae are overluminous for their radio luminosity.*

5.2. LINE RATIOS

We can use BPT diagrams (so called because they were pioneered by Baldwin, Phillips, and Terlivich, 1981) to investigate the line ratios exhibited by the cooling flow nebulae and compare them to the nebulae of a radio flux density selected sample of galaxies. In Figure 4 is shown a plot of the [NII] 6584 Å to Hα versus [OI] 6300 Å to Hα ratio as determined from long slit spectroscopy for the radio selected sampled of BHvB. The region of paremeter space on this diagram occupied by the cooling flow sources of HBvBM has been cross hatched. Symbols are plotted for each pixel location in each spectra where all relevant lines were detected.

Two points are apparent. First, as can be seen from this figure, line ratio variations are seen within individual sources, but these variations are much less than the variations seen from source to source. Phenomenologically, this suggests that the factors responsible for the line ratios (ionization mechanism/parameter, metallicity, temperature, continuum shape, ratio of ionizing photons to absorbing material, etc.) vary much more from source to source than they do within a given source. This point applies both to the cooling flow sources and the radio selected sample.

Second, it is clear that as a class, the radio galaxies in our sample exhibit a much greater variation in line ratios and thus form a more hetereogeneous sample than do the cooling flow sources. Thus, the cooling flow sources occupy a characteristic region of parameter space in the BPT diagrams, attesting to the relative uniformity of their spectra. *Cooling flow nebulae exhibit a characteristic spectrum.*

5.3. KINEMATICS

We can paramaterize the kinematics of the emission line nebulae in terms of σ the average velocity dispersion of the gas and Δ, a measure of the rotation velocity (or outflow/inflow velocity) of the gas. Δ is taken to be 1/2 the difference in the average velocity on opposite sides of the nucleus along the slit. This is a useful parameterization: a rotating disk of gas will have $\Delta/\sigma > 1$ while a turbulently supported nebula will have $\Delta/\sigma < 1$. In Figure 5, Δ/σ versus the [NII] to Hα ratio for the combined cooling flow and radio selected samples is plotted, where we have distinguished the cooling flow sources as filled squares. It is easy to see that the cooling flow sources have systematically high [NII] to Hα ratios (see 5.2 above) as well as systematically small values of Δ/σ. In fact, the cooling flow sources have systematically small absolute values of Δ as well.

Nebuale in cooling flow sources are kinematically distinguishable by the lack of organized velocity patterns in the gas. The nebulae are turbulently, and not rotationally, supported, and the nebulae have (as a class) very small absolute rotational velocities.

6. Evidence of a Cooling Flow – Emission Line Nebulae Connection

There are two strong pieces of evidence that there is indeed a connection between an x–ray inferred cooling accretion flow in a cluster and the presence of an optical emission line nebula. These are (1) the observation that Hα emission is only detected in the central dominant galaxies of clusters with x–ray inferred cooling accretion flows, and not in the galaxies at the centers of clusters without cooling accretion flows, and (2) the line luminosity of the nebulae correlates with the mass accretion rate of the cluster.

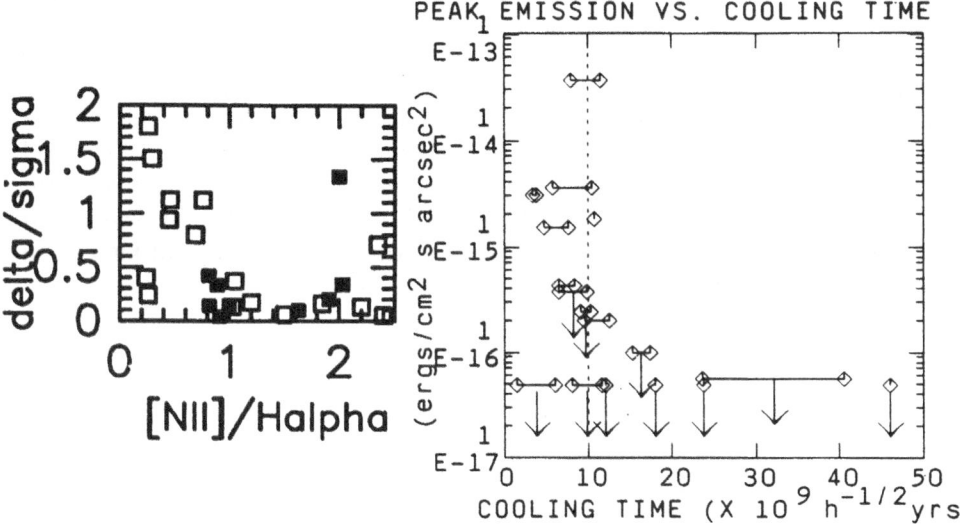

Fig. 5. Left. Δ/σ (the rotational velocity divided by the velocity dispersion) versus the [NII] to Hα ratio for the combined cooling flow and radio selected samples. The cooling flow sources are indicated by filled squares.

Fig. 6. Right. Plot of peak Hα surface brightness versus cooling time in the x-ray gas in clusters, taken from Hu 1988.

6.1. THE DETECTION OF Hα AND THE PRESENCE OF A COOLING FLOW

First, and foremost, optical line emission is detected only in clusters with x-ray inferred cooling flows; i.e., in clusters in which the central gas density is high enough that the cooling time of the gas is less than the 'Hubble time'. This is illustrated in Figure 6, reproduced from Hu (1988). In this figure the peak Hα surface brightness is plotted against the cooling time. The point made by Hu is that only clusters with a cooling time less than some amount (10^{10} years, or roughly the lifetime of the cluster) have detected line emission. However, it is also important to note that there are clusters with cooling times less than 10^{10} years with large x-ray inferred cooling accretion flows but which do not have detected Hα emission. Thus, a cooling time less than the lifetime of the cluster may be a *necessary* condition for Hα emission to be detected, but it is apparently not a *sufficent* condition.

There are two possible interpretations of this plot. The first and most straightforward is that the emission line nebulae owe their existence and/or their energization to a cooling accretion flow in the ICM. The second possibility is that the presence of a detectable emission line nebula in a central dominant galaxy is dependent primarily upon a second parameter, which is itself correlated with the presence of an x-ray inferred cooling accretion flow.

For example, the detectability of Hα emission may not actually be dependent on the cooling time in the hot gas but instead be dependent on a more basic physical property of the ICM, such as, for instance, the pressure in the hot ICM. We might expect such a dependence from the arguments presented in section 4.1. That is, an ionized filament of a given total mass and with a temperature of 10^4K, will have an emission measure (and hence a surface brightness) which is entirely dependent on the external pressure in which it is bathed (assuming, of course, that it achieves pressure equilibrium with that bath). Thus, the emission measure will be $EM \propto M_{cloud}^{1/3} P_{ext}^{5/2}$, and the detectability of the line emission will be a strong function of the external pressure exerted by the ICM.

6.2. CORRELATION OF $L_{H\alpha}$ WITH MASS ACCRETION RATE

Johnstone *et al.* 1987 and HBvBM have shown that the line luminosity of a cluster (as measured by say the Hα or Hβ emission lines) correlates with the x–ray inferred mass accretion rate. This is shown in figure 7 taken from HBvBM, where the Hα luminosity is plotted against the x–ray determined mass accretion rate for brightest cluster galaxies. The two parameters are correlated (at the 99.9% confidence level).

There is, however, a huge scatter in the correlation between line luminosity and mass accretion rate. At low mass accretion rates, there is roughly one order of magnitude scatter in the line luminosities. However, at high mass accretion rates (roughly 100 M$_\odot$ yr^{-1} and above), the scatter in the observed line luminosities spans the full range, i.e., three orders of magnitude! While it is undoubtably scientifically relevant that there is a correlation between the mass accretion rate and the line luminosity, it is also important to realize that the relationship between line luminosity and mass accretion rate is not at all a simple and direct one. Thus, one must explain the presence of clusters with the same (high) mass accretion rates but 1000 times less line luminosity.

7. Problems with The Cooling Flow – Emission Line Nebulae Connection

As outlined in the introduction, the cooling flow model, in its simplest form posits that the hot intracluster gas is cooling radiatively. Thermal instabilities in this gas lead to the formation of filaments. These filaments are sources of optical emission lines, as the gas in the filaments cools through 10^4K and electrons and atoms recombine. The detection of Hα filaments in cooling flow clusters provides apparent support for this interpretation. But, is this 'cooling flow gas' really what we are seeing when we observe Hα filaments in clusters?

When we look in detail at the properties of the emission line filaments we find that these properties are not at all what we would expect from filaments which have condensed quiescently out of a cooling accretion flow. That is, in gross terms, the nebulae don't look like instabilities in a cooling flow. The 'problems' with the view that the clouds condense out of the hot ICM are enumerated and detailed below.

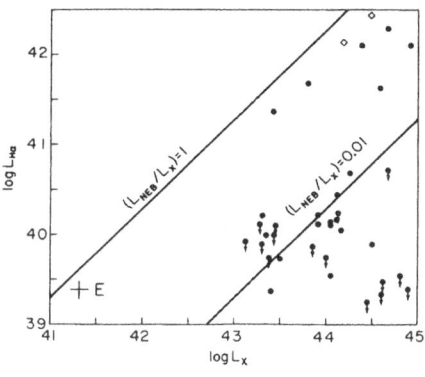 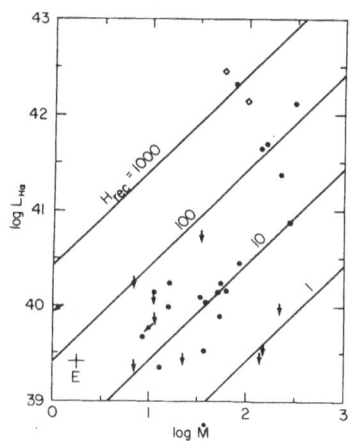

Fig. 7. Right. Plot of the log of the Hα luminosity versus the mass accretion rate, reproduced from HBvBM. Lines of constant H_{rec} (see text) are superposed.

Fig. 8. Left. Plot of the log of the Hα luminosity versus the log of the x-ray luminosity, reproduced from HBvBM.

7.1. THE EXISTENCE PROBLEM

Though early linear theories predicted that thermal instabilities should form easily in cooling accretion flows, more recent detailed theoretical work has shown that, in fact, that there is no *strong* linear thermal instability in a cooling flow atmosphere (see e.g., Balbus 1988, Balbus and Soker 1989; Tribble 1989; Loewenstein 1989; Tribble 1990). In laymen's terms this is because as the instability starts and the filament begins to condense, it falls inward under the pull of gravity. However, the density in the ICM increases with decreasing radius; thus the infalling filament reaches a point in the ICM where the external pressure balances the internal pressure; it is no longer overdense and the instability stabilizes.

7.2. THE MORPHOLOGY PROBLEM

The emission line nebulae show elongated filaments with unidirectional structure on scales of 10 kpc or more. Such organized, highly elongated, structures are not what one would naively expect would condense out of the ICM. What organizes the structure seen in the emission line nebulae? It may be that largescale magnetic fields exist. Since, cooling occurs more readily perpendicular to the magnetic field axis than parallel to it, largescale magnetic fields could produce highly elongated filaments. However, if largescale magnetic fields exist in cooling flow clusters (as may also be suggested by the high rotation measures observed in cooling flow clusters [see Ge and Owen 1991]), thermal conduction is not likely to be suppressed (as required by cooling flow theories). Alternately, the elongated nature of the filaments may indicate that the initial seed instabilities out of which the filaments grew were themselves elongated in shape. This might be the case if the seeds are largescale perturbations from galaxy motions, or if the filaments originate as cold gas which has been stripped or cannibalized from passing galaxies and retains its original coherence (e.g., Soker, Bregman, and Sarazin 1991).

7.3. THE LENGTH SCALE PROBLEM

Fits to the x–ray surface brightness profiles in the few cooling flow clusters which have been observed with high enough spatial resolution and signal to noise to allow a detailed analysis, indicate that the x–ray surface brightness does not rise as rapidly going into the center of the cluster as would be predicted by cooling flow theories in which all of the mass drops out in the center (e.g., Fabian *et al.* 1981; Canizares *et al.* 1983; Steward *et al.* 1984; Thomas *et al.* 1987) . Instead, the fits require that the mass accretion rate goes roughly linearly with radius. Thus, these fits require that cooling matter and cooled matter are distributed throughout the cluster cooling radius (typically several hundred kpc).

If the $H\alpha$ emission traces the inflow rate, and assuming, as required by the x–ray data, that the mass accretion rate is $\propto r$, then we expect $\mu(r) \propto r^{-1}$, $L(r) \propto r$, where $\mu(r)$ is the $H\alpha$ surface brightness and $L(r)$ is the $H\alpha$ luminosity. However, when one examines the radial distribution of $H\alpha$ emission in clusters, what is seen is that there is a much sharper cutoff in the surface brightness as a function of radius than predicted. This is a general feature of the emission line nebulae in clusters; the line emission is strongly concentrated to the inner 10 kpc. We can conclude from this that, if the line luminosity traces the mass deposition rate, then mass must drop out at the center. Alternately, we must conclude that the radial dependence of the line emission does not reflect the radial dependence of the mass deposition rate.

We can turn the question around and ask, if the ICM is cooling with a mass deposition rate proportional to radius, would we expect to see $H\alpha$ emission from the quiescently cooling gas? The answer, is no. For instance, we can calculate the surface brightness of $H\alpha$ we would expect from a 100 solar mass per year cooling flow in which the $H\alpha$ luminosity was spread over a region 100 kpc in radius. The expected surface brightness is $\sim 10^{-20} (1+z)^{-4}$ erg sec^{-1} cm^{-2} arcsec^{-2}, at least two orders of magnitude below the presently detectable limits. Thus, *we cannot detect quiescently cooling gas which is recombining at the x–ray determined mass accretion rate and with the x–ray determined spatial distribution.*

This calculation reminds us that, though instabilities can, in principle, exist in the ICM of all clusters, regardless of the cooling time in the hot ICM, in the absence of a gravitational focus for the gas and/or a source of energy to repeatedly reionize the gas, these instabilities will not be detectable as emission line nebulae.

7.4. THE VELOCITY PROBLEM

As discussed in section 4.7, the emission line filaments have line widths of several hundred to 1000 km sec^{-1}. What is the origin of these velocities? If the filaments condense out of a quasi-static ICM, then these velocities must represent the velocities attained by the filaments as they fall into the galaxy under the gravitational pull of the galaxy. However, as pointed out by Loewenstein and Fabian (1990) instabitilites which do not comove with the hot ICM will be quickly disrupted by drag forces and Rayleigh–Taylor instabilities. Thus we are forced to one of two conclusions. Either the hot ICM comoves with the 10^4K gas or the filaments have not condensed out of the hot ICM, but instead originate cold.

It is possible that the velocities of the optical filaments represent turbulent velocities and that the hot component of the ICM is itself turbulent. There are two (obvious) ways in which the ICM can acquire large scale turbulence, either via galaxy motions through the cluster or via merging of subclusters (e.g., Jaffe 1980; Miller 1986; O'Dea and Baum 1987b; Balbus 1988; Loewenstein and Fabian 1990). An alternate viewpoint would be that the velocities of the optical filaments do not represent turbulent velocities, but rather the inflow velocities of the cooling accretion flow. If this were correct, it would imply that there is trans infall of the cooling x-ray gas (e.g., Fabian, Nulsen, and Cannizares 1984; Voit and Donahue 1991, Tribble 1990) at distances of at least 10 kpc from the cluster center.

7.5. THE DUST PROBLEM

Evidence has now accumulated from several different wavebands and several different observational approaches that emission line nebulae in clusters frequently contain dust. These include (1) IRAS detections of roughly 1/2 of the central dominant galaxies in cooling flow clusters (Bregman *et al.* 1990; Hu 1988; but see also Bregman this volume), (2) reddening values ($E_{B-V} \sim 0.1 - 0.3$) determined from the ratio of the strengths of hydrogen recombination lines observed from the emission line nebulae (Hu 1988); (3) broad band optical colour maps showing dust lanes and dust patches in a large fraction of central dominant galaxies (McNamarra *et al.* 1991); (4) comparison of emission line images and broad band images of M87 and NGC4696, showing that there is substantial dust which is co-spatial with the line emitting gas (Sparks *et al.* 1989, Sparks private communication 1991).

It has further been pointed out that the infrared luminosities of this dust are in some cases truely staggering; $L_{IR} \sim 10^{43} - 10^{45}$ ergs sec^{-1}, rivaling the total x-ray luminosities of the cluster. There appears to be no correlation of L_{IR} with the presence of blue colours in the central dominant cluster galaxy, with the Hα luminosity, with the x-ray luminosity or with the x-ray derived mass accretion rate. The total masses in dust are difficult to estimate but can be expressed as $M_{DUST} \sim 0.3 - 3 \times 10^7 \times (T/30K)^{-5.9}$ M$_\odot$ (e.g., Bregman *et al.* 1990).

The 'problem' is this. The traditional view, based on straighforward physical arguments, is that the ICM should be dust free. This is because the sputtering time for dust in the ICM is short $t_{sputter} \sim 2 \times 10^7 (n_H/0.001)^{-1} (a/0.01\mu)$ yrs (Draine and Saltpeter 1979), relative to all other relevant timescales. However, this does not necessarily imply that the ICM will be dust free. What it means is that any dust present in the ICM, which is well mixed with the ICM, must have been injected in the last 10^7 years. Cool clouds which have condensed directly out of the hot ICM should be dust free, provided their collapse time is longer that $t_{sputter}$ and provided they have no internal dust source.

Theoretically, dust can be added to the ICM through stellar evolution at a rate of $\sim 0.15(L_B/10^{12}L_\odot)M_\odot \mathrm{yr}^{-1}$, for a gas to dust ratio of 100 (Faber and Gallagher 1976). Equilibrium is therefore reached with $\mathrm{M_{DUST}}/t_{\mathrm{sputter}} = 0.15$, or $\mathrm{M_{DUST}} \sim 10^7$ M$_\odot$. This calculation assumes that dust is effectively mixed with the hot ICM, and that the injection of dust due to stellar evolution is a continuous process.

We know that the filaments themselves are dusty in some sources and that there are dust lanes and patches in a large fraction of the central dominant galaxies in cooling flow clusters. Thus, dust which is localized to the inner ~ 10 kpc and which shows organized morphological patterns does exist. What is not yet clear is whether there is an additional dust component which is spread throughout the ICM. Interesting parameters to compare are the gas to dust ratio (by mass) in the ICM and in the filaments. Sparks *et al.* 1989 and Sparks *et al.* 1991 (in preparation) have measured gas to dust ratios *near* the Galactic value of 100 in NGC4696 and M 87 for the filaments. It is clear that the gas to dust ratio of the ICM as a whole is much higher than this (the total gas mass is $\sim 10^{14}$M$_\odot$, while the total dust mass is $\sim 10^7$M$\odot$). Even within the inner 10 kpc of a rich cluster, assuming all the dust in the cluster resides within this radius, the gas to dust ratio is well over one thousand. Thus, models in which the filaments condense out of the hot ICM must explain the dramatic increase in the dust to gas ratio seen going from the hot ICM to the $10^4 K$ filaments.

If there is a component of the dust which is distributed throughout the hot ICM, then dust which is stripped or ejected from cluster galaxies must become effectively mixed with the ICM, within 10^7 years. Alternately, the stripped dust must be contained within cold clouds, which themselves would need to be able to avoid evaporation (either because they are initially large, or because of tangled magnetic fields).

7.6. THE EMBARRASSMENT OF RICHES PROBLEM

Given an x-ray determined mass accretion rate, if we assume that each hydrogen atom emits one Hα photon as it cools and recombines, we can predict a total Hα luminosity expected from a cooling flow cluster. What we find is that the observed Hα luminosities exceed the predicted values, often by several orders of magnitude. This 'embarrassment of riches' problem has been parameterized by $\mathrm{H_{rec}}$. $\mathrm{H_{rec}}$ is defined to be the number of times each atom must be reionized to reproduce the observed Hα luminosity, assuming the rate at which atoms cool is given by the x-ray determined mass accretion rate. In a typical cluster, a value for $\mathrm{H_{rec}} \sim 10$ is required. In a subset of the clusters, $\mathrm{H_{recs}}$ as high as $100 - 1000$ are required. The required $\mathrm{H_{recs}}$ show a very wide dispersion, ranging from < 1 to ~ 1000 between clusters. Thus, the mass accretion rate is not a particularly good predictor of the Hα luminosity a cluster will show, particularly at *high* mass accretion rates (see also section 6.2, above).

The embarrassment of riches problem becomes even more embarrassing when we allow for the result from the x-ray observations that mass is dropping out over the entire cooling radius (typically a few hundred kpc). Then, within the radius at which the filaments are seen (typically only ~ 10 kpc), the number of atoms cooling at the x-ray derived mass accretion rate will be a factor of $\sim 10 - 25$ less than the number cooling over the entire cluster. The required H_{rec} will therefore be, on average, several hundreds (and not tens!). In the extreme cases, H_{recs} of ~ 10000 are required.

7.7. THE ENERGY SOURCE PROBLEM

From the above sections, we have learned that we are not simply seeing quiesently cooling gas when we observe emission line nebulae in clusters. The gas must be reionized, on average at least 10 times per atom, and in some sources as much as 100 - 10000 times per atom. The energy required to do this is, in most cases, a non-trivial fraction of the total energy extractable from the cooling flow.

The maximum value of H_{rec} retrievable from the flow is ~ 500, assuming that $(5/2)kT$ is released per particle cooling at the x-ray derived mass accretion rate. Thus, on average 2% of the total retrievable energy from the flow must be converted to line luminosity from gas at 10^4K, and in some cases 100%, or more, of this cooling flow luminosity must be converted to line luminosity to reproduce the observe Hα luminosity. However, since the x-ray derived mass accretion rate is observed to have a radial dependence roughly proportional to radius, and since the cooling flow filaments are concentrated to the inner 1/20 of the cooling radius, in actuality only 1/20 of the total energy retrievable from the cooling flow is available to the filaments. Thus, if the emission line filaments are energized by the cooling flow, they must intercept fully 1/2 of the total thermal energy emitted by the cooling flow within the inner 10 kpc! Further, it is clear, that in the most extreme cases, there simply is not any where near enough energy from the cooling flow within the radius at which the filaments are observed to energize the emission line filaments.

8. What Ionizes the Filaments

There are a host of possible energy sources available to power the emission line nebulae, and as we shall see in the discussion below, there is no obvious winner. Possibilities we shall consider are (1) photo-ionization by a nuclear (AGN) continuum, (2) photo-ionization by a population of (young?) hot stars, (3) mechanical (shock) heating, (4) heating via the radio source and (5) tapping the thermal energy of the ICM. Another possibility is that the filaments are energized when the intra-cluster magnetic field reconnects in the inner regions of the cooling flow (see Soker and Sarazin 1988). This process is poorly understood, at best, but simple calculations show that it will only be energetically important only if $\dot{M}$ is constant with radius (HBvBM). Yet another possibility is that the radiative decay of dark matter produces hydrogen ionizing photons (see Sciama 1988 and references therein).

8.1. PHOTO-IONIZATION BY AN AGN

Central dominant galaxies in cooling flow clusters are commonly active radio sources. Thus, it is logical to suppose that the nuclei of these sources produce an ultraviolet/x-ray continuum. The question is whether this continuum is strong enough to ionize the observed nebulae and whether the radial variation of line ratios is consistent with photo–ionization by a nuclear point source. The answer to both of these question appears, typically, to be no.

The nuclear ionizing continuum inferred from optical observations is typically one or two orders of magnitude too small to power the emission line gas (e.g., HBvBM). This does not immediately rule out nuclear photo-ionization, because the nuclear continuum may either be highly peaked in the ultraviolet, or the nuclear emission may be anisotropic, so that the nebulae are exposed to a higher nuclear flux than we observe.

However, there is another piece of evidence suggesting that nuclear photoionization is not the dominant ionization source for the cooling flow nebulae. That is, the line ratios as a function of radius are inconsistent with photo–ionization by a nuclear point source. (HBvBM; Johnstone and Fabian 1988). The argument goes as follows. If the gas is photoionized by a point source, then the ionization parameter (the ratio of the ionizing flux to the electron density) as a function of radius from the source is $U = Q_{nuc}/(4\pi r^2 n_e c)$, where Q_{nuc} is the number of ionizing photons per second emitted by the source, r is the radius from the source, and n_e is the electron density at radius r. If gas is photonionized by a nuclear source, than $U \times n_e \propto r^{-2}$. When we estimate the ionization parameter as a function of radius (by measuring line ratios which are sensitive to U), and we estimate the electron density as a function of radius (by measuring the [SII] ratio), we find that this relationship does not hold. The most straightforward interpretation is that the gas is not ionized by a point source and that the nebulae have, instead, a distributed ionization source.

However, since many central dominant galaxies in cooling flow clusters have radio loud AGN, they are likely to have a nuclear ionizing continuum which dominates the ionization within some radius (and some solid angle?). Thus it may be that photo–ionization by an AGN may somewhat alleviate the energetics problem in some of the cooling flow nebulae. However, (see section 5.1) it is important to realize that there are cooling flow clusters with high Hα luminosities, and only very weak (or undetected) radio sources (AGN). Thus, evidence from many sides shows that nuclear photoionization cannot be the whole answer.

8.2. PHOTOIONIZATION BY HOT STARS

If matter deposited by the cooling flow forms stars, and if this star formation does not produce exclusively low mass stars, then it may be that high mass stars are continuously produced in central dominant galaxies in cooling flow clusters. These high mass stars are likely to be surrounded by HII regions, and the question logically arises whether the emission line nebulae we see in cluster centers are in fact nebulosities which are photoionized by such young, hot stars.

Kennicut (1983) has noted the following simple empirical relation between star formation and Hα luminosity in spiral galaxies; $L_{H\alpha} \sim 10^{41} erg\ sec^{-1}\ \dot{M}_{disk}\ yr^{-1}$, where $\dot{M}_{disk}$ is the star formation rate with a disk IMF. If we apply this to the emission line nebulae in cooling flow clusters, we find that this relation translates into $H_{rec} \sim 4000$. Put more simply, this implies that (on average), only $\sim 4\%$ of material which is accreting at the x–ray determined mass accretion rate is required to form stars with a disk galaxy IMF in order to produce the Hα luminosities observed in clusters. Thus, energetically at least, ionization by massive ongoing star formation is a viable source for the emission line nebulae in clusters.

In addition to these 'energetic' arguments, there are several other observations which favour photo–ionization by stars. These are (1) there are extended regions of blue colours found in the inner 10 kpc of many cooling flow dominant galaxies (hereafter cfd) which may be associated with ongoing star formation (McNamara and O'Connell 1989 and references therein; Romanishin 1987; Norgaard-Nielsen 1990), (2) the strength of the 'D Break' [a measure of ongoing star formation, see e.g., Johnstone *et al.* 1987 for more details] correlates roughly with Hα luminosity in cfds (at the 97.5% confidence level, (3) If one computes the rate at which star formation with a disk IMF must be proceeding in order to produce the observed Hα luminosity and compares that rate with the rate of ongoing star formation with a disk IMF which is needed to produce the blue colours in the cfds, one gets roughly the same number (McNamara and O'Connell 1989; Johnstone *et al.* 1987).

While the arguments in favour of stellar photoionization are quite strong, there is, unfortunately, an equally long and compelling list of points against it. These are (1) the [OII] strength is independent of the strength of the blue colour anomalies in cfds (McNamara and O'Connell 1989) (2) there is no correlation of $L_{H\alpha}$ with blue colours, as measured by δUB, where δUB is purportedly a better indicator of the ongoing star formation rate than is the D break (McNamara and O'Connell 1989), (3) there are galaxies with strong Hα but small (or absent) uv excesses (blue colours), and (4) the line ratios exhibited by the nebulae are very different from the line ratios of normal spiral galaxies and of HII regions.

8.3. MECHANICAL HEATING

The term 'mechanical heating' is used here to denote any mechanism which converts kinetic energy into an energy source for the ionization of the emission line nebulae. The two obvious sources of this kinetic energy are (1) the turbulent energy of the ICM and (2) the kinetic energy of the outflowing radio plasma from the AGN at the cluster center. The latter is discussed in section 8.4 below, here only the kinetic energy of the ICM itself is considered. Within the ICM, the turbulent energy to be tapped may be either the turbulent energy of the cold and warm components of the ICM or the turbulent energy of the hot component of the ICM. The rate at which turbulent energy released by a given component of the ICM can be supplied to the emission line nebulae can be written as $M_{comp} \times v_{comp}^3/2r_{neb}$, where M_{comp} is the mass of the component within radius r_{neb}, v_{comp} is the turbulent velocity of the component and r_{neb} is the radial extent of the emission line nebulae.

In most clusters, there are no measurements of the mass and velocity width of the cold component of the ICM. However, in the Perseus cluster, both HI absorption and CO emission (see article in this volume by Jaffe and references therein) have been detected from cold gas in the central ten kpc of the central dominant galaxy, NGC1275, allowing estimates of the mass in cold gas and the turbulent velocity width of that gas to be made. Using the numbers presented by Jaffe, and the formalism above, the energy available from the dissipation of the turbulent energy of these cold clouds is only $\sim 20\%$ of the luminosity of the emission lines from the 10^4K gas. Thus, at least in the Perseus cluster, cloud–cloud collisions do not appear to be energetically capable of supporting the luminosity of the emission line nebula.

In order to estimate the rate at which turbulent energy is dissipated by the hot component of the ICM two assumptions are required; (1) the hot component is in pressure equilibrium with the warm component; this allows us to estimate the mass of the hot component within the region of the cluster within which the emission line nebulae is found (taken to be $4/3\pi r_{neb}^3$, where r_{neb} is the radial extent of the nebulae) and (2) the hot component comoves with the warm component; this allows us to assume that the velocity width of the hot component is the same as what is measured for the warm component. When the kinetic energy of the hot ICM is calculated in this way the results are; (1) the kinetic energy scales roughly with Hα luminosity (see HBvBM), and (2) conversion efficiencies of $\sim 100\%$ are required for the kinetic energy of the hot ICM to provide the energy source for the line luminosities from the nebulae.

Give that the energy is available, there are a host of possible mechanisms for translating kinetic energy into emission line luminosity. Among these are (1) cloud cloud collisions, (2) drag heating, and (3) repressurizing shocks. If the kinetic energy of the ICM is the energy source for the line emission from the 10^4K gas, then we might expect that the line ratios would be consistent with shock heating. This is indeed the case as discussed in Section 4.8. A further prediction of shock models might be that the strength of the temperature forbidden lines relative to the recombination lines would increase with the shock velocity. Crawford $et\ al.$ (1991) have shown that in two clusters, the [NII] to Hα ratio increases with increasing line width, and the surface brightness of the line increases as a strong power of line width. However, while some cooling flow clusters show this dependence, others do not. Combining the Crawford $et\ al.$ results with a similar analysis of line width versus line ratio for the seven sources in HBvBM gives the following picture. In a subset of cooling flow clusters there is a correlation of [NII] to Hα ratio with line width, suggesting that shock heating may indeed be the answer. In at least one of these clusters (A2597), the radio source and emission line nebulae are cospatial and of similar sizes, suggesting the possibility that perhaps the radio source (and not the ICM) is the source of the turbulent energy for this particular nebula.

In addition, however, care must be taken to separate a velocity dependence from a radial dependence. Since in general line widths and surface brightness are a strong fuction of radius, this is by no means an easy task. A radial decline in [NII] to Hα ratio would be predicted by, for instance, nuclear photo–ionization models. In addition, there are a number of clusters in which no correlation of line ratio with either distance or line width is seen. Thus, the situation is not simple.

8.4. THE RADIO SOURCE OF ENERGY

The radio source associated with the central dominant cluster galaxy may provide another source of energy for the emission line nebulae. Cosmic rays, magnetic fields, heat, and kinetic energy are transported via the radio jet from the nucleus into the ICM and these may all contribute to the heating and ionization of the emission line gas. Since Jaffe (this volume) presents a review of the radio properties of clusters, I just briefly outline the 'circumstantial' evidence connecting the emission line nebulae and the central radio sources in cooling flow clusters (Pedlar *et al.* 1990; O'Dea and Baum 1987a; Baum and Heckman 1987;Burns 1990; HBvBM; Baum and O'Dea 1991; Zhao, Burns, and Owen 1989).

1. The nebulae and the radio sources typically have similar size scales and are frequently cospatial (in projection).

2. There is evidence for a dynamical interaction between the radio source and the emission line gas in that the linewidths of the line emitting gas are larger in regions of bright radio emission.

3. The radio source can be energetically important. Typically, the total line luminosity from 10^4K gas is 10 times the radio luminosity. However, the radio luminosity is thought to be only a few percent of the kinetic and internal energy of the outflowing radio plasma. Thus, the radio source, on average, pours 10 times as much energy into the ICM as is lost by the 10^4K gas.

4. The radio luminosity correlates with the Hα line luminosity in central dominant galaxies.

5. The minimum energy estimate of the pressure in the radio emitting plasma is roughly equal to the thermal pressure of the emission line nebulae at the same radial distance.

6. Radio sources in cooling flows exhibit a host of unusual properties which suggest they are interacting strongly with the ICM.

Finally, it is important to realize that despite the evidence suggesting the presence of a connection between (1) cooling flows, (2) radio sources, and (3) emission line nebulae, there are cooling flow nebulae in clusters which have radio quiet central dominant galaxies and there are radio loud central dominant galaxies in clusters without cooling flows.

8.5. UTILIZING THE THERMAL ENERGY OF THE HOT ICM

The hot ICM contains $\sim 10^{14}$ solar masses of gas at $\sim 10^{7-8}$K. This is a vast thermal reservoir, which can, in principle, be tapped to supply energy for the emission line filaments.

A number of authors have considered the possibility that the emission line filaments are photoionized by the uv and x-ray emission of the cooling flow, that is by the the gas itself, as it radiatively cools from 10^7 to 10^4K. Unfortunately, the authors do not agree on the viability of this mechanism.

Voit and Donahue (1991) and Donahue and Voit (1991) have recently presented detailed models of the line ratios which would be emitted by a gas which is photoionized by the uv/x-ray radiation from the cooling plasma. They are able to correctly reproduce the line ratios exhibited by the nebulae, including the very high [NII] and [OII] to Hα ratios exhibted by some sources. Their models require clouds with column depths of $N_H = 10^{17} - 10^{21}$cm^{-2}, and they show that variations in cloud depth reproduce the variations in line ratios which are observed. Note that since the variation in line ratio is seen predominantly from source to source and not within individual sources, this implies that a given cluster will have an ICM with warm clouds with a fixed range of optical depths.

Thus, the models show that this scenario can reproduce the line ratios. The dispute over the feasibility of the models concerns the energetic requirements. The total energy available from the cooling gas is in most cases sufficient to provide the observed line luminosities (remember from section 7.6 that $H_{rec} \sim 500$ is obtainable from the cooling flow itself). However, that 'energy' must be focused in the region where the optical filaments exist (i.e., the inner 3–10 kpc) and the efficiency with which the radiation is converted to line luminosity must be very high, i.e., the covering factor of the gas must be very large. Observationally, the data suggest that neither of these constraints is met. That is, the x–ray data suggests that mass cools out over the entire cooling radius, and the high spatial resolution images of the emission line nebulae suggest that the filaments have only a small covering factor. Thus, the filaments most likely intercept only a very small fraction of the total radiation of the cooling gas and this model therefore faces serious energetic difficulties.

This is shown in a slightly different way in Figure 8, taken from HBvBM, where the log of the Hα line luminosity of the brightest cluster galaxy is plotted against the log of the x–ray luminosity of the cluster. A line indicating where the x–ray luminosity of the cluster equals the line luminosity from the nebula is superposed. It is clear that a subset of the clusters lie dangerously close to this line. For these clusters, models which seek to obtain the ionizing radiation for the emission line nebula from the x–ray emission from the cluster will be seriously strained. There is no significant correlation of x–ray and emission line luminosity.

There exists (at least) two alternate model for tapping the thermal energy of the hot ICM. One, proposed by Begelman and Fabian (1991), posits that turbulent mixing layers will be produced along the outer skins of cold clouds immersed in a hot, turbulent ICM. These layers will attain an intermediate temperature and ultraviolet and soft x–ray radiation from the layers may provide the photons which ionize the cold clouds. An alternate model is the thermal conduction model, which has been discussed by a host of authors who have reached differing opinions as to its validity. It has been championed most recently and most convincingly by Sparks *et al.* (1990) and independently by de Jong *et al.* (1990) (and see the review in this volume by Bregman for an elaboration of the model). In this model, thermal electron conduction causes heat to flow from the hot bath of the ICM into cool gas and dust which has been acquired in a merger or tidal interaction between the cfd and a gas–rich cluster member. An intermediate temperature layer forms between the cold gas and the hot ICM, and uv/x–ray radiation from this layer photo–ionizes the emission line gas (much as in the Voit model). Sparks *et al.* demonstrates that the model is energetically feasible and Donahue and Voit (1991) have shown that the model correctly reproduces the observed line ratios.

8.6. TRUTH TABLES

From the myriad possible energy sources for the emission line gas in cluster centers presented above, there are a few important points to note. First, most, if not all, of the possible mechanisms require a high conversion of the available energy into line luminosity (i.e., they are energetically strained). Second, if the process which 'energizes' the 10^4K gas heats the gas substantially above 10^4K, then lines from the intermediate temperature ($10^5 - 10^6$) gas should be as overluminous as is the Hα line. This would imply that lines such as the coronal lines of Fe X and Fe XIV should be easily detectable. This is not the case.

Third, there are exceptions to every rule. That is, there are emission line nebulae in central dominant galaxies without apparent ongoing star formation and in central dominant galaxies without associated radio sources. Conversely, there are cooling flow clusters which have no detected Hα emission but which have central dominant galaxies which have strong radio sources and/or which show blue colors and dust patches. Thus, each time we think we may have identified 'the energy source' (e.g, the AGN, or photoionization from young stars) we find an example where this interpretation cannot be correct.

The simplest interpretation of this confusing situation is that there is no one dominant source of energy for the emission line nebulae in cooling flow clusters, but rather that a number of different processes apply, any one of which can dominate in a given cluster. We are driven to this esthetically unpleasing solution (dubbed Oemler's Razor at this conference) as the only way to satisfy the energy and observational constraints. However, as always, it may be that we are 'missing something', and that there is indeed one dominant energy source which we have failed to correctly identify.

9. Final Remarks

The bottom line is this: the emission nebulae do not 'look' or 'behave' like filaments which have quiescently cooled out of an homogeneous, hot ICM with a cooling accretion flow, as envisioned in the traditional models. The implications are either that (1) the filaments originate from the hot ICM, but the standard models of the ICM and cooling flows must be modified, or (2) that the filaments do not originate in the hot ICM, but have, instead, been stripped cold from orbiting galaxies or acquired in a merger between the central dominant galaxy and a gas rich cluster member.

If the filaments do indeed condense out of the hot ICM, then we must understand: (1) the existence of dust in the filaments with gas to dust ratios and reddening curves similar to Galactic dust, (2) the large infrared luminosities of the clusters and the lack of correlation between the infrared luminosity and other properties of the ICM, (3) the coherent, elongated morphology of the filaments which can show structure on scales of tens of kpc, (4) the kinematics which show evidence for large scale turbulence, (5) the spatial distribution of the Hα emission, which is strongly concentrated to the inner ten kpc of the cluster, and (6) the energy source for the ionization of the gas. One implication is that the hot ICM itself may need to share the spatial and kinematic inhomogeneities of the 10^4K gas.

On the other hand, if the filaments originate as cold clouds stripped from gas rich galaxies, then the existence and properties of the dust, the filamentary morphology of the nebulae, and the spatial distribution of the emission can be easily understood.

Further, if as in the Sparks *et al.* (1989) model, the hot ICM transfers heat to the cold clouds via thermal conduction, then it may also be possible to understand the energization of the emission line gas and the large infrared luminosities of the clusters. To explain the correlation of emission line nebulae and cooling flows in clusters, we must add to this the prediction of Sparks *et al.* (1989) and de Jong *et al.* (1990) that the transfer of energy from the hot ICM to the cold clouds via conduction will lead to the cooling of the x–ray gas in the center of the cluster and thereby to the mimicing of the appearance of a cooling accretion flow.

This 'warming flow' scenario introduces a new set of problems which must be considered. First, x–ray observations allow the determination of the mass fraction of the hot ICM which is at a given temperature (for some temperature ranges). The measurements to date are fully consistent with gas cooling from the hot phase (10^8K) through the intermediate phases (10^6K) (Canizares *et al.* 1988; Mushotzky *et al.* 1988), as predicted by the cooling flow model. If this turns out to be a general property of clusters of galaxies, then it will be important to understand why a 'warming' (instead of cooling) scenario, should produce the same distribution of emission measures.

Second, cooling flows appear to be fairly ubiquitous in rich clusters (Edge *et al.* 1991) at the current epoch. Thus, for this scenario to be correct, tidal stripping and/or cannibalism of gas rich galaxies must be a common and ongoing phenomenon at the centers of rich clusters today. This would have important implications for our understanding of cluster and galaxy evolution.

Acknowledgements

SAB thanks Chris O'Dea for many helpful discussions and critical readings of the manuscript. SAB also thanks the Hubble Fellow Program for providing the Hubble Fellowship under which most of this work was completed. Partial support for this work was provided by NASA through grant HR-1001,01-90A awarded by the Space Science Telescope Institute which is operated by Associated Universities for Research in Astronomy, for NASA under contract NAS5-2655S.

References

Balbus, S. A., 1988, Ap. J., 328, 395.

Balbus, S. A., and Soker, N., 1989, Ap. J., 341, 611

Baldwin, J. A., Phillips, M. M., and Terlevich, R., 1981, Pub. A.S.P., 93, 5.

Baum, S. A., and Heckman, T. M., 1987, in Radio Continuum Processes in Clusters of Galaxies, Ed. C. P. O'Dea and J. M. Uson (Greenbank: NRAO), 119.

Baum, S. A., and Heckman, T. M., 1988a, Ap. J., 336, 680

Baum, S. A., Heckman, T. M., 1988b, Ap. J., 336, 702

Baum, S. A., Heckman, T. M., van Breugel, W., 1990, Ap. J. Suppl., 74, 389

Baum, S. A., Heckman, T. M., van Breugel, W., 1991, Ap. J., in press

Baum, S. A, Heckman, T. M., Bridle, A. H., van Breugel, W., and Miley, G. K., 1988, Ap. J. Suppl. 68, 833

Baum, S. A., and O'Dea, C. P., 1991, MNRAS, 250, 737

Begelman, M. C., and Fabian, A. C., 1990, MNRAS, 244, 26p

Binney, J., and Cowie, L., 1981, Ap. J., 247,464

Bohringer, H., and Fabian, A. C., 1989, MNRAS, 247, 1147

Bregman, J. N., McNamara, B. R., O'Cconnell, R. W., 1990, Ap. J., 351, 406

Burns, J. O., 1990, A. J., 99, 14

Canizares, C. R., Stewart, G. C., and Fabian, A. C., 1983, Ap. J., 272, 449

Canizares, C. R., Markert, T. H., Donahue, M. E., 1988, Cooling Flows in Clusters and Galaxies, ed., A. C. Fabian, Kluwer, 63

Cowie, L. L., and Binney, J., 1977, Ap. J., 215, 723

Cowie, L. L., Hu, E. M., Jenkins, E. B., and York, D. G., 1983, Ap. J., 272, 29

Crawford, C., Fabian, A., 1991, Clusters and Superclusters of Galaxies; Contributed Talks and Poster Papers, ed. Colless et al., (Cambridge: Nato), 69

Donahue, M., and Voit, G. M., 1991, Ap. J., in press

Draine, B. T., and Salpeter, E. E., 1979, Ap. J., 231, 438

Edge., A., Fabain, A. C., and Stewart, G. C., 1991, Clusters and Superclusters of Galaxies; Contributed Talks and Poster Papers, ed. Colless et al., (Cambridge: Nato), 35

Faber, S. M., and Gallagher, J. S., 1976, Ap. J., 204, 365

Fabian, A. C., and Nulsen, P. E. J., 1977, MNRAS, 180, 479

Fabian, A. C., Nulsen, P. E. J., and Canizares, C. R., 1985, Nature, 310, 733

Fabian, A. C., Nulsen, P. E. J., and Canizares, C. R., 1991, A & A Reviews, 2, 191 Nature, 310, 733

Fabian, A. C., Hu, E. M., Cowie, L. L., and Grindlay, J., 1981, Ap. J., 248, 47

Field, G. B., 1965, Ap. J., 142, 531

Ford, J., C., and Butcher, H., 1979, Ap. J. Suppl., 41, 147

Ge., J. P. and Owen, F. N., 1991, Clusters and Superclusters of Galaxies; Contributed Talks and Poster Papers, ed. Colless et al., (Cambridge: Nato), 121

Heckman, T. M., 1981, Ap. J., 250, L59

Heckman, T. M., 1987, IAU Symposium 121, Observational Evidence of Activity in Galaxies, ed. E. Khachikian, K. Fricke, and J. Melnick, (Dordrecht: Reidel), 421

Heckman, T., Baum, S., van Breugel, W., McCarthy, P., 1989, Ap. J., 338, 48

Hu, E. M., 1988, in Cooling Flows in Clusters and Galaxies, ed. A. C. Fabian, (Dordrecht: Kluwer), 73

Hu, E. M., Cowie, L. L., and Wang, Z., 1985, Ap. J. Supp., 59, 447

de Jong, T., Norgaard-Nielson, H. U., Jorgensen, H. E., and Hansen, L., 1990, A & A , 232, 317

Jaffe, W., 1980, Ap. J., 241,924

Johnstone, R. M., Fabian, A. C., and Nulsen, P. E. J., 1987, MNRAS, 224, 75

Johnston, R. M., and Fabian, A. C., 1988, MNRAS, 233, 581

Jones, C., and Forman, W., 1984, Ap. J., 276, 38

Kent, S. M., Sargent, W. L. W., 1979, Ap. J., 230, 667

Loewenstein, M., 1989, MNRAS, 238, 15

Loewenstein, M., and Fabian, A. C., 1990, MNRAS, 242, 120

Mathews, W. G., and Bregman, J. N., 1978, Ap. J., 224, 308

McNamara, B. R., and O'Connell, R. W., 1989, Ap. J., 98, 2018

McNamara, B. R., and O'Connell, R. W., 1991, Ap. J., submitted

Miller, L., 1986, MNRAS, 220, 713

Morganti, R., Ulrich, M.-H., Tadhunter, C. N., 1991, MNRAS, in press

Mushotzky, R. F., Hold, S. S., Smith, B. W., Boldt, E. A., Serlemitsos, P. J., 1981, Ap. J., 244, L47

O'Dea, C. P., and Baum, S. A., 1987a, Radio Continuum Processes in Clusters of Galaxies, ed., Chris O'Dea and Juan Uson (Charlottesville:NRAO), 141

O'Dea, C. P., and Baum, S. A., 1987b, AJ, 94, 1476

Osterbrock, D. E., 1989, Astrophysics of Gaseous Nebulae and Active Galactic Nuclei, University Science Books, Mill Valley, CA

Pedlar, A., et. al., 1990, MNRAS, 246, 477,

Rawlings, S., Saunders, R., Eales, S. A., Mackay, C. D., 1989, MNRAS, 240, 701

Romanishin, W., 1987, Ap. J., 323, L113

Romanishin, W., 1988, Cooling Flows in Clusters and Galaxies, (Dordrecht: Kluwer), 121

Sarazin, C. L., 1986, Rev. Mod. Physics, 58, 1

Sciama, D., 1988, Ap. J., 364, 549

Silk, J., 1976, Ap. J., 208, 646

Soker, N., Bregman, J. N., and Sarazin, C. L., 1991, Ap. J., 368, 341

Soker, N, and Sarazin, C. L., 1988, Ap.J., 327, 66

Sparks, W. B., Macchetto, F., and Golombek, D., 1989, Ap. J. 345, 153,

Steward, G. C., Fabian, A. C., Jones, C., and Forman, W., 1984, Ap. J., 285, 1

Thomas, P. A., Fabian, A. C., and Nulsen, P. E. J., 1987, MNRAS, 228, 973

Tribble, P. C., 1989, MNRAS, 238, 1247

Tribble, P. C., 1991, MNRAS, 248, 741

Voit, G. M., and Donahue, M., 1990, Ap. J., 360, L15

Zhao, J-H, Burns, J. O., and Owen, F .N., 1989, A. J., 98, 64

CLUSTERS AS GRAVITATIONAL LENSES

G. Soucail
Observatoire Midi-Pyrénées
14 Avenue E. Belin
31500 Toulouse
France

ABSTRACT. This paper reviews the present status of our understanding of gravitational lensing by clusters of galaxies. A spectacular manifestation of this phenomenon is the formation of highly distorted images of background sources (galaxies) known as gravitational arcs, but also the formation of more numerous small distorted images, the so-called arclets. Although this topic is still a recent one, some original new theoretical approaches have been developed and a lot of observations are performed on the largest telescopes through international collaborations, from which a large number of interesting results have already emerged. The redshift determination of more than 6 large arcs opens a new window in the study of high redshift field galaxies. The modeling of some of these arcs reinforces the idea that most of the dark matter in clusters is distributed in a smooth component outside the galaxies. The detection of more than 50 arclets in some rich clusters will enable one to better probe the mass content and the distribution of dark matter in clusters, as well as indications on the redshift distribution of the background galaxies.

1. Introduction

The story of gravitational lensing related to lensing by clusters of galaxies began in 1936-1937, when Einstein and then Zwicky predicted the formation of extended images and rings by "extragalactic nebulae". Then some isolated theoretical works were done, mainly by S. Refsdal in the 1960's, but nothing was observed until 1979, when the first multiple quasar 0957+561 was discovered (Walsh et al. 1979). In 1985-1987, the first observations of arcs in clusters of galaxies were announced (Soucail et al. 1987, Lynds and Petrosian 1986). Paczynski (1988) first guessed that these arcs could be manifestations of gravitational lenses by the clusters. The definitive confirmation of this assumption was the redshift measurement of the giant arc in Abell 370, a redshift about twice the cluster redshift (Soucail et al. 1988). A few month later, several "mini-arcs" were detected in the centre of Abell 370, interpreted as high redshift galaxies distorted by the cluster (Fort et al. 1988). This hypothesis was consistent with the detection by Tyson (1988) of a high density of very faint blue galaxies, supposed to be at high redshift. Furthermore, Tyson et al. (1990) reinforced this observation by detecting more than 50 such arclets in the cluster A1689, from which they tried to get a "direct" mapping of the dark matter. More than 15 clusters have now been identified with some manifestations

A. C. Fabian (ed.), *Clusters and Superclusters of Galaxies*, 199–218.
© 1992 *Kluwer Academic Publishers*.

of gravitational lensing on these background sources.

During the last 5 years of research on cluster lenses, the whole community of astronomers began to realise the potentialities of this subject in relation to the main topics in observational cosmology. First of all, lenses are sensitive to mass and are a useful for mapping the dark matter in clusters without any dynamical assumptions. So it is crucial to compare the results about the mass determination in clusters from lenses and from dynamical studies. One could also expect new important constraints on the localisation of the dark matter, especially using the arclets. Moreover, clusters of galaxies are used as gravitational telescopes to look at the very distant universe, and the sources of the arcs are typical examples of background field galaxies, seen with the magnification of the lens.

Before presenting the overall results obtained recently, I will rapidly recall some basic concepts of gravitational lensing which are useful for the understanding of the formation of rings, arcs and arclets in part 2. The giant arcs currently known from the modeling of the lensing configurations or their redshift measurements will be reviewed in part 3. We will also present the results concerning the arclets and the study of statistical lensing, applied to mapping the dark matter in cluster of galaxies (part 4) and possibly in the large-scale structures (part 5).

2. Basic concepts of gravitational optics

Several formalisms can introduce the basic equation of gravitational lensing, which relates the position of the image(s) for a given source to a given observing point (Figure 1). One of them is derived from Fermat's Principle, applied to the light propagating in an expanding inhomogeneous universe, in the framework of General Relativity (Schneider 1985, Blandford and Narayan 1986): " *The time of arrival of light at O, from S, is extremised for an actual ray*". It is a scalar formalism which expresses the time-delay of a light ray with respect to the direct ray (SO), as the sum of two terms. One is the geometrical time-delay

$$t_{geom}(\theta_I, \theta_S) = \frac{(1 + z_L) D_{OL} D_{OS}}{2c\, D_{LS}} (\theta_I - \theta_S)^2,$$

and the other one is the gravitational time-delay

$$t_{grav}(\theta_I) = -\frac{2(1 + z_L)}{c^3} \int \phi(\theta_I)\, ds,$$

where ϕ is the local Newtonian potential and the integral is performed along the ray. The distances involved are angular diameter distances.

Then one introduces the reduced parameters, assuming $c = G = 1$,

$$\tau = \frac{D_{LS}}{(1 + z_L) D_{OL} D_{OS}}\, t,$$

$$\Psi(\theta_I) = \frac{2 D_{LS}}{D_{OL} D_{OS}} \int \phi(\theta_I)\, ds.$$

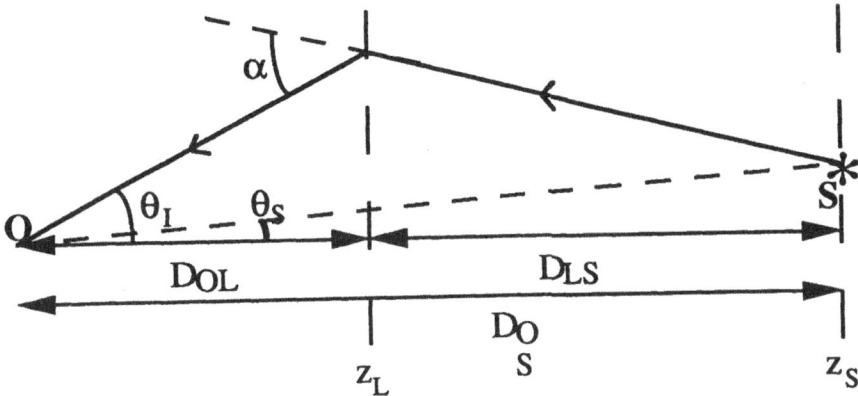

Fig. 1. Geometry of the problem in gravitational lensing. The lens is supposed to be thin and transparent, and the lengths are calculated in the small angle approximation.

$\Psi(\theta_I)$ is the two-dimensional potential, satisfying the 2-D Poisson equation: $\Delta\Psi = 2\Sigma/\Sigma_{cr}$. Σ is the projected surface density of matter, and Σ_{cr} is the critical surface density defined as

$$\Sigma_{cr} = \frac{c^2}{4\pi G}\frac{D_{OS}}{D_{OL}D_{LS}}.$$

Physically, Σ_{cr} can be interpreted as the surface density of matter of a uniform sheet which is just able to focus the beam from the source to the observer, with an infinite magnification.

This gives the expression of the reduced time-delay, for a given position of the source θ_S and for a position of the image θ_I:

$$\tau(\theta_S, \theta_I) = \frac{1}{2}(\theta_I - \theta_S)^2 - \Psi(\theta_I).$$

Fermat's Principle means that, for a given position of the source, the real position of the image(s) satisfies the equation $\partial\tau/\partial\theta_I = 0$, which expresses in a vectorial form:

$$\theta_S = \theta_I - \nabla_I\Psi(\theta_I) = \theta_I - \alpha(\theta_I).$$

α is the bending angle of the lens, and this equation is the "lens equation", which was introduced directly by Bourassa and Kantowski (1975) in its vectorial form (or equivalently with complex functions). It is important to note that both formalisms are equivalent, even if they have some different physical interpretations. Blandford and Narayan (1986) interpreted the basic lens equation as a mapping of the source plane on the lens plane, and they used the catastrophe theory to present a full understanding of the formation of multiple images. Finally, a full generalisation of Fermat's principle was also introduced by Kovner (1990), useful in the case of large scale inhomogeneities.

202

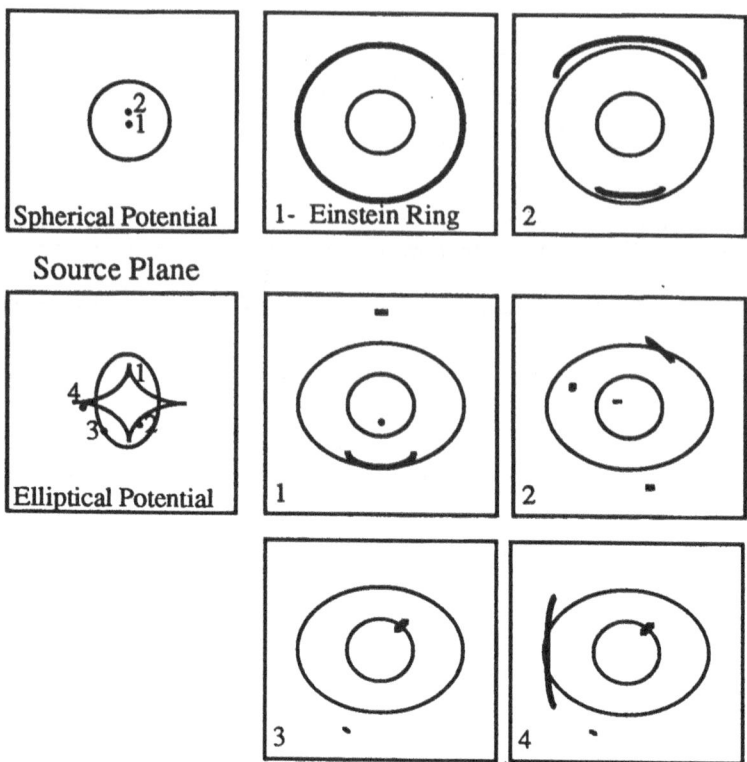

Source Plane

Fig. 2. The classification of images: the different types of images expected for a spherically symetric lens are displayed on the first line. Source 1 is on the axis of the lens and forms an "Einstein ring". Source 2 is slightly off-centred: two arcs are formed. The second line corresponds to a more general elliptical lens. Source 1 lies in a "cusp": 5 images are created, with 3 merging. Source 2 is on a "fold": from the 5 images, only 2 are merging. Source 3 forms radial images. Source 4 is on a "naked cusp" and creates only 3 merging images (from Grossman and Narayan, 1988).

Some of the properties of gravitational lensing are direct consequences of this lens equation:

— gravitational lensing is achromatic, in first order approximation.
— surface brightness is conserved (Etherington 1933).
— the magnification matrix is defined as $M_{ij}^{-1} = (\partial\theta_{Si}/\partial\theta_{Ij})$, so the "light amplification" corresponds to a "magnification" of the source, as well as a "distortion" of the image. The magnification factor is the inverse of the Jacobian of the transformation.
— magnification goes to infinity for images located on *critical lines* and sources on *caustics*.
— if the deflecting potential is non-singular, there must be an *odd* number of images (Burke 1981).

—   in the simple case of an isothermal sphere, the deflecting angle has a simple analytic expression: $\alpha(\theta_I) = 4\pi(\sigma^2/c^2)$, where σ is the line-of-sight velocity dispersion. This justifies the extensive use of this potential in many models of lenses.

An excellent review of these theoretical aspects of gravitational lensing was published by Blandford and Kochanek (1987). Also in Grossman and Narayan (1988) an interesting classification of the images is presented, illustrated in Figure 2. An entire and very good book about gravitational lensing is in press (Schneider, Ehlers, Falco 1991) which extensively describes all our knowledge about lensing, from both a theoretical and an observational point of view, including multiple quasars, arcs and arclets.

3. The giant arcs

There is no precise definition of what a "giant arc" is, mainly because of the large phenomenon's variety. The only characteristics which can have some meaning are geometrical, in the sense that the giant arcs are structures longer than 10-15 arcsec, with a width of 1.5" or less (most of them being unresolved, i.e. having a width smaller than the point spread function or seeing) lying in rich clusters of galaxies of intermediate redshift. Most of them have been observed in the optical range, with photometry and spectroscopy for the redshift determination. Some attempts to measure their near-IR fluxes are also in progress (Aragon-Salamanca and Ellis 1990). The scientific interests of studying giant arcs are multiple:

—   It is a powerful method of looking at very distant field galaxies, using the clusters of galaxies as gravitational telescopes. Multi-color photometry, coupled with a redshift measurement, are useful tools for studying the history of formation of galaxies at high redshift ($z \geq 2z_{cluster}$), especially because the selection effects are minimal in this sample.

—   The modeling of arcs is also an original way of getting some constraints on the cluster potential. The questions of the location of the dark matter in clusters, its distribution among or outside individual galaxies, clumps of matter, or the overall cluster can be addressed. But, as in the other techniques for the determination of masses, the problem is not simple, and is generally dependent on a large number of free or poorly constrained parameters. In any case, the redshifts of both the cluster (the lens) and the arc (the source) are important parameters for setting the geometric scales of the problem.

3.1. THE "CLUSTER ZOO"

Table I is a summary of what is currently known on the giant arcs, including their redshift when available, and the basic references. The reader should also refer to Figures 3 and 4 where most of the giant arcs are displayed, in order to see the variety of such a phenomenon!

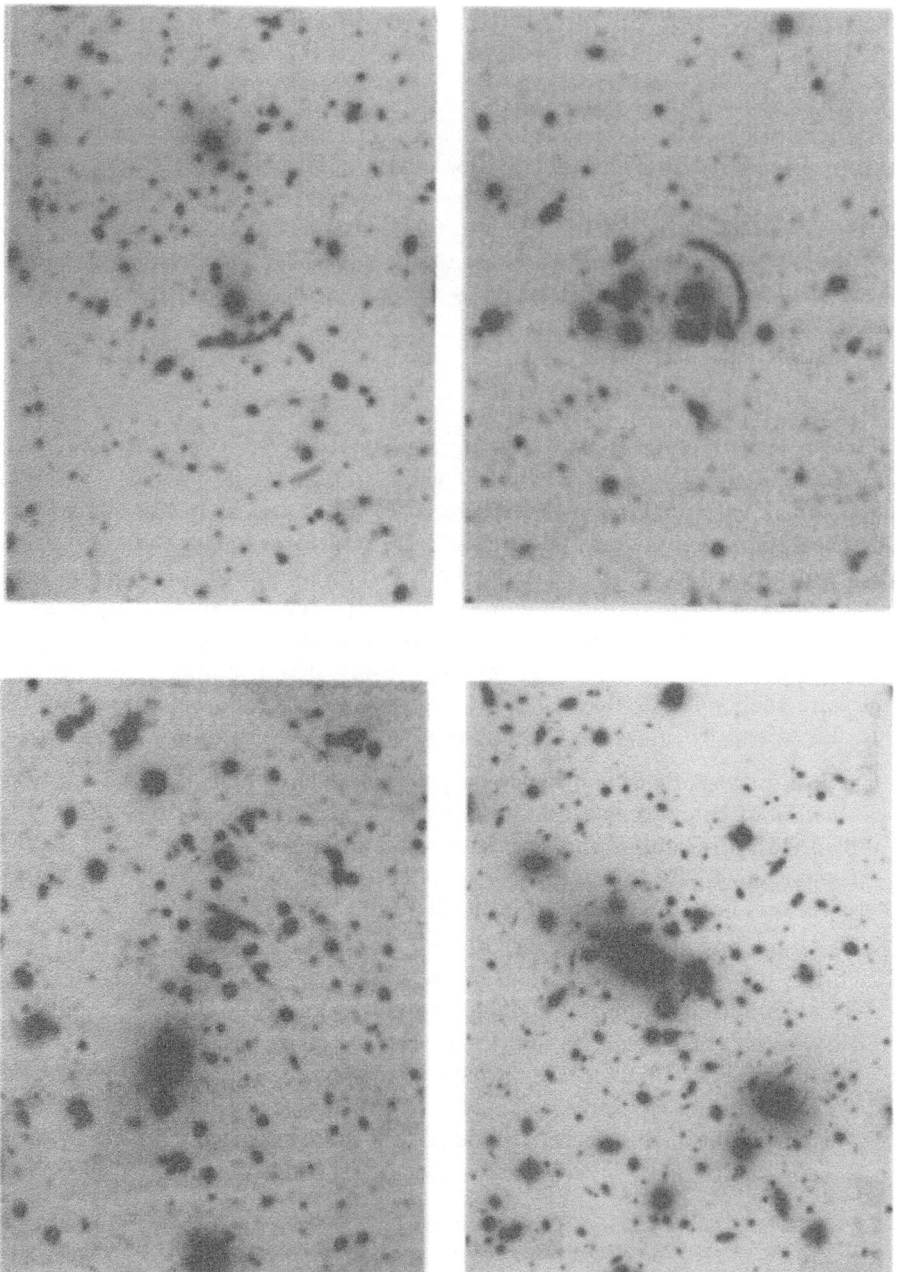

Fig. 3. Some examples of the most spectacular arcs. top left: the giant arc in Abell 370. Some arclets are visible, and A5 is on the bottom right of the image. (Fort et al., CFHT). top right: the circular arc in Cl2244–02 and some arclets (Fort et al., CFHT). bottom left: the straight arc in Abell 2390 (Soucail, Tyson, KPNO). bottom right: the arclets in Abell 2218 (Pello, Calar Alto, Spain).

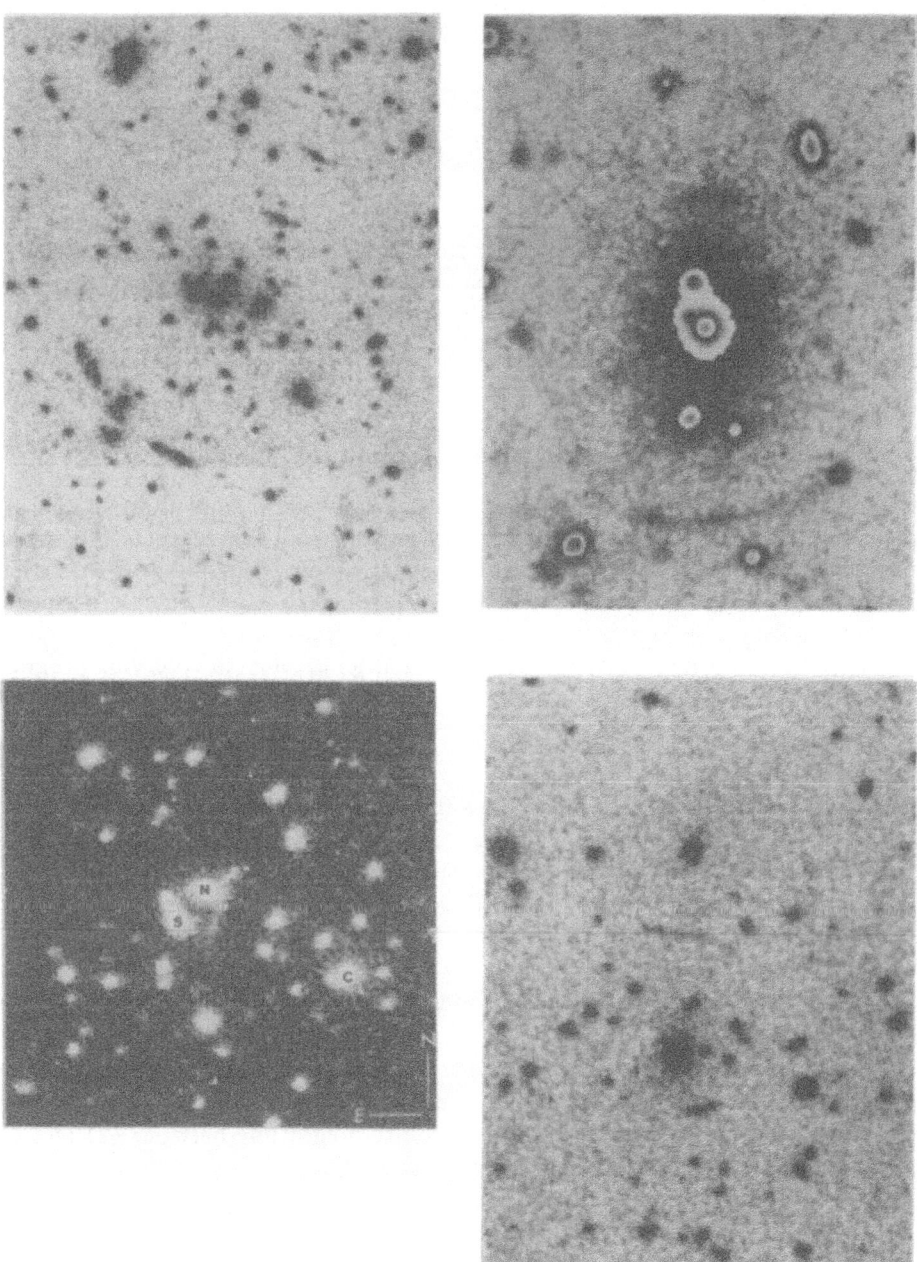

Fig. 4. Some examples of the most spectacular arcs (continued). top left: the splitted arc in Cl0024+17 (Fort et al., CFHT). top right: the double arc in Abell 963 (Lavery and Henry, 1988). bottom left: the arc in Cl0500–24 (Giraud, 1988). bottom right: two arcs in Cl0302+17 (Fort et al., CFHT).

TABLE I

Main properties of the giant arcs

Cluster	z_L	z_S	References
A370 (A0)	0.374	0.725	Soucail et al. (1988)
A370 (A5)		1.307 (?)	Mellier et al. (1991)
Cl 2244–02	0.331	2.237	Mellier et al. (1991), Lynds and Petrosian (1989)
A2390	0.231	0.913	Pello et al. (1991)
A963	0.206	0.77	Ellis et al. (1991)
Cl 0500-24	0.316	0.913	Giraud (1988, private communication)
Cl 0024+17	0.391	>1	Koo (1988), Mellier et al. (1991)
Cl 0302+17	0.42	—	Fort et al. (1991, in preparation)
A2218	0.171	0.702	Pello et al. (1988, 1991b in preparation)

3.2. SPECTROSCOPY AND REDSHIFT DETERMINATION

It was one of the observational challenges of recent years, since most of the arcs have a very low surface brightness, generally as faint as a few percents of the dark sky brightness. Nevertheless, low resolution spectroscopy combined with very long exposure times was successful in some cases, leading to a new sample of spectra of distant field galaxies. I will emphasise some examples, which illustrate both the difficulties of the measurement, and the potentialities of such an observing program.

3.2.1. The giant arc in Cl 2244-02. This arc, first detected by Lynds and Petrosian (1986), has been intensively observed by our group and others, to derive its redshift (Miller and Goodrich 1988, Lynds and Petrosian 1989, Mellier et al. 1991). The detection of an emission line at 3940 Å from data obtained at ESO (Chile) and CFHT after more than 10 hours of integration time, led us to identify it with Ly α. This identification was reinforced by the detection of some other absorption features such as CIV (1549 Å) and SiII (1304 Å). This arc has a redshift of 2.237, and is the distorted image of the most distant field galaxy currently identified (Figure 5).

Some properties of the source galaxy were also extracted from the spectroscopic data (Mellier et al. 1991):

- Corrected roughly from the magnification factor, the intrinsic magnitude of the galaxy is $B = 24.7 \pm 0.5$ and its surface brightness is about $\mu_B = 25.2$ mag arcsec^{-2}. Thus, this galaxy has an absolute magnitude between –21 and –22 (for $H_0 = 50$ km/s/Mpc), depending on the value of q_0 and on the k-correction that needs to be added. Its surface brightness would also be around 21.4, in the absence of any k-correction. These values are characteristic of rather bright galaxies in our neighbourhood, but not ultra-bright, compared with the high redshift radio-galaxies (Chambers et al. 1990).

- The equivalent width of the emission line identified as Ly α is 40 Å, corresponding to 12 Å in the rest frame. This gives an absolute Ly α luminosity of $1.5 \, 10^{42} \, ergs^{-1}$ and an estimate of the Star Formation Rate in the galaxy around $10 h_{50}^{-2} \, M_\odot yr^{-1}$.

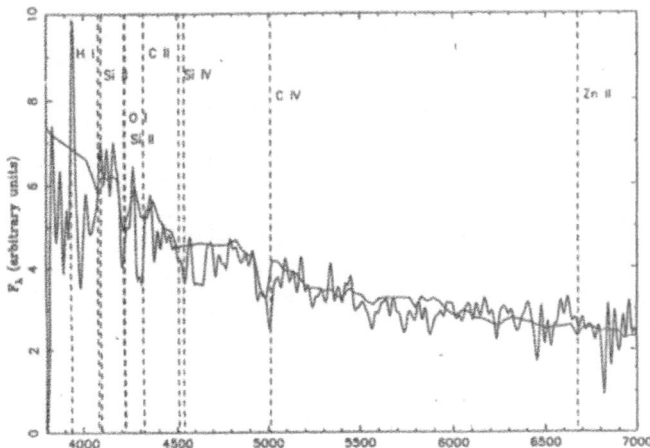

Fig. 5. Flux calibrated spectrum of the giant arc in Cl2244–02 obtained at ESO. A synthetic spectrum of a star-forming galaxy redshifted at $z = 2.237$ is superimposed, and some of the best identified lines are superimposed (from Mellier et al. 1991).

–   Note finally that the continuum of the spectrum observed in the optical range is indicative of a large fraction of very young stars, but there is no strong evidence for any nuclear activity in this galaxy.

3.2.2. The straight arc in Abell 2390.   This arc is peculiar by its geometry which presents an extremely high curvature radius, or equivalently a straight shape. The redshift, measured by Pello et al. (1991) at the WHT (La Palma), is 0.913, from the identification of the [OII] 3727 Å emission line and also consistent with the 4000 Å break, marginally detected. Note that the color index B–R=1.93 does not correspond to a very blue structure, even if it appears bluer than the galaxies of the cluster. Finally, the most striking result is the detection of a velocity gradient along the arc, corresponding to a velocity difference of 400 km s^{-1}. This gradient is either the signature of an internal rotation (from a disk?) of the source galaxy or is due to a close pair of interacting galaxies that are both lensed. It was used (with the first hypothesis) to attempt to derive a value of H_0 from the Tully-Fisher relation (Soucail and Fort 1991).

3.2.3. The arclet A5 in Abell 370.   This object is not exactly a giant arc, and will be discussed in the next section. In any case, its surface brightness is large enough to allow some spectroscopic measurements. The final spectrum which was obtained at ESO is flat, with some absorption features, but with no emission-line. The redshift identification which was proposed in Mellier et al. (1991) is based on the identification of 3 among 5 absorption lines: $z = 1.307$ seems presently the most probable redshift for this strange object, which also seems to be an extremely blue galaxy in the near IR (Aragon-Salamanca and Ellis 1990).

TABLE II

Available spectrophotometric data on the giant arcs. The intrinsic magnitude is roughly
evaluated by the ratio of the surface of the arc to the surface of a distant galaxy.

Cluster	B	B–R	μ_B	$B_{intrinsic}$	z_S
A370 (A0)	21.1	1.97	24.6	23.8	0.725
A370 (A5)	22.3	1.05	25.4	24.5	1.307
Cl 2244–02	21.4	0.75	25.3	24.7	2.237
A2390	21.9	1.93	25.3	24.6	0.913
A963 (North)	23.6	0.5	25.5	25.1	0.77
Cl 0024+17 (arc 3)	23.5	0.7	25	24.5	> 1
A2218 (arc H)	23.5	2.2	25	25	0.702

3.2.4. A new sample of very distant field galaxies ?. Table II sumarises most of
the spectrophotometric data about the arcs with an available redshift. These arcs
form a sample of spectra of field galaxies which by definition lie at a redshift larger
than 0.7 (in order to have a strong enough lensing effect from the clusters). This
bias is, on the one hand, very useful because it very easily selects high redshift field
galaxies, without any other strong selection biases, but on the other hand, it only
gives the high redshift end of the distribution of the galaxies within magnitudes
in the range $B = 24 - 25$. IR observations of most of them are also in progress in
collaboration with R. Ellis. The far-red stellar content is indicative of the oldest
stellar population, and consequently of the age of the galaxy. So this sample of
distant field galaxies must be increased to include its statistical properties in the
framework of galaxy formation and evolution (Figure 6).

3.3. SOME MODELINGS OF THE GIANT ARCS

The main goal when modeling a given arc is to derive some constraints on the gravitational potential of the deflector. Although models are useful for understanding
the overall shape of the cluster, they are generally limited when one is interested
in the influence of more local inhomogeneities of matter such as galaxies. Therefore
the modeling begins with the choice of a global potential function $\Phi_{2D}(x, y)$. Then
one has to solve the equation of the problem

$$\theta_S = \theta_I - \nabla_I \Phi(\theta_I)$$

for a grid of points in the image plane.

With this equation, one can reconstruct the source of the arc, and search for
additional images of the source. If the shape or size of the source does not match
the expected requirements, or if additional images which are undetected in the
real frames, appear in the model, one has to reject the model, or to modify the
parameters of the potential.

What are these parameters? Generally, a minimum number of 6 is required to
characterise a potential such as:

– the position of the centre of the potential (x_c, y_c).

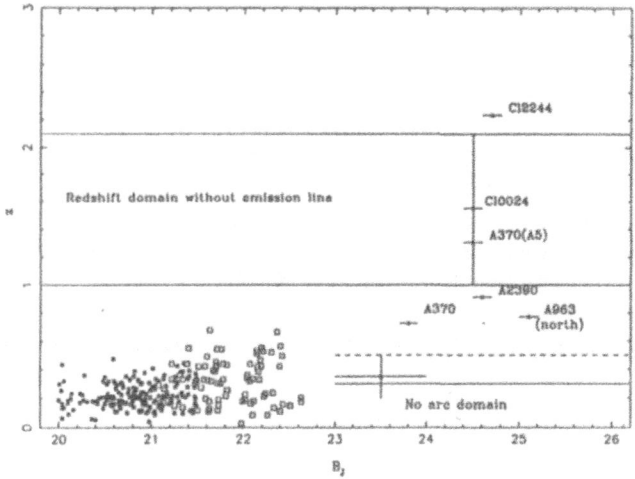

Fig. 6. Redshift-magnitude diagram derived from Table II. Data from two spectroscopic surveys are also plotted: Broadhurst et al. (1988, filled squares) and Colless et al. (1990, open squares). The cross represents the mean redshift of the sample from Lilly et al. (1991).

   –   an anisotropic term fixed by an ellipticity ϵ and a position angle θ (or a quadrupole term in the potential development).

   –   2 parameters to characterise the potential shape like a core radius r_c and the "strength" of the lens $f = \Sigma(0)/\Sigma_{cr}$.

One of the most useful potentials is the well-known elliptical pseudo-isothermal potential whose analytic form is:

$$\Phi_{2D}(x,y) = 4\pi \frac{\sigma^2}{c^2} \frac{D_{LS}}{D_{OS}} \sqrt{r_c^2 + (1 - \epsilon)x^2 + (1 + \epsilon)y^2}$$

where σ is the velocity dispersion of the cluster, also a characteristic of the strength of the lens.

3.3.1. Some models of A370. As this is the most observed cluster-lens, it is also the one in which most of the models were studied, especially after the first discovery of the giant arc, when the lensing hypothesis was not yet confirmed. Here, I will present only 3 models among those which were proposed, in order to compare their properties and differences.

   –   The first model, presented by Hammer (1987), is a multi point mass model. The mass of the galaxies is derived from their luminosity with a given M/L ratio. An additional mass is centred on the centre of the potential to represent the cluster dark component. The best fit of the arc is obtained with a total mass calculated inside a circle defined by the arc and centred on the cluster centre is then $2.3\,10^{14} M_\odot$. A significant fraction of this mass must also lie in the two giant galaxies which dominate the cluster.

– Another model, published by Grossman and Narayan (1989), includes a dark cluster component modeled by a nearly-isothermal sphere, with an elliptical term and a core radius r_c. The individual galaxies are represented by truncated isothermal spheres (weighted by their luminosity through the Faber-Jackson relation). Their main result is that the dark matter component must have a strong ellipticity (as large as 0.6) corresponding to an axis ratio of about 2. The best model has also a small core radius of 10 arcsec or 62 kpc, and the total mass inside the arc is $9\,10^{13} M_\odot$.

– Finally Mellier et al. (1990) tried an alternative in the morphological type of the source galaxy and they introduced a spiral galaxy to better reproduce the unusual shape of the giant arc (enlargement of one extremity and change of curvature). The model has two dark components (nearly isothermal potentials) more or less centred on the two giant galaxies plus the additional individual galaxies located along the arc. This kind of model is poorly constrained except that it must create a "naked cusp" in the source plane (Figure 2), and the spiral source must lie outside but near this cusp. In that case, the lens creates a stretching of the source image which is quite compatible with the shape of the arc.

Without describing them in detail, I would also cite other authors who developed some models for A370 and/or for other clusters: Kovner (1988,1989), Narasimha and Chitre (1988), Hammer and Rigaut (1989), Bergmann et al. (1990), Wambganss et al. (1989). In conclusion, all the models presented here have some common properties, such as the mass measured inside the radius of the arc: $1 - 2\,10^{14} M_\odot$ which can be immediately evaluated, assuming that the radius of the arc is the Einstein radius:

$$R_E = \sqrt{\frac{4GM}{c^2}\frac{D_{OL}D_{LS}}{D_{OS}}}$$

The mass-to-light ratio M/L_R is around 100, a value quite compatible with the dynamical value measured by Mellier et al. (1988) from optical data and redshift measurements: $M/L_R \simeq 60$. This indicates that most of the mass seen along this line of sight is located in the cluster A370, and that there is very little dark matter outside the cluster at other redshifts. Note anyway that gravitational lensing is sensitive, to a first order, only to the overdensities of matter along the line of sight!

Another result from the models is the necessity of including a dark component outside the galaxies, with a smooth distribution and a strong ellipticity (axis ratio of 1.5 to 2). Several authors have tested that it is not possible to reproduce the smooth shape of the arc if one puts all the mass in the galaxies, whatever the choice of the mass distribution. In particular, it is striking to note that at least 5 galaxies are located less than 5 arcsec in projection from the giant arc, and none of them strongly disturb the image of the arc.

But it is disappointing to see the number of models presented, and the difficulties of comparing them because most of them contain too many free parameters, and the shape of the potential is poorly constrained. A new analysis of what could be derived from both the giant arcs and the arclets is being investigated by various authors (see the contributions of Miralda-Escudé, Kochanek, Kneib in the proceedings of the Hamburg Conference on Gravitational Lenses, September 1991).

3.3.2. Modeling a "straight arc" in A2390. The main morphological properties of the giant arc in A2390, which have to be reproduced in any model are: a length of 15 arcsec, a width of 1.3 arcsec (the arc is resolved on high resolution images, Pello et al. 1991), and a nearly infinite curvature radius (i.e. larger than 10 arcmin !). Moreover, any model must be compatible with the velocity gradient observed along the arc.

Two groups have been working on this problem, but none of them were able to find any solution with a single dominant potential centred on the cluster centre, even with the addition of individual galaxies! In order to try to solve the problem, they had to include some kind of dark matter, not well related to the visible matter. In fact, a straight arc such as this one is a typical consequence of what is called a "marginal lens" (Kovner 1987), i.e. a cluster lens with parameters very close to critical values. In that case, large amplifications are possible without any splitting of the images, and the distribution of the light can create some very extended images.

— Pello et al. (1991) propose adding a clump of dark matter on the other side of the arc, in order to enlarge its curvature radius. It is justified by the fact that on the other side of the arc, there exists an overdensity of galaxies, although it is rather far away and not very rich. If this clump is responsible for the straightness of the arc, it should have a M/L ratio much larger than the global one for the cluster! On the other hand, the velocity gradient is explained in this case by a source disk-galaxy seen nearly edge-on and modeled by an elliptical contour.

— Kassiola, Kovner and Blandford (submitted) propose a new kind of cylindrical dark matter component. Correctly oriented, this potential should be able to create straight images without any fine tuning of the model. Moreover, in order to explain the velocity gradient, they use another hypothesis: the arc should be the image of two interacting galaxies close together near the caustics of the lens.

It is probably the most striking case in which gravitational lensing shows evidence for a dark matter distribution which does not follow the light distribution. It should be of particularly great interest to get a high resolution (smaller than 10 arcsec) X-ray map of this cluster, which is already known to be a strong X-ray emitter. This would add reliable information on the matter distribution in the cluster, especially in the cluster core. Also, ultra-deep imaging would give information on the distribution of arclets (see next Section), and consequently on dark matter distribution. Images with high resolution in many colors will also constrain the models, and possibly allow a reconstruction of the source, with a method similar to the one developed by Kochanek et al. (1989) for the radio-ring MG1131+0456.

4. The arclets

Again, there have been several tentative definitions of an arclet, but nothing very formal. I would propose defining an arclet as an elongated object, tangentially oriented with respect to the cluster centre, with an axis ratio $a/b > 2$. Moreover, arclets are most often faint and blue objects, i.e. $B > 24$ and $B - R < 1.5$. They have been immediately interpreted as gravitationally distorted images of the faint background population of galaxies (Fort et al. 1988, Tyson et al. 1990).

The best examples of such arclets are found in the clusters A1689 (Tyson et al. 1990) and A370, (Fort et al. 1988, 1991 in preparation) where more than 50 of them were detected. The statistics on these arclets in one cluster (position, distortion, orientation) is in principle a direct characteristic of the mass distribution in the cluster, provided one is able to invert the problem of gravitational lensing. Assuming that there exists a large population of faint background galaxies, with a uniform distribution, they form a distorted grid through the cluster of galaxies, and the distortion map is a direct "image" of the distribution of dark matter. Moreover, the sample of arclets found in clusters belongs to the background population, and the study of its properties, such as the distribution in color and magnitude or the number counts, will provide one of the best tools for understanding the evolution and the redshift distribution of the faint galaxies.

4.1. HOW TO OBSERVE ARCLETS AND WHERE?

In order to obtain a "good" statistics, one must detect a number of arclets as large as possible, i.e. lower the detection level at $\mu_B \geq 28$ mag arcsec^{-2}, knowing that at a limiting surface brightness of $\mu_B \geq 29$ mag arcsec^{-2} there are over 30 faint galaxies per arcmin^2 and per magnitude interval (Tyson 1988). Therefore techniques of ultra-deep photometry similar to those developed by Tyson. This necessarily implies very long total exposure times for each field (at least two filters are necessary for color information). As an example, it is interesting to note that at the CFHT Prime Focus with the RCA CCD, a pixel size of 0.41"/pixel and an integration time of 1 hour in the B-band, one reaches a detection level of $\mu_B = 26.3$ mag/"^{2} at a 3σ level (or equivalently with a S/N ratio of 3). To reach a detection level of 28 mag/"^{2} in the same conditions one needs a total integration time of 5 hours ! The survey of a large sample of clusters of galaxies, necessary to better understand the phenomenon, its occurence and its consequences, is consequently extremely demanding in telescope time. A large international collaboration was organised in order to coordinate observations of ultra-deep imaging in rich clusters. This includes the Toulouse Observatory (Fort, Le Borgne, Mathez, Mellier, Picat, Soucail), Bell Labs (Tyson, Bernstein), Princeton University (Turner, Guhathakurta, Miralda-Escudé), the University of Barcelona (Pello, Sanahuja) and also the University of Durham (Ellis, Aragon-Salamanca, Smail) mainly for IR data.

Moreover it is essential to observe in very good seeing conditions, better than 0.8" and with a sampling on the CCD of about 0.2–0.3 arcsec/pixel, in order to optimise both the detection and the separation between arclets and faint cluster galaxies. It is also suggestive that the background galaxies which are lensed by the

TABLE III
Temporary list of clusters in which arclets are detected

Cluster	Redshift	References	Comments
A1689	0.18	Tyson et al. (1988)	50 arclets
3C295	0.46	Tyson et al. (1988)	15–20 arclets
A2218	0.171	Pello et al. (1988, 1991 in preparation)	20–30 arclets
A370	0.374	Fort et al. (1988, 1991 in preparation)	60 arclets
Cl2244−02	0.329	Fort et al. (1991, in preparation)	20 arclets
A2390	0.231	Tyson, Soucail, Pello, in progress	20–30 arclets
Cl0024+17	0.39	Tyson, Turner, Fort, in progress	
A483	0.29	Picat, Mellier, Tyson, in progress	
A2397	0.224	Smail et al. (1991)	
A1942	0.22	Smail et al. (1991), Picat et al. in progress	
A2163	0.19	Soucail, Picat, in progress	
Cl0016+16	0.54	Smail et al. (1991)	< 10 arclets!

clusters have a small angular size, probably smaller than 0.5".

Table III lists the clusters in which at least "some" arclets have already been detected. Obviously this list is not exhaustive and probably more clusters could be added. But it is not always easy to compile different kinds of data observed by different observers on different sites !

One should also address the question of the number of arclets expected behind a given cluster. The starting point is the number of faint galaxies found in empty fields by Tyson, which is about 130 galaxies per square arcminute in the B magnitude range 23–27. Nobody currently knows what the redshift distribution of these galaxies is, and what fraction is at redshift larger than 0.7, a typical value above which lensing effects are sensible. On the contrary, studying the occurrence of arclets in a set of clusters similar in mass and richness, with a redshift distribution ranging from 0.15 to 0.5, would give interesting constraints: detecting a decrease in the number of arclets above a cluster redshift will give a limit on the maximum redshift of the background population. Up to now, the sample of well studied clusters is too small to derive any conclusions, but there are some suppositions that the background galaxies are hardly at a redshift larger than 1 or 1.3 (Fort, private communication). Note that another limit on the redshift of the background population was observed by Guhathakurta et al. (1990): from U deep imaging, they conclude that there least 93 % of the galaxies are below a redshift of 3.

The number of detectable arclets also strongly depends on the strength of the lens related to the dynamical state of the cluster. We can easily evaluate its dependence with the velocity dispersion of the cluster , assuming that the potential is represented by a singular isothermal sphere. In that case, the Einstein radius scales as

$$R_E = 4\pi \frac{\sigma_{los}^2}{c^2} \frac{D_{LS}}{D_{OS}}$$

and the ratio D_{LS}/D_{OS} does not depend strongly on the distance of the source provided this one is large enough (say at a redshift larger than 0.8). The number of arclets in a cluster is roughly proportional to the square of the Einstein radius, as most of them are located at a distance from the centre ranging from R_E to $2R_E$, and scales as:

$$N \propto \sigma_{los}^4$$

This strong dependence on the velocity dispersion favours the richest clusters and those with a high X-ray luminosity, in a redshift range of 0.15–0.5. At larger redshift, evolutionary effects become sensible, with probably a decrease of the velocity dispersion (a less evolved dynamical state for the cluster). Moreover, within the framework of Cold Dark Matter cosmology, Frenk et al. (1990) claimed that clusters with a velocity dispersion larger than 1500 km/s would be due to superimposition and selection effects. Good statistics of the occurrence of giant arcs and/or arclets in clusters will allow to test this hypothesis.

4.2. STATISTICAL LENSING: MAPPING THE DARK MATTER IN CLUSTERS

The pioneer work was done by Tyson et al. (1990) who first observed a large enough number of arclets in A1689 to introduce some statistics in their approach, and by Kochanek (1990) and Miralda-Escudé (1991a) who introduced new formulations of the problem. The questions adressed are:

— What is the relation between the distribution of arclets (shape, ellipticities) and the distribution of matter in clusters?
— How does it depend on other parameters such as the distribution in redshift of the sources, or their distribution in ellipticities?
— How does one optimise the observing conditions for such a study?
— How does one relate it to the light distribution, or other estimators of the mass distribution such as the X-ray profile ?

Different parameters of the cluster potential can be evaluated, the simplest one being the location of the centre of mass. Indications about the radial profile or the ellipticity of the potential can be tentatively evaluated. Also an estimate of the velocity dispersion can be given, but it needs to be compared with optical measurements.

4.2.1. Determination of the cluster centre. This point was first developed by Tyson et al. (1990) and Mellier et al. (1990) who introduced the notion of "distortion map": for a given point (X,Y) in the CCD image, one can calculate what they call "the net tangential alignment" defined as

$$\langle \frac{I_{TT} - I_{RR}}{I_{TT} + I_{RR}} \rangle_{blue\,objects}(X,Y)$$

where I_{TT} is the second order intensity weighted tangential moment and I_{RR} is the second order intensity weighted radial moment, calculated for each object bluer than a given value (for example, $B - R < 1.5$) and fainter than typically $B = 23$, with respect to the point (X,Y) set as the centre. The net tangential alignment is finally normalised with a random distribution of ellipticities for a similar sample of objects. The result of this calculation is a map identified with the distortion map, which characterises the distribution of the distorted objects. The maximum intensity in this map gives the location of the centre of the deflector, which can be determined with an accuracy of 10 arcsec or even better, mainly to the order of the average separation between the sources (Tyson and Seitzer 1988, Miralda-Escudé 1991a).

4.2.2. Constraints on the mass distribution. Some theoretical approaches were developed in the last two years to try to extract the best contraints on the deflecting potential from the distribution of arclets. Up to now, it has mainly been a theoretical analysis, as the application to real data is still a challenge. Kochanek (1990) and Miralda-Escudé (1991a) have both investigated how to relate the distribution of observed ellipticities and orientations of the arclets to the true distribution of shapes of the sources without distortion. The relation between the two is a function of the cluster potential, which can be determined with an accuracy depending on the number of arclets used in the statistics. Both authors demonstrate that one can find some statistical functions which do not depend too strongly on the unknown shape of the sources but which can constrain rather satisfactorily both the velocity dispersion of the cluster and the outer density profile (estimation of the cut-off radius). The problem is mainly limited by the statistical noise introduced by the averages on the distribution in ellipticities and redshift of the sources.

Instead of giving global properties for the cluster potential, Mellier et al. (1991) tried to directly extract some information on the radial profile, by taking their averages in radial bins from the cluster centre. With a large number of arclets per bin and enough bins one should, in principle, recover the radial profile of the potential with a resolution similar to that of the best X-ray maps. This approach seems powerful as it does not make any a-priori assumption on the potential, but is limited mainly by poor statistics. In any case, this work is still in progress, in order to compare the distribution of the dark matter with respect to the luminous matter (from which one measures the velocity dispersion).

In all cases, the difficulties are tied to the relatively small number of arclets. First, near the centre, the number of arclets is limited because most of them are deviated far enough, so it is difficult to constrain the core radius. Moreover, their detection is even harder because of the crowding effects, more sensible where the density of cluster galaxies is higher.

Second, far from the centre, the distortions are weaker and their measurement are depends more on seeing conditions, which tend to circularise every image. Another opportunity to measure the local surface density in the cluster rather far from the centre was proposed by Kochanek and Blandford (1991) for the few suspected examples of optical rings. They show that these rings should be detectable in some cluster-lenses and that they could set locally the potential shape. Such examples of more or less complete optical rings have been detected in the cluster A2218 (Pello, private communication).

Finally, from an observational point of view, most of the clusters are not spherical, and the arclet distribution is sensitive to the ellipticity of the potential, adding new parameters in the attempt to model the cluster potential. So it will be difficult to map the matter in clusters from the study of arclets, independently of other data. On the contrary, it seems that "good" X-ray maps with a resolution of 50–100 kpc (10–20" for intermediate redshift clusters) will be useful as complementary ingredients in the study of mass distribution in clusters, especially to constrain the core radius or the mass profile near the centre. This program is typically feasible with the new generation of X-ray satellites (ROSAT, GINGA, XMM...) although it also needs some spectral information on the temperature distribution of the hot gas.

4.2.3. Constraints on the redshift distribution of the background sources. In the statistical approach, one of the first assumptions is to neglect the effect of the redshift distribution of the sources provided they are far enough behind the cluster ($z > 0.8$). Indeed this is a second order effect which appears through the ratio D_{LS}/D_{OS}. But when the cluster mass distribution is well understood, one should be able to distinguish between sources at redshift of 1 and 2 (for example, for a cluster redshift of 0.5, the ratio D_{LS}/D_{OS} differs by 50 % between a redshift of 1 to 2).

More promising are the photometric results recently obtained by B. Fort and collaborators (in preparation) on the color distribution of the arclets. From a sample of 60 arclets candidates detected in A370, they show that the color distribution is quite compatible with the one from a comparison empty field. Moreover, from B,R and I photometry, they derive a "photometric redshift" which is between 0.8 and 1.4 for most of them. This indicates that a significant fraction of the population of faint background galaxies may be in this redshift range, and only a few of them at a larger redshift, but this preliminary result needs to be confirmed with observations in similar conditions of other clusters of galaxies.

In any case, the determination of this redshift distribution or, more roughly, of the mean redshift versus magnitude is a fundamental clue related to the problem of galaxy formation, one of the most crucial points in observational cosmology in 1991!

5. Future prospects and conclusions

During the last few months there have emerged, more or less independently, in different places the idea that statistical lensing could be used to investigate the weak gravitational distortion of the background sources by large-scale structures (Blandford et al. 1991, Miralda-Escudé 1991b, Bartelmann and Schneider 1991).

Blandford et al. (1991) show for example that at an intermediate redshift ranging from 0.3 to 0.8, the effect of these structures (superclusters, voids, walls... up to a scale of 100 Mpc) on the background population creates a correlated mean ellipticity of a few percents over an angular scale of 0.5 to 1 degree. Assuming that the density of the background galaxies is as high as $3\,10^5$ per sq. degree at a magnitude level of 28, that most of the sources have a redshift of about 3 and an angular size of 2", they demonstrate that the effect should be detectable with a survey over a field of view of a few square degrees. But if, as supposed by several authors, their mean redshift is around 1 and/or their angular size is about 0.8"(Lilly et al. 1991), it would be more difficult to detect, because of the circularisation of the images by the seeing. In any case, with the development of new large-area CCDs, it is time now to examine this observing program which will probe the large-scale distribution of matter.

I could not finish this review without a word about the determinations of the cosmological parameters from gravitational lensing, which is essentially sensitive to the large-scale geometry of the universe. Several propositions have emerged, but nothing concrete has been measured yet. First of all, Paczynski and Kovner (1988) immediately studied the possibility of detecting Supernovae in the source of a giant arc. If the arc is multiple-imaged, the time-delay between the SN event in the different images is about a few weeks or days, and it is well known that time-delays are simply related to H_0, but no SN event have been detected in any arc yet! Another attempt to measure the Hubble contant was proposed by Soucail and Fort (1991) but the results are still controversial since they strongly depend on the source reconstruction. However, high resolution images of some new arclets could lead to morphological information on the sources, useful for such purpose.

In principle, it should be easy to constrain q_0, with a well-understood lens in which two large arcs are found, with their redshifts measured. The scaling factor D_{LS}/D_{OS} is a function of q_0 only, provided the redshifts are known. This could be applied to the case of A370 for A0 and A5, but in practise, the lens is complex, and the uncertainties in the modeling are much larger than the weak dependence of the scaling factor on q_0. But as soon as it will be better constrained by the distribution of arclets, getting a new determination of q_0 will be a challenge.

Gravitational lensing in clusters of galaxies is indeed a very important and useful tool for cosmology, as it simultaneously probes the distribution of matter (in clusters and superclusters), the very deep universe through the magnification effects, and the geometry of the universe. Both observers and theoreticians are beginning to assimilate all the potentialities of this new class of gravitational lenses and it is certain that this topic will progress quite rapidly in the next few years.

Acknowledgements

I wish to thank B. Fort, G. Mathez and Y. Mellier for their help and comments during the writing of this review, as well as my other collaborators in Toulouse, namely M. Cailloux, J.P. Kneib, J.F. Le Borgne, R. Pello, and J.P. Picat.

References

Aragon-Salamanca A., Ellis R.S.: 1990, in Mellier Y., Fort B. and Soucail G., ed(s)., *Toulouse Workshop on Gravitational Lenses, Lecture Notes in Physics*, p. 288, Springer
Bartelmann M., Schneider P.: 1991, *Astron. Astrophys.* **248**, 349
Bergmann A.G., Petrosian V., Lynds R.: 1990, *Astrophys. J.* **350**, 23
Blandford R.D., Kochanek C.S.: 1987, in Bahcall J., Piran T. and Weinberg S., ed(s)., *Dark Matter in the Universe*, p. 131, World Scientific
Blandford R.D., Narayan R.: 1986, *Astrophys. J.* **310**, 568
Blandford R.D., Kovner I.: 1988, *Phys. Rev. A* **38**, 4028

Blandford R.D., Saust A.B., Brainerd T.G., Villumsen J.V.: 1991, *M.N.R.A.S.* **251**, 600
Bourassa R.R., Kantowski R.: 1975, *Astrophys. J.* **195**, 13
Broadhurst T.J., Ellis R.S., Shanks T.: 1988, *M.N.R.A.S.* **235**, 827
Burke W.L.: 1981, *Astrophys. J.* **244**, L1
Chambers K.C., Miley G.K., van Breugel W.J.M.: 1990, *Astrophys. J.* **363**, 21
Colless M.M., Ellis R.S., Taylor K., Hook R.N.: 1990, *M.N.R.A.S.* **244**, 408
Ellis R.S., Allington-Smith J., Smail I.: 1991, *M.N.R.A.S.* **249**, 184
Etherington I.M.H.: 1933, *Phil. Mag.* **15**, 761
Fort B., Prieur J.L., Mathez G., Mellier Y., Soucail G.: 1988, *Astron. Astrophys.* **200**, L17
Frenk C.S., White S.D., Efstathiou G., Davis M.: 1990, *Astrophys. J.* **351**, 10
Giraud E.: 1988, *Astrophys. J.* **334**, L69
Grossman S., Narayan R.: 1988, *Astrophys. J.* **324**, L37
Grossman S., Narayan R.: 1989, *Astrophys. J.* **344**, 637
Guhathakurta P., Tyson J.A., Majewski S.R.: 1990, *Astrophys. J.* **357**, L9
Hammer F.: 1987, in Bergeron J. et al., ed(s)., *Third IAP Astrophys. Meeting on High Redshift Objects*, p.467, Editions Frontières
Hammer F., Rigaut F.: 1989, *Astron. Astrophys.* **226**, 45
Kochanek C.S., Blandford R.D., Lawrence C.R., Narayan R.: 1989, *M.N.R.A.S.* **238**, 43
Kochanek C.S.: 1990, *M.N.R.A.S.* **247**, 135
Kochanek C.S., Blandford R.D.: 1991, *Astrophys. J.* **375**, 492
Koo D.C.: 1987, in Rubin V.C. and Coyne G.V., ed(s)., *Large Scale Motions in the Universe*, p.513, Princeton University Press
Kovner I.: 1987, *Astrophys. J.* **321**, 686
Kovner I.: 1988, in Kaiser N. and Lasenby A.N., ed(s)., *The Post-Recombination Universe, NATO ASI*, p. 315, Kluwer Academic Publishers
Kovner I.: 1989, *Astrophys. J.* **337**, 621
Kovner I.: 1990, *Astrophys. J.* **351**, 114
Kovner I., Paczynski B.: 1988, *Astrophys. J.* **335**, L9
Lavery R.L., Henry J.P.: 1988, *Astrophys. J.* **329**, L21
Lilly S.J., Cowie L.L., Gardner J.P.: 1991, *Astrophys. J.* **369**, 79
Lynds R., Petrosian V.: 1986, *Bull. Am. Astron. Soc.* **18**, 1014
Lynds R., Petrosian V.: 1989, *Astrophys. J.* **336**, 1
Miralda-Escudé J.: 1991a, *Astrophys. J.* **370**, 1
Miralda-Escudé J.: 1991b, *Astrophys. J.* , in press
Mellier Y., Soucail G., Fort B., Mathez.: 1988, *Astron. Astrophys.* **199**, 13
Mellier Y., Soucail G., Fort B., Le Borgne J.F., Pello R.: 1990, in Mellier Y., Fort B. and Soucail G., ed(s)., *Toulouse Workshop on Gravitational Lenses, Lecture Notes in Physics*, p. 261, Springer
Mellier Y., Fort B., Soucail G., Mathez., Cailloux M.: 1991, *Astrophys. J.* , 20 October issue
Mellier Y., Longaretti P.Y., Kneib J.P.: 1991, , in preparation
Miller J.S., Goodrich R.W.: 1988, *Nature* **331**, 685
Narasimha D., Chitre S.M.: 1988, *Astrophys. J.* **332**, 75
Paczynski B.: 1987, *Nature* **325**, 572
Pello R., Soucail G., Sanahuja B., Mathez G., Ojero E.: 1988, *Astron. Astrophys.* **190**, L11
Pello R., Le Borgne J.F., Soucail G., Mellier Y., Sanahuja B.: 1991, *Astrophys. J.* **366**, 405
Schneider P.: 1985, *Astron. Astrophys* **143**, 413
Smail I., Ellis R.S., Fitchett M.J., Norgaard-Nielsen H.U., Hansen L., Jorgenson H.E.: 1991, *M.N.R.A.S.* **252**, 19
Soucail G., Fort B., Mellier Y., Picat J.P.: 1987, *Astron. Astrophys.* **172**, L14
Soucail G., Mellier Y., Fort B., Mathez G., Cailloux M.: 1988, *Astron. Astrophys.* **191**, L19
Soucail G., Fort B.: 1991, *Astron. Astrophys.* **243**, 23
Tyson J.A.: 1988, *Astron. J.* **96**, 1
Tyson J.A., Seitzer P.: 1988, *Astrophys. J.* **335**, 552
Tyson J.A., Valdes F., Wenk R.A.: 1990, *Astrophys. J.* **349**, L1
Valdes F., Tyson J.A., Jarvis J.F.: 1983, *Astrophys. J.* **271**, 431
Walsh D., Carswell R.F., Weymann R.: 1979, *Nature* **279**, 381
Wambganss J., Giraud E., Schneider P., Weiss A.: 1989, *Astrophys. J.* **337**, L73

THE SUNYAEV-ZELDOVICH EFFECT

A.N. Lasenby
Mullard Radio Astronomy Observatory
Cavendish Laboratory
Madingley Road
Cambridge CB3 0HE
U.K.

ABSTRACT. The differential scattering to higher energies that occurs when energy from thermal electrons is injected into a gas of photons is called the Sunyaev-Zeldovich effect. As applied to the effect of clusters of galaxies upon the microwave background radiation, it promises to be a very useful cosmological tool, and is already being used to set bounds upon H_0 which need to be taken seriously. This review discusses the physical basis of the effect, its role in elucidating cluster physics, its relationship to other microwave background anisotropies, and the prospects for improved measurements of both its amplitude and two-dimensional structure via a new generation of instruments.

1. Introduction

The Sunyaev-Zeldovich effect (Sunyaev & Zeldovich, 1972) is a perturbation to the spectrum of the microwave background as it passes through the hot gas of galaxy clusters. It is expected to be a very useful diagnostic of the physical conditions inside clusters of galaxies, and in conjunction with information from other wavelengths, particularly X-rays, it holds the promise of providing estimates of the Hubble constant H_0 independent of the systematic effects present in existing optical determinations (although with some further systematics of its own). A key feature of the Sunyaev-Zeldovich (henceforth SZ) effect, is that for a given line integral of pressure through the cluster gas, the effect is independent of redshift. Thus measurements of 'blank fields' can in principle tell us a great deal about the hot gas content of clusters at epochs early enough for clusters to cover a significant fraction of the sky — such epochs are of course extremely difficult to probe optically, but the SZ effect is transmitted to us with undiminished amplitude.

The structure of this review will be to begin with a brief heuristic derivation of the effect, in order to to get the underlying physics in place, and in particular to stress the frequency dependence of the effect and contrast it with other cluster pertubations of the microwave background and with the signature expected from primordial perturbations associated with the development of structure in the very early universe. Then the status of the observations will be discussed — in particular

A. C. Fabian (ed.), *Clusters and Superclusters of Galaxies*, 219–239.

why it is so hard to detect; the best results available to date and the prospects for the future. This is followed by a section on the methods of finding H_0 and preliminary results from this work, mainly carried out by Mark Birkinshaw and collaborators at CfA. The work currently being carried out at Cambridge, U.K., aims to increase rapidly the number of clusters for which accurate SZ measurements are available, and to try to reduce some of the systematics in H_0 determinations by averaging over several clusters. The blank field measurements already carried out by Caltech, and which we wish to pursue at Cambridge are then discussed. In particular, it emerges from recent work by Dick Bond that the effects of the cosmological bias parameter will be very important here. The prospects for the measurement of the effect in *emission* at mm wavelengths will be mentioned — potentially this could be one of the most fruitful observational areas for the future. An overall moral which should emerge is that the SZ effect (and other cluster anisotropies) are potentially very useful tools, but are still extremely difficult to measure. Only a *multifrequency* approach will allow a final unambiguous separation of the various competing effects, and making the observations with a variety of instruments, and at a variety of wavelengths is vital.

2. The underlying physics

To start in understanding the SZ effect, consider the injection of energy, via Compton scattering, from an electron gas into an isotropic photon gas. We shall suppose the electron gas is non-relativistic and with no bulk motions, and that the radiation temperature T_{rad} is $\ll$ the electron temperature T_e. (In clusters we typically expect $T_e \sim 10^7$ to 10^8 K). With these conditions it is not immediately obvious there should be any effect — if the scattering is elastic, then one can't change the basically isotropic distribution of photons or electrons. It is true there is a 'Doppler Scattering' effect

$$\frac{\Delta\nu}{\nu} = \frac{v_e}{c} \times \text{angular factor}$$

where v_e is the velocity of the colliding electron, but this will average to zero, since the electron velocites are isotropic. (We mention this formula however since it will tell us what happens when there *is* a bulk motion.) The SZ effect arises as follows (and I must give acknowledgements to Pierre Salati for this particular way of looking at it). The scattering mechanism is Thomson scattering, (see Fig. 1) for which we know

$$\frac{d\sigma}{d\Omega} = \sigma_T \frac{3}{16\pi}(1 + \cos^2\theta),$$

i.e. there will be maximum scattering in the foward and backward directions. We therefore approximate the problem as 1-dimensional and consider in the comoslogical (photon gas) frame the head-on collision of a photon with energy $h\nu$ with an electron with a velocity $v = \beta c$ directed away from the incoming photon. In the electron rest frame the photon energy is transformed to $h\nu((1-\beta)/(1+\beta))^{1/2}$ and there is an equal scattering probabilty in the forward and back directions. In these directions there is no frequency shift, from the familiar Compton formula

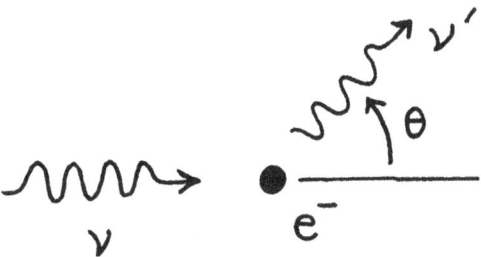

Fig. 1. Photon colliding with an electron initially at rest.

$$\lambda' - \lambda = 2\lambda_e \sin^2 \theta/2 \qquad (\lambda_e = \frac{h}{m_e c}).$$

Transforming back to the cosmological frame we find the photon has energy $h\nu(1 - \beta)/(1 + \beta)$ (i.e. *red*shifted) if its direction has been reversed in the encounter, or $h\nu$ if not. Now suppose the photon came from the other direction, so that initially in the photon's frame the electron velocity was directed towards it. In this case the same analysis yields a *blue* shift to an energy of $h\nu(1 + \beta)/(1 - \beta)$ if the photon direction is reversed and a constant energy of $h\nu$ if not. The mean frequency after collision for photons with momentum parallel to electron velocity is thus

$$\nu' = \nu \frac{1}{4} \left(2 + \frac{1 - \beta}{1 + \beta} + \frac{1 + \beta}{1 - \beta}\right) = \frac{\nu}{1 - \beta^2} \approx \nu(1 + \beta^2).$$

Thus there is a net second order effect.

What happens in the interval of time Δt? Restoring the 3-dimensional dependence upon scattering angle, the mean shift is

$$\langle \text{prob. of shift in } \Delta t \times (\beta \cos \theta)^2 \nu \rangle$$

averaged over the isotropic photon momentum distribution, i.e.

$$d\nu = \frac{1}{3} n_e \sigma_T c \Delta t \, \nu \frac{3kT_e}{m_e c^2}.$$

(Here $(n_e \sigma_T)^{-1}$ is the photon mean free path.) Thus

$$\frac{1}{\nu} \frac{d\nu}{dt} = \left(\sigma_T c n_e \frac{kT_e}{m_e c^2}\right).$$

The right hand side has units of $(\text{time})^{-1}$ and we write it as dy/dt. Integrating we obtain

$$y = \int_{t_i}^{t_f} \sigma_T c n_e \frac{kT_e}{m_e c^2}\, dt.$$

The quantity y is called the Comptonization parameter and here is worked out for the case of injection of energy from an initial time t_i to a final time t_f. Equating this to the time taken for a photon to traverse a cluster yields the final formula

$$y = \int \sigma_T n_e \frac{kT_e}{m_e c^2}\, dl$$

or $y = \frac{\sigma_T k}{m_e c^2} \times$ line integral of pressure through cluster. We comment straightaway that this dependence upon $n_e T_e$ promises to give complimentary information to that provied by X-ray emission, where the flux is determined by the integral of $n_e^2 T_e^{1/2}$. Now the equation

$$\frac{1}{\nu}\frac{d\nu}{dt} = \frac{dy}{dt} \qquad \text{implies} \qquad \nu(t) = \nu(0)\, e^y.$$

In the cluster case we have $\nu_{\text{out}} = \nu_{\text{in}} e^y$. This shift in frequency yields an apparent *deficit* in the Rayleigh-Jeans region of the CMB spectrum, and an increase in the Wien region (see Fig. 2). We have

$$\Delta I(\nu) = I\left(\frac{\nu}{e^y}\right) - I(\nu) \qquad \text{(generally)}$$

$$= 2\left(\frac{\nu}{e^y}\right)^2 \frac{kT_{\text{rad}}}{c^2} - \frac{2\nu^2 kT_{\text{rad}}}{c^2} \qquad \text{in the Rayleigh-Jeans region}$$

$$\approx I(\nu) \times (-2y) \qquad \text{for small } y.$$

For a rich cluster, then (taking the parameters of the Coma cluster as an example), $l \sim 3\,\text{Mpc}$, $n_e \sim 10^3\,\text{m}^{-3}$ and $T_e \sim 7 \times 10^7\,\text{K}$. Thus a typical y is 7×10^{-5}, making the assumption of its smallness in general well justified.

The result just quoted for the Rayleigh-Jeans (R-J) region is the correct one, as is the general idea of an decrement in the R-J region and an excess in the Wien region. However, our derivation has ignored several important physical factors, in particular that the photon energy $h\nu$ varies after collision in the electron frame; that the photons are not isotropic in the electron frame and that photons are bosons, the latter implying that there is stimulated emission also to consider. Putting all this together to find the correct equation yields the Kompaneets equation (Kompaneets, 1957). The general solution as a function of frequency, expressed in terms of R-J brightness temperature $(T_{\text{RJ}} = (\lambda^2/2k)I)$ is

$$\Delta T_{\text{RJ}} = \left(\frac{x}{e^x - 1}\right)^2 e^x \left(x \coth\frac{x}{2} - 4\right) y\, T_{\text{rad}},$$

where

$$y = \frac{k\sigma_T}{m_e c^2}\int n_e T_e\, dl,$$

as above and $x = h\nu/kT_{\text{rad}}$. This is what a heterodyne system would measure. If we were to measure *flux* rather than temperature, i.e. use a bolometer type system then

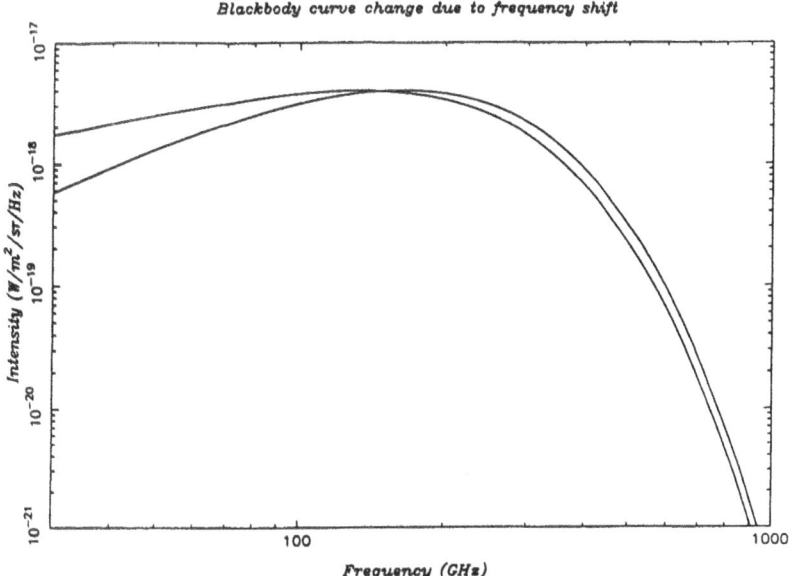

Fig. 2. The shift in the CMB spectrum caused by differential scattering to higher energies.

$$\Delta I \propto \nu^2 (\Delta T_{\mathrm{RJ}})_{\mathrm{equiv}} \propto x^2 \times \text{above.}$$

The importance of this is that the effect is thus much increased in the sub-mm region, where we are likely to be using bolometer techniques anyway. The relevant curves are shown in Figs. 3(a) & (b), which also show the same curves for a simple change in the black-body temperature of the CMB, such as would be relevant to primordial perturbations. (In general primordial perturbations, whether due to gravitational linkage (Sachs-Wolfe effect), doppler scattering of straight adiabatic compression all preserve the black-body nature of the CMB, merely shifting its apparent temperature.) One can see from Fig. 3(b) that observing the flux effect in three regions, centered at approximately 2mm, 1.2mm and 0.8mm say, would give a unique signature for the SZ effect, separating it quite distinctly from any anisotropies due to primordial pertubations. (Such an experiment, with simultaneous measurement at three wavelengths covering the SZ effect in emission, decrement and with a middle channel as atmospheric monitor, has been built by Chase, Joseph and Clements, and is currently being tested at Tirgo in the Swiss Alps, prior to deployment on the James Clerk Maxwell Telescope (JCMT), Hawaii.) Pursuing this point of different frequency dependencies further, we note that (as first pointed out by Sunyaev & Zeldovich, 1980; see also Rephaeli & Lahav, 1991) if a cluster has a systemic *radial* velocity then the bulk electron flow induces Doppler scattering with magnitude given by the bulk motion formula above

$$\frac{\Delta T}{T} \sim \frac{v_{\mathrm{pec}}}{c} \tau$$

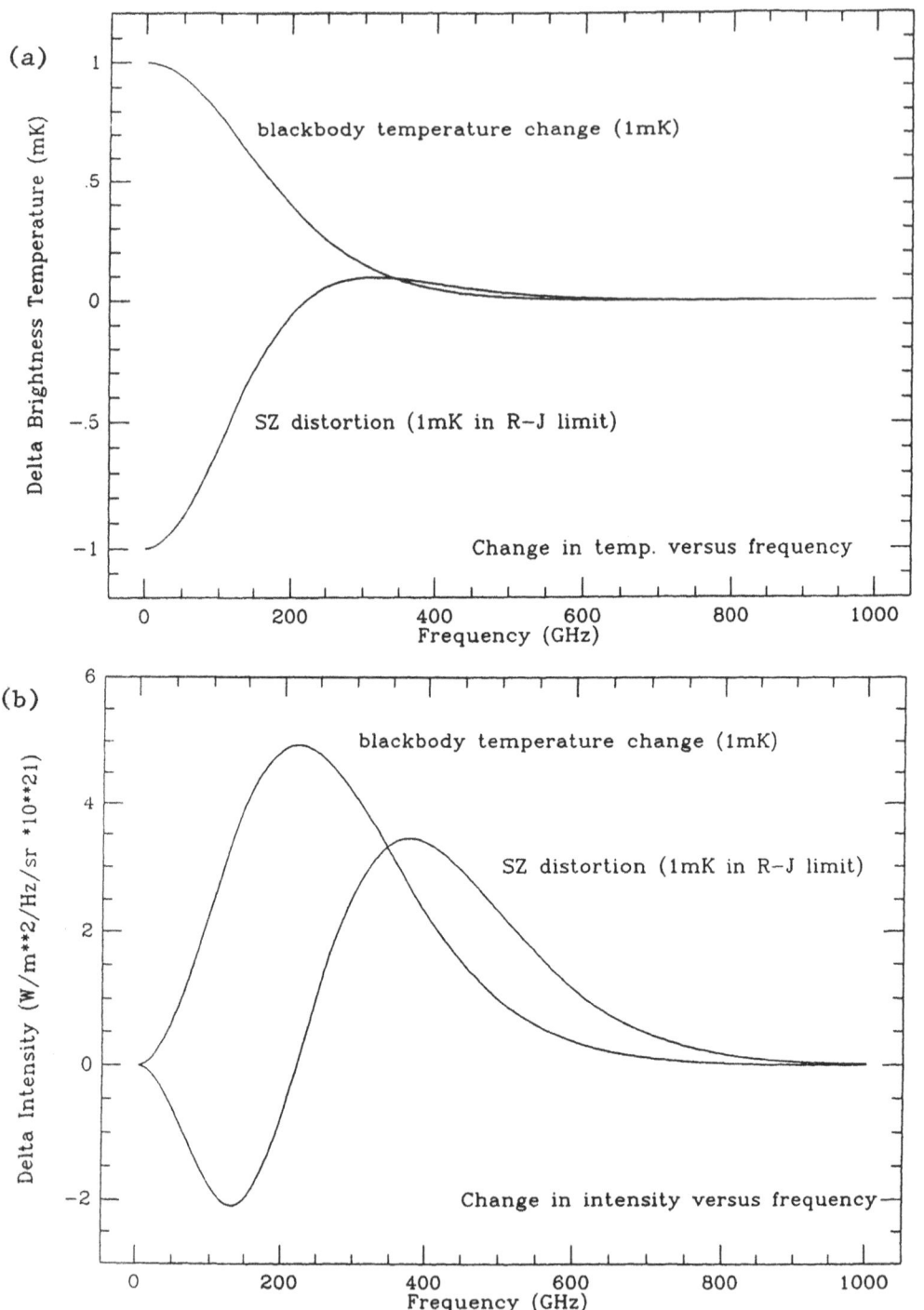

Fig. 3. (a) Plots of ΔT (brightness temperature) for a 1mK change in blackbody temperature (top curve) and due to the Sunyaev-Zeldovich effect with a Rayleigh-Jeans limit of 1mK. (b) same but for intensity distortion.

where τ is the optical depth to Thomson scattering through the cluster, $\tau \sim \int n_e \sigma_T \, dl$. The frequency dependence of this effect will be that of the black-body curves of Fig. 3, and so by observing at the peak of the black-body curve one can separate the direct 'thermal' SZ effect, which nulls here, from the kinetic or 'Doppler' one, which has a peak here (when measured in terms of flux). To see what the hopes are of being able to map the peculiar velocity field of clusters in this way, we take the τ appropriate to the Coma cluster, $\sim 1\%$, and assume a cluster velocity of $600 \, \mathrm{km\,s^{-1}}$, which may not be untypical of the higher velocites which could be expected. This yields a black-body temperature change of 0.05 mK. (A few clusters (see Rephaeli & Lahav) have velocites measured via the Tully-Fisher relation of up to $3000 \, \mathrm{km\,s^{-1}}$, giving an expected ΔT of 0.25 mK, but it seems impossible to believe at the moment that these velocities are real, rather than an artifact of the method.) For comparison, taking the Coma cluster as an example again, the direct SZ effect expected is 0.5mK, and the measured values in three clusters (see below) appear to be about 0.4 mK. Thus the mapping of the peculiar velocity field of clusters will be a difficult task, but the numbers do not look hopeless. Transverse motions of clusters also give a perturbation to the CMB via the 'moving lens effect' (see Birkinshaw & Gull, 1983, and correction by Gurvits & Mitrofanov, 1986), with a distinctive polarization signature, but for similar velocities the $\Delta T/T$ effect tends to be a few $\times 10^{-6}$ rather than few $\times 10^{-5}$ as for the thermal and radial kinetic effects.

3. Observations

There already exists an excellent review by Mark Birkinshaw (1990) in which he describes the history and current status of SZ measurements of clusters, so I will only pick out a few points here (see also Lasenby & Gull, 1985, for a description of mm attempts). The effect was first *claimed* to have been discovered by Parijskij (1972), in the Coma cluster, which we have been using as an example above. This was unlikely given the frequency of observation, and the fact that Coma has a strong cluster halo source, in addition to other radio sources near the centre. There were further claims in 1978 and 1981 for a detection in several clusters by Birkinshaw, Gull and Northover, using a dish at Chilbolton, U.K., at 10GHz, but again radio source confusion and possible systematic effects in the observations make it unlikely the the decrements seen (at about 1 mK amplitude) were real. The first reports of a detection that are widely believed, and seem to have withstood the test of further accumulation of data, are from Birkinshaw, Gull & Hardebeck (1984), using the 40m Owens Valley dish of Caltech, at 20GHz. This is the same telescope, frequency and receiver system as used by Readhead *et al.* (1989), for the intrinsic anisotropy work, for which a limit of $\Delta T/T \lesssim 1.8 \times 10^{-5}$, or $\Delta T < 50 \, \mu K$ is derived. The values for ΔT_{RJ} for the three clusters for which Birkinshaw, Gull & Hardebeck claimed a detection were $\sim 400 \, \mu K$. It might wondered therefore, why there is any difficulty or question about these measurements, given that the effects are about 10 times bigger than a 95% confidence blank sky limit which has been set with the same telescope. The answer lies in the (necessarily) different observing strategies adopted in the two cases. For the intrinsic anisotropy work, Readhead *et al.* were

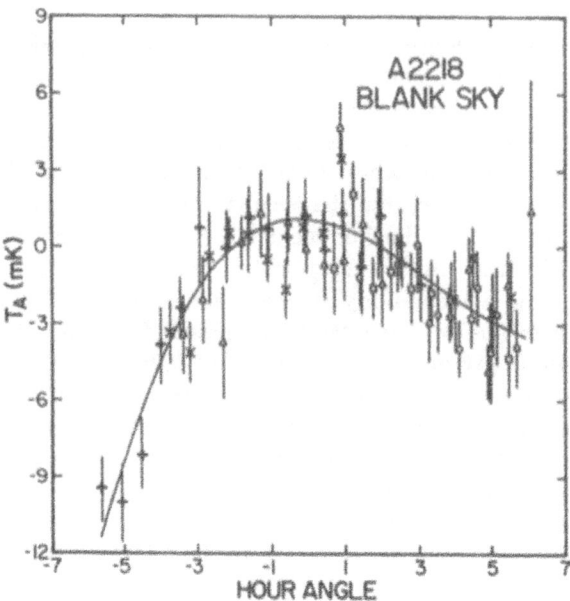

Fig. 4. Sytematic effects at the Kitt Peak 36' telescope, due to differential spillover interacting with local obstacles (from Perronod & Lada, 1979).

able to minimise telescope movement by working with fields very close to the North
Celestial Pole, and observing only near transit. For observations of specific clusters,
there is no such option, since the cluster must be followed around an appreciable
portion of the sky in order to build up sufficient sensitivity. This leads to two
problems. First, the combination of telescope on-off motions and beamswitching
used for both experiments in order to reduce atmospheric effects, can still leave
residual systematics due to incomplete subtraction of sidelobes of the two beams.
This is known as 'differential spillover'. It causes no problems if the telescope is
stationary, but as it moves around the difference pattern is swept over local obstacles
and horizon features, leading to Azimuth/Elevation dependent offsets, that can
mimic a true signal (see Fig. 4). A great deal of careful work is needed to eliminate
these, and in particular observing strategies are necessary in which only a small
proportion of the time is spent actually 'on-source', the rest of the time being
spent with the beams in control 'off positions'. (See e.g. Lasenby & Davies, 1983;
Birkinshaw & Gull, 1984.)

The second problem, is that as the source sweeps around the sky, the 'off'
beams themselves sweep over circular arcs, displaced from the cluster centre by the
beamthrow, which is 7' for the OVRO observations. If these 'reference arcs' contain
radio sources, these will produce negative dips in the central signal as the beams
pass over them, leading to a false impression of a decrement having been detected.
The only solution to this is an independent survey for the confusing radio sources.
To achieve sufficient sensitivity in such a survey requires the power of the Very
Large Array, but the area that needs to be covered for each cluster means that

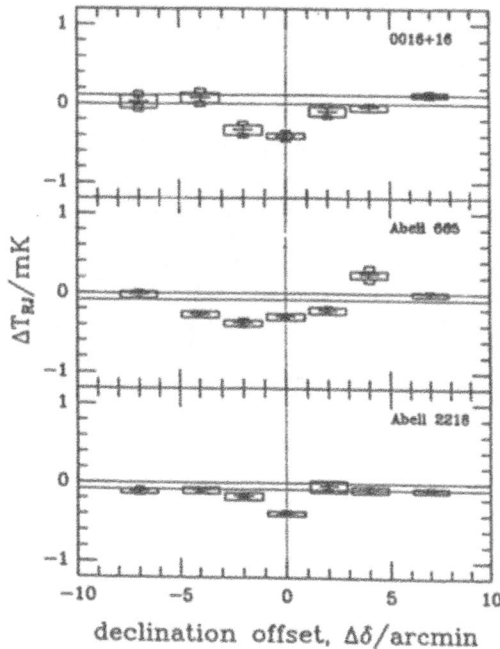

declination offset, $\Delta\delta$/arcmin

Fig. 5. N-S scans through the clusters definitely showing the SZ effect. The boxes are an estimate of maximum systematic error. (Courtesy M. Birkinshaw.)

several VLA fields need to be mosaiced together, requiring quite large amounts of observing time. In addition the results are dependent on an assumed spectral index from the VLA observing frequencies (5 and 8GHz) to the flux at 20GHz. This has now been carried out (e.g. Moffet & Birkinshaw, 1989) for the OVRO SZ observations, but uncertainty in radio source fluxes are still quite a large problem for interpreting the observations. The main fact which is now very convincing as regards the reality of the detections found at OVRO, is that over the course of 6 further years of observations, North-South declination cuts through the three clusters showing the effect — A665, A2218 and 0016+16 — show a profile which is in broad agreement with the SZ profile as expected from the X-ray measurements on these clusters (Fig. 5). For an isothermal cluster atmosphere, the HWHM of the SZ effect is expected to be about 4 times the X-ray core radius (see Lasenby & Gull, 1985). For the clusters 0016+16 and A665 this ties in well, and there is even a shift in the centre of the SZ dip in A665 which agrees with an independent redetermination of the X-ray centre (Birkinshaw, 1990). The SZ width of A2218 however looks rather narrower than expected. Residual problems with radio source contamination of the data cannot be ruled out.

So far the only other independent measurement of the SZ effect, for which there is a claim to see the effect unambiguously, and which has not been withdrawn, or at least for which there is no clear reason to doubt its validity, comes from Juan Uson, using the NRAO 140ft telescope, again at 20GHz. Unfortunately, the last reports from this work were at an early stage in the analysis of both the 140ft and OVRO data sets as regards the inclusion of data on radio source fluxes from the VLA

228

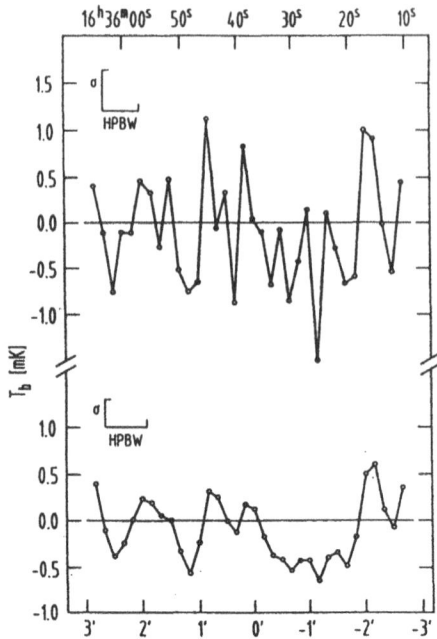

Fig. 6. Bonn stacked scans at 25GHz through the centre of A2218. Top, raw data; bottom, smoothed to 45″.

surveys, and the current status of the reported disagreements for the two clusters observed in common, 0016+16 and A665, (see Uson, 1986, and Birkinshaw, 1986) is not clear. The immediate prospects for independent measurements come from the Australia Telescope (though restricted to a different set of clusters of course) and at Cambridge using the upgraded 5km telescope (see below). The work with the 100m telescope at Bonn should also be mentioned in this category. A recent measurement in A2218 at 25GHz (Klein *et al.*, 1991) is interesting in showing that measurements at the sub-mK level can definitely be achieved, but the actual final stacked scans through the cluster centre are compatible with pure noise. A further problem is that the beamthrow used, of only 2′, would lead to a reduction of the effect through self-chopping. Using the Ginga determination of the electron temperature (McHardy *et al.*, 1990) of $T_e = 7$ keV, rather than the old Boyton *et al.* (1982) value of 24 keV as used by Klein *et al.*, yields an expected decrement after chopping of only 0.1mK, much too small to be seen in the final Bonn scan (Fig. 6). New multibeam systems at 30GHz are currently under development at Bonn however, which should solve these self-chopping problems and have higher sensitivity (Wielebinski, private communication), and the prospects for independent single-dish determinations of the SZ effect at Bonn look promising.

4. Determinations of H_0

The basis for the route to H_0 given a measurement of the SZ effect and X-ray information, is the differing dependencies of the X-ray emission and Comptonization parameter y upon the cluster electron density and pressure, already commented upon above. We have for the integrated X-ray flux

$$S_X \propto \int \frac{n_e^2 T_e^{0.5}}{D_L^2} dV,$$

where D_L is the luminosity distance to the cluster. Now what we know from observation are S_X; T_e (from the X-ray spectrum and iron line fits) and θ, the angular size of the cluster. Now $dV \propto (dl)^3$ and

$$dl = \frac{\theta D_L}{(1+z)^2} \quad \text{implies} \quad dV \propto D_L^3.$$

Thus

$$n_e^2 \propto D_L^{-1} \propto H_0,$$

(using $D_L \approx cz/H_0$, which is accurate enough for the distances involved).

The microwave decrement however obeys

$$\Delta T \propto \int n_e T_e \, dl,$$

i.e.

$$\Delta T \propto n_e D_L \propto H_0^{-0.5}.$$

Thus given a measured ΔT and the X-ray information, we can deduce H_0. Combining the Birkinshaw et al. (1984) result for A2218 with X-ray results from the GINGA satellite, McHardy et al. (1990) obtained a value of

$$H_0 = 24 \pm 10 \, \text{km s}^{-1} \, \text{Mpc}^{-1}.$$

This was initially disappointing, but more recently Birkinshaw, Hughes and Arnaud (1991) have performed a similar analysis for A665, yielding

$$H_0 = (40 - 50) \pm 12 \, \text{km s}^{-1} \, \text{Mpc}^{-1}.$$

The 40 to 50 here refers to the uncertainty in the zero level of the SZ determination. Looking at this more closely, Fig. 7 (courtesy M. Birkinshaw) shows the North-South declination scan through A665, with the dotted lines indicating SZ predictions based upon the best fit X-ray models. The two solid horizontal lines bounding zero are the estimate of where the true zero of the SZ effect really lies. This is a function of the accuracy of the nearby 'blank sky' determinations. With many levels of switching in the system it is not trivial to determine this zero, and in fact it accounts for the greatest remaining systematic uncertainty in the H_0 determinations. The X-ray models are parameterized by cluster core radius and the gas to galaxy scale height ratio, commonly called β. The χ^2 contours of a joint fit in the two-dimensional plane of these parameters, together with superimposed contours of H_0 are shown in Fig. 8. This is for a value of zero level at the end of the range delineated in Fig. 7, and therefore picks out a value of H_0 ($40 \, \text{km s}^{-1} \, \text{Mpc}^{-1}$) at the edge of the bounds reported above. This process has since been repeated for A2218 (see Hughes & Birkinshaw, 1991) yielding the scan plot shown in Fig. 9, and a value (expressed with symmetrical errors) of

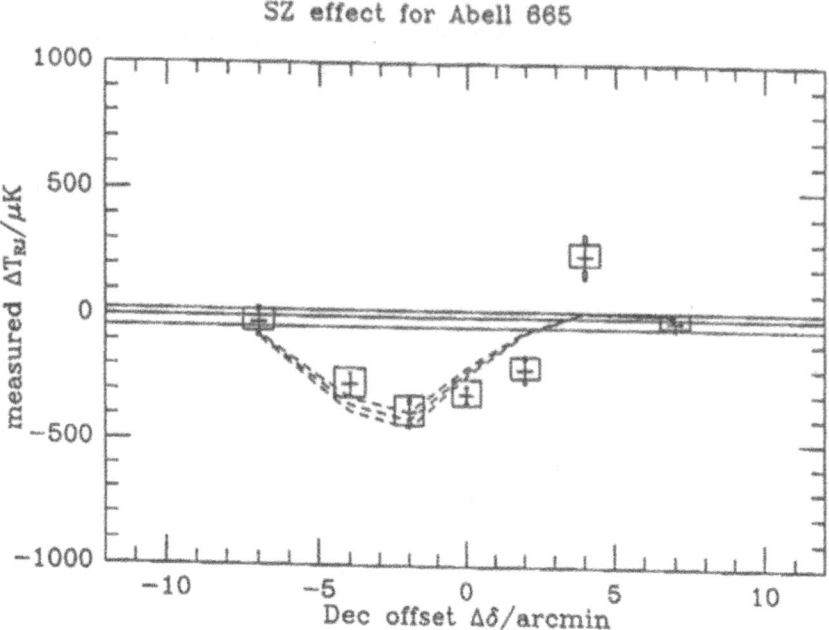

Fig. 7. N-S OVRO scan through A665, showing uncertainty in zero level, and best X-ray fits (dotted).

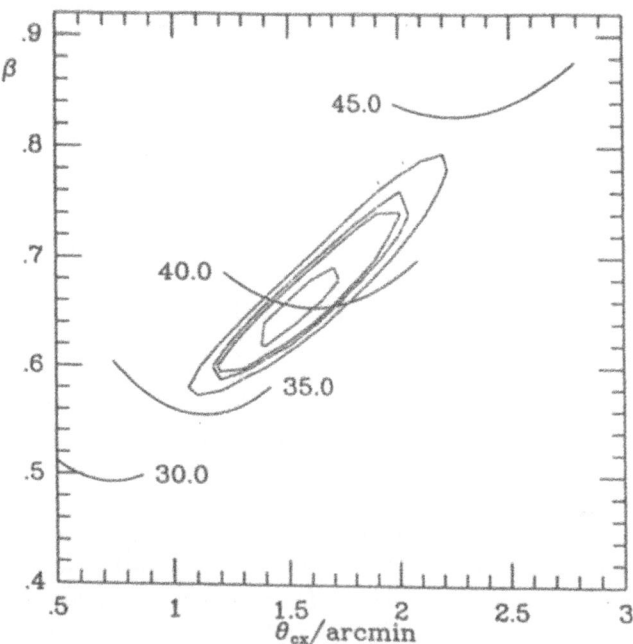

Fig. 8. Confidence level contours in the (β, θ_{CX}) plane for A665. Contours corresponding to different values of H_0 are shown superimposed.

SZ effect for Abell 2218

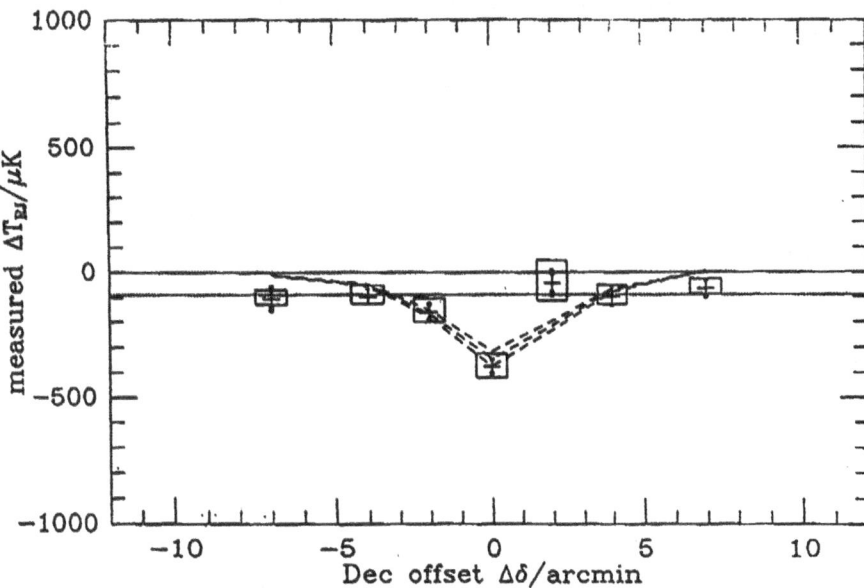

Fig. 9. N-S OVRO scan through A2218. Note the increased zero level uncertainty.

$$H_0 = 57 \pm 17 \, \text{km s}^{-1} \, \text{Mpc}^{-1}.$$

The dominant uncertainty in this is again the zero level offset, which spreads the range for the best χ^2 fit from $\sim 45 \, \text{km s}^{-1} \, \text{Mpc}^{-1}$ to $\sim 67 \, \text{km s}^{-1} \, \text{Mpc}^{-1}$. The reasons for the large discrepency with the McHardy *et al.* value are not clear, but possibly include the failure of the latter to take the actual SZ scan profiles, and the beam dilution effects properly into account when comparing with the X-ray observations. In this respect the latest Birkinshaw *et al.* H_0 determinations can be expected to be superior, since they are based upon the best data in seven years of SZ measurements at OVRO, rather than relying on the single points published in Birkinshaw *et al.* (1984), which is all McHardy *et al.* had available. Considering the Birkinshaw, Hughes and Arnaud numbers then, the major sources of error still expected to be present in the H_0 determinations are:-

1. Errors in the zero offset and width of the measured SZ effect: The former has already been discussed. The latter arises since some discrepent points (perhaps caused by variable radio sources, not adequately subtracted after the VLA surveys) are still present in the scans used, e.g. the point at +2' in Fig. 9.

2. Non-spherical symmetry of the cluster, e.g. prolate or oblateness: In transferring from angular size measured to a distance *through* the cluster (to predict the SZ effect), sphericity has to be assumed. Hughes (p.c.) estimates that for a single cluster, likely values of asphericity could affect the H_0 determination by up to $\pm 20 - 30\%$. The solution here will be to consider larger samples of clusters, in which the effects may be expected to average out.

3. Large scale substructure: This is similar to asphericity as a problem, except that independent verification of it is available to some extent from the optical galaxy distribution. Perhaps more importantly, higher resolution X-ray observations with ROSAT, and high resolution SZ maps with interferometers (see next section) will enable a much more detailed comparison of the X-ray and radio morphologies, and thereby corrections to be applied for this.

4. Small scale clumping: The different n_e dependencies of the X-ray flux and SZ decrement mean that the method is sensitive to the value of the 'clumping parameter'

$$\frac{\langle n_e^2 \rangle}{\langle n_e \rangle^2}.$$

However, the existence of small scale clumping will tend to make the method *over*estimate H_0! Thus in the context of resoving the dispute as to whether H_0 is closer to 50 or $100 \, \mathrm{km \, s^{-1} \, Mpc^{-1}}$ this is not a problem.

5. Observations at Cambridge

The above problems with the determination of H_0 highlight the need for a larger sample of clusters, observed with higher resolution in X-ray and radio. Apart from their use in finding H_0, two-dimensional maps of the SZ effect will be important in the comparison with maps of the dark matter content of clusters made via the gravitational lens distortion of background galaxies (see Soucail, these proceedings), not least because the $n_e T_e$ dependence of the SZ effect tends to probe the baryonic content of the outer regions of the cluster better than the strongly peaked X-ray distribution can be expected to do. Single dish methods, from which all the actual observations of the SZ effect discussed so far have come, have yielded (in the best case) one-dimensional cuts through 3 clusters after about 10 years work. In Cambridge, using the upgraded 5km Telescope, now called the Ryle Telescope, we hope to be able to achieve a sensitivity of $\sim 50 \, \mu\mathrm{K}$ (roughly that needed for a 10σ detection given that the peak ΔT_{RJ} will be less smeared in our beam) in 12 days per cluster, assuming 12 hours/day. The reason for this confidence in a much improved sensitivity and detection rate comes from the fact that we will be using an *interferometer*, for which nearly all of the systematic effects associated with single-dish work are not present. In particular, differential spillover, or its equivalents, does not arise. This means that much more of the observing time (nearly 100%) is devoted to actual observation on-source, and also one simultaneously achieves a two-dimensional map of the cluster area, rather than information at a single point. A further problem with single-dish observations, necessitating the twofold switching schemes mentioned in Section 2 and resulting in a fairly poor useful data rate, even at a good site such as Owens Valley, is the effect of the atmosphere. Fluctuations in atmospheric water vapour content passing by the antenna cause variations in measured temperature, with a timescale from seconds to minutes. An interferometer effectively switches at a much higher rate, and for baselines such as 18m (the minimum Ryle baseline) at 5 and 15GHz (the main operating frequencies) the atmospheric effects are reduced by orders of magnitude.

Given these tremendous advantages, the reason why interferometers have not

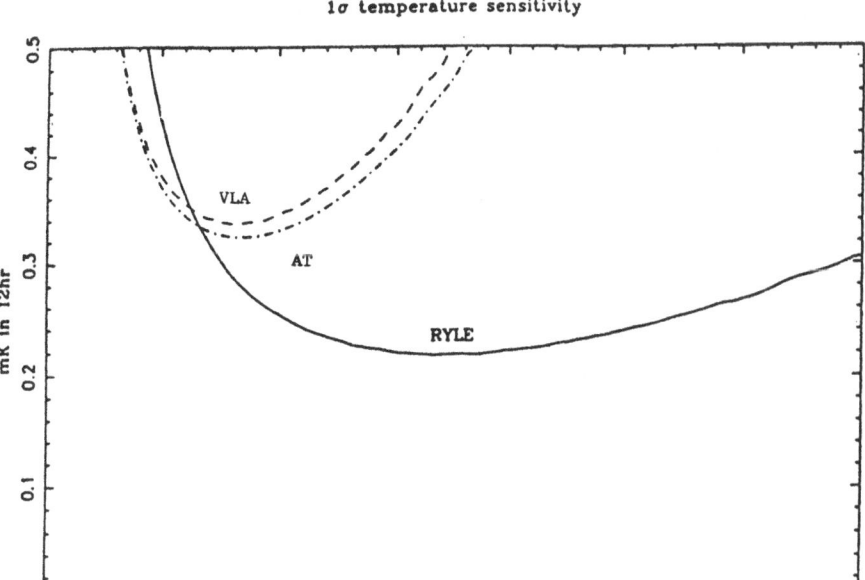

Fig. 10. Comparison of VLA, Australia Telescope (AT) and Ryle sensitivities as a function of SZ core size. All curves are calculated at 15GHz for a Dec 60° source.

been successfully used up to now in SZ work is twofold. Firstly, interferometers already in existence tend to have been built in order to map radio sources. The angular size scales of interest for these tend to be much smaller than the scales for the SZ effect, and therefore the interferometer will not have baselines small enough. This is so for the VLA for example, where despite tremendous flux sensitivity due to many large dishes, the minimum baseline achievable, 45m, is simply too large to give enough sensitivity on arcmin scales at frequencies where discrete source effects are small enough. In this context then, 'small is beautiful' and the Cambridge Ryle Telescope, with dishes half the size and a minimum baseline nearly three times smaller, does much better (see Fig. 10). The second reason is that although inter-ferometers eliminate certain single-dish systematics, such as differential spillover, they also have systematics of their own, such as dish-dish crosstalk, and correlator offsets. Both of these marred early attempts to use the VLA for SZ work. In up-grading the 5km telescope to the Ryle, we have attempted to eliminate many of these systematics at the outset, via the use of extra levels of phase switching for example, and offsetting the pointing centre from the electrical phase centre, so that correlator offsets accumulate outside the map area. In addition a number of other improvements have been introduced specifically for microwave background work: The telescope consists of eight 12.5m dishes mounted on an East-West baseline. It has been made about 20 times more sensitive than the old 5km telescope via installation of cooled receivers, and increasing the bandwidth by a factor of ~ 35, to 350MHz. Also a dense-pack mode, with several spacings of 18m and 36m has been introduced, and all the spacings are now correlated, which was not the case before. The main operating frequencies are 5 and 15GHz, with the higher frequency being expected to be most useful for SZ and primordial CMB work, since source confusion is much reduced there. 15GHz is expected to be in full operation starting January 1992, and serious SZ observations can start then. In the meantime we have been pursuing a programme of mapping of candidate clusters at 5GHz, including all Abell richness class R= 3, 4 clusters with good L_X/F_{5GHz}. Two example maps are shown in Figs. 11(a) & (b), together with overlaid contours of X-ray emission from the Einstein IPC. The maps are of the cluster A665, a definite SZ cluster, and 1558+41, which is a Gunn cluster at $z \sim 0.6$. The latter is included to illustrate the potential power of the SZ technique in reaching to large redshifts, since the decrement expected from 1558+41 will be of roughly the same amplitude as that from the much nearer ($z = 0.182$) cluster A665. The noise level on the maps is 115 μJy and 105 μJy respectively. This does not sound impressive by VLA standards, but the conversion required is that to brightness temperature. This is a function of baseline length; for example for the A665 map we have

$$\Delta T \sim 400 \left(\frac{d}{108\,\mathrm{m}} \right)^2 \mu\mathrm{K}.$$

and the angular scale involved can be got approximately from $\theta = \lambda/d$. E.g., on a 4′ scale, roughly corresponding to the decrement size in A665, this 5GHz map has a brightness temperature sensitivity of $\sim 90 \mu$K, which is beginning to be interesting from the point of view of measuring the SZ effect. However, these maps were taken before we recently introduced a further, and essential, level of phase switching, so that systematic offsets (of the dish-dish crosstalk type) have since been reduced by a factor of 10. Also, no source removal has been attempted for these maps as yet, and at 5GHz the removal of source effects to a level where the SZ effect could be unambiguously seen is difficult for A665, which has a soure as large as 26 mJy in the field. A key advantage of the Ryle, not exploited in this analysis yet, is the simultaneous availability of long baseline information, from baselines joining the dense-pack array to the three remaining outlying antennas. This will be vital at 15GHz, where possible source variablilty and the existence of flat and inverted spectra sources make it important to have simultaneous information at the actual frequency of interest. ROSAT and GINGA observations of our best candidates are being pursued, and for selected clusters we will also try to observe at mm wavelength using the SCUBA array (Cunningham & Gear, 1990) when that becomes available on the JCMT in 1993. This 100 element bolometer array working simultaneously at 400 and 800 μm (where the flux distortion peaks) has great potential for the accurate sub-mmm measurement of the SZ effect (see also Bond & Myers, 1991) due to the high sensitivity and throughput.

6. The integrated effect

The integrated SZ effect due to the combined action of many clusters at earlier epochs, where the covering fraction on the sky becomes significant, is of particular interest in constraining models of cluster gas evolution. Many predictions for this effect have been made, among them Schaeffer & Silk (1988), Bond (1988), Cole &

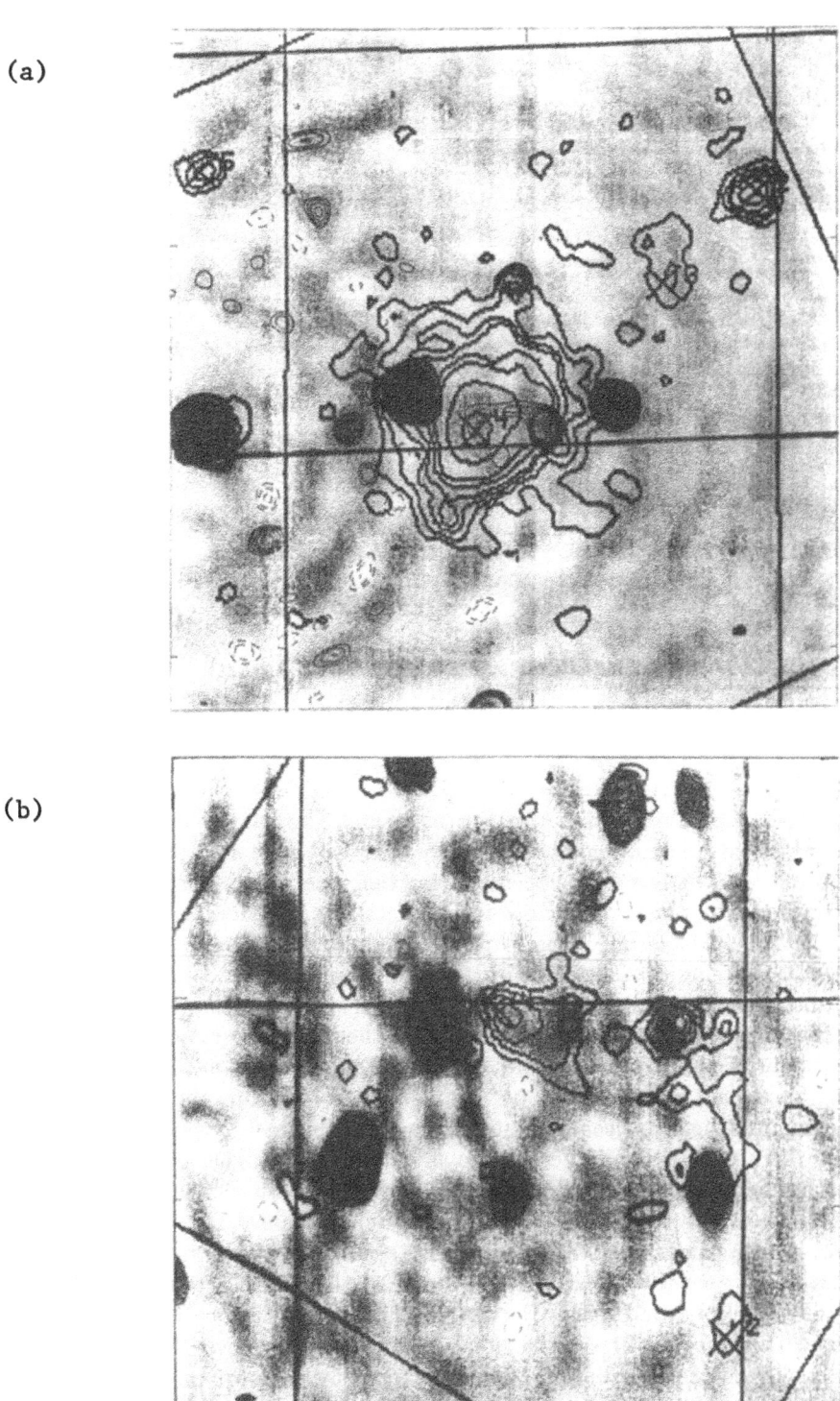

Fig. 11. (a) 5GHz Ryle map of A665. Map σ is $115\,\mu$Jy. Overlaid thick contours are for the Einstein IPC. (b) Same for the Gunn cluster 1558+41. Map $\sigma = 105\,\mu$Jy. Maximum source flux $= 12\,$mJy.

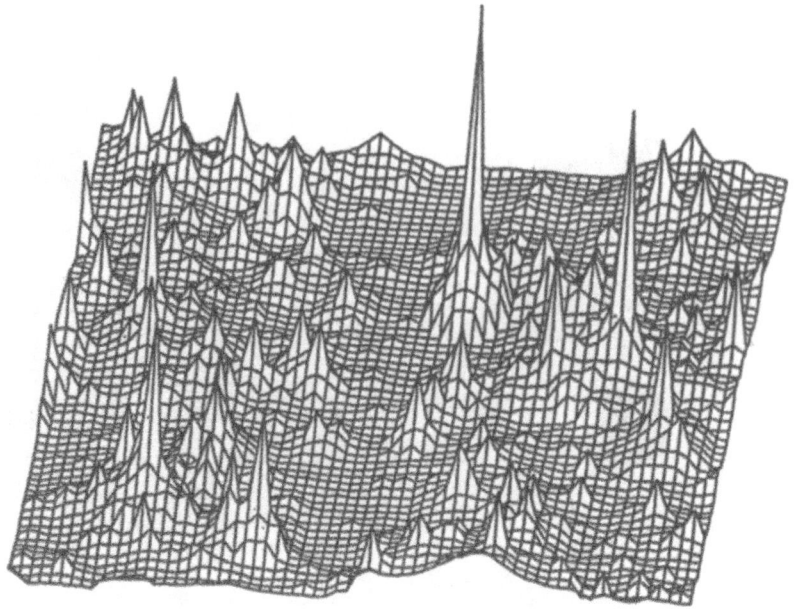

Fig. 12. Simulated sky map of the integrated SZ effect showing a 16×16 arcmin^2 region viewed with a resolution of 0.'25, and with temperature decrements plotted positively (Cole & Kaiser, 1988).

Kaiser (1988) — these were all semi-analytical — and Thomas & Carlberg (1989), who used numerical simulations. More recently Markevitch *et al.* (1991) have considered the question of the total *spectral* distortion due to all clusters, and will shortly present calculations on the anisotropy effect. Fig. 12 shows a typical result of the simulations of Cole & Kaiser, who like Thomas & Carlberg were working within the CDM scenario. A key feature to notice is that fluctuations produced are non-Gaussian, with hot-spots within a small core, surrounded by wider plateau, and areas inbetween with nothing. On the scale of the Readhead *et al.* (1989) small scale experiment ($\theta_{\mathrm{FWHM}} = 1.'8$), Cole & Kaiser would predict a $\Delta T/T$ of about 6.5×10^{-6}, about 1/3 of the 95% upper limit. This would bring the effect within striking distance of the Ryle deep blank field experiments which we plan, and (bearing in mind the non-Gaussianity, leading to the existence of higher peaks than the *rms* would imply) is well suited to testing against the results of the new Caltech RING experiment (Myers, 1991) in which 96 interlinked fields at Dec = $88°10'$ have been surveyed using the same 20GHz observing system on the 40m. The Thomas & Carlberg result for the Caltech scale is much smaller, with a peak near 10^{-6}, putting it effectively beyond reach. Part of the difference with the Cole & Kaiser results may be the different Press-Schechter normalization employed. Some detailed calculations by Bond & Myers (1991), using a 'hierachical peaks' model, give predictions for the integrated SZ effect as seen in the OVRO beam at 20GHz which for a bias parameter of $b = 1$ agree quite closely with the Cole & Kaiser results. The bias parameter in Cold Dark Matter, is a measure of the extent to which light traces

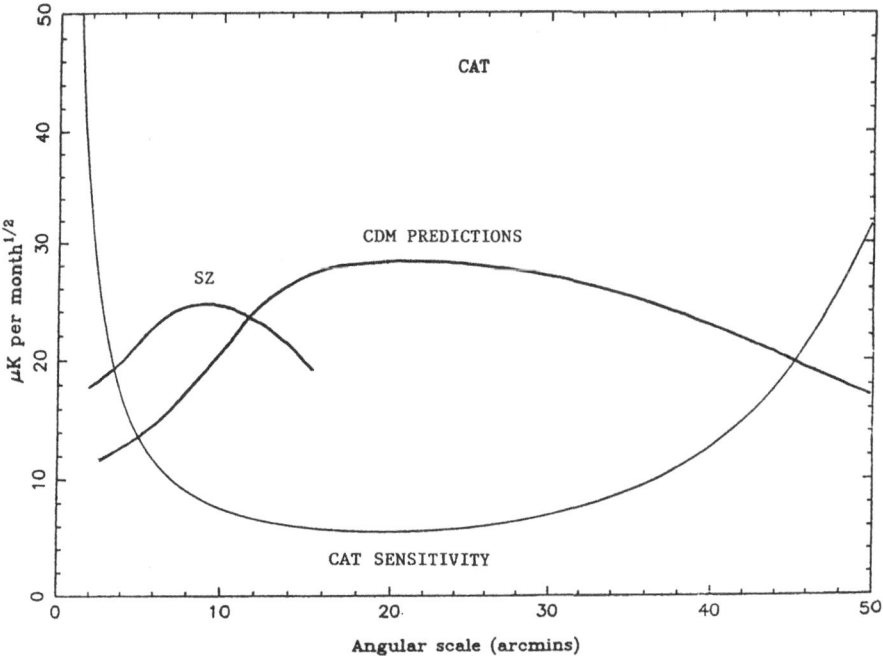

Fig. 13. The projected sensitivity in 1 month of the Cosmic Anisotropy Telescope (CAT) as a function of coherence angle, is shown in comparison with the expected CDM ΔT level and that due to the integrated SZ effect for bias value $b = 1$. (Calculations taken from Bond & Myers, 1991.)

the underlying mass distribution. $b > 1$ implies the light (i.e. the distribution of visible galaxies) is more clustered than the dynamically more important dark matter density perturbations. The results obtained (slightly increased from Bond & Myers, 1991 — Bond, p.c.) are an *rms* of $\Delta T/T = 9.0 \times 10^{-6}$ and a peak decrement of 4.4×10^{-4}, making the integrated effect quite plausible to search for in the Caltech data, or in Ryle blank fields. However, for an increase in b to 1.4, both the *rms* and peak values drop by a factor of 7. This surprising result is a reflection of the fact that in these models clusters are rare events at the tail of a Gaussian probability distribution. Thus even a small change in bias factor can change the mass function, and hence the predicted integrated SZ effect, dramatically. The angular scale at which these integrated SZ anisotropies will be most visible can be expressed in terms of a coherence angle (see e.g. Lasenby & Davies, 1988) and the amplitude of the effect plotted against angle in the same way as for primordial anisotropies. This yields an interesting comparison, shown in Fig. 13, in which standard CDM anistropy predictions are shown together with the curve for $b = 1$ integrated SZ effects. Also shown is the predicted sensitivity in one months' observation for a 3-element interferometer system we are building at Cambridge to probe the CMB on angular scales better matched to the predictions of CDM than can be reached with the Ryle Telescope (see Lasenby et al., 1991). Clearly, if the Bond & Myers predictions are correct, and if b is as small as 1, then SZ anisotropy is a possible contaminant to primordial CMB observations which must be bourne in mind even

238

on scales $> 10'$. (This is due to the extended envelopes of SZ that exist around the richest clusters in the hierarchical peak simulations.) In these circumstances, the only route to clear separation of the effects will be by multifrequncy studies (see Section 2).

7. Conclusions

So far, the SZ effect has not told us a great deal that we did not already know about the physics of the intracluster medium. Rather its measurement is currently at the level where agreement between the SZ observations and predictions based upon exisiting X-ray observations is an important supporting test. An overall normalization factor involved in checking this agreement contains the Hubble constant, and so one very important benefit that is beginning to flow from the SZ observations is a new route to determining H_0. I think that this work is now starting to get very exciting. For actually telling us something more detailed about the cluster medium, both in identified 'nearby' clusters, and via blank field observations which can constrain evolutionary models, I think that we still have a little way to go, but that a new generation of instruments just coming on-line will begin to produce important results, especially in conjunction with the new detailed X-ray maps we are getting from ROSAT, and hope to get from further X-ray satellites to be launched. Another frutiful overlap will be with methods that use the gravitational distortion of background galaxies to map the 'dark matter' content of clusters. (It is interesting that both the classic SZ clusters that are nearby enough to study in this way, A2218 and A665, show gravitational arclets (Soucail, p.c.), and also both contain low-frequency halo radio sources (Moffet & Birkinshaw, 1989).) Among the new instruments, the Ryle at Cambridge, the Australia Telescope and the multibeam developments at Bonn, have already been mentioned, as has the SCUBA array for sub-mm work, and the three-channel Chase and Josephs instrument. In addition, the planned NRAO mm array will contain 40 $\times 8$ m elements able to work at $\lambda 9$mm and shorter, and with a minimum separation of 10m (Owen, p.c.). This could provide an extremely powerful instrument for SZ work, and might be available from the late 1990's onwards. In space, the telescope most suitable for SZ work will be SIRTF. Overall, a key ingredient must be a multifrequency approach, not least in order to separate out the various competing effects that can lead to microwave background anisotropies.

8. Acknowledgements

The Ryle work at Cambridge is the result of continuing effort by many people, and I would particularly like to thank Richard Saunders (in overall charge of the project), Mike Jones and Dominic Lefebrve for their help and advice. My thanks also go to Mark Birkinshaw and Dick Bond for helpful conversations, advice and loan of figures.

References

Birkinshaw, M., 1986, in Radio Continuum Processes in Clusters of Galaxies, eds. O'Dea, C, & Uson, J.M. (Green Bank: NRAO), p261.

Birkinshaw, M., 1990, in The Cosmic Microwave Background: 25 Years Later, eds. Mandolesi, N. & Vittorio, N. (Dordrecht: Kluwer), p77.

Birkinshaw, M. & Gull, S.F., 1983. *Nature*, **302**, 315.

Birkinshaw, M. & Gull, S.F., 1984. *Mon. Not. R. astr. Soc.*, **197**, 571.

Birkinshaw, M., Gull, S.F. & Hardebeck, H.E., 1984. *Nature*, **309**, 34.

Birkinshaw, M., Gull, S.F. & Northover, K.J.E., 1978a. *Mon. Not. R. astr. Soc.*, **185**, 245.

Birkinshaw, M., Gull, S.F. & Northover, K.J.E., 1978b. *Nature*, **275**, 40.

Birkinshaw, M., Gull, S.F. & Northover, K.J.E., 1981. *Mon. Not. R. astr. Soc.*, **197**, 571.

Birkinshaw, M., Hughes, J.P. & Arnaud, K.A., 1991. *Astrophys. J.*, **379**, 466.

Bond, J.R., 1988, in The Early Universe, ed. Unruh, W.G. (Dordrecht: Reidel).

Bond, J.R. & Myers, S.T., 1991, in Trends in Particle Physics (Nov. 1990 UCLA), ed. Cline, D. (Singapore: World Scientific).

Boynton, P.E., Radford, S.J.E., Schommer, R.A. & Murray, S.S., 1982. *Astrophys. J.*, **257**, 473.

Cole, S. & Kaiser, N., 1988. *Mon. Not. R. astr. Soc.*, **233**, 637.

Cunningham, C.R. & Gear, W.K., 1990. ROE preprint.

Gurvits, L.I. & Mitrofanov, I.G., 1986. *Nature*, **324**, 349.

Hughes, J.P. & Birkinshaw, M., 1991, in Clusters and Superclusters of Galaxies (NATO ASI, July 1991), Contributed Talks and Poster Papers, eds. Colless, M.M., Babul, A., Edge, A.C., Johnstone, R.M. & Raychaudhury, S. (Cambridge: IOA), p73.

Klein, U., Rephaeli, Y., Schlickeiser, R. & Wielebinski, R., 1991. *Astr. Astrophys.*, **244**, 43.

Kompaneets, A.S., 1957. *Sov. Phys. JETP*, **4**, 730.

Lasenby, A.N. & Davies, R.D., 1983. *Mon. Not. R. astr. Soc.*, **203**, 1137.

Lasenby, A.N. & Davies, R.D., 1988 in Large–Scale Motions in the Universe, eds. Rubin, V.C. & Coyne, G.V., (S.J) (Vatican Press and Princeton), p277.

Lasenby, A.N., Davies, R.D., Watson, R.A., Rebolo, R., Gutierrez, C. & Beckman, J.E., 1991, in Observational Tests of Cosmological Inflation, eds. Shanks, T., Banday, A.J., Ellis, R.S., Frenk, C.S. & Wolfendale, A.W. (Dordrecht: Kluwer), p413.

Lasenby, A.N. & Gull, S.F., 1985, in ESO-IRAM-Onsala Workshop on (sub)mm Astronomy, eds. Shaver, P. & Kjar, K. (Munich: ESO), p137.

McHardy, I.M., Stewart, G.C., Edge, A.C., Cooke, B., Yamashita, K. & Hatsukade, I., 1990. *Mon. Not. R. astr. Soc.*, **242**, 215.

Markevitch, M., Blumenthal, G.R., Forman, W., Jones, C. & Sunyaev, R.A., 1991. *Astrophys. J.*, **378**, L33.

Moffet, A.T. & Birkinshaw, M., 1989. *Astr. J.*, **98**, 1148.

Myers, S.T., 1991. *Ph.D. Thesis*, Caltech.

Parijskij, Yu.N., 1972. *Astr. Zhurn.*, **49**, 1322.

Perronod, S.C. & Lada, C.J., 1979. *Astrophys. J.*, **234**, L173.

Readhead, A.C.S., Lawrence, C.R., Myers, S.T., Sargent, W.L.W., Hardebeck, H.E. & Moffet, A.T., 1989. *Astrophys. J.*, **346**, 566.

Rephaeli, Y. & Lahav, O., 1991. *Astrophys. J.*, **372**, 21.

Schaeffer, R. & Silk, J., 1988. *Astrophys. J.*, **333**, 509.

Sunyaev, R.A. & Zeldovich, Ya. B., 1972. *Comm. Astrophys. Space Phys.*, **4**, 173.

Sunyaev, R.A. & Zeldovich, Ya. B., 1980. *Mon. Not. R. astr. Soc.*, **190**, 413.

Thomas, P. & Carlberg, R.G., 1989. *Mon. Not. R. astr. Soc.*, **240**, 1009.

Uson, J.M., 1986, in Radio Continuum Processes in Clusters of Galaxies, eds. O'Dea, C, & Uson, J.M. (Green Bank: NRAO), p255.

LARGE-SCALE STRUCTURE WITHIN TEN THOUSAND KM/S

D. Lynden-Bell
Institute of Astronomy
The Observatories
Madingley Road
Cambridge CB3 0HA
U.K.

In the *Connaisances des Temps for 1783* published in Paris 1781, there is the first known reference to the Virgo Cluster written by Messier.

"The constellation of Virgo, especially the northern wing, is one of the constellations that includes most nebulae. This catalogue includes 13 numbered (M) 49, 58, 59, 60, 61, 84, 85, 86, 87, 88, 89, 90 & 91. All these nebulae appear to have no stars. They can be seen only in a very clear sky when they are close to the meridian. Most of these nebulae were given me by M. Méchain". On the basis of this entry, credit must be given to Messier and Méchain for the discovery of the Virgo Cluster. Their work certainly had an effect; only two years later William Herschel gave his first list of 1000 new nebulae in which he constantly refers to the list in *Connaisances* in order to discriminate between *new* and already-known nebulae. Herschel certainly knew of the great band of bright nebulae that we call the Supergalactic Plane and it was in part to see if this band continued into the Southern Sky that his son, John Herschel, spent three years mapping the southern nebulae from Cape Town. Dreyer, in mapping the galaxies in NGC and others at the end of the last century, knew the distribution was far from uniform; nevertheless Hubble's work at great depth showed an overall uniformity. Many of my generation were brought up to believe that apart from the great clusters, the galaxies were rather uniformly distributed. V.M. Slipher used the 36″ telescope of the Lowell Observatory in Flagstaff to measure redshifts of galaxies. By 1917 he had shown that most had redshifts. Eddington published Slipher's current list in the *Mathematical Theory of Relativity* 1922. There were 36 redshifts and 5 blueshifts: M31, M32, NGC 404, M33 & M82. From this data there were already indications that fainter galaxies had greater redshifts, but Hubble had to struggle to get even rough distances to them. We all know Hubble's law

$$v = H r \tag{1}$$

but few of us know how Hubble determined either his v or his r. Hubble's distances were based on luminosities and have since been found to be seriously inadequate,

A. C. Fabian (ed.), *Clusters and Superclusters of Galaxies*, 241–251.

but here we concentrate on his v. Hubble expected that the Sun and the Galaxy were both moving, so he actually wrote

$$v_H + \mathbf{v}_\odot \cdot \hat{\mathbf{r}} = H \, r \qquad (2)$$

where v_H is the heliocentric velocity of the Galaxy whose distance is r and $\mathbf{v}_\odot$ and H are constants to be solved for. With 4 constants to be solved for Hubble needed more than 4 galaxies with measured distances. In fact he had 17. The equation (1) could not be solved exactly, so he took the Hr over onto the LHS and minimised the sum of the squares of the residuals. Hubble thus found that the Sun moved with respect to the mean of nearby galaxies with the velocity

$$\mathbf{v}_\odot = 305 \ \text{km/s} \ \text{towards the direction} \ (71, -14) \ i.e. \ \ell^{\text{II}} = 71, \text{b}^{\text{II}} = -14 \,.$$

I have written the II explicitly because New Galactic Coordinates did not exist in 1929, so I have converted his direction. Hubble's velocity for the Sun is in pretty good agreement with modern ones. de Vaucouleurs' IAU standard is

$$300 \ \text{km/s to} \ (90, 0)$$

Much of this velocity is due to the motion of the Sun around the Galaxy. Some is due to the motion of the Galaxy toward Andromeda. There is only marginal evidence that the Local Group of galaxies moves relative to the mean flow of other galaxies, so we shall neglect that difference and shall not distinguish between the motion of the Local Group and the mean of the Hubble flow at the Local Group. Thus optical astronomers happily corrected galaxy velocities by $300 \sin \ell \cos b$ to put them into the frame of the Hubble flow. All this was rudely shaken by the measurements of anisotropy of the cosmic microwave background. First Conklin working with Bracewell measured a quite different direction and was disbelieved. Then Smoot *et al.* got an unequivocal result that other astronomers at first disliked but found they could not ignore. It did not go away like some bad dreams but rather refinement clearly established the result. There was a 0.1% dipole in the microwave background that was best explained by a motion of the Sun, not of 300 (90,0) but of 365 (267,50). The magnitudes were not far apart but the directions were 130° apart only 50° off opposite! This implies that the Local Group (and the mean of nearby galaxies) moves at 600 km/s toward (268,28). We shall now search for the cause of this motion. Three methods are available:

(i) Look for major non-uniformities in the density of galaxies over the celestial sphere.

(ii) Determine the gravity field at the Sun and the depth from which it arises.

(iii) Determine to what depth the streaming of galaxies persists by measuring deviations from pure Hubble flow.

Figure 1 shows Raychaudhury's plot of all galaxies with recessional velocities less than 350 km/s in the Local Group's frame. The galaxies are projected onto the Galactic Plane. The diagonal line of the supergalactic plane is very obvious. The forces on the Local Group due to these galaxies are described by dipolar and quadrupolar terms. The former generate a mean motion of the Local Group as

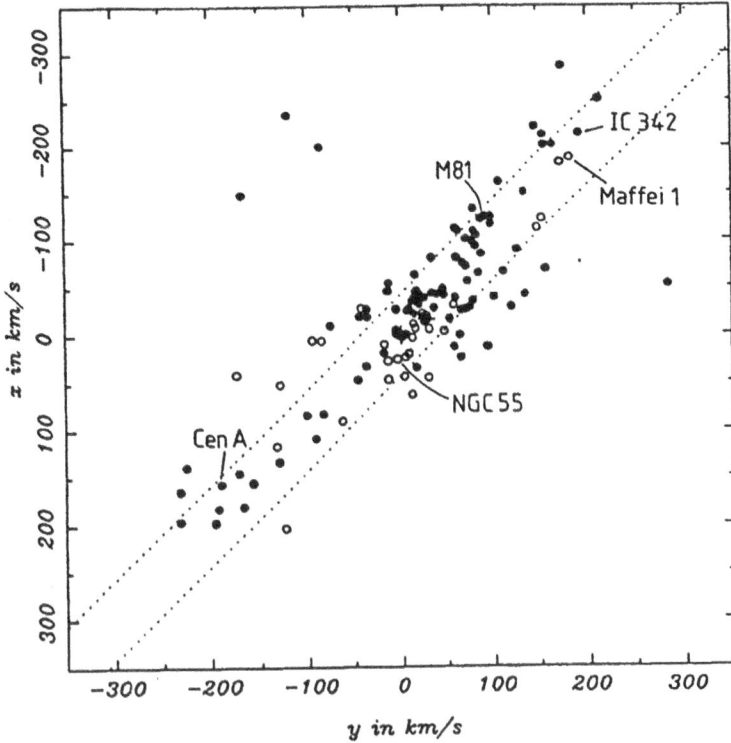

Fig. 1. All galaxies within 350 km/s of the Sun, projected on to the Galactic plane (filled circles for those above the plane, open circles for those below). It can be seen that they lie in a well-defined plane, known as the 'Coma-Sculptor disk' (the 1σ limits on the best-fit plane are superposed). This disk is aligned within 5° with the Supergalactic plane defined by de Vaucouleurs. The '+'s refer to the Galaxy (at 0,0) and M 31.

a whole which the latter provide tidal forces, torques *etc.* that affect the Local Group's internal dynamics.

Both gravity and light fall off as r^{-2}. Furthermore, heavier galaxies have more stars that give more light in proportion. Thus the net gravitational acceleration at the Local Group due to external galaxies should lie in the direction from which the net extragalactic light flux comes. This can be found by adding the *apparent* luminosities directed with the unit vectors to the galaxies. The quadrupolar terms are not quite so easy as the distance weighting in r^{-3} which is not the same as the fall-off of flux. The observed fluxes have therefore to be weighted with a further factor r^{-1}, so distances must be approximately known. Figure 2 shows plots of the optical dipole components P_x, P_y, P_z and the principal axes of the gravitational quadrupole as a function of the distance of the outermost galaxy considered. The dipole is still growing at the edge of the sample but the gravitational quadrupole, helped by its extra r^{-1} factor, converges more quickly. Figure 3 shows the directions of the torque on a pair of galaxies in the Local Group as a function of the direction

Galaxies within 12 Mpc

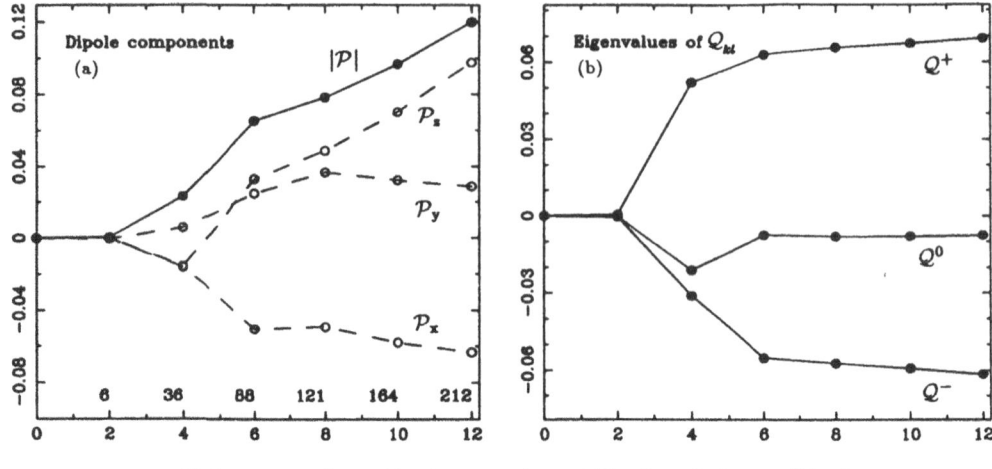

Distance r from the barycentre of the Local Group (Mpc)

Fig. 2. The behaviour of $\mathbf{P}$ and $\mathbf{Q}_{kl}$ with distance. This shows the cumulative behaviour of the Cartesian components of $\mathbf{P}$ and the eigenvalues of $\mathbf{Q}_{kl}$ in 2-Mpc bins from the barycentre of the Local Group. The numbers on the inner side of the lower boundary in (a) indicate the number of galaxies within the corresponding distance, which are the same in both (a) and (b).

of their separation. The arrows have zero length along the principal axes of the quadrupole.

One must distinguish clearly between this 'gravitational' quadrupole with its extra r^{-1} weight and the 'light quadrupole' obtained from the surface brightness $\sigma(\ell, b)$ of the extragalactic sky. One may expand

$$\sigma(\ell, b) = M + \mathbf{P} \cdot \hat{\mathbf{r}} + \hat{\mathbf{r}} \cdot \underline{\underline{Q}} \cdot \hat{\mathbf{r}} + \ldots$$

where $\hat{\mathbf{r}}$ is the unit vector $(\cos\ell \cos b, \sin\ell \cos b, \sin b)$. M is the light monopole or mean surface brightness, $\mathbf{P}$ is the light dipole, and the traceless tensor $\underline{\underline{Q}}$ describes the quadrupolar component of the light distribution over the sky. This $\underline{\underline{Q}}$ has no extra r^{-1} weighting so it is *not directly related* to gravity fields at the Local Group. It is, however, strongly influenced by major features of the extragalactic sky. Its negative principal axis lies towards the supergalactic poles while its positive axis lies along the tug-of-war line that joins the Centaurus Great-Attractor region to Perseus-Pisces.

To get good estimates of the Light Dipole and Quadrupoles one must add the contributions from galaxies up to at least ten thousand km/s in redshift. No complete galaxy catalogue of the sky goes this deep but the best one can do is to use Nilson's UGC in the North combined with Lauberts' ESO/Uppsala catalogue for the region $\delta < -17.5°$. This leaves a strip only catalogued in the less complete

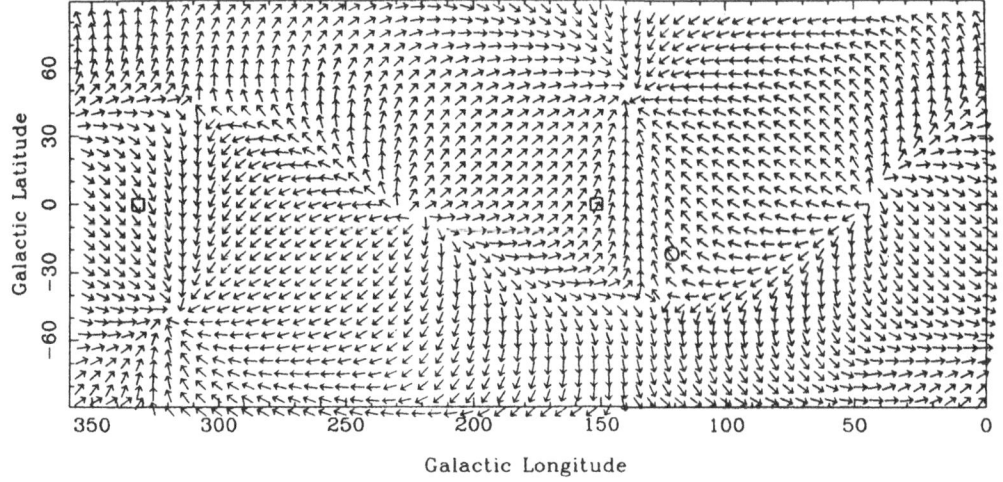

Fig. 3. The direction of the arrow at any point **r** on the sky shows the direction of the tidal force $\mathbf{Q}_0 \cdot \mathbf{r}$ on the Local Group due to nearby galaxies. The principal axes of $\mathbf{Q}_{kl}$ are easily seen. The present position of M31 is indicated by the open circle. Open squares represent the initial position of M31 as proposed by Gott & Thuan (1978).

MCG of Vorontsov-Velyaminov.

Figure 4 shows the resultant picture of the sky in celestial coordinates with the brighter galaxies given stronger symbols. The Virgo Cluster, lies on the Supergalactic Plane which passes on South through Centaurus where it broadens and strengthens in spite of its greater distance there. On the opposite side of the sky, Perseus-Pisces makes a remarkable Curved Arrowhead formation prominently visible in spite of its 5000 km/s distance.

Figure 5 shows the directions on the sky of the Sun's motion relative to nearby galaxies, the Sun's motion relative to the cosmic background and the Local Group's motion relative to the cosmic background. Superposed are the directions of the optical dipole labelled by numbers that give the distances in units of 1000 km/s out to which the dipole has been computed. This figure also shows the direction of the principal axes of the Light Quadrupole at the furthest distance out to which it can be computed and the dipoles as computed off the IRAS galaxies > 0.7 Jansky as determined by various authors (Table 1).

Figure 6 gives the growth in the magnitudes of the dipole and the principal components of the quadrupole as a function of depth. Overall, there is a good directional agreement between the dipoles and the Local Group's motion relative to the cosmic background. Most workers find that about 80% of the dipole comes from within 4000 km/s, thus the source of the Local Group's motion lies mainly in that range. Although the great Shapley concentration of galaxies at 14,000 km/s lies in Centaurus and contributes to the dipole, it is not a major contributor on present assessments, nor is it in the direction of the Local Group's motion, although

TABLE I
Optical, IRAS and X-ray Dipole Determinations

Date	Authors	Catalogue Data	λ	Method	N/100	ℓ	b	%	$\Omega/b^{\frac{1}{3}}$
1977	Smoot et al.	CMB	M	F	-	269	28	.1	-
1980	Yahil,Sandage, Tammann	RSA	O	F	12	217	78	-*	.2
1982	Davis,Huchra	CfA	O	F	12	220	69	-*	.2
1983	Shafer,Fabian	XRB	X	F	-	277	19	.5	-
1986	Yahil,Walker, Rowan-Robinson	IRAS	I	FB	80	248	40	-	.85
1986	Meiksin,Davis	IRAS	I	N	90	235	44	12	$\sim$.5
1987	Lahav	UGC,ESO, MCG> 1'.3	O	θ^2	169	227	42	49	.3
1987	Harmon,Lahav, Meurs	IRAS	I	F	90	273	31	34	-
1987	Villumsen,Strauss	IRAS	I	F	80	239	36	-	1.2
1988	Lahav,Rowan- Robinson & Lynden-Bell	UGC,ESO, MCG> 1'.3	O	θ^2	169	257	34	27	.15
		IRAS > .6Jy	I	FB	80	249	38	16	.8
1988	Strauss,Davis, Yahil	IRAS > 1.9Jy	I	N/r^2	20	231	48	-	.5
1988	Plionis	Lick counts	O	N	810	237	10	5†	-
1988	Lynden-Bell, Lahav	UGC,ESO MCG> 1'.03	O	θ^2	240	261	29	24	-
1989	Lynden-Bell	UGC,ESO}Mock	O	θ^2	240	275	26	41	.14
	Lahav,Burstein	MCG,ZCAT}Mean	O	θ^2	240	257	37	25	.3
1989	Lahav et al.	X-ray clusters	X	$L^{\frac{1}{2}}T^{-\frac{1}{4}}r^{-2}$	.53	258	76	-	-
1990	Miyaji,Boldt	XAGN	X	F	.7	299	35		
1990	Rowan-Robinson et al.	IRAS > .6Jy	I	N/r^2	21	237	43		.7
1991	Plionis, Valdarnini	Abell,ACO clusters	O	N$_g$/r^2	1.5	255	35	†	.06

* Sample too shallow to catch some of the dipole.
† Sample that concentrates on dipoles from very distant objects.

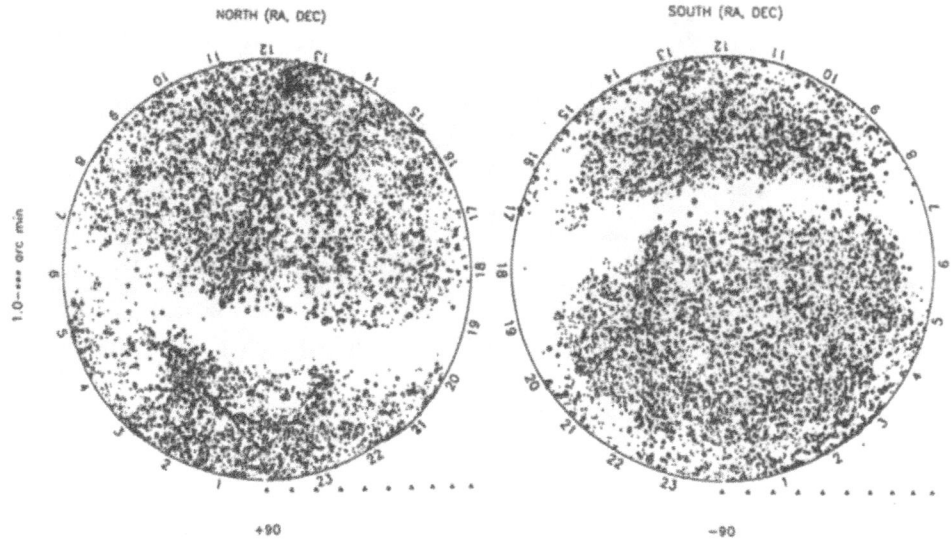

Fig. 4. Equal Area Projections of the North and South Celestial Hemispheres. The Northern galaxies are from the UGC catalogue. In the South, the galaxies are from the ESO ($\delta < -17.5°$), the UGC ($0° > \delta > -2.5°$) and MCG ($-2.5° > \delta > -17.5°$) catalogues. All galaxies are diameter coded.

the rather more broadly based mean peculiar velocity of all galaxies out to 4000 km/s lies quite close to it. The directions of the optical and IRAS dipoles are not in violent disagreement and their growths with redshift are quite similar, nevertheless the contrast in the optical sky is more than twice as great as the IRAS contrast. This has important consequences for the estimates of the cosmologists' Ω, the closure parameter. Optical estimates lie in the range 0.2–0.3 while IRAS estimates lie in the range 0.7–1.0. The derivation of these results is as follows.

As explained above the local acceleration due to gravity $\mathbf{g}$ is directly related to the extragalactic light flux $\mathbf{F}$ at the Local Group,

$$-\mathbf{g} = G \langle M/L \rangle \mathbf{F} \tag{3}$$

The average mass-to-light ratio must include dark matter and may be significantly larger than that of an individual galaxy. If $\langle \rho_L \rangle$ is the mean luminosity density produced in the universe and $\delta\rho_L$ the departure from that mean, then biased galaxy formation would lead one to expect a greater fraction of mass in galaxies where the density of matter was greater. If fluctuations in light are thus exaggerated as compared to fluctuations in the matter density by a factor b, then

$$\delta\rho_L/\langle \rho_L \rangle = b\,\delta\rho/\langle \rho \rangle . \tag{4}$$

b the bias factor would be unity if mass followed light exactly. When bias is included,

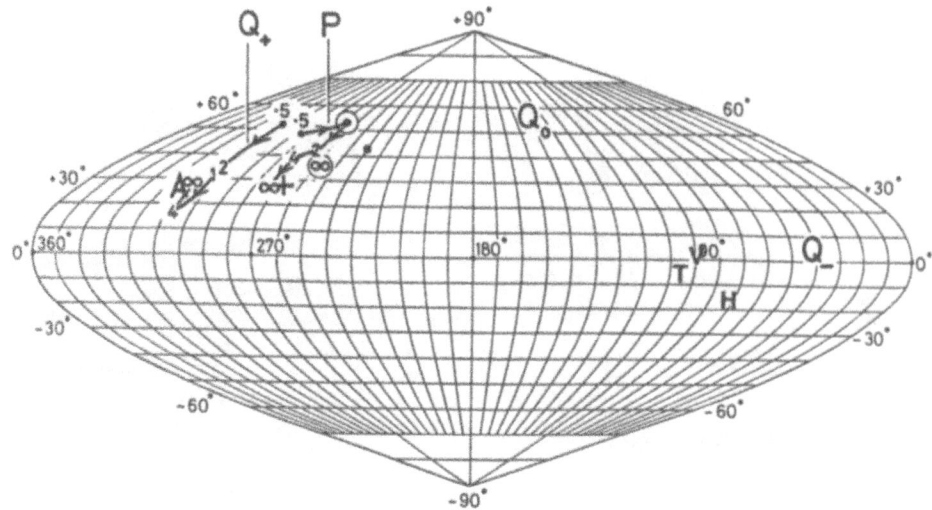

Fig. 5. Directions of solar motion in Galactic Coordinates.

H Hubble's determination relative to the best fitting Hubble flow of all nearby galaxies.

V de Vaucouleurs standard IAU direction of the Sun's motion relative to galaxies.

T The motion relative to the Local Group.

⊙ The Sun's motion relative to the Cosmic Microwave Background (CMB).

+ The Local Group's motion relative to the CMB.

• The direction of the IRAS dipole determined by Rowan-Robinson *et al.* Theory suggests this should coincide with +.

P and Q₊ show the directions of the optical dipole and the longest positive principal axis of the optical quadrupole as a function of the velocity limit $n \times 10^3$ km/s out to which galaxy light is summed. For P, ∞ should coincide with +.

∞ is the corresponding 'Mean sky' result.

equation (3) is modified to

$$-\mathbf{g} = \frac{G}{b}\left\langle \frac{M}{L} \right\rangle F = \frac{G}{b}\frac{\langle\rho\rangle}{\langle\rho_L\rangle}\mathbf{F} = -\frac{4\pi G}{3b}\frac{\langle\rho\rangle}{\langle\rho_L\rangle}\mathbf{P} \qquad (5)$$

where we used the relationship $\mathbf{F} = -\frac{4}{3}\pi \mathbf{P}$. Now Peebles has twice integrated the linear equations of motion of growing density perturbations in the expanding universe to give the relationship between the current velocity $\mathbf{v}$ and gravity $\mathbf{g}$ as

$$\mathbf{v} = \frac{H\Omega^{0.6}}{4\pi G\langle\rho\rangle}\mathbf{g} = \frac{\Omega^{0.6}}{3b}\frac{H}{\langle\rho_L\rangle}\mathbf{P}. \qquad (6)$$

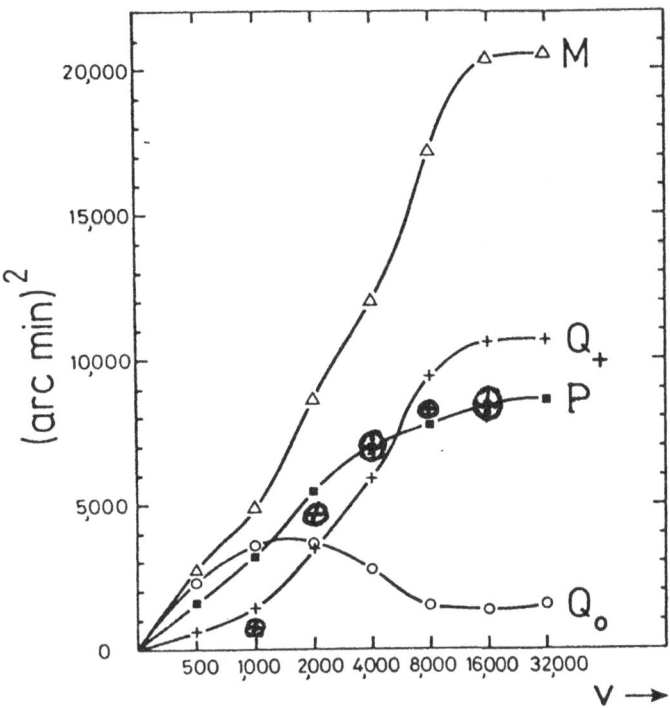

Fig. 6. The growth of the magnitudes of the monopole, dipole and the quadrupole's positive eigenvalues as functions of the depth of sample – measured by Hubble velocity. All quantities are measured in arcmin². Notice the strength of Q_0 up to 1000 km s⁻¹ and the saturation due to incompleteness beyond 12000 km s⁻¹. Q_- is the negative of the sum of Q_0 and Q_+. The symbol $\oplus$ shows the growth of the QDOT IRAS sample from Rowan-Robinson *et al.*, (1990).

Estimates of $\langle \rho_L \rangle$ are dependent on H because volumes scale as H^{-3} and luminosities corresponding to known magnitudes scales as H^{-2}. Thus their ratio $\langle \rho_L \rangle$ scales as H. Luckily we need $\langle \rho_L \rangle / H$ which can be determined without a knowledge of the distance scale. With $\mathbf{v}$, $\mathbf{P}$ and $\langle \rho_L \rangle / H$ determined, equation (6) can be used to determine $\Omega^{0.6}/b$. If b is taken to be 1 corresponding to unbiased galaxy formation, this determines Ω. Discrepancies between optical and IRAS determinations can of course be ameliorated by taking $b_{\mathrm{opt}} > b_{\mathrm{IRAS}}$ and there is some direct evidence for this from correlation functions, but the effect is not large enough to remove the present discrepancy completely. It may be the differences in the treatment of the missing band in the galactic plane that is the primary cause of the difference.

Streaming Motions

Various streaming motions of galaxies have been claimed starting with that determined by Rubin, Ford & Rubin, but one that seems now to be well established is a general streaming of nearby galaxies out to distances of a least 3000 km/s in the general direction of Centaurus, $\ell = 300$ b $= 18$. Most estimates of this motion are around 500 ± 100 km/s and although not far from the motion of the Local Group, the direction is quite distinctly closer to $\ell = 300$ than 268. The directional errors are about 10° . The direction is thus consistent with the concentration of galaxies in the plane of the sky nicknamed the Great Attractor (Fig. 6). Melnick & Moles pointed out that the remarkable Shapley concentration of galaxies and clusters lies behind the Great Attractor. Scaramella *et al.* suggested that a very significant contribution to the optical dipole might come from the Shapley concentration in spite of its 14,000 km/s redshift, but Raychaudhury's (1989) detailed assessment of the galaxies in his deep survey suggests that any such contribution is not more than 15%. The Samurai first found a uniform streaming (Dressler *et al.*, 1987), but later interpreted their data in terms of their Great Attractor model (Lynden-Bell *et al.*, 1988) which gave a better fit. Furthermore, that model allowed agreement with the conclusions of Aaronson *et al.* (1986) who showed that their distant clusters at 5000–8000 km/s were not streaming in the Cosmic Microwave Background's frame. Although Dressler & Faber found some evidence for a reversal of motion beyond 5000 km/s in the Great Attractor direction, there remained the possibility that the observed effect arose at least partly from Malmquist bias. Very recently Matthewson *et al.* have found a continued streaming out to beyond 8000 km/s in this general direction. This appears to be in conflict with both the IRAS redshift surveys and the growth of the optical and IRAS dipoles, both of which show convergence with only some 15% coming from beyond 4000 km/s. In assessing the significance of all this work one must realise that all *distance* estimates are beset by pitfalls and biasses that make them unreliable at large distances. Where there is contradiction, it is more likely that one of the assumptions underlying the distance estimates are false than that redshifts are wrongly measured or galaxies seriously miscounted. Lucey has already found one distant cluster for which the Tully-Fisher and the $D_n - \sigma$ methods produce a serious 2000 km/s disagreement. There is now a period of consideration in which the zero points of the different data sets must be seriously questioned, reviewed and standardised. Different data sets with different distance indicators agree in the nearby streaming and give similar streaming speeds and directions, but agreement at greater depth beyond 5000 km/s is evidently lacking and only the brave should take bets.

References

Aaronson, M., Bothun, G., Mould, J., Huchra J. Schramm, R.A. & Cornell, M.E. 1986, *Astrophys. J.* **302**, 536.

Dressler, A. & Faber, S.M. 1980, *Astrophys. J.* **354**, 13.

Dressler, A., Faber, S.M., Burstein, D., Davies, R.L., Lynden-Bell, D., Terlevich, R.J. & Wegner, G. 1987, *Astrophys. J. Let.* **313**, L37.

Herschel, W. 1786. *Phil. Trans. R. Soc.* **76**, 457.

Lucey, J.R., Bower, R.G. & Ellis, R.S. 1991, *Mon. Not. R. astr. Soc.* **249**, 755.

Lynden-Bell, D., Faber, S.M., Burstein, D., Davies, R.L., Dressler, A., Terlevich, R.J. & Wegner, G. 1988, *Astrophys. J.* **326**, 19.

Lynden-Bell, D., Lahav, O. & Burstein, D. 1989, *Mon. Not. R. astr. Soc.* **241**, 325.

Mathewson, D.S., Ford, V.L. & Buchhorn, M. 1991, preprint.

Messier, C. 1781, *Connaisances des Temps 1783*, Paris.

Melnick, J. & Moles, M. 1987, *Rev. Mex. Astr. Astrophys.* **14**, 72.

Raychaudhury, S. & Lynden-Bell, D. 1989, *Mon. Not. R. astr. Soc.* **240**, 195.

Raychaudhury, S. 1989, *Nature* **342**, 251.

Rowan-Robinson, M. *et al.* 1990, *Mon. Not. R. astr. Soc.* **247**, 1.

Rubin, V.C., Ford, W.K. & Rubin, J.S. 1973, *Astrophys. J. Let.* **183**, L111.

Scaramella, R., Baiesi-Pillastrini, G., Chincarini, G., Vettolani, G. & Zamorani, G. 1989, *Nature* **338**, 562.

Tully, R.B. & Shaya, E.J. 1984, *Astrophys. J.* **281**, 31.

Yahil, A., Sandage, A. & Tamann, G.A. 1980, *Astrophys. J.* **242**, 448.

SUPERCLUSTERS AND LARGE-SCALE STRUCTURE

G. Chincarini
Università di Milano
and Osservatorio Astronomico di Brera
Milan – Italy
L. Guzzo
Osservatorio Astronomico di Brera
Milan – Italy
R. Scaramella
Osservatorio Astronomico di Roma
Monteporzio Catone – Italy
G. Vettolani
Istituto di Radioastronomia CNR
Bologna – Italy

and

A. Iovino
Osservatorio Astronomico di Brera
Milan – Italy

ABSTRACT. We briefly highlight a few selected, controversial issues and some aspects of the (yet) elusive concept of supercluster(ing) within the description and study of the large–scale structure of the universe. Of particular interest is the distribution of clusters of galaxies and radio galaxies on the plane defined by the Local Supercluster. It seems, furthermore, that an empirical analysis of clusters and galaxies populating the plane shows either walls (large scale structures containing no clusters) or superclusters (defined by clusters) in the location of the first peaks of the pencil beam redshift surveys. The cluster redshift distribution in a new wide–angle survey strikingly follows the location of the peaks over a large angle, but shows also other directions with no evidence for periodicity. We discuss also the point of the statistical characterization of clustering on the scales of superclusters, following recent developments on the scaling properties of galaxies at large separations. Finally, we look at the interplay between galaxy morphology and luminosity on one side, and the supercluster environment (local density) on the other. We discuss recent evidences showing that galaxies in dense environment tend to be slightly more luminous.

1. Introduction

The boundaries of the "Cosmos" have been enlarging with the advancement of knowledge, and as a consequence its properties and the (alleged) distribution of matter in it have changed accordingly. In all the historical models we know, local properties were assumed as global and what we knew seemed to be all that was

A. C. Fabian (ed.), Clusters and Superclusters of Galaxies, 253–274.

254

"The whole world is made of sand" –
the scientifically sound extrapolation
of an ant living in the Gobi desert

Fig. 1. The main questions: Where are we (a), and are we sure we are giving an unbiased answer to this question (b)? (Reproduced from Börner 1988, with permission from the author and Springer-Verlag)

arond us. The two cartoons reproduced in Fig. 1 from Börner (1988) summarize quite well the main question to be answered and the trap to be avoided. Where are we really located and are we comfortable with our answer? In other words, do we observe a large enough sample of the Universe to estimate its global properties and to understand the fundamental properties of the distribution of matter? and how do we define a fair sample? or which are the parameters which we measure that identify the Universe as a whole?

While the discussion of the past achievements is left to the History of Science, it is quite interesting to read, with the knowledge we have today, a lecture Oort gave in 1958 on the "Distribution of Galaxies and the Density in the Universe". At that time the most accurate catalogue available was the Shapley and Ames (1932) compilation, fairly complete to $m_{pg} = 13.0$. Within this limiting magnitude the distribution of galaxies in the Virgo region is dominant. Referring to the distribution of galaxies Oort states: *"One of the most striking aspects of the Universe is its* inhomogeneity *... even a superficial inspection of Figs.1 and 2* (here he refers to the two polar cap projections published in Shapley and Ames, 1932) *shows that the large majority of the galaxies in this part of the Universe are limited together in larger or smaller* Structures *and that there is no such thing as a regular* field *on which the structures we see are superimposed ... Some astronomers believe that the major part of the features which I have described forms some sort of* superstruc-

ture which they have called the **Supergalaxy**. *Whether or not this is a useful concept, I do not know. There is also* no clear deviation *between the average expansional velocity of this region and that of the Universe at large ...* "

We will not certainly take the extreme point of view that Oort was already aware of what was demonstrated later on with redshift samples at larger depth; there is no doubt, however, that he was on the right track. His description is very close indeed to the picture we have today. Obviously, due to the lack of redshift surveys, he could not have evidence of the local perturbations of the Hubble flow caused by density excesses, in spite of the fact that some of his work on galaxy formation would call for peculiar velocities on rather large scales. Voids were not yet known (see the review article by Rood 1988).

Cosmography is the result of galaxy catalogs, redshift surveys and statistical analysis. The modern knowledge on the distribution of galaxies is based on the early work by Hubble, Shapley, Zwicky, Abell, Shane and Wirtanen and on the statistical approach pioneered by Neymann and Scott. In the seventies it became also possible, thanks to the image tube technology, to measure reasonably large samples of redshifts using a limited amount of telescope time. [1] In this period various superclusters were detected and defined as large unrelaxed agglomerates of galaxies. Regions devoid of galaxies, one of the most important topological feature of the Universe, were for the first time discovered and defined (Rood 1988).

The work by Peebles and collaborators introduced and developed a statistical and theoretical machinery to understand the observations . Hauser and Peebles (1973) evidenced that clusters of galaxies and galaxies are not equivalent tracers of the distribution of matter in the Universe. The gravitational instability theory (see Peebles 1980) naturally leads to a hierarchy of structures with density contrast over the mean density which, on average, decreases with larger and larger scales. The largest structures, today recognized as such, are modest enhancements ($\Delta n/n = \mathcal{O}(a\ few)$) in the number of galaxies over the mean density.

The mean density is measured as the mean density of the particular sample. For a "fair" sample this should coincide with the mean density of galaxies in the Universe. One of the most important issues in Observational Cosmology lies in the experimental verification of such hypothesis (see section 5 for further discussion on this concept).

2. Superclusters

We are nowadays familiar with the appearance of the structures which we call Superclusters. In Fig. 2 one of the first detected as such, the Coma–A1367 Supercluster is shown. We refer to Geller and Huchra (1988), Haynes *et al.* (1984) for the map of other regions of the sky. At first glance the definition is very empirical, since we have the visual impression of contiguous objects defining a structure with

---

[1] The exposure time needed to obtain the spectrum of a galaxy using a two stages RCA Image Tube was, depending from the central surface brightness of the galaxy, about 5 to 15 minutes and limited by the brightness of the sky background. Such exposure times should be compared with the exposures used by Humason, Mayall and Sandage in 1956, when a spectrum needed up to 10 hour exposure.

256

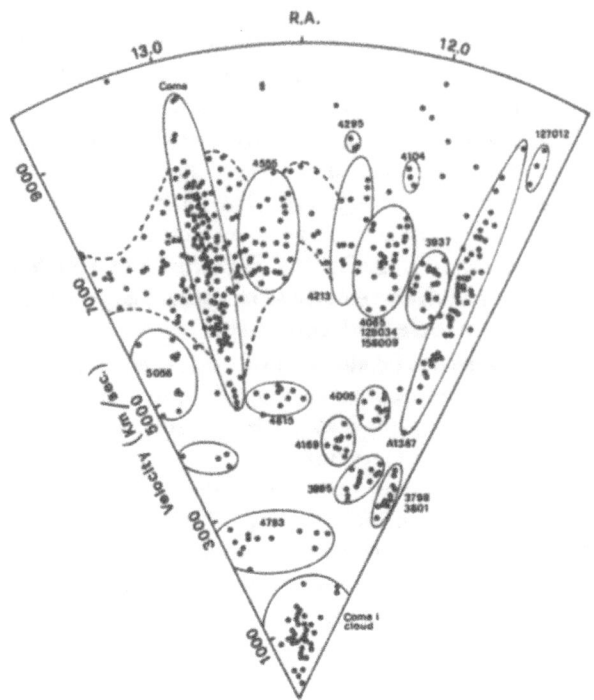

Fig. 2. Wedge diagram of the Coma–A1367 region (adapted from Jaffe & Gavazzi 1986).

a rather irregular morphology. On the other hand, using various algorithms (e.g. percolation), the definition of the structure can be put on a rather objective basis. In these kind of diagrams, as well as in the statistical studies, we use the redshift as the third coordinate (depth) and squeeze one of the angular coordinates in order to have a two dimension plot (in Fig. 2 the declination has been compressed). Could this cause any misunderstanding? The cluster virial velocities [2] for instance, distort considerably the shape of a cluster in the radial direction; note again in Fig. 2 the elongation along the line of sight of both Coma and A1367. A further example of possible spurious features due to the compression in one coordinate is represented in Fig. 3, which shows three declination sections always in the region of Coma. Here a spontaneous question arises: could peculiar velocities severely distort the picture we derived so far about voids and superclusters? Soon after the discovery of the coherence of peculiar motions, concern about spurious topology, and the reality

[2] We would like to point out again the historical inconvenience of calling the cluster redshift elongations "the fingers of God". The term was introduced in astronomy by Bart Bok in connection with the 21 cm maps of the Milky Way. As it is well known the mapping of the spiral arms of the Galaxy is based on a dynamical model since we have no way of determining the distance of the HI clouds. In the early maps, and due to the inaccurate dynamical model of the Galaxy, a few elongations in the distribution were visible. Prof. Bok remarked that that was the finger of God pointing at us and indicating we were doing something wrong.

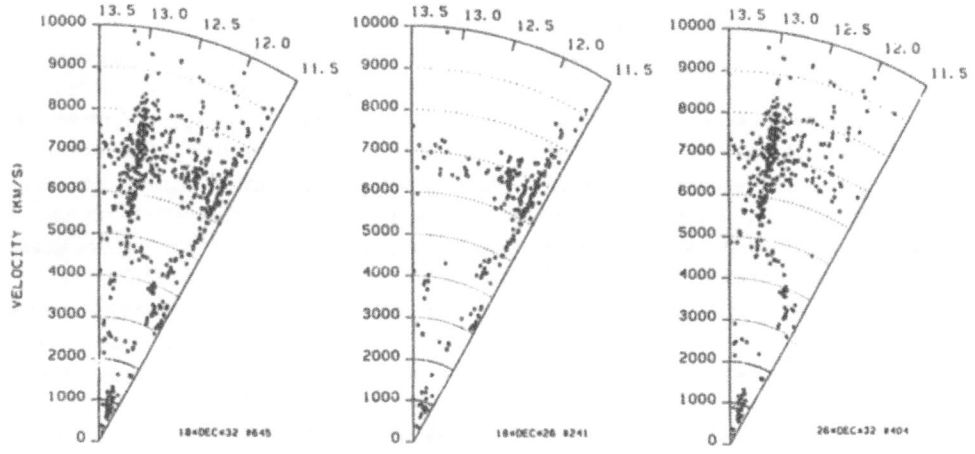

Fig. 3. Three declination slices in the region of the Coma–A1367 complex: note how the spurious elongation produced by the velocity dispersion in the two clusters contribute to the definition of the void borders, but are in reality on different planes (courtesy of G. Gavazzi).

and extent of some voids, was expressed by Burstein (1986) and Kaiser (1987). A quantitative result on voids has been published by Geller and Huchra (1989). By measuring the parameters of the Tully–Fisher relation for 40 objects on the near and far edge of a nearby void located between 4500 and 9500 $km\ s^{-1}$ and with R.A. between 13 and 14 hours, they find a ratio of distances $= 2.08 \pm 0.05$ and a ratio of velocities 2.19 ± 0.05, consistent with small distortions of the Hubble flow. In fact, one expects that in a median "slice" of a dynamically expanding circular void, the boundaries will appear in redshift space as an ellipse with major axis in radial direction, the amount of the distortion being proportional to the outward peculiar velocity of the edges of the void itself.

The difference between redshift space and real space has been also accounted for, and simulated statistically, in various contexts, especially in the computation of the two point correlation function or in the study of the infall pattern around single clusters (see Regös and Geller 1991 for a recent application). From these evidences, therefore, it seems justified to continue to use in general the redshift as a measure of distance for studies on the large scale distribution of galaxies. However, both for a detailed topology and to measure accurately the peculiar motions on large scales, we need to map large regions of the sky using intrinsic distance indicators. Since the best ones available today have a typical uncertainty of about 15 %, we should try hard to find more accurate methods.

Shape and size of superclusters have been often discussed in the literature, but with very small (in the number of SC's) database. The largest structures could be either the extreme of a hierarchical statistical distribution, or the physical result of a particular process of formation favoring particularly large structures, and perhaps only some of them. The largest structure so far detected is the Great Wall (Geller and Huchra 1990). Its dimensions are about 170 h^{-1} Mpc along the right ascension, 60 h^{-1} Mpc along the declination and only $5-10$ h^{-1} Mpc in depth. It is interesting that also the Perseus-Pisces supercluster seems to be so thin in depth (Chincarini et al. 1983). Does this fact represent an important real information on the flattening of some structures, as it seems to be the case, or are we affected by detection and/or measure bias? Could we easily detect a great wall or a Perseus-Pisces supercluster if it were seen along the line of sight? It is in fact possible that the small thickness is partly due to the effect of peculiar velocities; if galaxies are moving toward the spine (i.e. the region of highest density) of the structure, then the redshift range along the line of sight is smaller than the distance range, because of the velocity caustic in radial direction caused by the pancake collapse. An interesting point concerns the level of random velocities present within the wall itself. In fact, collisionless pancakes tend to thin during their evolution (Szalay and Silk 1983), and also tend to yield roughly the same level of final peculiar velocities, even for a wide range of random initial peculiar velocities ("temperature") of the infalling particles.

The effect due to the gravitational pull generated by the main supercluster structure is illustrated in Fig. 4 for 81 galaxies with measured 21 cm line width in the Perseus-Pisces region (Baffa et al. 1991). Here the peculiar velocity has been computed using the infrared Tully-Fisher relation. It is plotted against the Hubble velocity predicted according to the model by Han and Mould (1990). It is clearly seen how the nearby objects (smaller V_{pred}), have a positive peculiar velocity, while farther objects lying beyond the supercluster, have a negative peculiar velocity. According to these preliminary results the structure is shrinking along its spine.

A further crucial question is if these long thin structures are **caSUally** or **caUSally** connected.

The overdensity of a supercluster is naturally a function of the definition of the supercluster itself since within the supercluster the galaxy number density goes from the density of the clusters cores ($\Delta n/n \sim \mathcal{O}(10^2)$) to that of a void (e.g. $\Delta n/n \simeq -3/4$ for Bootes, Strauss and Huchra 1989). Various figures referring to the average overdensity for ill defined boundaries have been given, however, in the literature (see also Oort 1983). We generally have $\Delta n/n$ between 1 and 6 (Postman et al. 1986 derive a value of ~ 5 for the intracluster field in Cor Bor). The Great Wall has an overdensity, with respect to the mean density of the survey, of about 5 (Geller & Huchra 1989).

Saunders et al. (1991) map the Universe out to 100 h^{-1} Mpc using an all-sky sparse redshift survey of galaxies detected by IRAS. Their algorithm to detect superclusters and voids works well in detecting known and new structures. The maximum density contrast measured in this survey is of about 5.5.

On the theory side, Superclusters can impose tight topological and physical constraints on the cosmological models for galaxy and cluster formation. At the present their existence is difficult to explain in current theories with lack of large scale power. As discussed elsewhere in these proceedings the cluster-cluster two

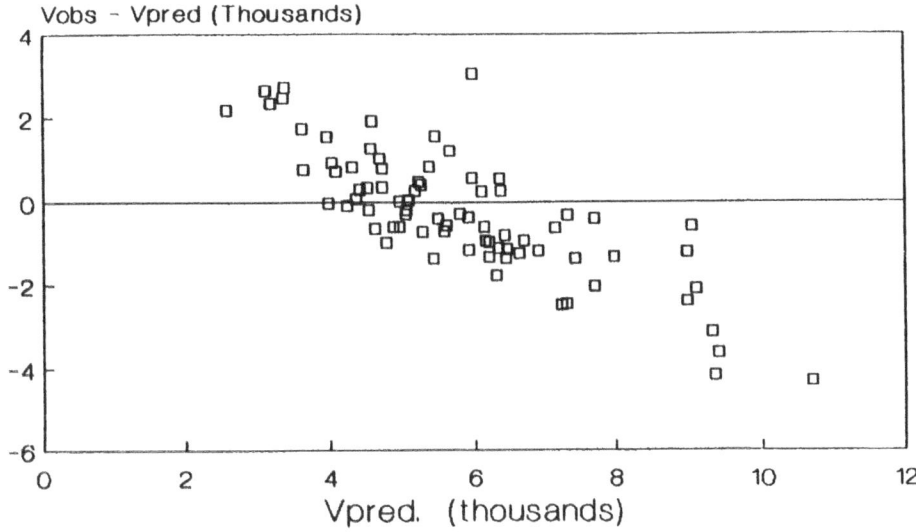

Fig. 4. The peculiar velocity field around the Perseus–Pisces supercluster. V_{pred} (kms^{-1}) is the velocity distance predicted according to the local group model by Han & Mould (1990), which takes into account local anomalies and the effects of Virgo and the Great Attractor (from Baffa *et al.* 1991).

point correlation function has a similar slope as the function measured by galaxies, but larger amplitude by a factor of 9 to 25 (more likely 16) according to different authors (Bahcall and Soneira 1984, Huchra *et al.* 1990, Sutherland and Efstathiou 1991), where the differences are partly due to different methods of data reduction, which follow different assessments of biases in the cluster catalogs, and partly due to different cluster densities in the samples examined. In fact, it seems to be present a scaling of the correlation amplitude as a function of number density of objects (see Bahcall 1988).

Superclusters defined by an agglomerate of clusters of galaxies should somewhat coincide with the agglomerates defined by galaxies since clusters seem to be the high peaks of the density enhancements as defined by galaxies (the reverse is not necessarily true, but it would be quite contrived for large volumes with large overdensities). Clustering of clusters seems to involve much larger sizes, however, so that regions defined by overdensities in the cluster distribution will quite likely contain regions of galaxy underdensity, therefore diluting the contrast (see Bahcall 1988 for a model of the correlation functions from this perspective). Also, a simple form of biasing (Kaiser 1984, Bardeen *et al.* 1986) predicts an exponential enhancement in the ratio of overdensities as observed in clusters with respect to that of galaxies. These points should be better understood to fully comprehend the relation between clustering of clusters and clustering of galaxies. To accomplish this we need of course both deeper and more extensive redshift surveys and (Bayesianly

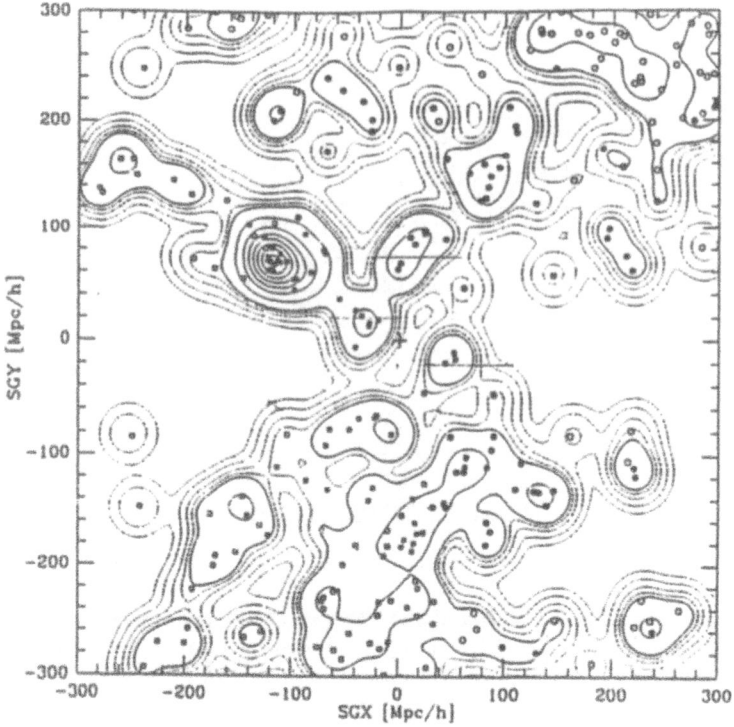

Fig. 5. A thin slice centered on the SGX-SGY plane. Circles denote Abell clusters, squares ACO clusters. Solid points have measured redshifts (from Zamorani *et al.* 1991).

?) "objective" cluster catalogs, selected in X–rays (e.g. ROSAT), or through algorithms on large galaxy samples which are now in the production stage (Lumsden *et al.* 1991; Nichol 1992; Maddox *et al.* 1991).

3. The Supergalactic Plane

At the IAU meeting in Benjing (1983) it was pointed out that the large scale clustering of clusters detected by Tully (1986, 1987) was striking partly because of its size, which is however of the same order of magnitude of the structure detected by Batuski & Burns (1985), but mostly because of its alignement with the Supergalactic Plane (Chincarini & Vettolani 1987), which is defined in terms of the *local* galaxy enhancement. If true, the preference of clusters to reside in a very large superplane could by itself hardly be due to mere Poissonian chance, although the signal from the 1987 data (essentially north Abell clusters) would still be compatible (with frequency ∼ 10%) with a moderate level of large scale power, consistent with the known two point function of clusters (Postman *et al.* 1989). In Fig. 5 we reproduce the distribution of clusters in a thin slice (50 h^{-1} Mpc thick and centered on the supergalactic plane) as seen in projection on the supergalactic plane itself (Zamorani *et al.* 1991). With a larger sample, which now includes also the southern cluster catalog (ACO), the significance of the SG plane has considerably grown (Scaramella & Tully , these proceedings; Tully *et al.* 1991). The overdensity

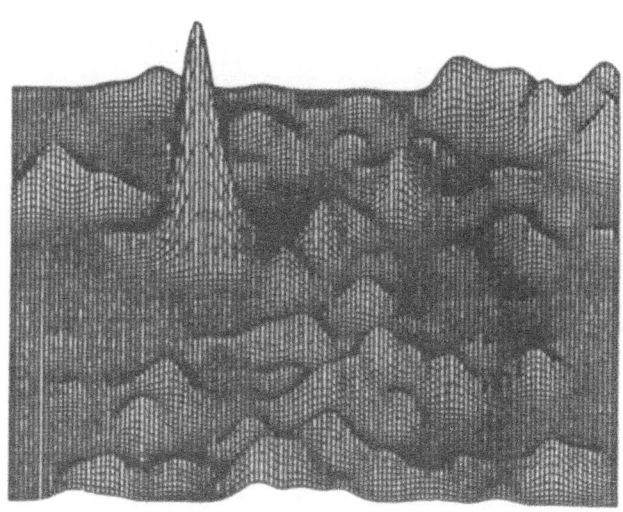

Fig. 6. A 3-D view of the density field described by the cluster distribution of Fig. 5.

of the varius structures is better visualized in Fig. 6. Almost all the known important features within $100\ h^{-1}\ Mpc$ lie on the plane, as it does the "Giant Attractor" (or Shapley concentration), which is the dominant agglomerate of clusters within $200\ h^{-1}\ Mpc$ (Scaramella *et al.* 1989), and helps the Great Attractor (Lynden–Bell et al 1987) to win the gravitational tug–of–war at our position against the opposite pull of the Perseus–Pisces supercluster (cf. Yahil 1987).

Shaver (1991) has recently looked at the space distribution of radio sources. He finds that within a scalelength of about $60\ h^{-1}\ Mpc$ the radio sources are preferentially located on the supergalactic plane, (Fig. 7). The plane, originally defined by the distribution of galaxies in the Virgo Supercluster, seems indeed to be better defined by clusters and radiosources (Fig. 8). This means that either the observed radio sources are born preferentially in clusters, and thus simply reflect the distribution of their parent structures, or, less likely, the environment of the supergalactic plane favours the formation of this kind of objects. It is not obvious, however, what could make the supergalactic plane peculiar in this respect to strongly influence either formation processes or evolution of galaxies. Also recent work by Peacock (these proceedings) confirms the preference for nearby radiosources to lie in the SG plane, while no effect is present in his sample for large distances.

If all these were not coincidences, as it seems to be, then the supergalactic plane of the local supercluster would merely be a detail of a phenomenon which occurred on a much larger scale.

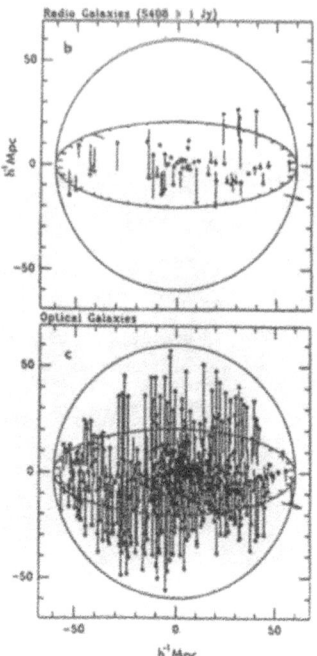

Fig. 7. The space distributions of optical and radio galaxies after Shaver (1991). The arrows show the direction towards the Great Attractor.

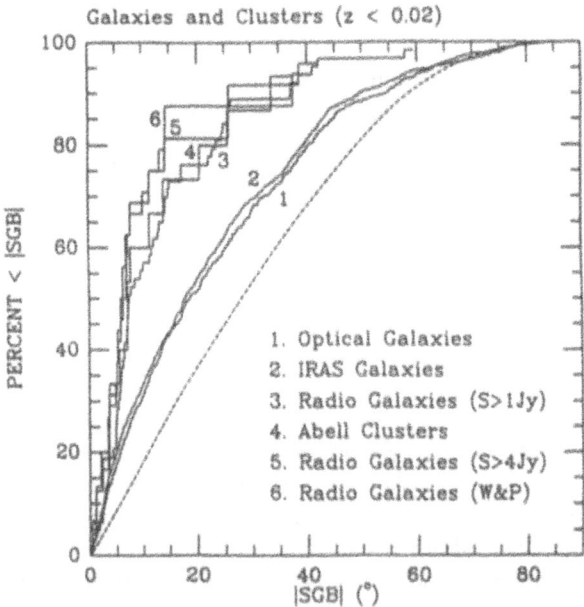

Fig. 8. Integral distributions in supergalactic latitude for several samples of galaxies. The dashed curve is the expected distribution for a random sample with $|b| > 10°$ (from Shaver 1991).

4. Superclustering at Large z

"Any coincidence", said Miss Marple to herself,
*"is always worth noticing. You can throw it away
later if it is only a coincidence"*.

Agatha Christie

Very little, if anything, is known about the Large Scale Structure at high red-shifts, ($z \geq 0.4$). Most of the evidence we have is indirect as far clusters and/or galaxies are concerned. Clustering has been detected in the distribution of quasars (Iovino & Shaver 1988). Bahcall and Chokshi (1991) explain the quasars clustering as a consequence of the clustering of the parent objects: poor groups and clusters. QSO's located in small groups (about 10 galaxies) and Radio Loud QSO's located in richer environment (about 30 galaxies). This is of course strongly dependent on the assumption that quasars are indeed located in clusters (see Yee, these proceedings). Another intriguing point is that raised by West (1991), who finds that, for comoving separations up to about 45 h^{-1} Mpc the radio major axes of high red-shift sources exhibit a tendency to point in the direction of neighboring quasars and radio-galaxies. This anisotropy is interpreted (see also these proceedings), as evidence of structure on large scale at early epochs.

Perhaps the most innovative and provocative modern surveys have been the deep pencil beam surveys (Broadhurst *et al.* 1990, BEKS hereafter; Szalay *et al.* 1991). These data (Fig. 9) show not only the presence of strong clustering at large z (the median redshift of the survey is z=0.3) but also give evidence for a 128 h^{-1} Mpc periodicity in the radial direction, which might be indicative of a global cellular structure. While future studies must directly verify this very important topological indication, do we have any other hint of this periodicity or, at least, are other data in agreement with the findings of BEKS? The North–South beam, which happens to lie in the Supergalactic plane, detects quite a strong signal at the redshift of the Great Wall. Bahcall (these proceedings) shows that her 1984 catalog of superclus-ters gives a posteriori evidence that indeed nearby peaks of the pencil beam redshift distribution broadly coincide with the location of known superclusters. These are indications that we might be able to identify the peaks with known (or yet to be discovered) galaxy strucutures. Tully *et al.* (1991) and Scaramella & Tully (these proceedings) notice, always a posteriori, that a few peaks coincide either with walls connecting superclusters (the Great Wall and the Cetus Wall) or with superclus-ters (see Fig. 10), and that there is a hint of large structures orthogonal to the beam direction. These, however, seem not to be large perturbations to the overall distribution. An even clearer evidence along this line is provided by the recently completed redshift survey of a sample of clusters from the new (automatically com-piled) Edinburgh Cluster Catalogue (ECC in the following) (Guzzo *et al.* 1989; Lumsden *et al.* 1991; Guzzo *et al.* 1991). The histogram of the redshift distribution

264

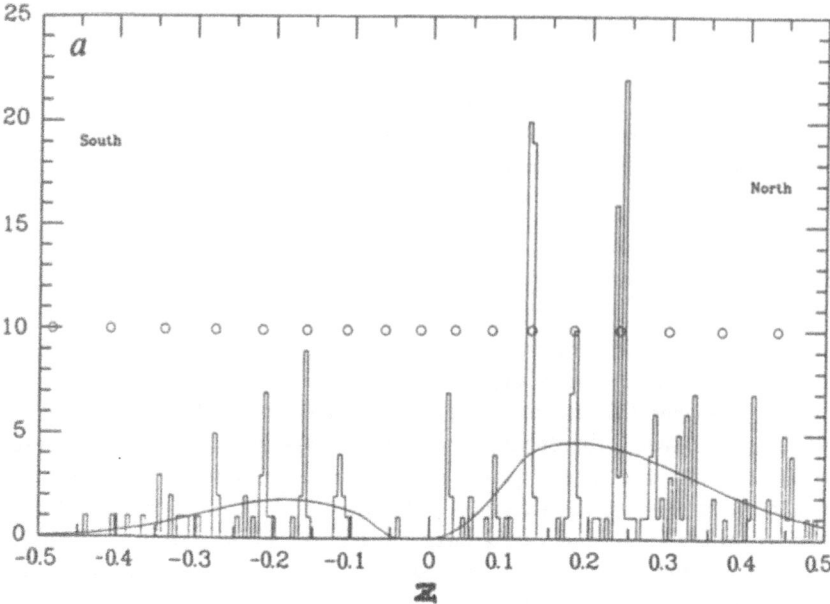

Fig. 9. Redshift histogram of BEKS North–South faint probe. Circles indicate the best–fit constant comoving separation of 128 h^{-1} Mpc for $q_o = 0.5$.

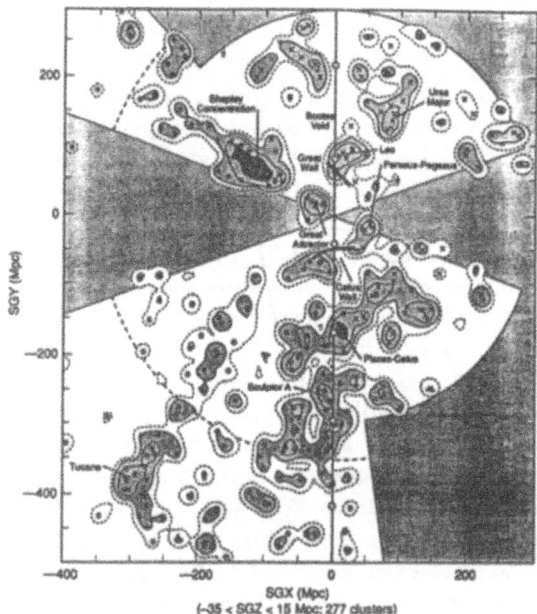

Fig. 10. The projection on the SGX–SGY plane of a slab 50 h^{-1} Mpc thick. The BEKS probe line is passing vertically through the origin, and 6 cycle peaks are shown as knots in line (from Tully et al. 1991).

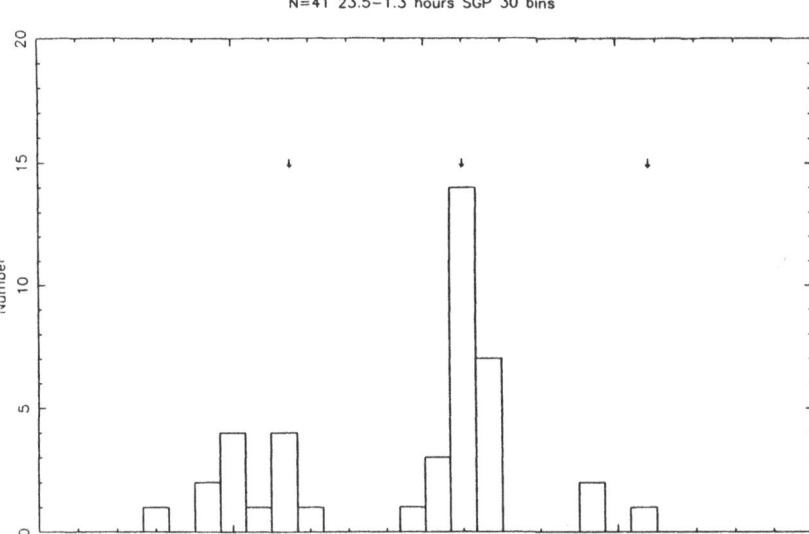

Fig. 11. Redshift distribution of clusters from the ECC redshift survey in a $\sim 20 \times 20$ square degree area centered on the BEKS SGP pencil beam. Arrows show the 128 h^{-1} Mpc BEKS expected peak positions (from Guzzo *et al.* 1991).

for a region of 20 × 20 square degrees surrounding the BEKS South Galactic Pole (SGP) pencil beam is shown in Fig. 11. The two maxima of the histogram coincide very well with the first two peaks (marked by arrows in the figure) of BEKS. The largest peak is due to a a large concentration of clusters at $z \simeq 0.11$ extending for about 100 h^{-1} Mpc, visible in the cone diagram of the whole survey in Fig. 12.

From these results it seems that BEKS are mostly sensitive to very large structures, as superclusters or walls[3]. From Fig. 10 we see that in correspondence with the BEKS peak beyond the Bootis void, at about 250 h^{-1} Mpc distance, there is not known concentration of clusters (supercluster). However, two overdensities are visible on both sides, so it is very likely the presence of a bridging wall, and a deeper survey could well detect it. The reasoning is, in spite of the absolutely weak statistics, very appealing: superclusters are connected by walls and the periodicity is the result of a network composed by superclusters and walls. Indeed, such complex network arises naturally in N–body simulations with large scale power, but does not usually have periodic appearence and characteristics. Periodic structures in narrow probes have been shown to arise sometimes if the large–scale distribution of galaxies follows a topology well described by a Voronoi tessellation (Icke & Van de Weygaert 1987, Van de Weygaert 1991; but see Williams *et al.* 1991).

---

[3] Superclusters in general might be defined through the distribution of clusters, while walls seem to be confined to a mixture of galaxies and groups. Are they different structures?

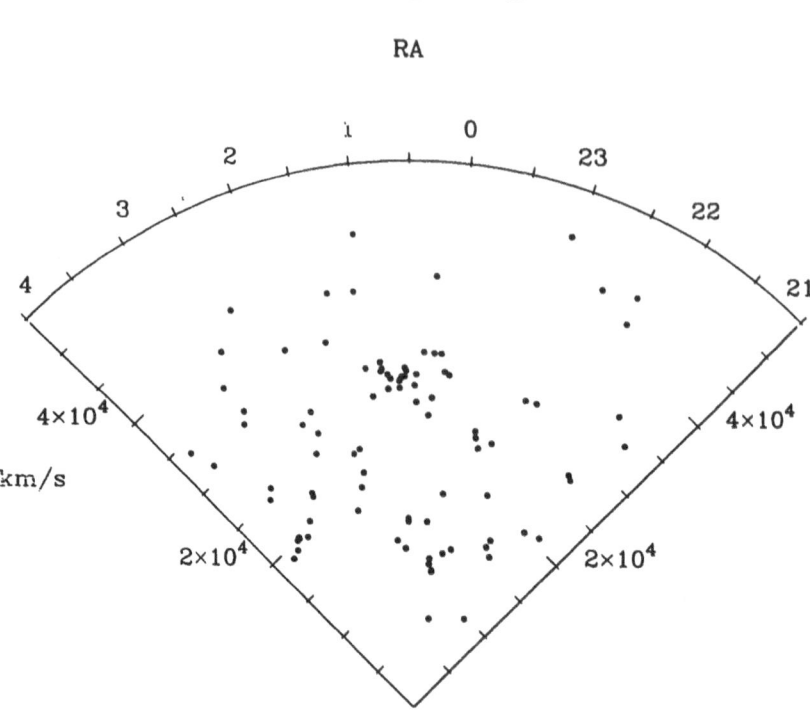

$$21 \quad < RA < 4$$
$$-45 \quad < DEC < -20$$

RA

Fig. 12. The whole redshift sample of clusters from the ECC, showing in the central part the two concentrations of clusters giving rise to the corresponding BEKS peaks.

5. Statistical Measures of Superclustering

It is of considerable interest to quantify the statistical properties of clustering on the scale of superclusters, given the fact that fluctuations of these masses have $\delta\rho/\rho < 1$, and thus should still be evolving following linear theory. This means that their statistical properties could be possibly traced back to the recombination epoch, to put severe constraints on the power spectrum $P(k)$ describing the fluctuations at that time.

As discussed elsewhere in these proceedings, the most popular statistics used to quantify the clustering of galaxies has been the autocovariance or two-point correlation function, that in its continuous definition can be expressed as (Peebles 1980)

$$\xi(r) = \frac{\langle n(\mathbf{x}) \cdot n(\mathbf{x} + \mathbf{r}) \rangle}{\langle n \rangle^2} - 1$$

where the averages are performed over the whole solid angle and over the sample. $\xi(r)$ represents the excess probability over random of finding an object at distance r from a given one. In principle, this requires the knowledge of the mean density of the Universe, which provides the expected number for a random distribution to

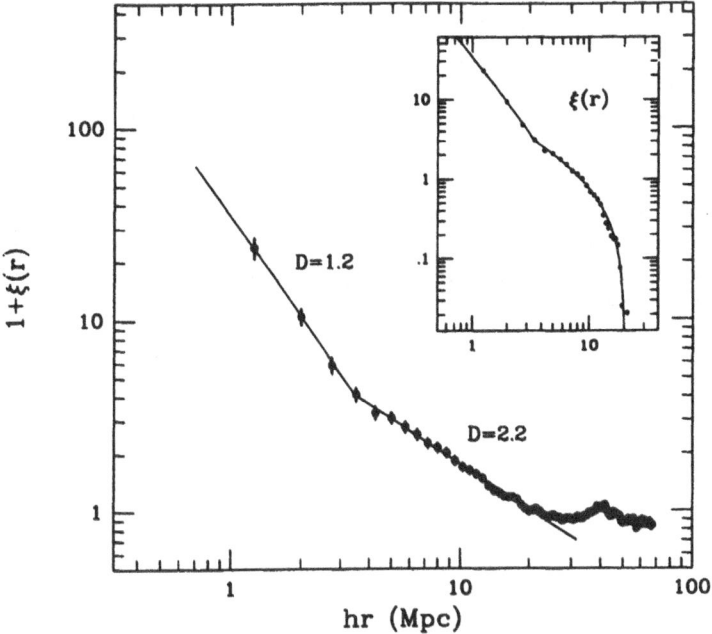

Fig. 13. $g(r) = 1 + \xi(r)$ for a volume–limited subsample of the Arecibo redshift survey with $M \leq -19$, containing 1092 galaxies. Error bars are bootstrap resampling estimates. The inset shows $\xi(r)$ and the solid line corresponds to the fit of two power laws to $g(r)$. Note the accurate fit of the shoulder feature (from Guzzo *et al.* 1991b).

which $\xi(r)$ is normalized. In practice, the mean density is estimated from the sample itself, even directly or using more sophisticated techniques (e.g. Davis & Huchra 1983). The hope is that the sample is large enough (i.e. it is a *fair sample*), so that clustering is washed out when averaging to estimate the mean density. This is close to be true if the typical scales of clustering are smaller than the sample size. This problem is also referred to as the *integral constraint* on $\xi(r)$, discussed in detail by Peebles (1980), and from a different perspective by Pietronero (1987). Its net effect is in general that of introducing an underestimate of the amplitude and correlation length of $\xi(r)$, i.e. an overnormalization of the correlation function. Pietronero (1987) has suggested to characterize clustering in terms of different functions, like the *conditional density* $\Gamma(r)$, i.e. the mean density in a shell at distance r from an object chosen at random. However, it is highly desirable to keep any estimator as close as possible to $\xi(r)$, which provides a powerful connection of statistics with the dynamics of clustering (e.g. Davis & Peebles 1977). It can be shown (Guzzo *et al.* 1991b) that the function $g(r) = 1 + \xi(r)$ allows to overcome the above problem, while still remaining close to the traditional description of clustering. In particular, the *scaling* properties of clustering are correctly evidenced by $g(r)$ also in the weak clustering regime. The clustering analysis by Guzzo *et al.* (1991b) in terms of $g(r)$, evidences the advantages of this description. In Fig. 13 it is reported a plot of $g(r)$ for a volume–limited subsample of the Perseus–Pisces redshift catalogue. Note the two clear power–law ranges, evidence for a simple fractal–like scaling that can be characterized in terms of a *correlation dimension* D_C, being $g(r) \propto r^{D_C - 3}$ (see e.g. Mandelbrot 1982). Galaxy clustering is therefore characterized by $D_C \simeq 1.2$ and $D_C \simeq 2.2$, respectively for scales smaller or larger than the transition radius $r_b \simeq 3.5 \ h^{-1} \ Mpc$. This means that on large scales density goes as $\sim r^{-0.8}$, which might represent a problem, e.g. for some infall models that assume as standard a $\sim r^{-2}$ law on all scales. It is somewhat surprising that this behaviour of $1 + \xi(r)$ was noticed already in 1984 by Dekel & Aarseth in the CfA survey data, even if only in terms of *topological* dimension, i.e. as evidence for pancake–like structures $(D \sim 2)$ on large scales. From Fig. 13 it is also clear the flattening of $g(r)$ for $r > 30 \ h^{-1} \ Mpc$, evidence for homogeneity on these scales. The fact that $g(r)$ goes below 1 at this point could either be true evidence for anticorrelation, or more probably a residual effect of the integral constraint. Note how for $g(r)$ this residual uncertainty on the mean density represents only a shift in the log–log plane, leaving unscathed the real flattening scale, while it strongly affects the break scale of $\xi(r)$ (inset of Fig. 13). The two scaling ranges find a natural interpretation as related to the transition between large–scale linear density fluctuations and nonlinear clustering at smaller separations. In this picture, the large–scale $D_C \simeq 2.2$ range should still be close to the form of $P(k)$ at recombination, and it can be shown (Guzzo & Branchini 1991) how this corresponds to a power spectrum close to the form $P(k) \propto k^{-2}$ down to wavenumbers $k \sim r_c^{-1} \ h \ Mpc^{-1}$, where r_c is the homogeneity scale shown by $g(r)$. Preliminary results from N–body simulations starting with such a kind of power spectrum at recombination (Branchini *et al.* 1991) show that it is easy to end up with a $g(r)$ with the observed 'double–slope' shape, matching both large–scale clustering and rms velocities at large separations.

6. Some Effects of Clustering on Large Scale

The process of clustering implies the creation of a variety of environments by which the formation and evolution process of the single galaxy could be affected. Dressler (1980) was the first to break with the old and coarse description by selecting, in the description of the population of a region of the sky, a parameter unique to cluster and non cluster galaxies: the density, Fig. 14.

While phenomena like the HI deficiency, Fig. 15 (Haynes *et al.* 1984) and distribution of shapes for the rotation curves of galaxies (Di Stefano *et al.* 1990, Whitmore 1990), could be characteristics only of clusters of galaxies, it has been shown by Postman and Geller (1984) and Giovanelli *et al.* (1986), that the correlation between density and morphology holds at any density, from the very sparse outskirts of superclusters to the innermost regions of clusters of galaxies. Recently, Iovino *et al.* (1991) have investigated the correlation between galaxy luminosity and the density of the environment using the Arecibo redshift survey. Previous analyses on a smaller version of the catalogue had given apparently contradictory results (Haynes *et al.* 1984; Chincarini 1988). Further work in this direction has then been done also by Scaramella (1988), Hamilton (1988) and Vall-Gabaud *et al.* (1988). The main point of the new analysis is to disentangle the effects due to the fact that the type content is a function of density. A different mixing of morphological types, coupled to a morphology dependent luminosity function, could originate spurious

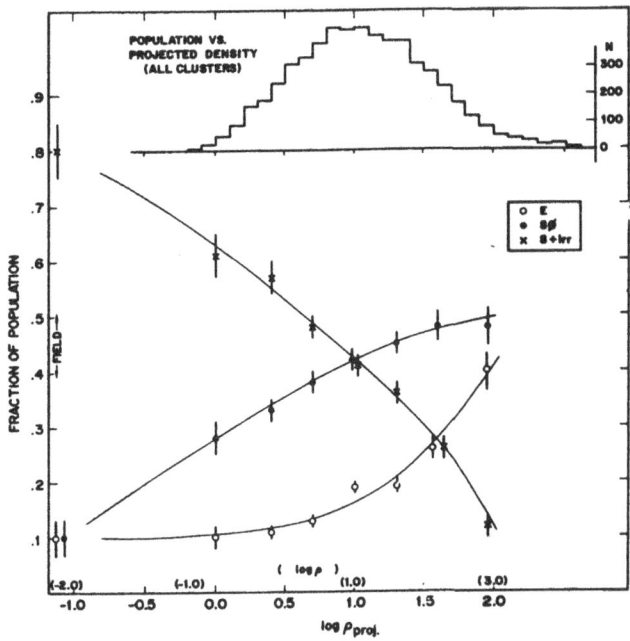

Fig. 14. The morphology–density relation for clusters of galaxies, after Dressler (1980).

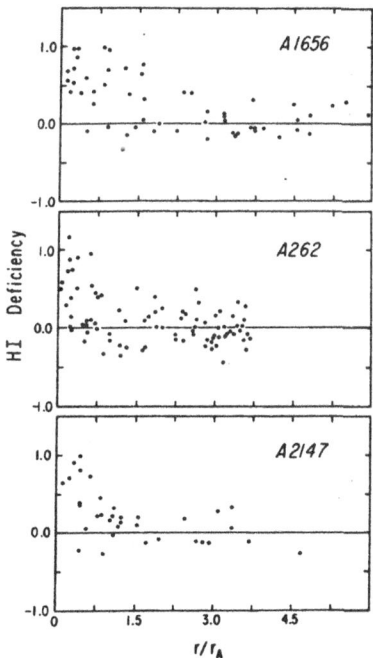

Fig. 15. HI deficiency as a function of radial distance in three clusters. Radial distance is scaled by the Abell radius r_A of each cluster (from Haynes *et al.* 1984).

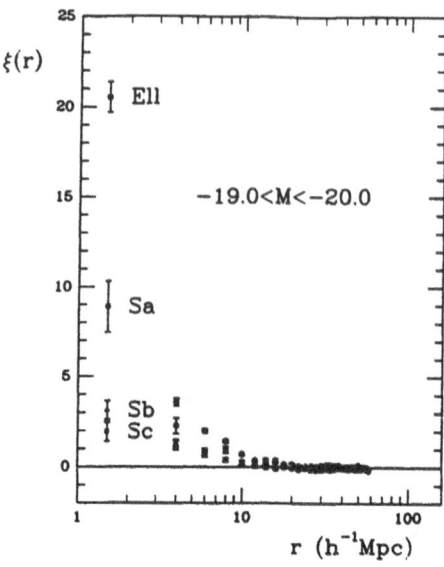

Fig. 16. The two–point spatial correlation function for a subsample of the Arecibo redshift survey as a function of the morphological type from Iovino *et al.* (1991).

dependence of luminosity on density. That galaxy clustering is a function of morphological type and indipendently of luminosity is illustrated in Fig. 16 where the autocorrelation function has been computed for galaxies in the narrow luminosity bin between $M = -19.0$ and $M = -20.0$. To test for luminosity and avoid spurious effects or large statistical correction, a rather narrow velocity cut-off has been choosen: $4000 \ km \ s^{-1} \leq v \leq 6000 \ km \ s^{-1}$. This makes it impossible to divide the sample into morphological classes, and then subdivide each class in one magnitude luminosity bins, while still having enough objects per class to allow an autocorrelation analysis. It has been preferred, therefore, to test if the absolute magnitude of the objects in pairs at small separation is significantly brighter than that of objects in pairs at large separation. This shows, with an high statistical significance, that the more luminous galaxies are typically more clustered than the fainter ones. In other words, luminosity seems to depend on the local density indipendently from the morphological type. This is further shown by plotting the mean value of the normalized number of neighbours as a function of the absolute magnitude, and for each morphological type (Fig. 17). If we choose a bin in magnitudes and we read the graph from the top to the bottom, the number of neighbours decreases for each morphological type. This is the autocorrelation function for that type. On the other hand the amplitude is larger for early morphological types, that is the segregation in morphology evidenced in Fig. 16. Reading each section for a selected morphology, we see that the most luminous galaxies have a larger number of neighbours, which means that in denser regions of space we tend to find brighter galaxies. This result seems to show rather convincingly that both morphology and luminosity are

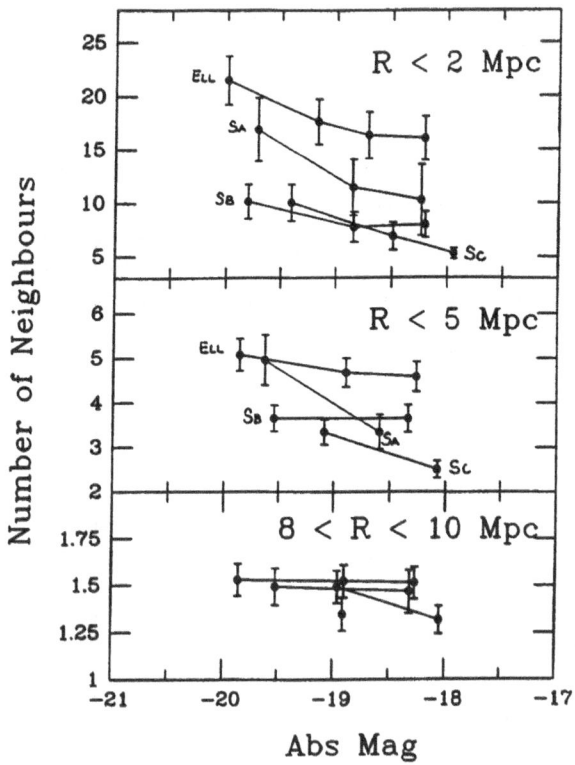

Fig. 17. The normalized number of neighbours within a distance R as a function of absolute magnitude and morphological type. For $R < 5\ h^{-1}\ Mpc$ the number decreases both as a function of type (morphology segregation) and luminosity (luminosity segregation) from Iovino et al. (1991).

two parameters which depend, indipendently from each other, from the density. However, for a different conclusion see Saslaw and Crane (1991).

7. Concluding Remarks

In the following we will simply mention a few points that should be more deeply understood in the near future.

— More effort should probably be directed in understanding what the distribution of objects on the supergalactic plane is telling us and how deep is the disagreement among observations of different tracers and the available theories or numerical models.

— How far shall we look for masses causing large scale motions and/or is the velocity field a better measure of the large scale matter distribution? What is the coherence scale of the flow?

— Since many years we are chasing a "fair sample" The definition of it is however operational, and it is a function of the questions we are trying to answer.

Is it a sample large enough that the mean values of the luminosity and density reflect the mean luminosity and density of the Universe? Large enough to measure the geometry of space, H_o and q_o, free of any local effect? Large enough that we feel confident to represent correctly the distribution of the various clustering scale lengths? Large enough that we might pin down evolutionary effects? Large enough to be a fair sample of the Universe!? It might well be that some of its properties can be determined only by considering the Universe as a whole.

— A network of walls and superclusters has been proposed to explain the rather simple but constraining information given by the pencil beam observations. How did all this come about and, now that so many groups have sophisticated codes for N-body simulations, is it possible to have eventually also some prediction power?

— Clustering at high redshift is one of the future most important fields of research if we want to understand how galaxies and large-scale structures formed and evolve. At the moment we have only indirect informations which, once they are put on more solid grounds, will be nevertheless fundamental to our knowledge. Here two future project could allow a break-through: a) the Survey Telescope proposed by the Chicago–Princeton consortium, and b) the wide–field X–ray telescope (WFXT), an endeavour proposed to the American and Italian space agencies (NASA & ASI) following an idea by Giacconi.

— How important are the properties of galaxies which are related to the environment to the understanding of the global picture?

— Ultimately the properties of the Universe must be described in a quantitative way. It is not yet clear which is the most appropriate statistics and the role of the interplay between models and statistics: how strongly the latter is uncompromised by the former?

References

Bahcall, N.A., 1986 Ap.J. 302, L41

Bahcall, N.A., 1988, Ann. Rev. Astron. & Astrophys. 26, 631

Bahcall, N.A., & Soneira, R.M., 1983, Ap.J. 270, 20

Bachall, N.A., & Chokshi, 1991, preprint

Bachall,N.A., & Soneira, R.A., 1984, Ap.J. 277, 27

Baffa, L., Chincarini, G., Henry, D., Manoussoyanaki, J., & Salinari, P., 1991, in preparation

Bardeen, J.M., Bond, J.R., Kaiser, N., & Szalay, A.S., 1986, Ap.J., 304, 15

Börner,G., 1988, "The Early Universe", Springer–Verlag

Branchini, E., Guzzo, L., Provenzale, A., & Governato, F., 1991, in preparation

Broadhurst, T.J., Ellis, R.S., Koo, D.C., & Szalay, A.S., 1990, Nature 311, 726.

Batuski, D.J., & Burns, J.O., 1985, Ap.J. 299, 5

Burstein, D., 1986, private communication

Chincarini, G., 1988, in "Origin, Structure and Evolution of Galaxies", Fang Li Zhi ed.

Chincarini, G., Giovanelli, R., & Haynes, M.P., 1983, A. & A. 121, 5

Chincarini, G., & Vettolani, G., 1987, in "Observational Cosmology", A. Hewitt, G. Burbidge, and F. Li Zhi eds., p.275

Davis, M., & Huchra, J.P., 1983, Ap.J. 254, 437

Davis, M., & Peebles, P.J.E., 1977, Ap.J. Suppl. 34, 425

Dekel, A., & Aarseth, S.J., 1984, Ap.J. 283, 1

Di Stefano,A., Rampazzo, R., Chincarini, G., & de Souza, R., 1990, A. & A. Suppl. 86, 7

Dressler, A., 1980, Ap.J. 236, 351

Faber, S.A., Burstein, D., 1988, in "Large Scale Motions in the Universe", V.C. Rubin ang G.V. Coyne eds., Princeton University Press, p. 256

Geller, M.J. & Huchra, J.P., 1988, in "Large Scale Motions in the Universe", V.C. Rubin ang G.V. Coyne eds., Princeton University Press, p. 3

Geller, M.J. & Huchra, J.P., 1989, Science 246, 897

Giovanelli, R., Haynes, M.P., and Chincarini, G, 1986, Ap.J. 300, 77

Guzzo, L., & Branchini, E., 1991, in preparation

Guzzo, L., Nichol, R.C., Collins, C.A. & Lumsden, S.L., 1989, The Messenger 60, 45.

Guzzo, L., Nichol, R.C., Collins, C.A. & Lumsden, S.L., 1991, Ap.J. Letters, submitted

Guzzo, L., Iovino, A., Chincarini, G., Giovanelli, R., & Haynes, M.P., 1991b, Ap.J. Letters, in press

Hamilton, A.J.S., 1988, Ap.J. Lett. 331, L59

Han, M., & Mould, J., 1990, Ap.J. 360, 448

Hauser, M.G., & Peebles, P.J.E., 1973, Ap.J., 185, 757

Haynes, M.P., & Giovanelli, R., 1988, in "Large Scale Motions in the Universe" V.C. Rubin & G.V. Coyne, Princeton University Press, p. 31

Haynes, M.P., Giovanelli, R. & Chincarini, G., 1984, Ann. Rev. Astron. Astrophys., 22, 445

Huchra et al. 1990, Ap.J. 365, 66

Icke, V., & van de Weygaert, R., 1987, A. & A. 184, 16

Iovino, A., and Shaver, P.A., 1988, Ap.J. Letters 330, L13

Iovino, A., Giovanelli, R., Haynes, M.P., Chincarini, G., & Guzzo, L., 1991, Ap.J., submitted

Jaffe, W., & Gavazzi, G., 1986, A.J. 91, 204

Kaiser, N., 1984, Ap.J. 284, L9

Kaiser, N., 1987, MNRAS 227, 1

Lumsden, S.L., Nichol, R.C., Collins, C.A., & Guzzo, L., 1991, MNRAS, submitted

Lynden-Bell et al. 1988, Ap.J., 326, 19

Maddox, S., 1991, in "The distribution of Matter in the Universe", G. Mamon ed., Kluwer, in press

Nichol, R.C., 1992, Ph.D. Dissertation, University of Edinburgh

Oort, J.H., 1958, in "Onzième Conseil de Physique", Institut International de Physique Solvay

Oort, J.H., 1983, Ann. Rev. Astron. & Astrophys. 21, 373

Peebles, P.J.E., 1980, "The Large Scale Structure of the Universe" Princeton University Press, Princeton

Pietronero, L., 1987, Physica 144 A, 257

Postman, M. & Geller, M.J., 1984, Ap.J. 281, 95

Postman, M., Huchra, J.P., Geller, M.J., & Henry, P.J., 1985, A.J. 90, 1004

Postman, M., Spergel, D.N., Sutin, B., Juszkiewicz, R., 1989, Ap.J., 346, 588

Regös, O., & Geller, M.J., 1991

Rood, H.J., 1988, Ann. Rev. Astron. & Astrophys. 26, 245

Rowan-Robinson et al. 1990, MNRAS 247, 1

Saslaw, W.C., & Crane, P., 1991, ESO preprint

Saunders, W., et al. , 1991, Nature 349, 32

Scaramella, R., 1988, Ph.D. Thesis, S.I.S.S.A., Trieste.

Scaramella, R., Vettolani, G., Zamorani, G., 1991 Ap.J. Letters 376, 1

Scaramella, R., Baiesi-Pillastrini, G., Chincarini, G., Vettolani, G., Zamorani, G., 1989, Nature 338, 562

Shapley, H., & Ames, A., 1932, Harvard Annals 88, 43

Shaver, P.A., 1991, Austr. J. of Phys. in press

Strauss, M., & Davis, M., 1988, in "Large Scale Motions in the Universe", V.C. Rubin ang G.V. Coyne eds, Princeton University Press, p.256

Strauss, M., & Huchra, J.P., 1989

Sutherland, W.J., & Efstathiou, G.P., 1991, MNRAS 248, 159

Szalay, A.S., & Silk, J., 1983, Ap.J. 264, L31

Szalay, A.S., Ellis, R.S., Koo, D.C., Broadhurst, T.J., 1991, in "After the First Three Minutes" S. Holt ed., in press

Tully, R.B., 1986, Ap.J., 303, 25

Tully, R.B., 1987, Ap.J., 323, 1

Tully, R.B., Scaramella, R., Vettolani, G., Zamorani, G., 1991, Ap.J.,in press

Valls-Gabaud, D., Alimi, J.-M., & Blanchard, A., 1989, Nature 341, 215

Van de Weygaert, R., 1991, MNRAS 249, 159

West, M.J., 1991, Ap.J., in press

Willik, J.A., 1990, Ap.J. Letters, 351, 5

Williams, B., Peacock, J.A., & Heavens, A.F., 1991, MNRAS, in press

Whitmore, B.C., 1990, in "Clusters of Galaxies", Oegerle, W.R., Fitchett, M.J. & Danly, L. eds. (Cambridge: Cambridge University Press) p. 139, and references therein.

Yahil, A., 1988. in "Large Scale Motions in the Universe", V.C. Rubin ang G.V. Coyne eds., Princeton University Press, p. 219

Zamorani, G., Scaramella, R., Vettolani, G., & Chincarini, G., 1991, in proceedings of Ringberg workshop "Traces of Primordial Structure", H. Böhringer ed., in press

DISTRIBUTION AND PROPERTIES OF SUPERCLUSTERS

NETA A. BAHCALL
Astrophysical Sciences
Princeton University
Princeton, NJ 08544-1001

ABSTRACT. The distribution and properties of superclusters and their relation to the large-scale structure of the universe are discussed. Different types of observations are reviewed including the spatial distribution of galaxies, clusters of galaxies, narrow pencil-beam surveys, and quasars. A consistent picture regarding supercusters and the large-scale structure they trace appears to be emerging from the above observations. A network of large-scale superclusters, up to ~ 100 - 150 Mpc in scale, is suggested. The supercluster network surrounds low-density regions of similar scales, suggesting a "cellular" structure of the universe. ($H_0 = 100$ km /s/ Mpc is used throughout.)

1. Introduction

The existence of large-scale systems of galaxies - superclusters - has been known for over half a century. Shapley (1930) noticed a large remote "cloud of galaxies" in Centaurus, which he estimated to be ~ 14 times more distant than the "Coma-Virgo Cloud A". Recent observations reveal that this system, known as the Shapley Supercluster, is a large structure (~ 50 Mpc in size), that is rich and dense in clusters of galaxies (Raychaudhury *et al.* 1991). Zwicky, in 1937, noticed the very large galaxy concentration in Pisces, that also encompasses several clusters. Abell (1958) was the first to notice that rich clusters of galaxies were themselves clustered into second order clustering, i.e., superclusters. The scales of the above superclusters reached tens of Mpc.

The terminology of superclusters generally implies a large-scale concentration of galaxies, considerably larger than a typical rich cluster of ~ 1 - 2 Mpc radius. The specific definition of superclusters may vary however according to the method of selection. Some superclusters refer to a rich cluster of galaxies with a large, extended halo of galaxies (e.g., Virgo supercluster). Others may refer to the clustering of clusters of galaxies which may extend to ~ 100 Mpc scale (e.g., Corona Borealis supercluster; Abell 1958, Bahcall and Soneira 1984). Still other superclusters may be detected by the gravitational influence of their large masses on the peculiar velocities of nearby galaxies (e.g., the Great Attractor; Lynden-Bell *et al.* 1988). In either case, superclusters are now known to be very large systems (~100 Mpc), irregular in shape, with no well-defined boundaries.

What are the properties of superclusters? What is their shape and how are they distributed in space? While detailed answers to these questions are not yet available, a great deal has been learned about superclusters in the last decade or so. Since large superclusters have a rather low space density, a large survey is needed in order to answer the above questions with accuracy. Such surveys are planned for the coming decade.

A. C. Fabian (ed.), Clusters and Superclusters of Galaxies, 275–292.

In this paper I summarize the main properties and distribution of superclusters as determined from different types of observations: from the spatial distribution of galaxies, clusters of galaxies, narrow pencil-beam surveys, as well as the distribution of quasars and AGNs. A consistent picture regarding superclusters and the large-scale structure they trace in the universe appears to be emerging from the above observations. A network of large-scale superclusters, up to ~ 100 - 150 Mpc in scale, is suggested. The supercluster network surrounds lower density regions of similar scales, suggesting a "cellular" structure of the universe, similar to the "pancake" model discussed by Zeldovich and collaborators.

The following topics will be addressed in the paper:
•Section 2: Supercluster Properties
•Section 3: Superclusters surrounding Voids
•Section 4: Superclusters and Pencil-Beam Surveys
•Section 5: Geometrical Models of Superclusters
•Section 6: The Cluster Correlation Function
•Section 7: Quasars and AGNs in Superclusters
•Section 8: Conclusions

2. Supercluster Properties

Some of the early redshift surveys of galaxies have already revealed that superclusters are extended regions in the galaxy distribution that are flattened or filamentary in shape. Gregory and Thompson (1978) obtained a redshift survey of galaxies in the direction of the Coma cluster. They find the large, flattened Coma supercluster (Fig. 1), which is

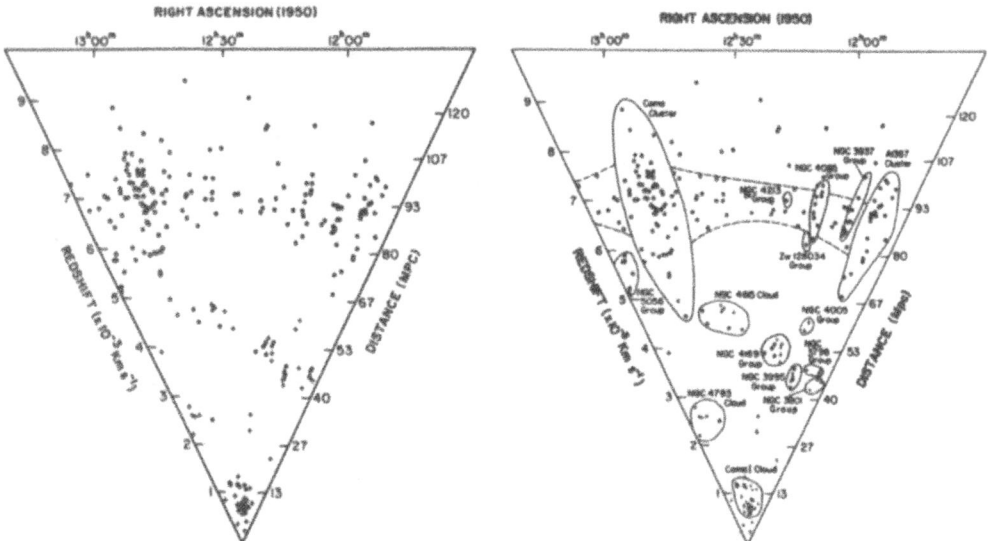

Figure 1. Redshift-cone diagram of the Coma supercluster (Gregory and Thompson 1978). The elongated supercluster is part of the Great-Wall. The supercluster borders a low-density region ("void").

part of the recently named Great-Wall, extending to at least ~ 40 Mpc. The supercluster appears to surround a large under-dense region of comparable size. Additional surveys by Gregory *et al.* 1981, and Chincarini *et al.* 1981 yielded similar results in the Hercules and Perseus superclusters. More recent galaxy redshift surveys (Giovanelli *et al.* 1986, de-Lapparent *et al.* 1986, da Costa *et al.* 1988) reveal similar large-scale superclusters surrounding low density regions.

Large-scale superclusters have been traced very successfully also by rich clusters of galaxies (Abell 1958, Bahcall and Soneira 1984). Like mountain-peaks tracing extended mountain-chains, so do the rich clusters trace the large-scale superclusters.

A complete catalog of superclusters - defined as clusters of clusters of galaxies - was constructed by Bahcall and Soneira (1984; hereafter BS84) from a complete redshift sample of rich Abell (1984) clusters to $z \leq 0.08$. The catalog identifies all superclusters with $z \leq 0.08$ located at $|b| \geq 30°$ and $\delta > -27°$ that include richness class $R \geq 1$ or $R \geq 0$ clusters at a spatial density enhancement $f \geq 20$ times larger than the mean cluster density. A map of the supercluster catalog for different f values is presented in Figure 2.

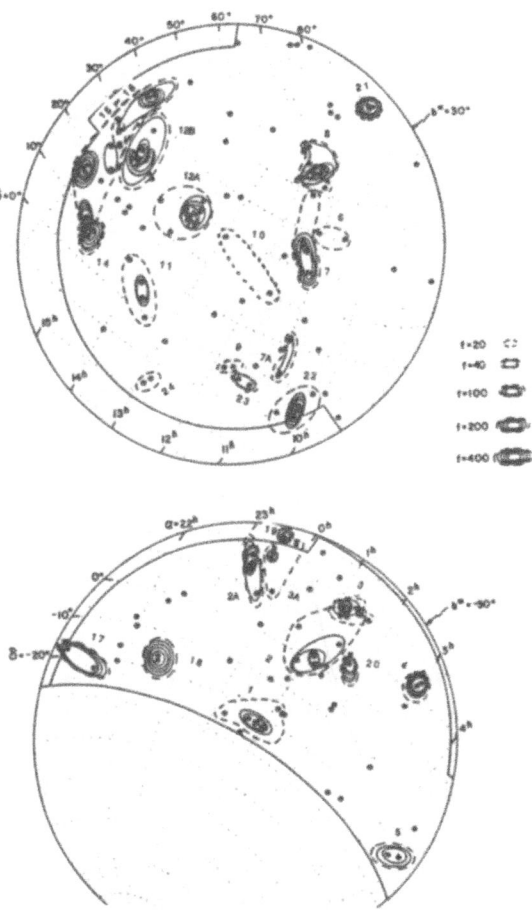

Figure 2. Projected contour map of the Bahcall-Soneira (1984) superclusters. The density contours represent the spatial density enhancement of the superclusters for $f \geq 20$.

A total of 16 superclusters are cataloged for $R \geq 1$ and $\underline{f} \geq 20$, and 26 superclusters for R ≥ 0 and $f \geq 20$. Some properties of the Bahcall-Soneira superclusters are summarized below. The mean density of superclusters is $\sim 10^{-6}$ Mpc^{-3}, with an average mean supercluster separation of ~ 100 Mpc. The number of clusters per supercluster varies from 2 to 15 for the $f \geq 20$ superclusters; the average number of clusters per supercluster is approximately three. The superclusters appear to contain a large fraction of all clusters: $\sim 54\%$ at $f \geq 20$. Comparisons with random catalogs show that this fraction is considerably higher than expected by chance. This indicates that most of the high overdensity superclusters are real physical systems of the largest scale yet observed. The linear size of the largest observed superclusters are ~ 150 Mpc (e.g., Corona Borealis, at $\sim 15h + 30^\circ$). These superclusters appear to be elongated in shape. The fractional volume of space occupied by the superclusters is very small: $\sim 3\%$ at $f \geq 20$, and decreases rapidly with f.

A redshift-cone diagram of the superclusters in the declination slice $\delta = 0^\circ$ - 40° is presented in Figure 3.

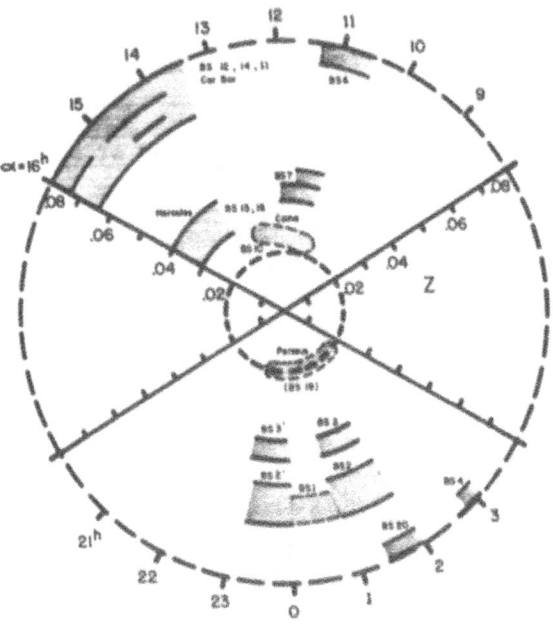

Figure 3. Redshift-cone diagram of the Bahcall-Soneira superclusters in the $\delta = 0^\circ$ - 40° slice. (See Bahcall 1991 for more details.) The Coma-Hercules supercluster union is the Great Wall.

The superclusters are numbered by their cataloged number, and some are also listed by their popular common name (such as Coma, Hercules, and Corona Borealis superclusters). The typical mean separation of the superclusters, ~ 100 Mpc, is apparent in the diagram. We shall also see below (§3) that the superclusters appear to surround large, low-density regions such as the Bootes void as well as underdense regions seen in pencil-beam surveys. (For example, the Bootes void of Kirshner et al. 1981, is located in the region between the Hercules and Corona Borelais superclusters; Fig. 3.) The minimum extent of the superclusters (i.e., the size containing the high density $f \geq 20$ region of clusters, not counting any extended halos of the superclusters), appear to be ~ 150 Mpc for the largest systems (eg., Corona Borealis).

How do these superclusters compare with the structures found by galaxy redshift surveys? In Figure 4, I superimpose the supercluster contours from Figure 3 on top of the cumulative galaxy redshift map from the CfA survey (Geller and Huchra 1989), plotted on the same scale. It is clear that the superclusters identified by the clustering of clusters highlight well the main large-scale systems seen in the galaxy survey in the overlap region. In particular, the union of the Coma and Hercules superclusters of Figure 3 constitute the "Great-Wall" seen in the CfA survey, as well as in the Gregory and Thompson survey of Figure 1. The Great-Wall is thus a merging of two BS superclusters, with a total extent of ~150Mpc and thickness of $\lesssim$ 10 Mpc. The supercluster extent in declination is not yet certain, but appears to be $\gtrsim$ 70 Mpc. This extent and flattened shape is comparable to the other large superclusters in the BS catalog; for example, the Corona Borealis supercluster is another such Great-Wall considerably greater and richer than Coma-Hercules. It is located behind a large void in Bootes; the void is surrounded by the Hercules supercluster in front and Corona Borealis in the back. This comparison of the galaxy and cluster distribution indicates that the large-scale structure traced by both galaxies and rich clusters is consistent with each other; both find the same superclusters. While the rich clusters are most efficient in finding the largest-scale structures, the galaxies are essential for tracing the small-scale connectedness to the larger scales. Combining both galaxy and cluster redshift survey information is thus a powerful tool in surveying large volume samples for superclusters.

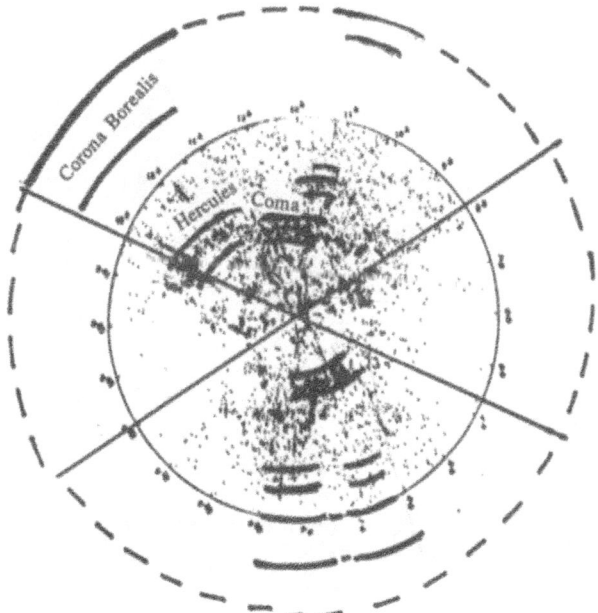

Figure 4. The Bahcall-Soneira supercluster contours of Fig. 3 superpsoed on the CfA galaxy redshift distribution. The BS superclusters highlight the main galaxy superclusters in the CfA survey. The Great-Wall is the union of the Coma-Hercules superclusters.

A rather different method of finding superclusters is that used by Lynden-Bell *et al.* (1988) utilizing peculiar velocity information to infer the existence of large massive superclusters such as the Great Attractor. The estimated mass of the Great Attractor, ~ 5 x 10^{16} $M_\odot$ (Lynden-Bell *et al.* 1988) is comparable to that of the large Bahcall-Soneira superclusters. The Great Attractor does not appear however to contain rich clusters.

In summary, we see that clusters, galaxies, and velocity fields (as well as pencil-beam surveys; §4), appear to trace similar superclusters. These superclusters are the largest systems yet observed. Their sizes extend to ~ 150^2 x 20 Mpc^3, and their mass is estimated to be ~ 2 - 10 x 10^{16} $M_\odot$ (e.g., Bahcall 1988). This mass is comparable to the mass of ~ 20 - 50 rich clusters. The superclusters, Great Walls, and Great Attractors appear to all be similar systems. There are some indications that the supercluster distribution is not random. Bahcall and Burgett (1986) suggest positive correlations among superclusters on scales ~ 100 - 150 Mpc (Figure 5). Uncertainties, however, are large; a considerably larger cluster survey is needed in order to obtain a more accurate determination of the supercluster correlation.

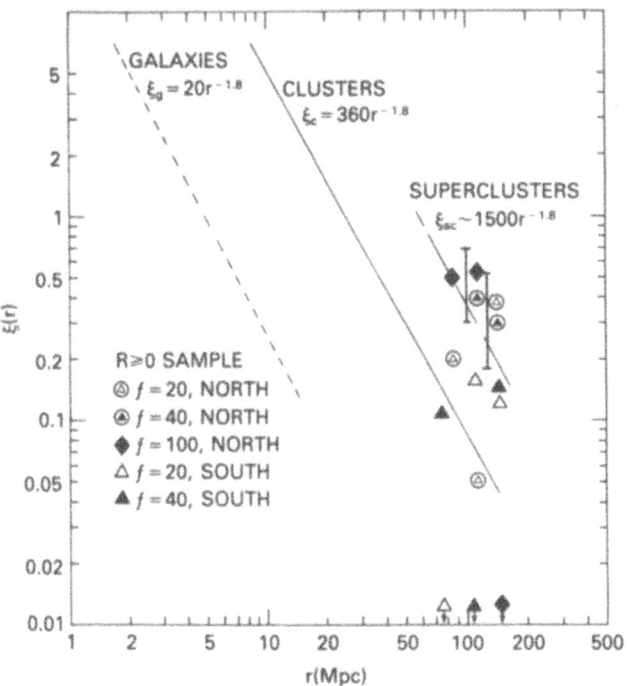

Figure 5. The spatial correlation function of superclusters (Bahcall and Burgett 1986). Detection of correlations in this sample are possible only for scales $50 \lesssim r \lesssim 200$ Mpc.

3. Superclusters Surrounding Voids

The BS84 supercluster catalog was used by Bahcall and Soneira (1982) to study the area around the large, ~ 60 Mpc diameter void of galaxies in Bootes detected by Kirshner *et al.* (1981). The largest, densest superclusters are located near and around the area devoid of galaxies. The Bootes void, at ~ 14.5h + 50°, is located near superclusters BS12 (Corona Borealis) and BS15+16 (Hercules) (~ 100 Mpc away in projection; see Figure 2 and BS84). In the redshift-cone diagram of Figure 3, the void fills the space between Hercules (part of the Great-Wall), and Corona Borealis (the next Great-Wall). It is interesting to note that the overdensity of galaxies observed by Kirshner *et al.* (1981) on both redshift sides of the void, at $z \simeq 0.03$ and $z \simeq 0.08$, coincide in redshift space with these two surrounding superclusters (Figure 6). This suggests that the large superclusters surround the galaxy void (at $z \simeq 0.05$), and that the tails of their galaxy distributions account for the overdensities observed $\gtrsim 100$ Mpc away by Kirshner *et al.* This connection provides a strong indication of large halos (~ 150 Mpc) to rich superclusters.

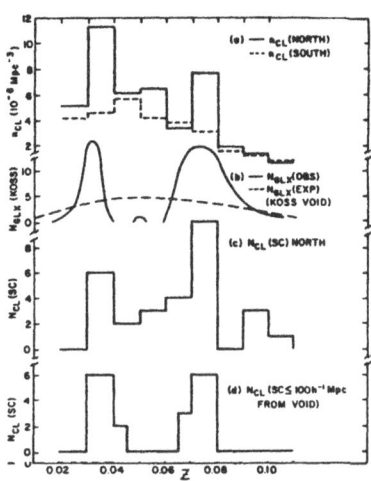

Figure 6. The frequency distribution of Abell clusters (N_{cl}), and galaxies in the Bootes direction (N_{glx}), as a function of redshift (Bahcall and Soneira 1982). The Hercules supercluster at $z \simeq 0.03$ and the Corona Borealis supercluster at $z \simeq 0.08$ appear to surround the void at $z \simeq 0.05$.

Previous observational evidence (Gregory and Thompson, 1978, Gregory *et al.* 1981, Chincarini *et al.* 1981) together with these results, as well as similar conclusions regarding comparisons with pencil-beam surveys (§4) and large galaxy redshift surveys (§2) suggest that galaxy voids are generally associated with surrounding galaxy excesses; the bigger the void, the stronger may be the related excess (see also §4 and §5).

4. Superclusters and Pencil-Beam Surveys

Recent observations of the redshift distribution of galaxies in narrow (~ 40 arcmin.) pencil-beam surveys to $z \lesssim 0.3$ (Broadhurst *et al.* 1990; hereafter BEKS) reveal a highly clumped and apparently periodic distribution of galaxies. The distribution features peaks of galaxy counts with an apparently regular separation of 128 Mpc, with few galaxies between the peaks. What is the origin of this clumpy, periodic distribution of galaxies? What does it imply for the nature of the large-scale structure and the superclustering properties discussed above? Bahcall (1991) investigated these questions observationaly, by comparing the specific galaxy distribution with the distribution of known superclusters (§2).

The pencil-beam peaks in the galaxy distribution are unlikely to originate in the main regions of rich clusters of galaxies. The small size of clusters (~ 1 - 2 Mpc radius), combined with their low space density (~ 10^{-5} Mpc^{-3}), yield a much lower probability than observed (by a factor of ~ 50) of clusters being intersected by a narrow beam survey.

Bahcall (1991) showed that the observed galaxy clumps originate from the tails of the large-scale Bahcall-Soneira superclusters. When the narrow-beam surveys intersect these superclusters, which have an average mean separation of ~ 100 Mpc (§2), the distribution of galaxies reported by BEKS is reproduced.

The redshift distribution of the superclusters in the δ = 0° - 40° slice (Figure 3) is plotted as a histogram (shaded area) in Figure 7 (with total count normalized to the number of clusters per supercluster). This distribution is superimposed on the narrow-beam galaxy distribution of BEKS (as presented in their preprint for the combined beam surveys, corrected for the observed selection function). It is apparent from Figure 7 that the supercluster distribution and the BEKS galaxy distribution are essentially identical for z ≲ 0.1. It indicates that the galaxy clumps observed in the pencil-beam survey originate from these superclusters as the beam crosses the superclusters' surface. The main superclusters that contribute to the clumps are indicated in Figure 7. The first northern clump originates from the Coma-Hercules supercluster (= the Great-Wall); the second northern clump is mostly due to the large Corona Borealis supercluster (BS12) (see also Figure 3). The first southern clump is due to the tail of the Perseus supercluster, and the second, rather broad clump is due to a few superclusters in this region.

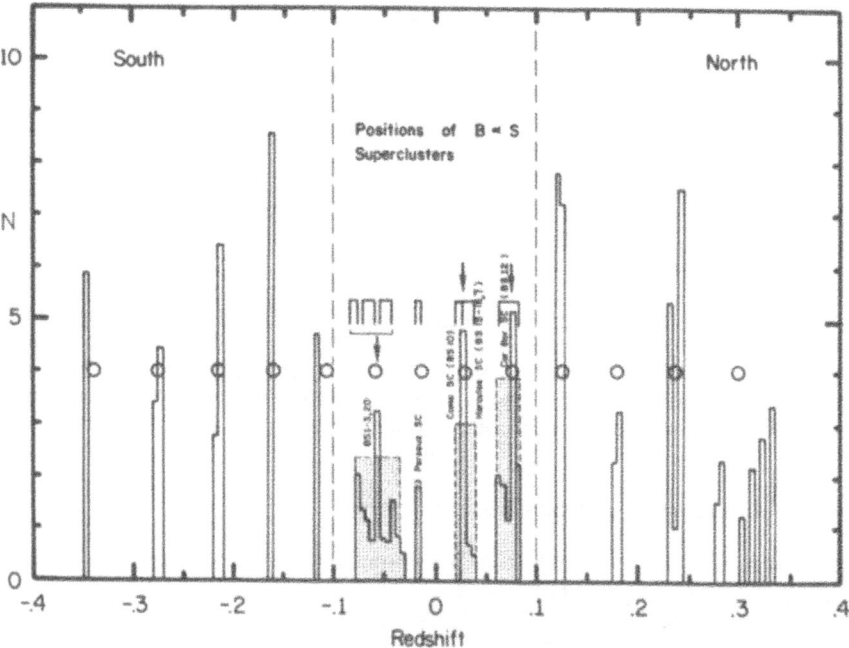

Figure 7. Histograms of the redshift distribution of the Bahcall-Soneira superclusters in the slice δ = 0° - 40° (Fig. 3) (shadded area), for z ≲ 0.1, superposed on the BEKS galaxy distribution (corrected for selection; BEKS preprint). The specific location and names of the BS superclusters are marked (Bahcall 1991).

The narrow-beam survey of BEKS is directed toward the north and south galactic poles. Some of the BS superclusters coincident with the BEKS peaks are located at projected distances of up to ~ 50 - 100 Mpc from the poles. This suggests, similar to the Bootes void analysis (§3), that the high-density supercluster regions are embedded in larger surfaces, ~ 100 Mpc in size, and that these large pancake-like structures surround large underdense regions (or voids).

The observed number of clumps and their typical mean separation are consistent with the mean number density of superclusters and their estimated average extent. The BEKS mean separation of ~ 130 Mpc corresponds to an average supercluster cross-section of ~ (90 Mpc)^2; this is comparable to the observed extent of superclusters (~ 100 x 100 x 20 Mpc; §2 – 3).

The narrow widths of the BEKS peaks are consistent with, and imply, the flatness of superclusters. From simulations of superclusters and pencil-beam surveys, Bahcall and Miller (1991) find that the distribution of the observed peak-widths is consistent with that expected of randomly placed superclusters with ≲ 20 Mpc width (and ~ 100 - 150 Mpc extent). See Figure 8.

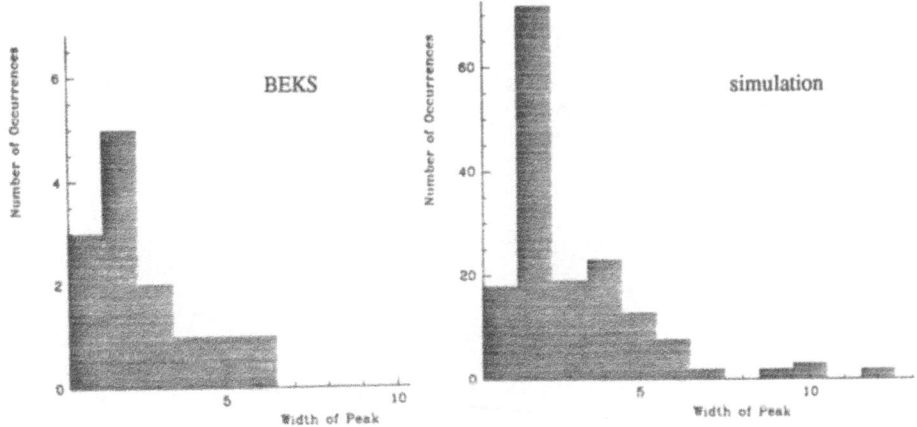

Figure 8. Histogram of the width distribution of galaxy peaks in the BEKS (left) and simulated (right) surveys. The simulated survey contains superclusters of dimensions 150 x 150 x 20 Mpc^3 (Bahcall and Miller 1991). The peak-widths are in 10 Mpc units.

The BS superclusters exhibit weak positive correlations on scales ~ 100 - 150 Mpc (§2). This implies that the superclusters, and thus their related galaxy clumps, are not randomly distributed but are located in some weakly correlated network of superclusters and voids, with typical mean separation of ~ 100 - 150 Mpc. This picture is consistent with statistical analyses of the BEKS distribution (eg. Kurki-Suonio 1990, Ikeuchi and Turner 1991, Park and Gott 1991, Bahcall and Miller 1991), as well as with the observational data presented in sections 2 and 3.

Based on the above picture, what would we expect future narrow-beam surveys in different directions to yield? Figures 2,3 and 7 suggest that surveys in different directions will continue to yield a highly clumped distribution of galaxies, not necessarily periodic, but with typical mean separation of ~ 100 - 150 Mpc. The positions of the clumps will change somewhat with direction. Large changes of beam direction (≳ 150 Mpc) should yield more significant variation in peak positions. The apparent periodicity in the galaxy distribution will be greatly reduced when pencil-beam surveys in various directions are combined.

5. Geometrical Models of Superclusters

The observational data described above (§1 - 4) suggest a "cellular" structure in the universe (e.g. a Zeldovich "pancake" model), in which large-scale flattened superclusters surround low-density regions. Such a model was simulated by Bahcall, Henriksen and Smith (1989), where galaxies were placed on surfaces of randomly placed shells, and clusters were placed at shell intersections. It was found that such a "cellular" model produced cluster correlations that are consistent with observations, showing the large increase in correlation strength (§6) from galaxies to clusters (Figure 9). The results are not very sensitive to the exact parameters used. The model galaxy correlations are also consistent with observations, even showing the tail of weak positive correlations at large separations recently reported by the APM survey (Maddox *et al.* 1990). These results suggest that the observed strong cluster correlation function may be due to the global geometry in which clusters are positioned on randomly placed shells or similar structures; the typical structure size is best fit with a radius of ~ 20 Mpc. Similar simulations based on the explosion model for shell formation were also carried out by Weinberg *et al.* (1989) with similar results.

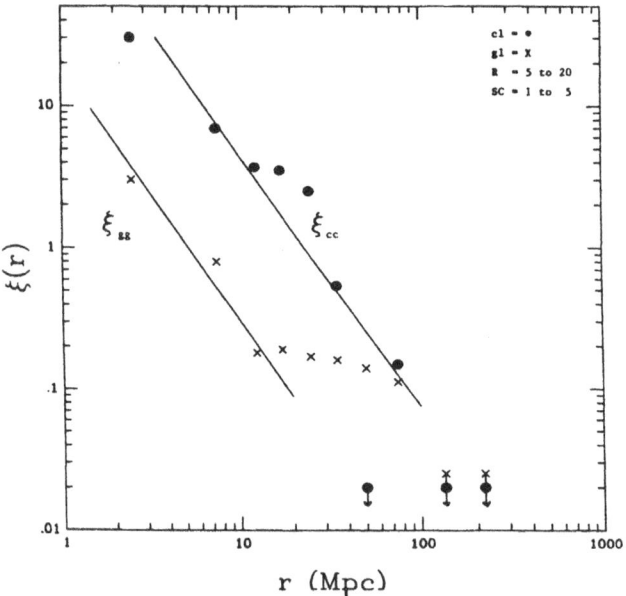

Figure 9. Shell-model correlation functions for clusters (dots) and galaxies (crosses), and their comparisons with observations (ξ_{cc} and ξ_{gg} lines). (Bahcall, Henriksen and Smith 1989.)

6. The Cluster Correlation Function

A large fraction of clusters ($\gtrsim 50\%$) are located in superclusters (§2). For comparison, only ~ 5% of galaxies are located in rich clusters. The clumpiness of such a large fraction of clusters is the cause of the strong correlation function observed among clusters. The cluster correlation function is stronger than the galaxy correlation function by a factor of ~ 15 (Bahcall and Soneira 1983, Bahcall 1988); the correlations yield, respectively,

$$\xi_{cc}(R \geq 1) \simeq 300 \, r^{-1.8} \qquad \text{Clusters} \ (R > 1) \qquad (1a)$$

$$\xi_{gg} \simeq 20 \, r^{-1.8} \qquad \text{Galaxies} \qquad (1b)$$

Many different samples and catalogs of clusters have now been analyzed, all yielding consistent results with (1) above (Klypin and Kopylov 1983, Shectman 1985, Postman *et al.* 1986, Bahcall *et al.* 1986, Huchra *et al.* 1990, Lahav *et al.* 1989 for X-ray selected clusters, West and van den Bergh 1991 for cD selected clusters, Postman *et al.* 1991). A composite of some of the data is shown in Figure 10.

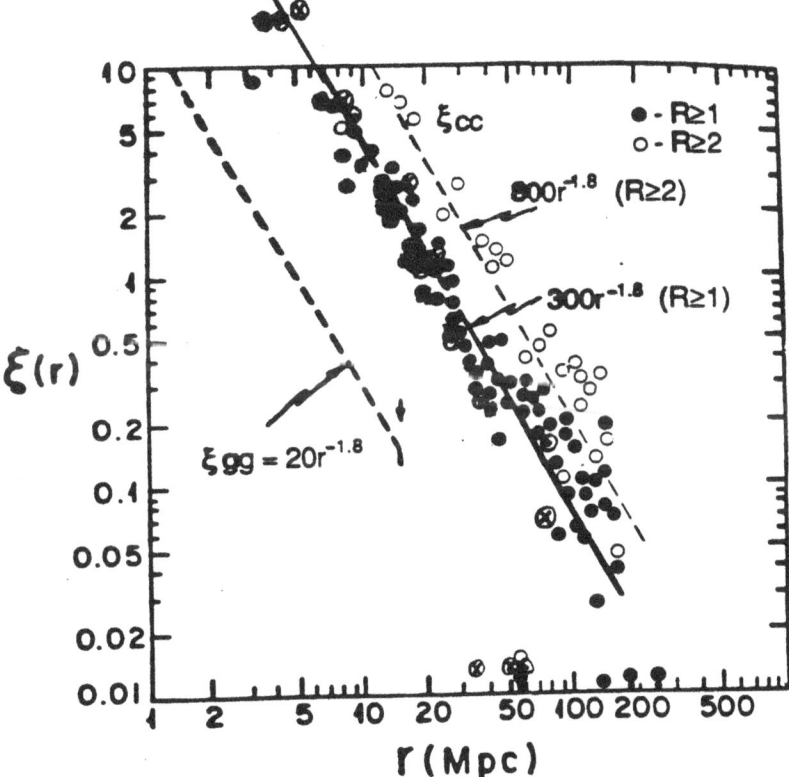

Figure 10. A composite of the spatial correlation function of clusters from different investigators and different cluster samples, including X-ray and cD clusters.

It is important to emphasize that *all* observational determinations of the correlations of rich clusters, for richness class $R \geq 1$, yield correlation scales that are in the range

$$r_0 \simeq 22 \pm 2 \, \text{Mpc} \qquad\qquad R \geq 1 \, \text{Clusters} \qquad\qquad (2)$$

(where $\xi (r) = A r^{-1.8} = (r/r_0)^{-1.8}$). This includes different catalogs (Abell, Zwicky, Shectman), as well as X-ray selected clusters and cD selected clusters. The correlation results do not appear to be significantly influenced by systematics or projection effects.

It has also been shown (Bahcall and Soneira 1983, Bahcall 1988) that the cluster correlation function is richness-dependent: the correlation amplitude increases with the richness of the galaxy clusters. This richness dependence for clusters of various richness classes is presented in Figure 11.

A relation that approximately describes this dependence is given by (Bahcall 1988)

$$\xi_i (N_i) \simeq (20 \, N_i^{2/3}) \, r^{-1.8} \qquad\qquad (3)$$

where N_i is the Abell (1958) defined richness of the cluster i and $\xi_i (N_i)$ is the correlation function of clusters of richness N_i. This richness-dependent correlation appears to hold well. It is of interest to note that the newly determined cluster correlation function of the APM survey (Dalton *et al.* 1991) is consistent with the richness-dependent cluster correlations; their correlation scale of ~ 13 Mpc is consistent with that expected for the poorer richness threshold of the APM clusters (Bahcall and West 1991).

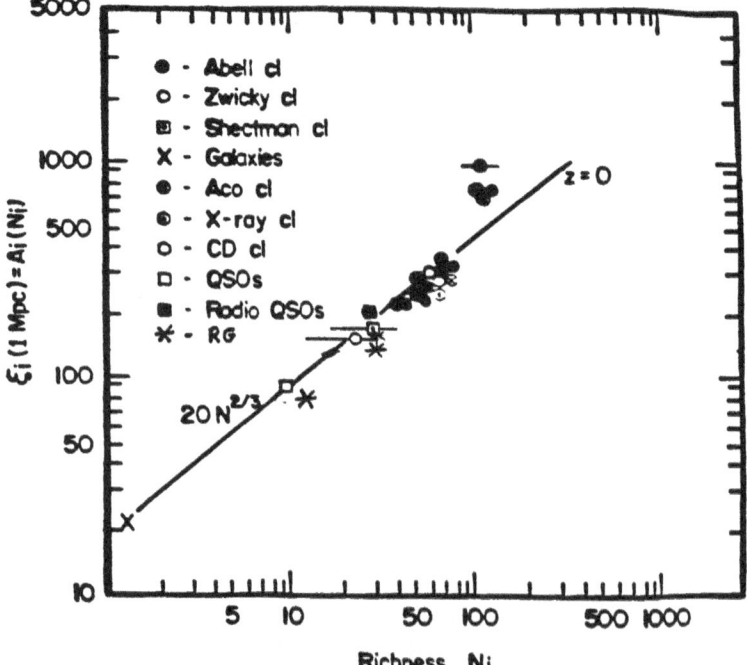

Figure 11. The richness-dependent cluster correlation function. Data points include different samples and catalogs of clusters, as well as X-ray and cD clusters. Quasars and radio-galaxies, represented by their parent-groups (§7), are also shown.

A second dependence of the cluster correlations is observed as a function of the mean space density, n (or mean separation, d $\propto$ n$^{-1/3}$) of the clusters (Figure 12). This dependence, $\xi \propto$ d$^{1.8}$, yields a *universal* dimensionless correlation function, when normalized by the mean separation of clusters (Szalay and Schramm 1985, Bahcall 1988):

$$\xi_i \, (d_i) \simeq 0.2 \, (r/d_i)^{-1.8} \qquad (4a)$$

or, equivalently, the correlation-scale is approximated by:

$$r_{0,i} \simeq 0.4 \, d_i \qquad (4b)$$

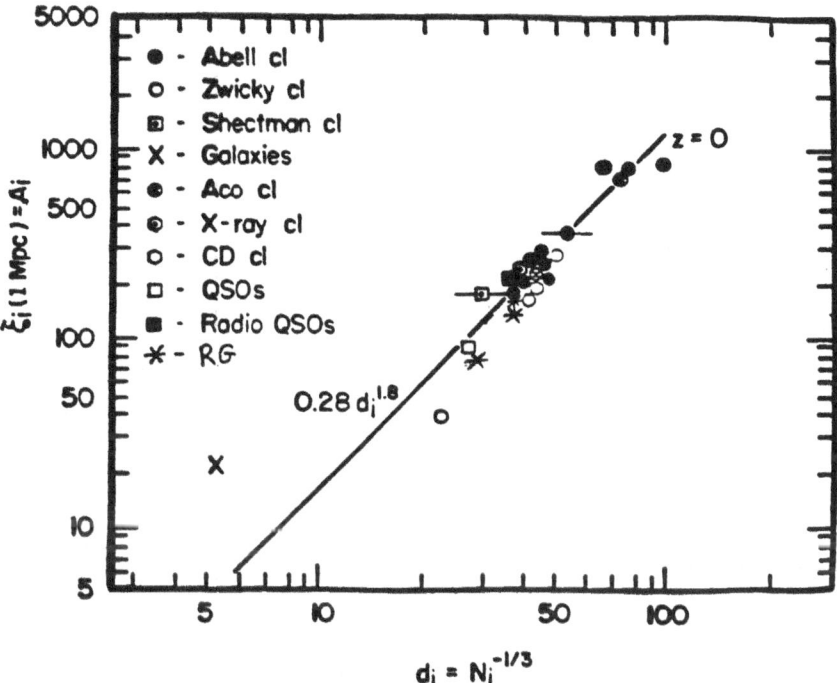

Figure 12. The *universal* dimensionless cluster correlation function. All other notations are the same as in Fig. 11.

Again, this dependence appears to hold well for all systems studies so far. The new APM clusters, with a mean density *four* times larger than that of the R $\geq$ 1 clusters (2.4 x 10^{-5} Mpc^{-3} versus 0.6 x 10^{-5} Mpc^{-3}, respectively), fit well this *predicted* relation (i.e., d(APM)/d(R $\geq$ 1) $\simeq$ 4$^{-1/3}$ $\simeq$ 0.63, and thus r$_0$ (APM) $\simeq$ 0.63 x r$_0$(R $\geq$ 1) $\simeq$ 0.63 x 22 Mpc $\simeq$ 13.8 Mpc, as is indeed observed by the APM).

The universal dimensionless cluster correlation function (eq. 4) is consistent with a fractal structure in the distribution of clusters. This may simply reflect the "cellular" geometry discussed in the previous sections in which clusters placed at "cell" intersections represent a similar, scale-invariant structure when normalized by their mean-separation (e.g. §5).

7. Quasars and AGNs in Superclusters

7.1 QUASARS

Observations over the last several years reveal that the distribution of quasars in space is not uniform, but is rather strongly correlated (Iovino and Shaver 1991; Shanks *et al.* 1988; Kruszewski 1985; Fang *et al.* 1985; Chu and Zhu 1988; Crampton *et al.* 1989). The correlation function of quasars is determined to be stronger than that of bright galaxies (1b), but weaker than the correlation of the richest clusters (1a). The quasars therefore trace large-scale structure in the universe in an intermediate manner between galaxies and rich clusters. Some large groups - or superclusters - of quasars have also been reported (Clowes and Campusano, 1991, Crampton *et al.* 1989); these findings are consistent, qualitatively, with the positive quasar correlations discussed above.

What is the origin of the observed quasar correlation and its implied large-scale structure? Bahcall and Chokshi (1991a) investigated the data and suggest that the quasar correlations and superclusters may reflect the same large-scale structure traced by groups and clusters of galaxies provided the quasars are preferentially located in these high density systems.

Using observational studies of the galaxy environment around nearby quasars to $z \lesssim$ 0.7 (Yee and Green 1987, Boyle *et al.* 1988, Ellingson *et al.* 1991), Bahcall and Chokshi (1991a) estimated the mean richness of the average parent-group around the quasars. They find that optically selected quasars are located in small groups of average richness $\sim$ 10L* galaxies (as compared with $\simeq$ 65L* galaxies for richness R = 1 clusters). The optically selected quasars have the same correlation function as expected for these small groups using the richness-dependent cluster correlation function (§6; see Figure 13).

Radio selected quasars are found to be located in richer groups of $\sim$ 30L* galaxies on average at $\bar{z} \sim 0.6$, having the stronger correlations expected for these richer groups (Figure 13).

The quasar correlations thus agree well with the universal richness-dependent cluster correlation function, as well as with the universal dimensionless cluster correlation (Figure 12), provided the quasars are in groups of the average richness observed above. This suggests that the quasar correlations are due to the groups in which they are located, thereby displaying the same large-scale structure traced by their parent groups. The agreement of the quasar correlations with the universal relations provides a unified model for large-scale correlations for galaxies, clusters, and quasars.

The recent observations of large groups - superclusters - of quasars ($\gtrsim$ 100 Mpc in size) are consistent with this picture. According to this scenario, quasars inhabit groups or clusters of galaxies which themselves trace the large superclusters detected to scales of $\sim$ 150 Mpc (§ 1 - 6). The quasars therefore highlight the same superclusters.

It is also of interest to note that groups and clusters of galaxies, as well as the large-scale structure they trace, are suggested to exist at the high redshifts observed for the quasar correlations, i.e., $z \lesssim 2$. This fact may help discriminate among various cosmological models which predict different formation times for clusters and superclusters.

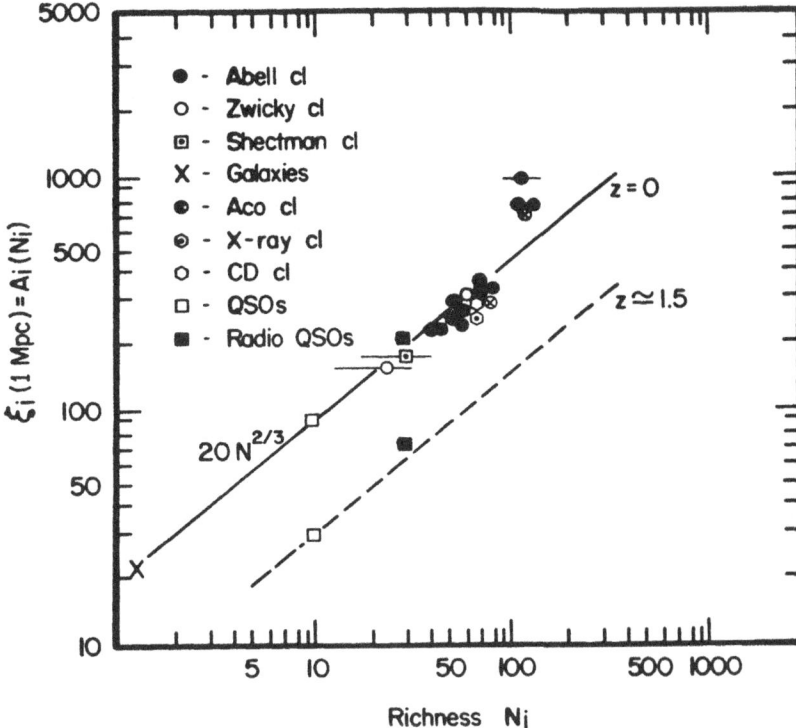

Figure 13. The richness-dependent cluster correlation function for z = 0 and z ≃ 1.5 (assuming a stable clustering model), and the optical and radio - QSO fits (Bahcall and Chokshı 1991a).

7.2 RADIO-GALAXIES

Radio-galaxies, like quasars, appear to be strongly clustered in space (Peacock and Miller 1988; Peacock and Nicholson 1991). Intermediate power radio-galaxies are observed to be clustered more strongly than individual galaxies but weaker than rich clusters of galaxies. The radio-galaxies therefore do not trace randomly the general distribution of galaxies.

Bahcall and Chokshi (1991b) investigated the richness of the environment around the radio-galaxies from works of Hill and Lilly (1991) and Prestage and Peacock (1988), and compared it with the observed correlation strengths. The correlation-scale of the radio-galaxies are $r_0 \simeq 15$ Mpc for the sample of FRI galaxies with log $P_{2.7} = 23 - 24$ W Hz^{-1} sr^{-1} (Peacock and Miller 1988) and $r_0 \simeq 11$ Mpc for a mixed sample of FRI + FRII galaxies with log $P_{1.4} = 22.5 - 24.5$ W Hz^{-1} sr^{-1} (Peacock and Nicholson 1991). These scales are both larger than the average galaxy correlation scale of r_0 (galaxies) $\simeq 5$ Mpc (Groth and Peebles 1977), but smaller than the rich clusters, $R \geq 1$, correlation scale of $r_0(R \geq 1) \simeq 22$ Mpc (eq. 2).

The median richness of the radio-galaxies parent-groups is found to be (Bahcall and Chokshi 1991b) ~ 33L* galaxies for the FRI sample, and ~ 13L* galaxies for the mixed sample. These richnesses of the parent-groups are plotted in Figure 11 versus the observed radio-galaxy correlation strength. The data appears to be consistent with the richness-dependent cluster correlation function, as well as with the universal dimensionless correlations (Figure 12), when the mean-separation of the parent-groups is used.

The above suggests that, like the quasars, the radio-galaxy clustering arises from their preferential location in galaxy groups. Radio-galaxies, and quasars, may thus be a good tracer of superclusters in the universe, especially at intermediate to high redshifts.

8. Conclusions

The following conclusions are suggested in this paper.

1. Large-scale superclusters are observed to scales of ~ 150 Mpc in the distribution of galaxies , clusters of galaxies, and probably quasars and AGNs. The same superclusters are traced well by galaxies and by rich clusters.

2. The superclusters appear to be flattened systems, with dimensions of up to ~ 150^2 x 20 Mpc^3; their mean space density is low: ~ 10^{-6} Mpc^{-3}, and their mean separation is ~ 100 Mpc.

3. Great-Walls, Great-Attractors, and the generic Superclusters are all similar structures with different names. They appear to surround large under-dense regions of comparable sizes.

4. Superclusters are the main origin of the galaxy peaks observed at ~ 100 - 150 Mpc intervals in narrow pencil-beam surveys. The peaks originate when the narrow beam crosses the large-scale superclusters.

5. It is suggested that superclusters are not randomly distributed in space but rather are weakly correlated on large scales.

6. A network system of superclusters is suggested by the data; a "cellular", or Zeldovich "pancake" type model may provide an approximate representation of the observations. Understanding the detailed topology of the structure will require considerably larger redshift samples of galaxies and clusters than currently available.

7. It is suggested that quasars and radio-galaxies also follow the supercluster structures reasonably well.

8. A richness-dependent cluster correlation function and a universal dimensionless cluster correlation appear to represent well the available data for galaxies, groups, and clusters, as well as quasars and radio-galaxies. The predictive power of these relations has succeeded, since new data appear to be consistent with these predictions.

9. A unified picture is emerging regarding superclusters and large-scale structure using different observations as tracers: galaxies, clusters, pencil-beam surveys, velocities, quasars, and radio-galaxies.

10. Larger samples that will be available over the next decade will help clarify the emerging picture of superclusters, and will address in more detail the topology and scale range of the observed structures. Theoretical models will then tell us, hopefully, how it all started.

REFERENCES

Abell, G.O. (1958) *ApJ. Suppl.* **3**, 211.
Bahcall, N.A. (1988) *Ann. Rev. Astron. Astrophys.* **26**, 631.
_____(1991) *ApJ.* **376**, 43.
Bahcall, N.A. and Burgett, W.S. (1986) *ApJ. Letters* **300**, L35
Bahcall, N.A. and Chokshi, A. (1991a) *ApJ. Letters* (in press).
_____(1991b) *M.N.R.A.S.* (submitted).
Bahcall, N.A., Henriksen, M.J. and Smith, T.E. (1989) *ApJ. Letters* **346**, L45.
Bachall, N.A. and Soneira, R.M. (1982) *ApJ. Letters* **258**, L17.
_____(1983) *ApJ.* **270**, 20.
_____(1989) *ApJ.* **277**, 27.
Bahcall, N.A. and Miller, N. (1991) *ApJ.* (to be submitted).
Bahcall, N.A. and West, M. (1991) *ApJ.* (to be submitted).
Boyle, B.J., Shanks, T. and Yee, H.K.C. (1988) "Large Scale Structure of the Universe" IAU Symp. 130, p.576 (J. Audouze *et al.* eds.).
Broadhurst, T.J., Ellis, R.S., Koo, D.C. and Szalay, A. (1990) *Nature* **343**, 726.
Chincarini, G., Rood, H.J. and Thompson, L.A. (1981) *ApJ. Letters* **249**, L47.
Chu, Y.Q. and Zhu, X.F. (1988) *A.A.* **205**, 1.
Clowes, R.G. and Campusano, L.E. (1991) *M.N.R.A.S.* **249**, 218.
Crampton, D., Cowley, A.P. and Hartwick, F.D.A. (1989) *ApJ.* **345**, 59.
da Costa, L.N. *et al.* (1988) *ApJ.* **327**, 544.
Dalton, G.B., Efstathiou, G., Maddox. S.J. and Sutherland, W.J. (1991) preprint.
de Lapparent, V., Geller, M.J. and Huchra, J.P. (1986) *ApJ. Letters* **302**, L1.
Ellingson, E., Yee, H.K.C. and Green, R.F. (1991) *ApJ.* **371**, 49.
Fang, L.Z., Chu, Y.Q. and Zhu, X.F. (1985) *Astrophys. Space Sci.* **115**, 99.
Geller, M.J. and Huchra, J.P. (1989) *Science* **246**, 897.
Giovanelli, R., Haynes, M. and Chincarini, G. (1986) *ApJ.* **300**, 77.
Gregory, S.A. and Thompson, L.A. (1978) *ApJ.* **222**, 784.
Gregory, S.A., Thompson, L.A. and Tifft, W. (1981) *ApJ.* **243**, 411.
Groth, E. and Peebles, P.J.E. (1977) *ApJ.* **217**, 385.
Hill, G.J. and Lilly, S.J. (1991) *ApJ.* **367**, 1.
Huchra, J.P., Henry, J., Postman, M. and Geller, M.J. (1990) *ApJ.* **365**, 66.
Iovino, A. and Shaver, P. (1991) A.S.P. Conference Series, in press.
Ikeuchi, S. and Turner, E.L. (1991) *M.N.R.A.S.*, (in press).
Kirshner, R.P., Oemler, A. Jr., Schechter, P.L. and Shectman, S.A. (1981) *ApJ. Letters* **248**, L57.
Klypin, A.A. and Kopylov, A.I. (1983) *Sov. Astron. Letters* **9**, 41.
Kruszewski, A. (1985) preprint.
Kurki-Suonio, H., Mathews, G.J. and Fuller, G.M. (1990) *ApJ. Letters* **356**, L5.
Lahav, O., Edge, A.C., Fabian, A.C. and Putney, A. (1989) *M.N.R.A.S.* **238**, 881.
Lynden-Bell, D., Faber, S., Burstein, D., Davies, R., Dressler, A., Terlevich, R. and Wegner, G. (1988) *ApJ.* **326**, 19.
Maddox, S.J., Efstathiou, G., Sutherland, W. and Loveday, J. (1990) *M.N.R.A.S.* **242**, 43.

Oort, J. (1983) *Ann. Rev. Astron. Astrophys.* **21**, 373.

Park, C. and Gott, J.R. (1991) *M.N.R.A.S.* (in press).

Peacock, J.A. and Miller, L. (1988) "Optical Surveys for Quasars" A.S.P. Conference Series **2** (Osmer *et al.*, eds.).

Peacock, J.A. and Nicholson, D. (1991) preprint.

Postman, M., Geller, M.J. and Huchra, J.P. (1986) *A.J.* **91**, 1267.

Postman, M., Huchra, J.P. and Geller, M.J. (1991) *Ap.J.* (in press).

Prestage, R.M. and Peacock, J. (1988) *M.N.R.A.S.* **230**, 131.

Raychaudhury, S., Fabian, A.C., Edge, A., Jones, C. and Forman, W. (1991) *M.N.R.A.S.* **248**, 101.

Shanks, T., Boyle, B.J. and Peterson, B. (1988) "Optical Surveys for Quasars" A.S.P. Conference Series **2**, 244 (Osmer *et al.*, eds.).

Shapley, H. (1930) Harvard Obs. Bull. No. **874**,9.

Shectman, S. (1985) *Ap.J. Suppl.* **57**, 77.

Szalay, A. and Schramm, D.N. (1985) *Nature* **314**, 718.

Weinberg, D., Ostriker, J.P. and Dekel, A. (1989) *Ap.J.* **336**, 9.

West, M. and van den Bergh, S. (1991) *Ap.J.* **373**, 1.

Yee, H.K.C. and Green, R.F. (1987) *Ap.J.* **319**, 28.

GALAXY CLUSTERS AROUND QUASARS

H.K.C. Yee
Department of Astronomy
University of Toronto
Toronto
Ont. M5S 1A6
Canada

ABSTRACT. A number of systematic surveys of fields around quasars have been carried out in the past decade to study their global environments. This paper summarizes some of the results, especially those pertaining to galaxy clusters around quasars. Quasars are found in environments significantly richer than those of average galaxies. However, at low redshift ($z < 0.4$), few, if any, are located in rich clusters. Nevertheless, a significant fraction of optically bright radio-loud quasars at $z \sim 0.6$ are found to be at or near the centers of galaxy clusters of Abell class 1 or richer. This can be understood if there is a much more rapid evolution of quasars in rich clusters compared with those in poorer environments. The evolution interpretation is confirmed by a survey of faint quasars in which lower luminosity quasars at lower redshift are also found in rich clusters. Possible candidates for the end products of this rapid evolution of quasars in rich clusters are radio galaxies in rich clusters at low redshift. By combining data from quasars and low-redshift radio galaxies, it is shown that they can both be consistently described by a single luminosity function and evolution law. Possible explanations for the more rapid evolution of quasar activity in clusters are discussed. Preliminary results from multi-object spectroscopy of galaxies around quasars are described. In general, clusters hosting bright quasars appear to have relatively low velocity dispersions, suggesting that the dynamical state of the cluster plays a role in the evolution of these quasars. Furthermore, interactions and mergers are important mechanisms in the triggering and maintenance of quasar activity.

1. Introduction

Quasars, being the most easily detectable objects in the distant Universe by virtue of their high luminosity, provide us with an unique glimpse into the Universe at an early cosmological time. In the standard paradigm, quasars are the active nuclei of galaxies at the cosmological distance indicated by their redshift. This, along with the tendency for galaxies to cluster, makes quasars excellent markers for locating less luminous and more ordinary objects at great distances, allowing us to study the properties and evolution of galaxies and galaxy clusters up to a redshift of one and possibly beyond.

The study of quasar environments, a term coined in the early 1980's, originated as an attempt to confirm the cosmological interpretation of quasar redshift. The logic is that, if quasars are at the cosmological distance indicated by their redshift,

A. C. Fabian (ed.), Clusters and Superclusters of Galaxies, 293–309.

then galaxies at a similar redshift should be detected. The early efforts consisted of deep photographic imaging of low-redshift ($z < 0.3$) quasars and very low-resolution spectrophotometry of detected companion galaxies. Gunn (1971) found galaxies with redshifts similar to those of the quasars in the fields of Ton 256 ($z = 0.131$) and PKS 2251+11 ($z = 0.323$). He also noted that PKS 2251+11 is situated in a group of galaxies. Oemler, Gunn, and Oke (1972) showed that 3C 323.1 ($z = 0.264$) is situated at the outskirts of a Zwicky cluster of similar redshift, 6'.5 from the central cD galaxy. Stockton (1978) was the first to complete a systematic spectroscopic search for galaxies associated with quasars with the aim of confirming the cosmological interpretation of quasar redshift. He found that 8 out of 27 quasars in his sample have at least one companion having the same redshift as the quasar. Systematic investigations of quasar nebulosity using photographic imaging were also carried out by Wyckoff et al. (1981) and Hutchings et al. (1981).

The advent of new electronic digital area detectors such as SIT and CCD cameras in the early 1980's had a large impact on the observation of the environment of quasars. Many systematic imaging surveys of quasars were carried out, with most of them motivated by the study of the nebulous component of quasars (e.g, Malkan 1984, Gerhren et al. 1984, Smith et al. 1987). Almost all these surveys noted the apparent large number of companion galaxies to the quasars and the possibly distorted morphologies of the host galaxies. Such evidence is taken as indication that galaxy interactions and mergers are important mechanisms in triggering quasar activities. Imaging programs with the study of associated galaxies in the global environment of quasars as the prime objective were carried out by Yee and Green (1984, 1987), Hintzen (1984), Tyson (1986), Ellingson, Yee and Green (1991), and Hintzen et al. (1991). Similar imaging studies of the environments of radio galaxies have been conducted by Yates, Miller and Peacock (1989) and Hill and Lilly (1991).

In this paper, I will concentrate on results pertaining to the global environments of quasars, and in particular, galaxy clusters associated with quasars. Section 2 describes the method used to measure quantitatively the richness of quasar environment. In Section 3, I review the results from different imaging surveys over the past decade. Section 4 summarizes the evidence for the evolution of the environment of radio-loud quasars. The possible connections to radio galaxies are discussed in Section 5. In Section 6, I present a recent analysis of the data, modelling the rapid evolution of quasar activity in rich clusters. The implications of these luminosity function models are also discussed. New results from multi-object spectroscopy of cluster galaxies in quasar fields are presented in Section 7. Section 8 discusses the enviornment of radio-quiet quasars. And a summary is presented in Section 9. In this paper, $H_o = 50$ km s^{-1}Mpc^{-1} is used throughout.

2. The Quantitative Measure of Richness of Quasar Environment

Because a prohibitively large amount of telescope time is required to measure the redshifts of a substantial fraction of galaxies in fields of a large number of quasars, statistical methods using direct images are used to derive quantitative information concerning the richness of the environment and the properties of the associated galaxies themselves. Due to the lack of information on the actual spatial distribution of galaxies in the quasar fields, the process is a complex one, involving many assumptions. However, it will be shown that multi-object spectroscopy data from a limited number of quasar fields indicate that these statistical estimates of richness are accurate.

To determine the richness of quasar environment, we first measure the excess galaxy counts in the quasar fields. From the observations, one needs to extract catalogs of photometric magnitudes and positions of the galaxies detected. Furthermore, reliable completeness limits, and well-determined background galaxy counts as a function of magnitude are also required. These data are necessary for estimating the excess number of galaxies in the quasar field as a function of magniutde. Due to the small field size of most digital detectors, the background galaxy counts are generally obtained in separate control fields, well away from the quasars.

Determining a simple excess, however, does not convey any quantitative information concerning the richness of the association. To derive the absolute richness of the associated galaxies, we need to correct for two sampling effects of galaxy counting: different sampling depth in the magnitude of the galaxy luminosity function (LF), and different metric areas imaged around the quasar. These effects arise because the quasars are at different redshifts and observations are made under varying conditions, often with different instruments. Thus, one needs to determine or assume an average LF and an average spatial distribution of the associated galaxies in order to correct the excess counts and derive a measure of the true richness of the environment.

The parameter that we adopted to measure the richness of quasar environment is the amplitude of the quasar-galaxy covariance function, B_{gq}. Longair and Seldner (1979) first developed this method based on the galaxy-galaxy covariance function formalism to study the environment of radio galaxies. Yee and Green (1984 and 1987, hereafter YG87) adopted this method with the additional step of using self-consistent models of cosmological parameters, galaxy LF's, and background galaxy counts.

The method for determining B_{gq} is summarized briefly here; more details can be found in Longair and Seldner (1979), YG87, and Ellingson, Yee and Green (1991). We can represent the spatial distribution of galaxies around the quasar in excess over the average background density by a power law which is called the quasar-galaxy spatial covariance function $\zeta(r)$:

$$\zeta(r) = B_{gq}r^{\gamma}. \tag{1}$$

From imaging data, we can only measure this over-density projected onto two dimensions. The directly observable quantity is the angular-covariance function $\omega(\theta)$ with amplitude A_{gq}:

$$\omega(\theta) = A_{gq}\theta^{1-\gamma}, \tag{2}$$

which measures the anugular distribution of excess galaxies as a function of angular distance, θ, from the quasar. The assumption of spherical symmetry is used in relating the power-law indices of the two functions. It can be shown easily that the galaxy counts within a circle of angle θ is related to A_{gq}, the amplitude, by:

$$A = \frac{N_{tot} - N_g}{N_g} \frac{3 - \gamma}{2} \theta^{\gamma - 1}, \tag{3}$$

where N_{tot} is the total number of galaxies counted, and N_g, the expected background galaxy counts YG87.

By assuming spherical symmetry, a universal LF for galaxies, and a clustering structure independent of look-back time, we can relate the measureable A amplitude to the B amplitude via (Longair and Seldner 1979):

$$B = \frac{AN_g(m)}{I_\gamma} \frac{1}{\Psi[M(m,z)]} \left(\frac{(1+z)^2}{D} \right)^{(3-\gamma)}, \tag{4}$$

where D is the luminosity distance of the quasar; $\Psi[M(m,z)]$, the integrated luminosity function of galaxies at absolute magnitude $M(m,z)$; and I_γ, an integration constant dependent on γ. (Note the typographical correction of this formula from that in both YG87 and Ellingson, Yee and Green 1991).

For consistency, we adopt $\gamma = -1.77$, the canonical value for the galaxy-galaxy covariance function (e.g., Seldner and Peebles 1978). The average radial distribution of galaxies around quasars is found to be consistent with this slope (YG87).

The added complication of cosmological models and galaxy LF evolution is addressed by using parameters that are self-consistent. YG87 derived several such models from their data. In brief, this is achieved by deriving the shape of the galaxy LF as a function of redshift from the excess galaxies associated with the quasars. This in principle couples the evolution of the galaxy LF to the adopted cosmological parameters in a consistent manner. Note that by deriving the LF from the actual excess count data, we avoid a significant systematic uncertainty in the computation of B_{gq}. The LF is then normalized by integrating the LF in magnitude and z to fit the background galaxy counts. It is found by using such consistent sets of parameters, the choice of q_o has a very small effect on the B_{gq} values even at $z \sim 0.6$.

As reference, we use the galaxy-galaxy covariance amplitude, B_{gg}, of 67.5 Mpc$^{1.77}$ (Davis and Peebles 1983). B_{gq}/B_{gg} is then the ratio of the richness of the galaxy environment of the quasar as compared to that of the average galaxy. Also, Prestage and Peacock (1988, and see Prestage Peacok 1989 for an erratum) estimated the galaxy covariance amplitudes for the centers of galaxy clusters to be 360, 645, and 945 Mpc$^{1.77}$, for Abell richness classes 0, 1 and 2, respectively. Thus, using these calibrations, we can in principal estimate the Abell richness class of the environment of a quasar by determining the excess galaxy counts in its field.

3. Quasars in Galaxy Clusters

Although many investigators noted that there is an apparent excess of faint galaxies around quasars in their studies of quasar fuzz (e.g., Hutchings et al. 1984, Malkan 1984), the first systematic quantitative survey of galaxies associated with quasars using a digital detector was carried out by Yee and Green (1984). This survey, using a SIT camera on the Palomar 1.5 m, consists of two samples of optically bright quasars – a radio-loud sample brighter than $m \sim 17.5$ mag, and the Palomar-Green sample (Green, Schmidt and Liebert 1986) of optically selected quasars brighter than ~ 16.5 mag. This survey shows conclusively that there is a significant enhancement in galaxy density around quasars. Moreover, the distribution of the angular covariance amplitudes is found to be a function of redshift in that there is no significant excess of galaxy counts in fields of quasars with $z \gtrsim 0.5$. Since the completeness limit for the detection of galaxies in that survey is Gunn $r \sim 21.5$ mag, equivalent to first-ranked galaxies at $z \sim 0.5$, this dependence of the angular covariance amplitude on z is consistent with the cosmological interpretation for the quasar redshift. The data from this survey are of sufficient depth to obtain reliable B_{gq}'s for nearby quasars out to $z \sim 0.3$. It is found that, on average, quasars tend to be situated in regions of higher-than-average galaxy density – typically in small groups, and a few in poor clusters. However, with the exception of one or two out of a sample of 56 (e.g., Ellingson et al. 1989), none are found situated in regions with galaxy density comparable to that at the centers of Abell class 1 clusters. This is in accordance with the "conventional wisdom" that quasars are not found in rich clusters.

A follow-up deeper imaging survey of essentially the same sample of quasars with $0.3 < z < 0.65$ was carried out using a CCD detector at the Steward 2.3 m telescope by YG87. The average completeness magnitude of $r \sim 22.7$ allows B_{gq} of quasars with redshift up to 0.65 to be determined. For $0.3 < z < 0.5$, both radio and optical quasars are found to be situated in environments similar to those from the SIT survey with $< B_{gq}/B_{gg} >$ of 2.8 and 2.0 for the two samples, respectively. However, for a small subsample of 9 radio-loud quasars with $0.55 < z < 0.65$, we found $< B_{gq}/B_{gg} > \approx 8.0$, a surprising and significant increase in the average richness of the environment. Over half of the quasars in that subsample inhabit regions with galaxy densities equivalent to Abell class 1 clusters or richer.

Yee and Green (1987) interpreted this apparent change in the environment of radio-loud quasars as an indication that there has been a strong evolution in the optical luminosity of radio-loud quasars situated in rich clusters. Quasars in clusters may have dimmed by several optical magnitudes between redshifts of 0.6 and 0.3, causing them to be excluded from an apparent-magnitude limited sample at lower redshift; whereas quasars in poor environment have suffered a much less drastic evolution, and hence dominate the sample at $z < 0.5$.

In another systematic imaging survey, Hintzen and collaborators also found some radio-loud quasars situated in rich environments. They set out specifically to search for the most efficient way of identifying quasars hosted by rich clusters by imaging radio-loud quasars that have distorted morphologies. The rationale is that radio sources in dense environments will be more likely to be distorted (Hintzen, Boeshaar and Scott 1981). Compared to a control sample of classical "clean" doubles, Hintzen (1984) found that quasars with distorted morphologies have a richer average environment. Detailed study of one of these quasar, 3C 275.1, demonstrated that the quasar is embedded in a large nebulosity which may be interpreted as an incipient cD galaxy in the center of a rich cluster at $z = 0.57$ (Hintzen and Romanishin 1986). However, YG87 pointed out that this correlation between galaxy cluster environment and radio morphology may not be definitive owing to the fact that the distorted morphology sample has a higher average redshift than the control sample. Furthermore, most of the quasars found in rich clusters by YG87 have relatively symmetric Fanaroff-Riley (FR) II class morphologies; i.e, similar to classical doubles.

4. Evidence for the Evolution of Quasar Environments

Motivated by the results of these surveys, my collaborators and I expanded the imaging survey of quasars to increase the sample size, and to cover both a larger redshift and luminosity range. The CTIO imaging survey of Parkes quasars is intended to cover a wider range of properties of radio-loud quasars to search for correlations between environment and quasar properties. Preliminary results from a subset of bright radio-loud quasars at $0.5 < z < 0.7$ confirm the increase in B_{gq} found for quasars at $z \gtrsim 0.5$. Figure 1 plots the histograms of B_{gq} for the sample in the two redshift bins. The distributions are different at the 99% confidence level.

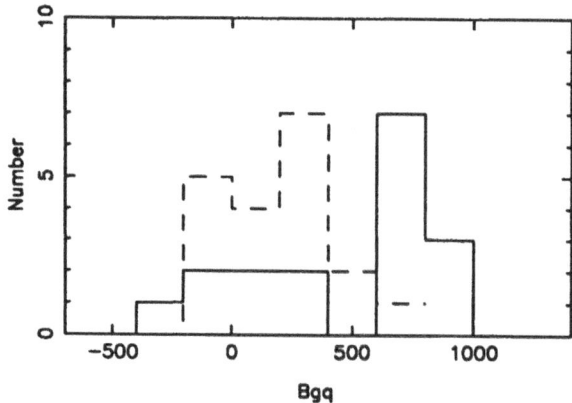

Figure 1. Histograms of the distribution of B_{gq} for optically bright quasars ($m < 17.5$) for $z < 0.5$ (dotted line), and $z > 0.5$ (solid line).

An equally important step in confirming this evolutionary interpretation is to investigate the environments of optically faint quasars. If the scenario of drastic dimming of quasars in rich clusters in the recent epoch is correct, we should then expect fainter radio-loud quasars at lower redshift to be located in the centers of rich clusters. Furthermore, such observations will allow us to estimate the approximate time scale for the disappearance of these quasars. Thus, a survey of faint quasars with magnitudes between 17.5 and 19.5, and $0.3 < z < 0.6$, was carried out using the Steward 2.3 m and KPNO 2.1 m telescopes (Ellingson, Yee, and Green 1991). Samples of 31 radio-loud quasars and 32 radio-quiet quasars with $0.3 < z < 0.6$ were imaged to a similar depth as that of the bright quasars survey of YG87.

Quasars in the faint sample are found to have environments similar to those from the brighter quasars in the same redshift range, i.e., in poor groups and clusters. However, a small number of radio-loud quasars at $z < 0.5$ with absolute magnitudes fainter than -25 are found situated in regions with $B_{gq} > 500$, representing environments as rich as Abell class 1 clusters. This is consistent with the scenario that quasars situated in the centers of rich clusters have dimmed between redshift of 0.6 and 0.4. A powerful way to illustrate the difference in the evolution of radio-loud quasars in clusters and poorer environments is shown in Figure 2. The quasars are divided into two samples according to their environment: those with $B_{gq} > 500$ which are assumed to be located in clusters as rich as Abell class 1, and the ones with $B_{gq} < 500$. The figure plots the absolute magnitudes of the quasars, M_{qso}, versus redshift for the objects in the two different environments. Here, we see that the rich environment sample has a very different distribution from that of the small B_{gq} sample. The former shows a drastic dimming in M_{qso} as a function of redshift. A 2-dimensional K-S test indicates that the two samples are different at the 98% confidence level. This difference in the distribution demonstrates conclusively that the increase in the average B_{gq} for radio-loud quasars at higher redshift is not due to selection effects in the quasar sample. We have chosen our sample without prior knowledge of the environment of the quasars, hence, whatever selection effect suffered by one sample would also affect the other. Moreover, this also shows that the rich cluster environments that we observed at $z \sim 0.6$ is not an artifact of a radio-power selection effect (arising from the strong radio-power-redshift correlation for a flux limited sample). In the combined sample, quasars in rich clusters are no longer confined to a narrow redshift range as a selection effect due to radio-power would predict.

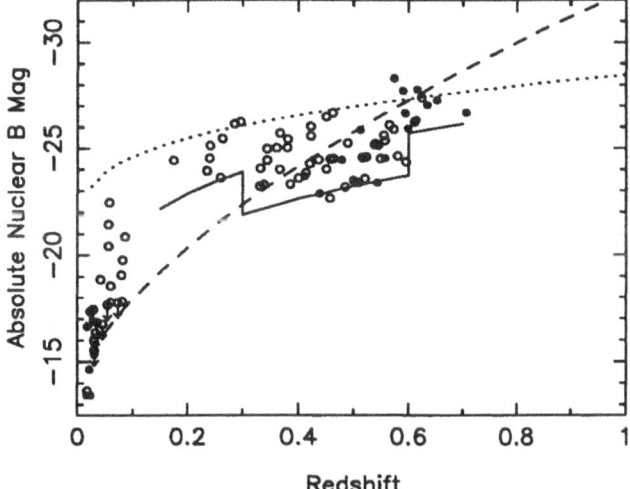

Figure 2. Plot of M_{qso} vs z. Objects situated in environments with $B_{gq} > 500$ (Abell class 1 or richer) are plotted as solid dots, while those in poorer environments are shown as open dots. The solid line represents the redshift and apparent magnitude sample limits of the quasar sample. Objects with $z > 0.15$ are from the quasar samples while those in the radio galaxy sample have $z < 0.1$. The dotted line represents the envelope model predicted by the Boyle et al. (1987) quasar LF and evolution, while the dashed line represents the Bolye et al. LF with an evolution 5 to 6 times more rapid.

5. The Environments of Radio Galaxies

Longair and Seldner (1979) showed that the properties of low-redshift radio galaxies are correlated with their environment. By correlating 3CR radio galaxies having $z < 0.1$ with the Lick galaxy count catalog, they found that FR I (e.g., head-tail morphology) objects tend to inhabit richer environments and that they are typically quiescent optically. On the other hand, FR II's, typical classical double radio sources, are situated in poorer environments, and have much more prominent optical emission lines. Similar investigations of higher redshift radio galaxies were conducted by Hill and Lilly (1991) and Yates, Peacock, and Miller (1990) using CCD images. At higher redshift ($z \gtrsim 0.4$), their results show that radio galaxies with FR II radio morphologies are also found in rich clusters. (Recall that the morphologies of many of the quasars in clusters found by YG87 are FR II.) Thus, it is likely that some kind of evolution of radio galaxies in rich clusters may have also taken place. Hill and Lilly (1991) suggest that the increase in intracluster gas may be a possible explanation for such evolution.

This scenario of rapid fading of quasars in clusters at $z \sim 0.5$ leads to an interesting question: Where are the remnants of these once bright quasars that used to be the central objects in rich clusters? One possible candidate is the relatively weak FR class I radio galaxies at low redshift which, as Longer and Seldner noted, are often the brightest members of nearby rich clusters and are normally considered optically quiescent. To study these objects, high signal-to-noise ratio spectra of the nuclei of a small sample of 8 3CR radio galaxies in cluster environment were obtained by Yee and De Robertis (1989) at CFHT using a relatively narrow slit of 1″ to isolate the nuclear spectrum. This, combined with careful removal of the stellar component, allows us to search for extremely low-level activities in the nucleus. We found that contrary to the original classification of them as "absorption line" galaxies, they all have weak Seyfert-like emisssion-line spectra, including broad Hα. As an example, detailed analysis of the spectra and the environment of 3C 465, the central cD galaxy in Abell 2634, has been carried out by De Robertis and Yee (1990). It is shown that 3C 465 has a nuclear spectra with widths of Hα and S[II] emission lines over 1000 km s^{-1}, similar to those of typical AGN. Furthermore, careful decomposition of the Hα line demonstrates that there is a weak broad Hα component with a width of about 4500 km s^{-1}. Using the line luminsoities, we estimate the magnitude of the "mini-quasar" in the nucleus of 3C 465 to be ~ -15.7 mag. Hence, these radio galaxies may represent some of the faintest examples of active nuclei occurring in the centers of rich clusters, and possibly the end-product of the rapid evolution seen at earlier epochs.

6. The Rapid Evolution of Quasar Activity in Cluster Environments

6.1. LUMINOSITY FUNCTION MODELS

The evidence that many low-redshift radio galaxies in clusters, either cD's or first-ranking galaxies, contain a weak quasar-like nucleus suggests a link between them and the bright quasars found in similar environments at higher redshift. To test this idea, we recently combined data from quasars and low-redshift radio galaxies to demonstrate that these two classes of objects can be described using a single LF and evolution law (Yee and Ellingson 1992, also see Ellingson and Yee, 1991) .

In addition to the quasars, we plot in Figure 2 a sample of 3CR radio galaxies with $z < 0.1$. The sample is essentially that used by Longair and Seldner (1979) in their study of the relationship between environment and optical properties of low-z radio galaxies. In quantifying the environment of each galaxy, we use B_{gq}'s derived by Yee and De Robertis (1992) whenever possible. These values are computed from CCD images obtained at KPNO 0.9 m with a field of $6' \times 6'$, using the same procedure as that for the quasars. In a few cases, where no imaging data are available, the B_{gq} values obtained by using Lick galaxy counts by Longair and Seldner (1979) are used. To determine M_{qso}, a decomposition of the nuclear and galaxy components is necessary since galaxy light dominates in almost all objects. For the majority, the nuclear magnitudes are obtained from the decomposition of multi-chanel spectroscopy data in Yee and Oke (1978). For objects in rich clusters observed by De Robertis and Yee (1992), the nuclear components are estimated using measured Hβ or Hα fluxes and the relationship between line and nuclear luminosities for AGNs obtained by Yee (1980). For objects with no discernable nuclear component and not observed by De Robertis and Yee, a lower limit is estimated based on the detection limit of Hβ in the spectrophotometry data of Yee and Oke.

Figure 2 illustrates that the radio galaxies also show a dependence of nuclear activity on the environment. This is in fact just another way of expressing the original result of Longair and Seldner (1979) in which they found that radio galaxies with no emission lines are situated in richer environments. It is also apparent from Figure 2 that the radio galaxy nuclear luminosities may be consistent with the extrapolation of the quasar data.

In order to analyse this quantitatively, taking into account the LF of active objects and volume selection effects, we constructed what we called "envelope models" on the M_{qso}-z plane. Given a LF and an evolution law of this LF, one can predict the appearance of the $M_{qso} - z$ diagram of an apparent-magnitude limited sample of the objects. A particularly simple attribute to model is the isodensity contours in the $M_{qso} - z$ diagram. These are clearly a function of the LF (which measures the space density of the objects as a function of luminosity) and the volume sampled at each redshift. These models can be constructed by solving for $M(z)$, the absolute magnitude as a function of redshift, at which the space density is such that a constant number of objects in the sample is expected in the space element of the redshift bin. An envelope model is simply the contour of the distribution of points in $M_{qso} - z$ diagram that represents the magnitude at which one would expect statistically to see one object. These models have the advantage that only the brightest objects at each redshift bin are used for comparison, and hence effects due to incompleteness in the fainter end of the LF are not critical. On the other hand, such models are not necessarily unique since the solution involves an integral of the LF; i.e., different combinations of LF and evolution may produce the same result. Thus, these models are best used to demonstrate consistency of the data with certain parametrization of the LF and the evolution, and are not intended as "best-fitting" models.

In Figure 2, two such models are shown. The dotted line represents the envelope expected for a parametrization of quasar LF and evolution derived by Boyle et al. (1987) for optically-selected quasars. Here, we use their best fitting model with $q_o = 0.1$, i.e., a two-power-law LF with slope $\beta = -3.5$ for $M < M^*$ and $\beta = -1.25$ for $M > M^*$, with $M_o^* \equiv M^*(z = 0) = -22.1$ in the B band. The evolution is parametrized by a power-law dependence with redshift for M^*:

$$M^*(z) = -2.5\kappa_L \log(1 + z) + M_o^*, \tag{5}$$

with $\kappa_L = 3.7$. The envelope model has been arbitrarily normalized at $z = 0.65$. This model matches the upper envelope of both quasars and radio galaxies rather well, suggesting that the parametrization of the LF for optically-selected quasars is also a reasonable fit for radio-loud quasar, and that radio galaxies and radio-loud quasasrs can be described by a single LF and evolution model. Note that with such a steep slope of the LF, the effect of rapid change of density with luminosity dominates the behavior of the envelope contour. A change of a factor of two in sample volume produces a change of only ~ 0.3 mag in the envelope. Thus, this model is insensitive to the completeness level of the sample, and we can therefore combine the radio-loud quasars and radio galaxies with relative impunity.

Also plotted in Figure 2 is a model which best describes the upper envelope of quasars and radio galaxies in rich environments. Using the same parametrization, it is found that a κ of 20 is required to match the rapid drop in the luminosities of the brightest active nuclei in rich clusters. This represents evolution 5 to 6 times more rapid than that of objects in poor environments. Note that, as in the case for objects in poor environments, the quasars and radio galaxies again can be both described using a single LF model.

An alternative way to model this steeper drop of the envelope magnitude with redshift without resorting to more rapid evolution is to change the slope of the LF. Two such models are plotted in Figure 3. Here the the evolutionary parameter κ is set to zero, and the bright-end slope of the luminosity function, β, is set to -1.50 (dashed line) and -1.25 (dotted line). It is clear that this family of models cannot match the data from both the quasar and radio galaxy samples simultaneously. The model that describes the magnitude envelope of the rich-environment quasars drops too steeply at low redshift to account for the radio galaxies in rich clusters. Given this, and the unrealistically shallow slope of the required LF, the rapid-evolution model is preferred.

We can also use these envelope models to predict the luminosities of quasars at higher redshift. In Figure 4, we plot as crosses the remaining known quasars in the northern hemisphere from the 3C, 4C, and Parkes catalog whose environments have not been determined. Again, the standard LF model of Boyle et al. (1987) fits the data extremely well, out to $z \sim 1$. However, simple extrapolation of the models that describe the envelope of quasars in rich clusters predicts the existence of much brighter quasars at $z \gtrsim 0.7$. These are not seen. This suggests that some significant change in the evolution of quasars in rich clusters at $z \sim 0.7$ has occurred.

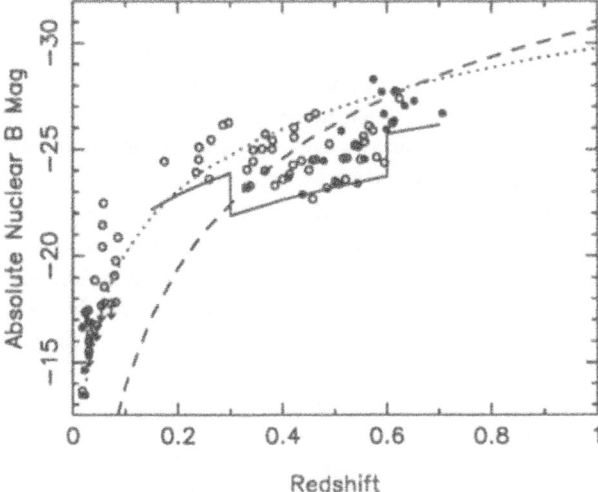

Figure 3. Same as Figure 2, but with LF models of different bright-end power-law slopes. The dashed line represents an envelope model with a LF with the bright-end slop of –1.25; and the dotted line, a bright-end slope of –1.50. Note that with these flat-slope LF's, we cannot fit the quasars and the radio galaxies in rich environments simultaneously.

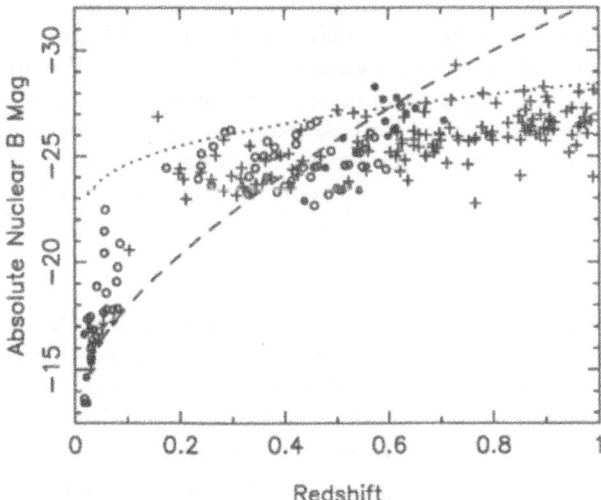

Figure 4. Same as Figure 2 but with quasars from the 3C, 4C, and Parkes samples for which we have no environment information plotted as crosses (+). Note that the extrapolation of the rapid evolution model predicts an envelope much brighter than the data at $z > 0.7$.

6.2. IMPLICATIONS

The data presented provide forceful evidence that the evolution of quasar activity is a stong function of their environment. A likely explanation for the rapid evolution of AGN located in the centers of rich clusters of galaxies is that the decline in quasar activity is in response to a change in some physical property in the cluster core. There are several possible scenarios we can explore.

Given the preponderance of the evidence for interactions and mergers as possible mechansims for triggering quasars, the evolution of quasars may well be expected to be tied to the dynamical evolution of galaxy systems. The timescale of about 1 Gyrs of the observed evolution is similar to dynamical timescales in these cluster cores, strengthening this interpretation. One possible scenario is that this rapid evolution is caused by an increase in the cluster core velocity dispersion as the cluster virializes. The efficiency of galaxy encounters is strongly dependent on their relative velocity; virialization of cluster galaxies should therefore be accompanied by a decline of quasar activity in these environments (e.g. Roos 1985, DeRobertis 1985). Small groups of galaxies, which have low velocity dispersion at the present epoch and a long dynamical time scale, would evolve much slower, allowing the quasars to have a longer lifetime. Carlberg (1990) has applied such a meger scenario to explain the general evolution of the quasar population. In the next section, preliminary results pertaining to the dynamical state of clusters associated with quasars will be presented.

Alternatively, the rapid evolution of quasars in clusters may be connected to the evolution of the intracluster gas of a rich cluster as it virializes. Stocke and Perrenod (1981) used such a mechanism to explain the general evolution of quasars. In this scenario, the gas content of the host galaxy of the quasar or the interacting neighbors plays a role in the fuelling of the quasar. The cluster gas density increases as the cluster evolves, producing more effective stripping. This in turn starves the quasar and causes the rapid evolution. Such scenario is supported by recent reports of very steep evolution of the x-ray LF of clusters even at the relatively low redshift of 0.2 (Edge et al. 1990, Gioia et al. 1990). However, direct observation of the gas contents of associated clusters using X-ray instruments such as ROSAT is required to test such scenario.

Another model, based on the hypothesis of cooling flows in rich clusters as fuel for the central engine, is also possible. Here, the disruption of cooling flows due to mergers of clusters into richer clusters is used to explain the rapid dimming of quasars in clusters (Fabian and Crawford 1991). For a more detailed discussion on this scenario, see the article by Fabian in these proceedings.

The evolution of quasar in clusters appears to have taken a significant turn at $z \sim 0.7$. It is clear from Figure 4 that the rapid evolution of quasars in clusters does not exptrapolate beyond $z \sim 0.7$. This indicates that $z \sim 0.7$ may be a special epoch in cluster or quasar evolution. One possible speculation is that individual quasars always evolve rapidly, at the time scale of cluster virialization. The ensemble of quasar population in rich environments evolves slower at $z > 0.7$ as there are always new sites becoming available. However, at $z \sim 0.7$, the birthrate of quasars in rich clusters declines, (perhaps due to a drop in the birthrate of clusters), causing the ensemble to evolve at or near the individual rate, and producing the very rapid decline of quasar population in clusters after $z \sim 0.7$ that we observe.

7. Spectroscopic Observations of Clusters Associated with Quasars

In order to delineate the relationship between cluster environment and the evolution of the quasars situated in them, we have been carrying out a program of spectroscopic observations of faint galaxies in the fields of quasars. First results based on the bright quasars in the YG87 sample are reported in Ellingson, Green and Yee (1991). Observations were obtained in fields of 17 quasars using multi-object spectroscopy at the KPNO 4m. Of the 127 redshifts determined, a total of 44 were considered as "associated". These data are combined with additional data from the literature, giving a sample with 88 associated galaxies in 38 quasar fields. Composite velocity distributions were derived by assuming that the quasar is at the dynamical center of the cluster, so that velocities relative to the quasar in different fields can be combined. This allows the inclusion of fields where only a few galaxies associated with the quasars were observed, and also minimizes the impact of any individual velocity distribution. Figure 5 illustrates the composite velocity distribution for the whole sample of 88 galalxies. In all combinations of samples, e.g., the whole sample, radio-loud, radio-quiet, rich environment ($B_{gq} > 500$), and poor environment subsamples, etc., velocity dispersions around 450 km s^{-1} are obtained. This is considerably lower than the average of ~ 850 km s^{-1} (Colless, 1988) expected for current epoch Abell 1 richness clusters. This result strongly suggests that cluster hosting bright quasars are in a dynamically young state.

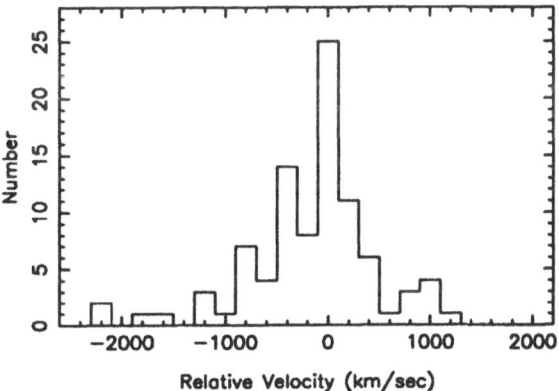

Figure 5. Composite velocity distribution of 88 galaxies associated with bright quasars based on data from the KPNO multi-object spectroscopy survey and the literature (Ellingson, Green and Yee 1991).

In a more recent survey, we have obtained spectra of galaxies in five additional clusters associated with quasars using the PUMA multi-object spectrogaph at the CFHT (Ellingson and Yee, 1992). Four of these are from the faint-quasar sample with $0.4 < z < 0.6$ of Ellingson, Yee, and Green (1991). This set of data was obtained with the purpose of comparing the dynamical states of clusters associated with quasars having different luminosities at different redshifts. In an evolution interpretation, these objects are in different evolutionary states. From these fields, 141 galaxy redshifts were measured; of which 44 are found to be associated with the 5 quasars. We combined this data set with that from Ellingson, Green and Yee (1991) and divided them into two subsets, based on the absolute magnitude and redshift of the quasars. The first subset includes galaxies from fields surrounding bright quasars with $z \sim 0.6$ and the second subset from galaxies associated with fainter quasars at $z \sim 0.4$, corresponding to two positions on the envelope model shown in Figure 2. In addition, a comparison sample was constructed using low redshift Abell class 1 clusters associated with cD radio galaxies using data from Zabludoff et al. (1990).

Preliminary analysis shows that the velocity dispersions of galaxies associated with bright quasars at high redshift ($z \gtrsim 0.5$) are significantly lower than normal Abell clusters at low redshift, (~ 500 vs ~ 800 km s^{-1}), in agreement with the models discussed above. Clusters associated with fainter quasars are found to have an dispersion of ~ 650 km s^{-1}, intermediate between the two, implying that the evolution of AGN in rich clusters is accompanied by an increase in velocity dispersion. At present, the large uncertainties in the velocity dispersions do not warrant a quantitative comparison with theoretical models of quasar activity in virializing clusters (e.g. Roos 1985), but these results are in qualitative agreement with the models.

With the well-controlled CFHT spectroscopic sample, we can also compare the association statistics with those estimated from simple galaxy counting. At this point, we have sufficient data for only two quasar fields: 3C 215 at $z = 0.4134$, and 5C 2.10 at $z = 0.4773$. Preliminary analysis indicates that the excess galaxy counts agree well with the spectroscopic data. For both fields, we were able to obtain redshifts for 80% of the galaxies brighter than $r = 21.5$ mag that were observed spectroscopically. (In each field, about half of the galaxies to this limit were observed.) Defining a sample within 1 Mpc of the quasar to this magnitude limit, we have 10 and 8 associated galaxies in the 3C 215 and 5C 2.10 fields, respectively. These represent 37 and 53 % of all the galaxies observed in each field. Galaxy counting indicates excesses of 64 ± 10 and $40 \pm 15\%$, respectively, over the same area and to the same magnitude, in general agreement with the spectroscopic data. These values indicate that these two quasars are situated in clusters of richness Abell class 1. Based on the spectroscopic characteristics of the associated galaxies, we also estimate the blue galaxy fractions for the two clusters to be 0.33 ± 0.16 and 0.25 ± 0.18, respectively. Thus, the blue galaxy fractions are similar to those expected from spectroscopic investigations of rich galaxy clusters at similar redshifts conducted by Gunn and Dressler (1988) to study the Butcher-Oemler effect (Butcher and Oemler, 1985).

8. The Evironment of Radio-Quiet Quasars

In Yee and Green (1984 and 1987), a marginally significant difference was found between the environments of low-redshift radio-loud quasars and radio-quiet quasars in that the former are situated in slightly richer environment than the latter. With the addition of the new data from the considerably larger sample of the faint-quasar survey, Ellingson, Yee and Green (1991) concluded at a 99% confidence level that the average environment of radio-loud quasars is richer than that of radio-quiet quasars. This implies that the existence of radio emission from quasars is a strong function of the environment. The result is consistent with the suggestion from investigations of the fuzz component of quasars that the host galaxies of radio-loud quasars are more similar to elliptical galaxies, whereas those of radio-quiet quasars are more like spiral galaxies (e.g., Malkan 1984, Smith et al. 1987). The mophology-density relationship of galaxies (e.g., Dressler, 1980) can then natually account for the difference in environments between the two types of quasars.

An additional difference between radio-loud and radio-quiet quasars is that there is no evidence that the preferred sites of radio-quiet quasars have undergone the same change at $z \gtrsim 0.5$ as found for radio-loud quasars. However, this conclusion is based on a small number of quasars as there is a shortage of bright optically-selected quasars in the redshift range between 0.5 and 0.8 due to selection effects in quasar surveys. To remedy this situation, an imaging survey of faint optically-selected quasars with $z < 0.9$ from the AAT fibre survey (Boyle et al. 1987) has been completed using the La Palma Issac Newton 2.5 m Telescope. Preliminary results have been reported in Boyle, Shanks, and Yee (1988); and a more complete analysis will appear in Yee, Boyle, and Shanks (1992). With a small sample of 8 relatively faint optically-selected quasars (absolute magnitudes ~ -24), we found the average B_{gq} computed by counting galaxies within a radius of 500 kpc to be similar to that obtained from the optically-selected quasars in the YG87 and Ellingson, Yee and Green (1991) samples. This confirms that at $z \sim 0.6$, radio-quiet quasars, unlike some radio-loud ones, are not found in centers of rich clusters. However, the CCD camera at INT has a total field of about $6' \times 4'$, allowing one to examine the environment of these quasars out to a radius as large as 1.5 Mpc at $z \sim 0.6$. We found the surprising result that the average B_{gq} computed by counting galaxies in a radius of 1 Mpc around the quasars increases by a factor of almost 3. This finding is also manifested by the average radial distribution of galaxies derived with the quasars as centers. This distribution shows a significant bump in excess galaxies around 1 Mpc away from the quasars. The simplest interpretation of this excess is that some of these faint radio-quiet quasars are preferentially situated at the edges of rich clusters, whereas a significant fraction of radio-loud quasars at $z \gtrsim 0.5$ are found at or near the centers of rich clusters. At this point, it is not possible to say whether there is any evolutionary effect because in the surveys of Yee and Green (1984 and 1987) and Ellingson, Yee and Green (1991), the environments of the low-redshift comparison sample were only imaged out to a radius of 250 to 500 kpc radius due to the small field sizes and relatively low redshift. It will be of great interest if the environment of the low-redshift sample can be re-examined with a CCD camera with a much wider field.

9. Summary

The study of quasar environments has produced many fruitful results. Using principally direct imaging data with proper background galaxy count corrections and reasonable assumptions, we are able to derive quantitative information concerning the environment of quasars and their associated galaxies over an interesting redshift range. We learn that quasars at low redshift in general are located in regions of higher-than-average galaxy density. However, at higher redshift ($z \sim 0.5$) a significant fraction of radio-loud quasars are found in the centers of very rich clusters. The appearance of powerful quasars in clusters at these higher redshifts can be understood in terms of a much more rapid evolution of quasars in clusters. This evolution scencario is strengthened by combining data from both quasars and radio galaxies. It is found that many low-redshift radio galaxies which are first ranked members of rich clusters contain a "mini-quasars". The level of optical activity in radio galaxies and quasars can be parametrized by a single LF, with the objects in rich environments requiring 5 to 6 times more rapid evolution than those in poor environments.

By studying galaxy clusters associated with both bright and faint (and perhaps dying) quasars, we may be able to isolate specific conditions that are favored by quasar activity and hence gain insights into the mechanisms of quasar formation and sustenance. Possible mechanisms that can explain the rapid evolution of quasars in rich clusters include the evolution of the dynamical state, the gas content, or the cooling flow rate in rich clusters over a time scale of one to two Gyrs. Preliminary evidence from multi-spectroscopy observations of galaxies in clusters associated with quasars suggests that quasars tend to inhabit clusters and groups with low velocity dispersion. this supports the scenario that interactions and mergers play an important role in sustaining quasar activity.

10. Acknowledgement

I wish to thank my collaborators, including B. Boyle, M. De Robertis, E. Ellingson, R. Green, C. Pritchet, and T. Shanks, for allowing me to present the new results before publication. A special thanks is due for E. Ellingson who has been my main collaborator in many of these projects and also for her many helpful and important comments on the manuscripts.

References

Boyle, B. J., Fong, R., Shanks, T., and Peterson, B. A. 1987, *M. N. R. A. S.*, **227**, 717.
Boyle, B. J, Shanks, T., and Yee, H. K. C. 1989, in *The Proceedings of the IAU Symposium 130: Large Scale Structure and Motions of the Universe*, eds. J. Adouze et al., (Dordrecht: Reidel), page 577.
Butcher, H., and Oemler, A. 1978, *Ap. J.*, **226**, 559.
Carlberg, R. G. 1990, *Ap. J.*, **350**, 505.
Colless, M. M. 1988, Ph. D. thesis, Kings College, Cambridge University, Cambridge.
Davis, M., and Peebles, P. J. E. 1983, *Ap. J.*, **267**, 465.
De Robertis, M. M. 1985, *A. J.*, **235**, 351.
De Robertis, M. M. and Yee, H. K. C. 1990, *A. J.*, **100**, 84.
De Robertis, M. M. and Yee, H. K. C. 1992, in preparation.

Dressler, A. 1980, *Ap. J.*, **236**, 351.

Edge, A. C., Stewart, G. C., Fabian, A. C., and Arnaud, K. A. 1990, *M. N. R. A. S.*, **245**, 559.

Ellingson, E., Green, R. F., and Yee, H. K. C. 1991, *Ap. J.*, in press.

Ellingson, E. and Yee, H. K. C. 1991, in *Workshop on the Sapce Density of Quasars*, ed. D. Crampton et al., in press.

Ellingson, E. and Yee, H. K. C. 1992, in preparation.

Ellingson, E., Yee, H. K. C., and Green, R. F. 1991, *Ap. J.*, **371**, 49.

Ellingson, E., Yee, H. K. C., Green, R. F., and Kinman, T. D. 1989, *A. J.*, **97**, 1539.

Fabian, A. and Crawford, C. 1991, *M. N. R. A. S.*, in press.

Gehren, T., Fried, J., Wehinger, P. A., and Wyckoff, S. 1984, *Ap. J.*, **278**, 11.

Gioia, I. M., Henry, J. P., Maccacaro, T., Morris, S. L., Stocke, J. T., and Wolter, A. 1990, *Ap. J.*, **356**, L35.

Green, R. F., Schmidt, M., and Liebert, J. 1986, *Ap. J. Suppl.*, **61**, 305.

Gunn, J. E., 1971 *Ap. J.*, **164**, L113.

Gunn, J. E., and Dressler, A. 1988, in the *Proceedings of the 5th Erice Workshop on Towards Understanding Galaxies at Large Redshift*, ed. R. G. Kron and A. Renzini, (Dordrecht: Kluwer), page 227.

Hill, G. and Lilly, S. J. 1991 *Ap. J.*, **367**, 1.

Hintzen, P. 1984, *Ap. J. Suppl.*, **55**, 533.

Hintzen, P., Boeshaar, G., and Scott, J. 1981, *Ap. J.*, **246**, L1.

Hintzen, P., and Romanishin, W. 1986, *Ap. J.*, **311**, 1.

Hintzen, P. Romanishin, W., and Valdez, F. 1991, *Ap. J.*, **366**, 7.

Hutchings, J. B., Crampton, D., Campbell, B., and Pritchet, C. 1981, *Ap. J.*, **247**, 743.

Hutchings, J. B., Crampton, D., and Campbell, B. 1984, *Ap. J.*, **280**, 41.

Longair, M. S. and Seldner, M. 1979, *M. N. R. A. S.*, **189**, 433.

Malkan, M. 1984, *Ap. J.*, **287**, 555.

Oemler, A., Gunn, J. E., and Oke, J. B., 1972, *Ap. J.*, **176**, L47.

Roos, N. 1985, *A. A.*, **104**, 218.

Prestage, R. M. and Peacock, J. A. 1988, *M. N. R. A. S.*, **230**, 131.

Prestage, R. M. and Peacock, J. A. 1989, *M. N. R. A. S.*, **236**, 959.

Seldner, M. and Peebles, P. J. E. 1978, *Ap. J.*, **225**, 7.

Smith, E. P., Heckman, T. M., Bothun, G. D., Romanishin, W., and Balick, B. 1987, *Ap. J.*, **306**, 64.

Stockton, A. 1978, *Ap. J.*, **223**, 747.

Stocke, J. T. and Perrenod, S. C. 1981, *Ap. J.*, **245**, 375.

Tyson, A. J. 1986, *A. J.*, **92**, 601.

Wyckoff, S., Wehinger, P. A., and Gehren, T. 1981, *Ap. J.*, **247**, 750.

Yates, M., Peacock, J. A., and Miller, L. 1989, *M. N. R. A. S.*, **240**, 129.

Yee, H. K. C. 1981, *Ap. J.*, **241**, 894.

Yee, H. K. C., Bolye, B. J., and Shanks, T. 1992, preprint.

Yee, H. K. C. and De Robertis, M. M. 1989, in *The Proceedings of the IAU Symposium 134: Active Galactic Nuclei*, eds. D. Osterbrock and J. Miller, (Dordrecht:Reidel), page 457.

Yee, H. K. C. and De Robertis, M. M. 1992, in preparation.

Yee, H. K. C., and Ellingson, E. 1992, preprint submitted to *Ap. J.*.

Yee, H. K. C. and Green, R. F. 1984, *Ap. J.*, **280**, 79.

Yee, H. K. C. and Green, R. F. 1987, *Ap. J.*, **319**, 28 (YG87).

Yee, H. K. C. and Oke, J. B. 1978, *Ap. J.*, **226**, 753.

Zabludoff, A., Huchra, J. P., and Geller, M. J. 1990, *Ap. J. Suppl.*, **74**, 1.

X-RAY EVOLUTION OF CLUSTERS OF GALAXIES

J. Patrick Henry
Institute for Astronomy, University of Hawaii
2680 Woodlawn Drive
Honolulu, Hawaii 96822
USA

ABSTRACT. I review the evidence for the evolution of the X-ray luminosity of clusters of galaxies with cosmic epoch. Such evolution has been found in two independent studies. I discuss the constraints that this evolution places on the primordial density fluctuation spectrum.

1. What Properties of Clusters Can Be Measured In X-rays?

X-ray observations of extragalactic objects are always photon starved, due to the low fluxes of the objects and the (usually) short life of any given X-ray observatory. Hence, it is useful to understand the minimum number of photons required for a "good" measurement of some interesting properties of clusters. For the luminosity, about 25 net photons must be obtained in order to meet the 5σ criteria of Murdoch, Crawford, and Jauncey (1973). However things rapidly become harder. Approximately 1000 photons are required to measure an X-ray temperature so that there are only 2 clusters with redshift greater than 0.2 with a measured temperature (A483, Arnaud *et al.*, 1991; A963, Yamashita, 1991). Approximately 5000 photons are needed to make an image. Consequently only 3 clusters with redshifts greater than 0.2 have images obtained with instruments capable of resolving them (A963, Henry *et al.*, 1991; 3C295, Henry and Henrikisen, 1986; Cl0016+16, White, Silk, and Henry, 1981). Therefore, it is only possible to obtain samples which are large enough and distant enough to investigate luminosity evolution and I will restrict myself to this topic for the remainder of this paper.

Because clusters have a very wide range of X-ray luminosities, greater than a factor of 100, it is necessary to study their X-ray luminosity functions (XLFs) as a function of redshift in order to search for luminosity evolution. I show two typical low redshift XLFs in Figure 1 (Piccinotti *et al.*, 1982; Edge *et al.*, 1990) In this Figure and elsewhere in this paper (except Section 5), I use $H_o = 50$ km s^{-1} Mpc^{-1} and $q_o = 0.5$. The XLFs are usually characterized by a power law: $N(L_{x,44}(E_1,E_2)) = K[L_{x,44}(E_1,E_2)]^{-\alpha}$, where $L_{x,44}(E_1,E_2)$ is the luminosity in the energy band E_1 to E_2 in units of 10^{44} ergs s^{-1}, K is the normalization with units of Mpc^{-3}(10^{44} ergs s^{-1})$^{\alpha-1}$, and α is the power law index. The power law fit to the Edge *et al.* data is also shown in Figure 1. Thus luminosity evolution manifests itself through changes in K or α with redshift.

2. Why Are Evolution Measurements of Clusters Interesting?

Clusters provide information on the properties of the universe on scales of order 40 Mpc, which is a very large scale. This is the size that is required to accumulate a mass of 10^{15} M$_o$, the mass of a

311

A. C. Fabian (ed.), Clusters and Superclusters of Galaxies, 311–322.

rich cluster, if the universe has closure density. Thus the properties of clusters can yield information on the properties of the mass density field of the universe on these scales, and, in particular, evolution of clusters can be related to evolution of the mass density field. The nature of this field is a fundamental unknown in cosmology.

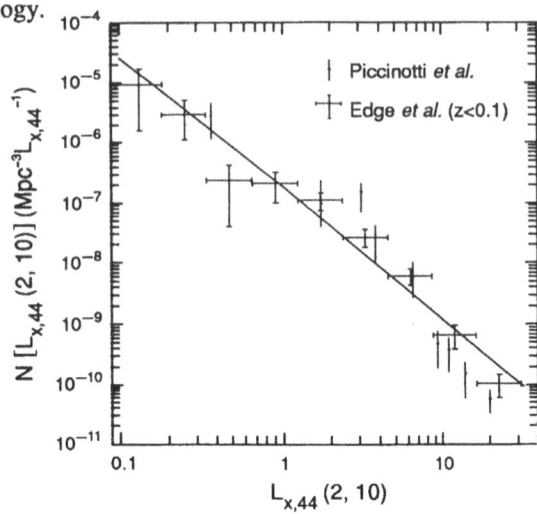

Moreover, extracting the information about the mass density field from cluster observations is relatively straightforward and is quite sensitive to the observed quantities. The simplicity arises because the mass density field on these scales is still undergoing linear evolution, ie the rms value of the mass fluctuations are less than the average mass density. Of course the clusters themselves are non-linear, having collapsed, but the underlying mass field is still linear. I give an outline of the linear theory in Section 5. Because clusters have collapsed from a linear density field, they must be the rare 3σ and 4σ objects capable of reaching the critical overdensity needed to locally halt the expansion of the universe. The properties of clusters are thus very sensitive to the mass density field, since they are on the exponential tail of some distribution. Section 5 provides a few more details.

Figure 1. Two low redshift cluster X-ray luminosity functions with the best fitting power law to the Edge et al. (1990) data overlaid.

Finally, X-ray observations of clusters are "clean" in the sense that projection effects are minimized and object selection is done entirely by objective criteria. The X-rays come from deep potential wells, that is from real physical objects not from objects in projection. Since the X-ray luminosity is proportional to roughly the third power of the richness (Bahcall, 1977) any projection is further minimized because the luminosities of two small objects do not add up to that of a single large one, but the galaxy counts may. Until quite recently all catalogs of clusters have been selected by human eyeball examining plates. The selection criteria of such a process are difficult if not impossible to quantify. Hence the conclusions draw from studies of these catalogs have been suspect at some level. In contrast, X-ray catalogs are generated by automated searches through photon event files, which if nothing else has the advantage that the selection criteria may be understood by everyone.

3. Early Studies Using Optically Selected Clusters

The first instrument capable of detecting the X-ray emission from clusters of galaxies at cosmological distances was the Einstein Observatory, launched in 1978 (Henry et al., 1979). The Einstein Observatory was the first imaging X-ray telescope useful for celestial objects that was placed in orbit. Consequently, like most telescopes, it was pointed at preselected targets. Since no previous experiment had been able to detect distant clusters, these objects were selected from the lists of clusters that were known circa 1978. The clusters were almost exclusively optically selected.

No evolution was seen in these early studies. Henry et al. (1982) showed that the slope of the cluster X-ray luminosity function was the same at a redshift of 0.5 as it was at low redshift (Figure 2a), and Henry and Lavery (1984) showed that the X-ray luminosity function of Abell clusters at a redshift of 0.25 was the same as that at low redshift (Figure 2b). Work at moderate redshift by Kowalski, Ulmer, and Cruddace (1983) using the non-imaging HEAO-1 A1 experiment, which scanned the entire sky and was not limited to preselected targets, also showed no evolution. They found that the volume emissivity of a sample of 31 Abell clusters with average redshift of 0.07 was the same as that of a sample of 38 clusters with average redshift 0.17.

It is likely that all this early work was dominated by selection effects. If the more distant clusters are richer, in order to be visible against the background of faint galaxies, then the more distant clusters are also more X-ray luminous, from the luminosity-richness effect. Thus optical selection may counter the trend discovered by later X-ray surveys that distant clusters tend to be less luminous on average. I describe these new results next.

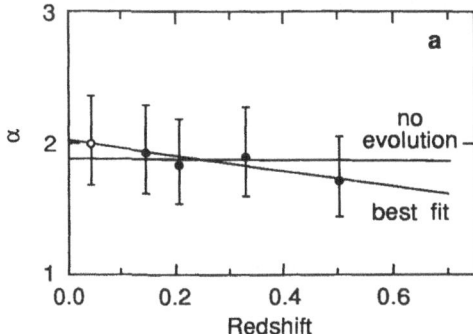

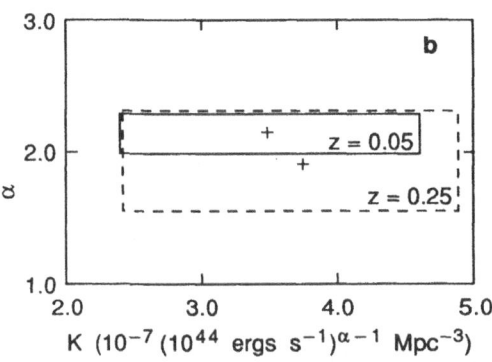

Figure 2. Lack of evolution in the cluster X-ray luminosity function of predominantly optically selected objects. a. The power law index of the luminosity function vs. redshift. The best fitting straight line of the form $\alpha = a + bz$ is overlaid. b. The $\pm 1\sigma$ confidence regions for the Abell cluster X-ray luminosity function power law index and normalization.

4. X-ray Selected Samples of Clusters

X-ray selected samples of clusters have finally become large enough and deep enough that evolution is detectable, though barely. There have been two simultaneous studies published to date (Edge et al. 1990; Gioia et al. 1990a) which find evolution in the same sense.

4.1. HIGH FLUX LIMIT, ALL-SKY SURVEY

Edge et al. (1990) present a sample of 46 clusters with the following selection criteria: $F(2,10) \geq 1.7 \times 10^{-11}$ ergs cm^{-2} s^{-1} and $|b^{II}| \geq 20°$. This sample is compiled mostly from the HEAO-1 A2 and Ariel V all-sky surveys conducted with non-imaging instruments. Confusion effects resulting from the rather broad beams of these two surveys were reduced through the Einstein and EXOSAT experiments which both had 0.75° fields of view.

The evidence for evolution is found in the difference between the logN-logS distribution for low and high luminosity clusters. This difference is significant at the 3σ level. The distribution for

$L_x(2,10) < 8 \times 10^{44}$ ergs s^{-1} is consistent with that expected for a Euclidean distribution of sources down to a flux of 1.7×10^{-11} ergs cm^{-2} s^{-1}, while the distribution for clusters with luminosities greater than that was inconsistent in the sense that there are too few low flux clusters (see Figure 3). Since it is difficult to imagine a selection bias based on luminosity, flux is what is detected, Edge *et al.* concluded that their sample was missing high luminosity clusters. Since high luminosity clusters are predominantly at high redshift, the clusters are being lost because of evolution. There are a factor of three fewer clusters at z ~ 0.15 with $L_x(2,10) > 6 \times 10^{44}$ ergs s^{-1} than there are at zero redshift.

There are two major deficiencies with this work. First, the high redshift sample is small, only 3 clusters are at redshifts greater than 0.1. Second, the claimed complete flux limit is lower than that of any of the surveys used to compile the sample (1.7×10^{-11} ergs cm^{-2} s^{-1} claimed *vs.* ~ 3.0×10^{-11} ergs cm^{-2} s^{-1} for the HEAO-1 A2 and Ariel V surveys). The estimated number of clusters missed due to incompleteness introduced by lowering the flux limit is about 10 while the number of clusters lost because of evolution is 9. The similarity of these two numbers is worrisome. However, there is no obvious reason why all the clusters lost due to incompleteness will be at high redshift because the redshifts involved are quite small.

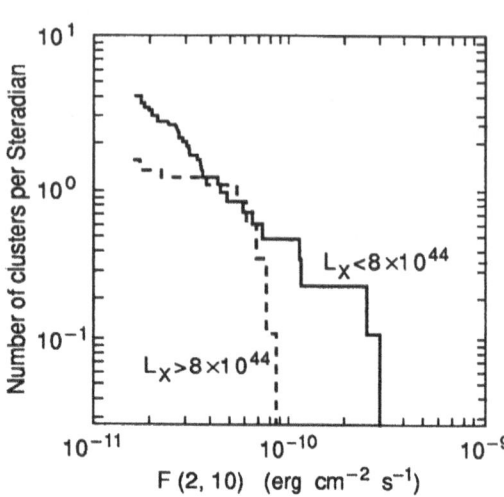

Figure 3. The logN-logS distribution for low and high luminosity clusters from the Edge *et al.* (1990) sample.

4.2. LOW FLUX LIMIT, SMALL SOLID ANGLE SURVEY

Gioia *et al.* (1990a) present a sample of 67 clusters with the following selection criteria: $F(0.3,3.5) \geq 1.3 \times 10^{-13}$ ergs cm^{-2} s^{-1} in a 2.4x2.4 arcmin^2 detect cell, $\delta \geq -40^\circ$, $z \geq 0.14$, and $|b^{II}| \geq 20^\circ$. Note the differences between this survey and the previously described one. It is in a different, softer, energy band and it is approximately 100 times deeper. Only ~20% of the clusters in this sample are Abell clusters. Such a low fraction agrees with the preliminary analysis of the ROSAT survey data (Boehringer, this conference).

The depth described above comes at a price however. The clusters are from the Einstein Extended Medium Sensitivity Survey (Gioia *et al.*, 1990b) which is comprised of those sources detected serendipitously in Einstein Observatory pointings. Since the Medium Survey was conducted with an imaging instrument some of the flux of every cluster is resolved out, the amount of which is a function of the apparent size and surface brightness of the cluster and hence a function of at least its redshift. This situation means that the total flux of the cluster must be corrected for that outside the detect cell, which can only be done if the redshift is known. Thus any quantity requiring the flux of the clusters, such as the logN-logS distribution, can only be obtained after the redshifts are obtained. To correct for the lost flux Gioia *et al.* (1990a) adopted the β model for the cluster surface brightness distribution (see Henry *et al.*, 1992 for additional details). In this model the surface brightness as a function of angle from the cluster center, $I(\theta)$, is given by

$$I(\theta) = I_0[1 + \theta^2/\theta_0^2]^{1/2-3\beta}$$

The model parameters adopted are $\beta = 2/3$ and $D_A*\theta_0 = 0.25$ Mpc where D_A is the angular diameter distance (see Henry *et al.*, 1992 and Jones, this conference, who finds an average value of 0.23 for $D_A*\theta_0$ in a sample of 158 clusters with β set to 2/3). Using this assumed surface brightness and the redshift of the cluster, the observed flux in the detect cell may be corrected to arrive at the "true" total flux and luminosity. Such corrections are typically a factor of 2.5. Gioia *et al.* also made corrections for the K dimming and for absorption by the Milky Way, but these corrections are small effects. The resulting XLFs in three redshift shells are shown in Figure 4.

I compare in Figure 5 the lowest redshift EMSS XLF with the low redshift XLF from Piccinotti *et al.* (1982) and the Edge *et al.* (1990) XLF which exhibits the evolution. Qualitatively, all three are in agreement, with the EMSS XLF somewhat higher. It is necessary to assume a spectrum in order to make a quantitative comparison, because the EMSS XLF is in a different energy band than are the other two. Under the assumption that all clusters emit a 6 keV thermal spectrum, the normalization of the EMSS XLF is a factor of 1.5 ± 0.5 larger than the Piccinotti *et al.* luminosity function. The power law exponents agree to within 2%. That is, the lowest redshift EMSS XLF agrees with previous work within the errors up to its highest luminosity of 5×10^{44} ergs s^{-1} or, in other words, *there is no strong bias of the type discussed by Pesce et al. (1990) resulting from the selection of clusters by imaging observations.*

The EMSS sample also contains evidence for luminosity evolution which is shown in Figure 6. There are fewer high luminosity high redshift clusters as exhibited by the higher power law index for the higher redshift sample. The difference between the two XLFs is significant at the 3σ level. The direction of the evolution found by both of the studies described here is in the same sense, but

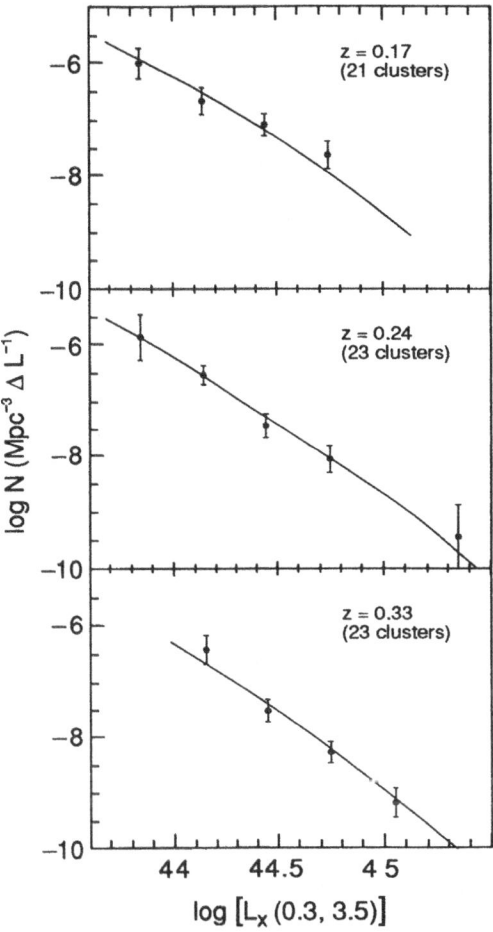

Figure 4. The X-ray luminosity functions for clusters of galaxies in three redshift shells from the EMSS sample. The best fitting theoretical luminosity functions are overlaid.

316

the EMSS data can not address the speed reported by Edge *et al.* (1990) since they only detect an effect above a luminosity of 6 x 10^44 ergs s^-1 whereas the EMSS low redshift sample only extends to 5 x 10^44 ergs s^-1.

The major deficiency in the analysis of the EMSS sample is the correction for the finite source size which is probably too simplistic. Clusters are not all the same size, and even if they were, there is no reason to suppose that the size does not evolve. After all, the luminosity does. Not withstanding the deficiencies in both samples discussed here, it is encouraging that they both find evolution in the same sense. It is probable that *the XLF of clusters really does evolve*.

I close this section with a prediction for ROSAT all-sky survey. Based on the EMSS XLF and its evolution, Evrard and Henry (1991) find that there will be approximately 300 clusters in the 314 deg^2 region around the North Ecliptic Pole (the region of greatest depth) above a flux in the 0.1 to 2.4 keV band of 9 x 10^-14 ergs cm^-2 s^-1. Only 10% of these clusters will have redshifts greater than 0.4 with no clusters at redshifts greater than 1. If there is no evolution then the redshift distribution will be dramatically different with many clusters at redshifts beyond 0.5. The ROSAT survey and its follow up optical identification program will put the results described here on a much more secure statistical basis.

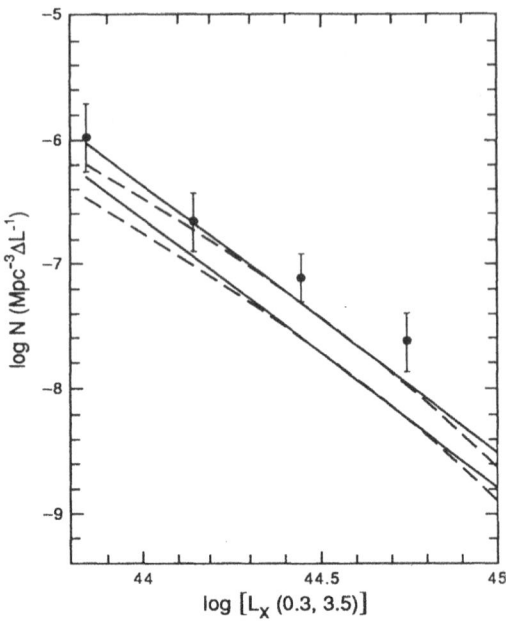

Figure 5. The X-ray luminosity function for the lowest redshift shell EMSS clusters (0.14 ≤ x ≤ 0.2) compared to the X-ray luminosity functions of Piccinotti *et al.* (solid lines, indicating the error range) and Edge *et al.* (dashed lines).

5. What Does It All Mean?

I may not succeed in answering the above question, but I hope to be able to describe a model which explains the observed X-ray evolution of clusters and some of their other properties as well. The sign of the observed evolution means that clusters almost certainly formed as the result of mergers of smaller objects, *ie* through a hierarchical process. There is a "relatively" simple theory which then connects observables with the fundamental theoretical idea, the mass density field of the universe. In this section I will outline this theory and describe what constraints the X-ray observations place on it. Here the Hubble constant is $H_0 = 100$ h km s^-1 Mpc^-1.

5.1 OUTLINE OF A HIERARCHICAL THEORY OF CLUSTER FORMATION

I start with the Press-Schechter (1974) mass function, N(M), which gives the number of collapsed objects of mass M per unit volume and per unit mass. The fundamental assumption of this mass function is that the fluctuations of the mass density field about its average value are described by a Gaussian probability distribution with dispersion $\sigma_\rho(r)$. Now the mass of a cluster is very difficult to observe, so the temperature or luminosity functions must be derived from the mass function. They come from the chain rule:

$$N(X) = N(M) \, dM/dX \qquad (1)$$

where X is the observable luminosity or temperature. One now needs a temperature-mass or luminosity-mass relation in order to proceed.

The temperature-mass relation comes from considering the spherical collapse of a top hat density fluctuation normalized by Evrard's (1990) numerical calculations.

$$kT = 6.4 \text{ keV } (1+z) \, (hM_{15})^{2/3} \qquad (2)$$

M_{15} is the mass of the cluster in units of 10^{15} M_0. There is little that can be done to assess the validity of this relation since its whole reason for being is because the masses of clusters are not well known. However, for the Coma cluster, $kT = 8.2$ keV which, from the above relation, implies $hM_{15} = 1.5$ whereas Hughes (1989) finds a hM_{15} of 1.0 ± 0.5 within $2.5h^{-1}$ Mpc. This mass is at least consistent with equation (2).

For the luminosity-mass relation I use the same spherical collapse model normalized by the *observed* luminosity-temperature relation at low redshift. After correcting for the finite band of the EMSS data I find (see Henry *et al.*, 1992 for the details).

$$h^2 L_{x,44}(0.3, 3.5) = 1.8 \, (1+z)^{3.15} \, (hM_{15})^{8/5} \qquad (3)$$

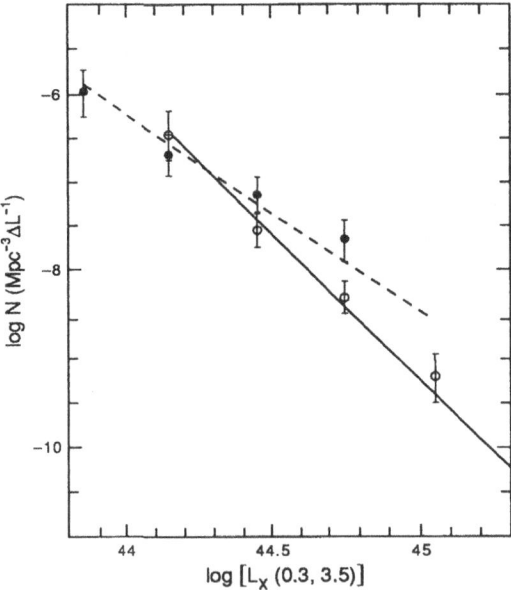

Figure 6. The X-ray luminosity functions for clusters in the EMSS sample for its lowest and highest redshift shells.

It is customary and convenient to work with the spatial Fourier transform of the fluctuations in the mass density field normalized by the average mass density, $\delta(k)$. The Gaussian assumption of the Press-Schechter formulation implies that $\sigma_\rho(r)$ is given by the square of $\delta(k)$ (usually called the power spectrum) which we assume is given by

$$|\delta(k)|^2 = 1/(k_o^3 V_u) \, (k/k_o)^n \qquad (4)$$

where V_u is some large volume within which the statistical properties of the mass density field are the same as in any other such large volume. Thus the power spectrum is parameterized by k_o and n and the object of what follows is to constrain the possible values of these two parameters by observations. The much discussed cold dark matter model has an n of about -1 on cluster scales.

318

Many times the normalization of the power spectrum is described by the bias parameter, b, which is defined to be $1/\sigma_\rho(8h^{-1}Mpc)$, but such a description is not convenient for this work.

The process I want to follow is shown schematically by

$$\delta(k) \longrightarrow \sigma_\rho(r) \longrightarrow N(M) \begin{array}{l} \longrightarrow N(L) \longrightarrow >< \longrightarrow Data \\ \longrightarrow N(kT) \longrightarrow >< \longrightarrow Data \end{array}$$

How well does the above analytic calculation agree with the numerical N-body results? I show in Figure 7 a comparison between the calculations of Frenk *et al.* (1990) for the cold dark matter model and the temperature function calculated according to the method outlined here for the case n = -1.2. Since the numerical work did not include hydrodynamics, I have converted their velocity dispersions to temperatures using a value of 1.2 for the ratio $\mu m_p\sigma^2_{DM}/kT$ where μ is the mean molecular weight (0.6), m_p is the proton mass, and σ^2_{DM} is the dark matter velocity dispersion. The value adopted for this ratio is consistent with equation (2). Figure 7 shows that the cumulative difference between the two results is very small for kT ≥ 3 keV but that it reaches ~40% at 2 keV. Since nearly every rich cluster known has a temperature greater than 3 keV, we conclude that the simple model presented here is adequate to test existing data against fluctuation spectra with indices near -1.

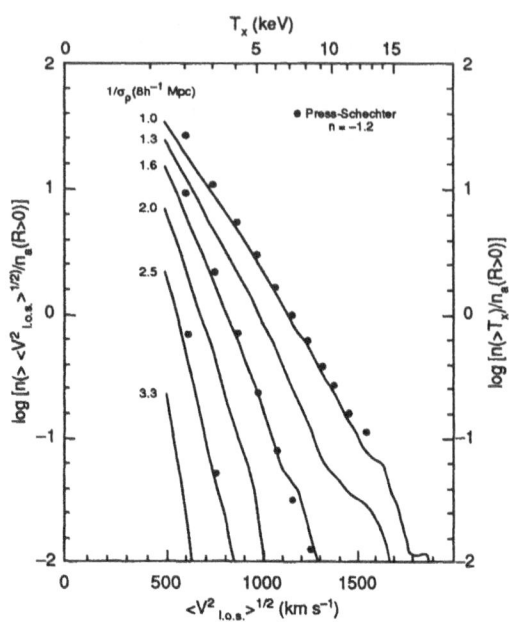

Figure 7. Comparison of numerical N-body calculations (solid lines) with the analytic model described here with n = -1.2 and normalization parameterized by b (solid dots at integral values of the temperature). The integral distributions normalized by the density of Abell clusters with richness class one and greater are plotted.

5.2 CLUSTER LUMINOSITY EVOLUTION

Inserting equation (3) into equation (1) and using the mass density dispersion calculated from equation (4) in the Press-Schechter mass function yields the luminosity function and its evolution which has been fit to the EMSS cluster sample. The best fitting functions are shown in Figure 4 and the constraints on the mass fluctuation spectrum from this fit are shown in Figure 8. Although the errors are correlated, as is shown in Figure 8, these constraints may be roughly stated as $n = -(2.1^{+0.3}_{-0.2})$, $k_0 = (0.029^{+0.008}_{-0.013})$ h Mpc^{-1}, which together imply b = $1.64^{+0.08}_{-0.11}$. The alignment of the error ellipse approximately along the lines of constant rms density dispersion (or b) yields an extremely sensitive measurement of this quantity.

The normalization is what is required by most numerical models in order to obtain the large scale structure observed (Carlberg and Couchman 1989; Park 1990). However the best fitting index n disagrees with that predicted by cold dark matter (CDM) models. CDM has been very successful in explaining many features of the large scale distribution of galaxies but it does appear to have too little power on cluster scales (cf Efstathiou *et al.* 1990). The larger index required by the EMSS data is just another reflection of this deficiency.

5.3 CLUSTER TEMPERATURE FUNCTION

The sample used is defined by $F(2,10) \geq 3 \times 10^{-11}$ ergs cm^{-2} s^{-1} and $|b^{II}| \geq 20^{0}$. This sample is that of Piccinotti *et al.* (1982) corrected for source confusion by Lahav *et al.* (1989) and Edge *et al.* (1990). There are 25 clusters with redshifts ≤ 0.09. All but one cluster has a temperature measurement, all but two with the same experiment. See Henry and Arnaud (1991) for the details. A power law representation of the temperature function obtained from these data is

$$N(kT) = 1.8^{+0.8}_{-0.5} \times 10^{-3} h^3 Mpc^{-3} keV^{-1}$$
$$(kT)^{-4.75 \pm 0.5}.$$

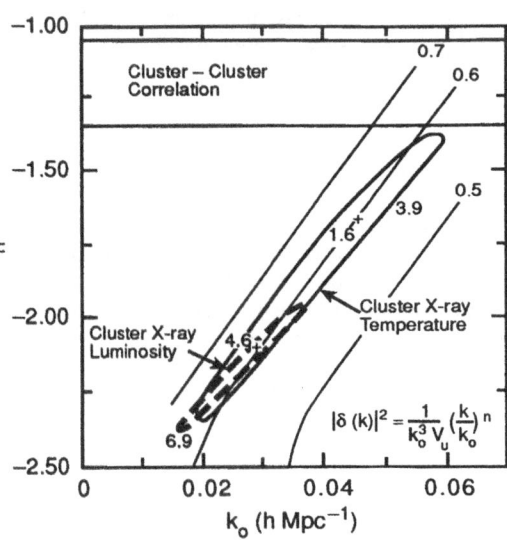

Figure 8. Constraints on the mass density fluctuation power spectrum characterized by a power law index n and normalization k_o. The contours labeled X-ray temperature and luminosity are at the 68% confidence level given by the χ^2 values indicated. Constraints from the cluster-cluster correlation are also shown. The light diagonal lines are the loci of constant values of $\sigma_\rho(8 \ h^{-1} \ Mpc)$.

In the same manner as before, inserting equation (2) into equation (1) and using equation (4) to calculate the mass density dispersion gives the temperature function. Fitting this function to the data yields the constraints on the mass fluctuation spectrum shown in Figure 8. They can be roughly described as n = -$(1.7^{+0.65}_{-0.35})$, $k_o = (0.045^{+0.015}_{-0.025})$ h Mpc^{-1}, which together imply b = 1.69±0.06. As may be seen in Figure 8, these values agree with those determined from the luminosity evolution but are less restrictive because the sample size is about a factor of three smaller. This agreement between two independent samples measuring different quantities must be indicating that the simple theory outlined here has some validity.

5.4 CLUSTER-CLUSTER CORRELATION FUNCTIONS

Bahcall and Soneira (1983) and Klypin and Kopylov (1983) were the first to obtain the cluster-cluster correlation function for Abell clusters. The difference between that function and the galaxy-galaxy correlation function provided some of the initial impetus to the consideration that the light from galaxies might be a biased representation of the mass distribution.

In an exactly analogous way to the Press-Schechter derivation of the mass function, the correlation function for collapsed objects of equal masses may be shown to be (Kashlinsky, 1987)

$$\xi_{MM}(r) = \delta_c^2/\sigma_\rho(r)^2 \, \xi_{\rho\rho}(r) \tag{5}$$

where $\xi_{MM}(r)$ and $\xi_{\rho\rho}(r)$ are the correlation functions for collapsed objects of equal mass M and for the underlying density field respectively, and δ_c is the critical overdensity of a mass perturbation which is just sufficient to cause it to collapse. This equation assumes that $\xi_{\rho\rho}(r)$ is << 1 and exhibits the biasing that occurs, ie the two correlation functions are not equal. With the power spectrum from equation (4), the above becomes

$$\xi_{MM}(r) = (r/r_0)^{-(3+n)} \tag{6a}$$

with

$$r_0 = 9.5h^{-1} \text{ Mpc } [0.32 \, 2^n(-n)(1-n)(3-n)]^{1/(3+n)} \, (1+z)^{-1} \, (hM_{15})^{1/3} \tag{6b}$$

the power law exponent of ξ_{MM} is often denoted γ. Equations (6) show explicitly that the correlation function is independent of the normalization of the power spectrum but that it does depend weakly on the mass of the objects being considered.

There have been three recent determinations of the cluster-cluster correlation function: Huchra et al. (1990), (HHPG); Postman, Huchra, and Geller (1992), (PHG, from whom I quote the results of their statistical sample, uncorrected for contamination); and Dalton et al. (1992), (DEMS). The properties of these three samples are given in Table 1 where R is the Abell richness class of the clusters in the sample.

TABLE 1. Properties of Cluster-Cluster Correlation Function Determinations

	PHG	HHPG	DEMS
Catalog	Abell	Abell	APM
Clusters	208	132	190
z	≤ 0.08	≤ 0.24	0.02-0.12
R	≥ 0	≥ 0	≥ 0.5
γ	-1.9 ± 0.2	-1.5 ± 0.3	-1.9 ± 0.3
r_0 (h^{-1} Mpc)	21 ± 5	21 ± 7	13 ± 1

The APM catalog is from the Automatic Plate Measuring machine scans of 185 UK Schmidt plates from which clusters were extracted by computer. It is one of the first optical cluster catalogs selected by objective methods. The Abell (1958) catalog has long been the main source of clusters, but it was selected by Abell's eyes and brain. The PHG and HHPG samples are nearly disjoint (having only 15 clusters in common), yet their correlations are remarkably the same. This agreement implies that the correlation function of Abell clusters is robust. Although the power law index, γ, of the APM clusters agrees with that determined for the Abell clusters, the scale length disagrees. This disagreements occurs despite the fact that approximately half of the APM clusters are also ACO clusters, the southern counterparts of the Abell catalog (see Abell, Corwin, and Olowin, 1989). Since the above sample is the first objectively selected sample of clusters in the optical, it is not known how robust the result is. It is important pioneering work which needs to be repeated in another region of the sky.

Taking the above results at face value, the constraints on the mass fluctuation spectrum are, after averaging the three indices since they are the only parameters which provide a constraint, $n = -(1.19 \pm 0.15)$, $k_0 =$ anything. This value of n is plotted in Figure 8. It clearly disagrees with the X-ray results, but it is in agreement with CDM. At the moment, it is not clear to me what this disagreement means.

With the above value of n and the scale lengths in Table 1, the characteristic mass of clusters is obtained from Equation (6b): $hM_{15} = 5^{+4}_{-3}$ and $1.2^{+0.5}_{-0.4}$ for the Abell and APM samples respectively. Either of these masses is a plausible value for a rich cluster. Finally Equation (6b) may provide an explanation for the increasing correlation of more massive objects noted by Bahcall (1988; see also this conference). With n as given above, (6b) becomes $r_0 = 12h^{-1}$ Mpc $(hM_{15})^{0.33}$. This relation should be compared with Equation 15 in Bahcall (1988): $r_0 = 35$ Mpc $(M_{15})^{0.28}$. The approximate agreement seems suggestive, particularly since her mass scale may be different from the one used here. Her Equation (15) implies that the mass of a R = 1 cluster is $0.3 \times 10^{15} M_0$, about an order of magnitude less than above.

6. Remember This Stuff

To summarize:

The X-ray luminosity of clusters of galaxies evolves in the sense that there are fewer high luminosity clusters in the past. This evolution implies that hierarchical formation (merging of subclusters) is likely to be the formation process. All of the X-ray data are consistent with a simple hierarchical model which requires a different shape mass spectrum than that expected from cold dark matter but with a normalization which agrees with other estimates. However recent measurements of the cluster-cluster correlation function are consistent with CDM (and thus inconsistent with the X-ray results). Spatially resolved X-ray temperature measurements of clusters of galaxies will permit the mass of clusters to be measured accurately which will then allow a more stringent test of the model discussed here.

I would like to thank Prof. J. Truemper and the ROSAT group at MPE for their hospitality during my sabbatical year at MPE. This work has received financial support from NASA grants NAG5-1256 and NAG5-1446 and NSF grant INT 8912660.

7. References

Abell, G.O. 1958, *Ap. J. Suppl.*, **3**, 211.
Abell, G.O., Corwin, H.C., and Olowin, R. 1989 *Ap. J. Suppl.*, **70**, 1.
Arnaud, M., Lachieze-Rey, M., Rothenflug, R., Yamashita, K., and Hatsukade, I. 1991, *Astron. Astrophys.* **243**, 56.
Bahcall, N.A. 1977, *Ap.J.*, **217**, L77.
Bahcall, N.A., and Soneira, R.M. 1983, *Ap.J.*, **270**, 20.

322

Bahcall, N.A. 1988, *Ann. Rev. Astron. Astrophys.*, **26**, 631.

Carlberg, R.G. and Couchman, H.M.P. 1989, *ApJ.*, **340**, 47.

Dalton, G.B., Efstathiou, G., Maddox, S.J., and Sutherland, W.J. 1992, *MNRAS*, in press.

Edge, A.C., Stewart, G.C., Fabian, A.C., and Arnaud, K.A. 1990, *MNRAS*, **245**, 559.

Efstathiou, G., Kaiser, N., Saunders, W., Lawrence, A., Rowan-Robinson, M., Ellis, R.S., and Frenk, C.S. 1990, *MNRAS*, **247**, 10p.

Evrard, A.E. 1990, in Clusters of Galaxies, eds W.R. Oegerle, M. Fitchett, and L. Danly, (Cambridge: Cambridge University Press), p. 287.

Evrard, A.E., and Henry, J.P. 1991, *Ap. J.*, in press.

Frenk, C., White, S.D.M., Efstathiou, G., and Davis, M. 1990, *ApJ.*, **351**, 10.

Gioia, I.M., Henry, J.P., Maccacaro, T., Morris, S.L., Stocke, J.T. and Wolter, A. 1990a, *ApJ.*, **356**, L35.

Gioia, I.M., Maccacaro, T., Schild, R.E., Wolter, A., Stocke, J.S., Morris, S.L., and Henry, J.P. 1990b, *Ap. J. Suppl.*, **72**, 567.

Henry, J.P., Branduardi, G., Briel, U., Fabricant, D., Feigelson, E., Murray, S., Soltan, A., and Tananbaum, H. 1979, *ApJ.*, **234**, L15.

Henry, J.P., Soltan, A., Briel, U.G., and Gunn, J.E. 1982, *ApJ.*, **262**, 1.

Henry, J.P., and Lavery, R.J. 1984, *ApJ.*, **280**, 1.

Henry, J.P., and Henriksen, M.J. 1986, *ApJ.*, **301**, 689.

Henry, J.P., and Arnaud, K.A. 1991, *ApJ.*, **372**, 410.

Henry, J.P., Gioia, I.M., Maccacaro, T., Morris, S.L., Stocke, J.T., and Wolter, A. 1992, *Ap. J.*, in press.

Huchra, J.P., Henry, J.P., Postman, M., and Geller, M.J. 1990, *ApJ.*, **365**, 66.

Hughes, J.P. 1989, *ApJ.*, **337**, 21.

Kashlinsky, A. 1987, *ApJ.*, **317**, 19.

Klypin, A., and Kopylov, A. 1983, *Sov. Astr. Letters*, **9**, 41.

Kowalski, M.P., Ulmer, M., and Cruddace, R.C. 1983, *ApJ.*, **268**, 540.

Lahav, O., Edge, A C., Fabian, A.C., and Putney, A. 1989, *MNRAS*, **238**, 881.

Murdoch, H.S., Crawford, D.E., Jauncey, D.L. 1973, *ApJ.*, **183**, 1.

Park, C. 1990, *MNRAS*, **242**, 59p.

Pesce, J.E., Fabian, A.C., Edge, A.C., and Johnstone, R.M. 1990, *MNRAS*, **244**, 58.

Piccinotti, G., Mushotzky, R.F., Boldt, E.A., Holt, S.S., Marshall, F.E., Serlemitsos, P.J., and Shafer, R.A. 1982, *ApJ.*, **253**, 485.

Postman, M., Huchra, J.P., and Geller, M.J. 1992, *ApJ.*, in press.

Press, W.H., and Schechter, P. 1974, *ApJ.*, **187**, 425.

White, S.D.M., Silk, J., and Henry, J.P. 1981, *ApJ.*, **251**, L65.

Yamashita, K. 1991, private communication.

EVOLUTION OF CLUSTERS OF GALAXIES

Nick Kaiser
CIAR Cosmology Program
CITA University of Toronto
60 St. George Street
Toronto Ontario
Canada

ABSTRACT. In hierarchical models for structure formation like CDM, the mass scale of clustering evolves rapidly and clusters in the past are predicted to be less massive, but denser and more numerous. The indication from optical searches is that the cluster population is roughly stable in that high velocity dispersion clusters seem to be about as abundant at $z \simeq 0.5$ as at the present while X-ray surveys show a strong negative evolution, there being many fewer high-luminosity clusters in the past compared to the present. Using self-similar scaling laws I find that the neutral evolution in the optical agrees quite well with what is expected, but that the negative X-ray evolution is very hard to reconcile with the idea that state of the gas in clusters is the result of gravitational clustering. One way to remedy this discrepancy is to assume that the gas we now see in clusters was heated at some earlier epoch by non-gravitational processes. If one discards the optical observations, an alternative is to invoke primordial fluctuations with a much redder spectrum, and I discuss how one might discriminate between these theories with future observations.

1. Introduction

In theories like the cold dark matter model, structure is thought to form by gravitational instability; overdense regions detach from the cosmological expansion, turn around and form condensed, virialised objects, this process proceeding hierarchically. In this picture, rich clusters live at the high mass end of the present day mass spectrum of condensed objects and, as such, provide important constraints on theory since in these objects we can both 'see' the dark matter through its gravitational influence and also plausibly believe that the cluster properties are primarily a result of initial conditions rather than astrophysics.

In models like CDM the hierarchical growth of structure is very rapid; the characteristic scale of clustering (this concept will be made more precise presently) grows roughly as $M_* \propto (1 + z)^{-3}$, and one might expect this to be quite noticeable in deep cluster surveys. One problem which faces us is that we must test the theory—which really predicts the structure in the dark matter—against observations of 'luminous tracers'; either X-ray emitting gas or bright galaxies. Neither of these is free from problems. With galaxies, we may reasonably expect that the

A. C. Fabian (ed.), Clusters and Superclusters of Galaxies, 323–330.

velocity dispersion reflects the depth of the confining cluster potential well, but there may be bias (Carlberg and Dubinski, 1991). If galaxies are more concentrated in clusters, either because galaxy formation was biased towards dense regions or because of dynamical friction, then the galaxy velocity dispersion will be less than that of the dark matter, and we will underestimate the total mass by some factor which depends on the rather uncertain dark matter profile. A second problem with optically detected chusters is that they may be subject to projection effects (Frenk, et al. 1990), and this may bias the observed richness distribution. In using X-ray emission there are analogous problems in that the gas temperature at a given radius will depend both on the dark-matter potential well depth and on the details of the gas density profile, and only very few clusters have the detailed temperature and surface brightness profiles necessary to solve for the mass profile. Indeed, in most of the data where we can probe evolution we have little or no information on the gas temperature and we must rely on the total X-ray luminosity which is an even less secure guide to the true cluster potential. One real advantage of X-ray observations though is that they are much less inflicted by projection effects.

In very rough terms, X-ray observations of clusters tell us that the present day 'temperature function' is very steep; $N(> T) \propto T^{-4}$ for temperatures of a few to ten keV, and that there seems to be a cut-off at about 10 keV (Edge et al. 1990). A rather similar picture emerges from the Abell cluster richness distribution; the inferred temperature function is somewhat shallower, though this difference may well be attributed to projection effects, and there again appears to be a cut-off at the high mass end. These gross properties would seem to be quite compatible with hierarchical pictures for formation of structure. A sharp high mass cut-off in the spectrum is to be expected in theories with Gaussian initial fluctuations, and the details of the temperature function seem to be compatible with that derived from N-body simulations (see Simon White's contribution to these proceedings) once suitably normalised.

The observational indications for evolution are more puzzling. In optical searches we see little evolution of the temperature function; the abundance of high velocity dispersion clusters seems much the same at $z \sim 0.5 - 1$ as it is today (Gunn, 1990), yet the X-ray cluster luminosity function (Edge et al. 1990; Gioia et al. 1990) shows strong negative evolution (i.e. fewer bright clusters in the past).

In the following, I shall argue that the neutral evolution of the temperature function as inferred from optical observations (albeit rather roughly) is quite compatible with hierarchical models with a spectrum like the CDM model—the strong negative evolution of the mass-function in these models notwithstanding—but that the strong negative evolution in X-rays is a serious problem if we assume, as is customary, that the present state of the gas in clusters is determined by gravitation clustering alone (with the gas being shocked on infall in order to come to hydrostatic equilibrium). It should be stressed that the optical observations are highly uncertain, but the a high abundance of high velocity dispersion clusters at earlier times is supported by observations of gravitational lensing (e.g. Tyson, 1991).

One way to help reconcile the X-ray observations with theory is to adopt a 'redder' primordial fluctuation spectrum. This is the model discussed by Pat Henry in his contribution to these proceedings, but I will argue that this would have deleterious effects on the optical predictions. An alternative is to appeal to some non-gravitational energy input to the gas in the past—perhaps from galaxy formation—and I will show that a crude model incorporating these effects seems to work quite well.

2. Self-Similar Clustering

There are three primary techniques which have been used to calculate the evolution of clusters. One approach is numerical simulations incorporating hydrodynamics (e.g. Evrard 1990). A second approach is to use the Press-Schechter (1974) formula or peaks statistics (Bardeen *et al.* 1986) to relate the cluster mass-spectrum to the statistics of the primordial mass fluctuation field. The third approach is to exploit the approximate self-similarity of the clustering expected in models like CDM.

As has been noted many times, the spectrum of mass fluctuations in CDM is quite well approximated by a power law with spectral index $n \simeq -1$ on the scale of clusters. With power-law initial conditions, the rms density fluctuations scale with mass as $\langle (\delta\rho/\rho)^2 \rangle^{1/2} \propto M^{-6/(n+3)}$. Now in linear theory, the density fluctuations evolve with time as a power law: $\delta\rho/\rho \propto a \propto t^{2/3}$, and so the initial conditions satisfy a scaling symmetry in that a change in the initial time is equivalent to a change in the mass scale. As things go non-linear the evolution is very complicated, but since the final configuration depends only of the ratio of the final to intial time (if we live in an Einstein de Sitter universe at least), the scaling symmetry is preserved in the non-linear regime, and the clustering pattern at any given time is an exact scaled replica of the present pattern. This property can be used to provide a useful test of calculations of the clustering of collisionless particles. What is less widely appreciated is that the same scaling applies to the gas, provided that one neglects cooling and non-gravitationally induced heating.

To convert this scaling into predictions for the mass function of clusters is fairly straightforward. Suppose, for concreteness, that we define a cluster of scale R to be a region where the density in a sphere of radius R exceeds some given multiple of the background density. We can define a cumulative mass fraction $f(> R)$ to be the fraction of mass in clusters larger than R, and we can define a *characteristic radius* R_* such that some conveniently chosen fraction of mass resides in clusters bigger than R_*. It is relatively easy to show that the characteristic radius evolves as $R_*(a) \propto a^{-(5+n)/(3+n)}$, and the evolution of the cumulative mass fraction is simply that $f(> R, a)$ is just a universal function of $R/R_*(a)$, or equivalently $f(> R, a') = f(> R R_*(a')/R_*(a))$. We can also define a characteristic mass $M_* \propto \bar{\rho} R_*^3$, and this has the familiar evolution law $M_* \propto a^{6/(n+3)}$, and one can readily calculate other characteristic quantities such as temperature or density. There is also a characteristic comoving number density of objects $N_* \propto 1/M_*$. From the cumulative mass fraction (as a function of R or M or whatever) it is straightforward to calculate say the differential mass function and we then have, for instance, $dn(M; a')/dM = (N_*(a')/N_*(a))^2 dn(M M_*(a')/M_*(a); a)/dM$.

An important distinction between self-similarity and the other calculational methods is that it does not predict the actual form of the mass function—and cannot therefore discriminate between theories with the same spectral index but with different statistics for the initial fluctuations—but it does allow one to predict the mass or temperature at any previous time from observations at a present epoch.

3. Temperature Function

The Edge *et al.* cumulative temperature function $N(> T)$ is roughly a power-law $N(> T) \propto T^\gamma$ with $\gamma \simeq -4$, but steepens above $T \simeq 10$ keV. Using the self-similar approximation we can easily convert this to the temperature function at any epoch z simply by scaling $N(> T) \propto (1 + z)^{6/(n+3)}$ and $T \propto (1 + z)^{(n-1)/(n+3)}$. On a log-log plot this is just a translation by

$$\Delta \log N = 6/(n + 3) \log(1 + z), \tag{1}$$

$$\Delta \log T = (n - 1)/(n + 3) \log(1 + z). \tag{2}$$

The change of number density with time at constant T is

$$\frac{\partial \log N}{\partial \log(1 + z)} = \frac{6 + \gamma - \gamma n}{n + 3}. \tag{3}$$

For $\gamma = -4$ there is no evolution for $n = -0.5$ and the evolution with redshift is positive or negative for n greater than or less than 0.5 respectively. This behaviour is shown graphically in figure 1. The solid line is the current epoch temperature function and the other lines the predicted temperature function at $z = 0.5$ for the spectral indices shown. If we interpret the Gunn *et al.* result as saying that the abundance at fixed T has not *increased* by more than about a factor 2 since $z = 0.5$ (though given the difficulties involved in finding high redshift clusters the data may well be compatible with a large decrease), my reading of the figure is that there is no problem with spectral indices $n > -1$, but that substantially 'redder' spectra than CDM, say $n < -1.5$ (which might seem favourable from the point of view of adding extra large-scale power), are ruled out. Note that spectra with $n \gtrsim 0$ give positive evolution, but the curves bunch together so it will be difficult to discriminate between such models using the temperature function evolution. On the other hand, if future surveys do show strong positive evolution all these models will be excluded.

4. X-ray Luminosity Function

As discussed above, if the gas started out bound to small clumps of dark matter and has evolved only under the influence of gravity (i.e. no heating save from gravitationally induced motions, no cooling), the gas and dark matter in the clusters should both evolve self-similarly, so we can use techniques very similar to those used in §2.1 to predict how the LF should appear at high z. I will assume that the luminosity is given by $L \propto M\rho T^{1/2}$ (i.e. as relevant for a broad band detector) in which case the characteristic luminosity varies with redshift as $L_* \propto (1 + z)^{(7n+5)/(2n+6)}$, and the translation in the N, L plane is

$$\Delta \log L = (7n + 5)/(2n + 6) \log(1 + z), \tag{4}$$

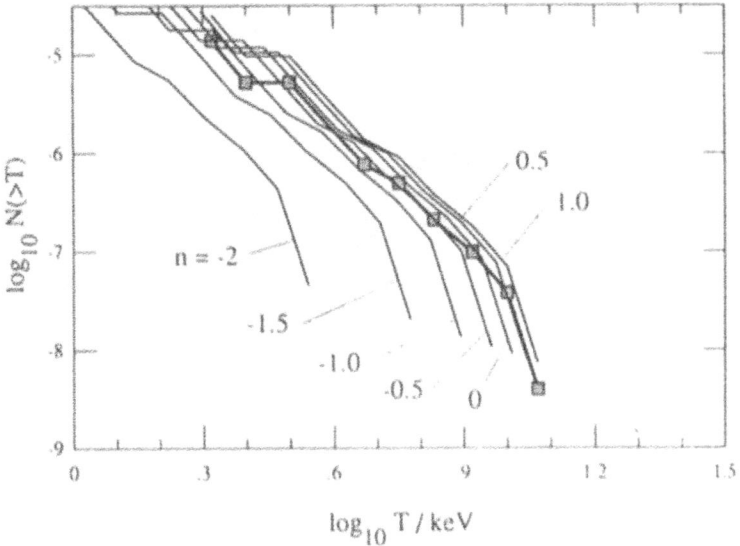

Fig. 1. The symbols are the cumulative temperature function from Edge *et al.* 1990. Narrow lines are the temperature function predicted at $z = 0.5$ for self-similar models with a variety of spectral indices. Curves for spectra with $n \gtrsim -1$ bunch together with very little apparent evolution. Redder spectra predict negative evolution which would be hard to reconcile with the abundance of high z clusters found by Gunn and collaborators, but models with $n \gtrsim -1$ appear to survive this test. One visible distinction between the bluer models is that the 'break' in the luminosity function (i.e. the last but one point) moves by different amounts, but the spread in T_* is rather small so this would be difficult to measure.

$$\Delta \log N = 6/(n+3)\log(1+z). \tag{5}$$

The evolution actually observed in in clear conflict with most of the models plotted in figure 2. For example, Edge *et al.* find that the abundance of high ($\gtrsim 10^{45}$ erg/s) clusters is down by about a factor 3 from the present epoch value even at redshifts $0.1-0.2$. The EMSS results suggest a somewhat less rapid rate of evolution, but for the brightest clusters the evolution is still very rapid, and has the opposite sign to what is predicted for most of the models shown in figure 2. The exception is for red spectra with $n \simeq -2$, but these models would not fit the lack of evolution seen optically.

5. Cluster Formation with a Preheated IGM

While the self-similar method is at best approximate, and there is also substantial uncertainty in the observations, there is clearly a big problem with the self-similar

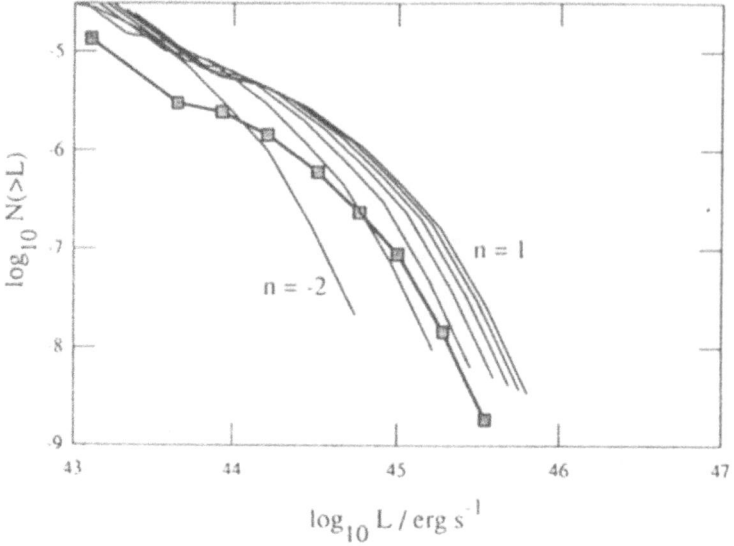

Fig. 2. The cumulative luminosity function predictions for models in which both the gas and the dark matter evolve self-similarly. The symbols are from Edge *et al.* 1990, and the narrow lines show the LF at $z = 0.2$. As with the temperature function in figure 1, the models with $n \gtrsim -1$ bunch together, but here with strong positive evolution. This is in the opposite sense to the strong evolution seen in the Edge *et al.* sample. Only models with very red spectra can reproduce the evolution seen, but, as shown in figure 1, these predict too much negative evolution of the temperature function.

X-ray evolution for an $n \simeq -1$ model.

One plausible modification to the model is to assume that the gas was heated, perhaps at the time of galaxy formation, by supernova or UV radiation. With such a modification, the gas will not be able to become concentrated in the shallow potential wells at early times due to the entropy barrier, but as time goes by the gas will cool adiabatically and the dark matter potential wells will deepen and the gas will become more and more concentrated and hence more luminous in X-rays and in this way the evolutionary trend for the X-ray LF is reversed.

A rather crude implementation of this idea (Kaiser, 1991) is to assume that the gas is initially on a single adiabat, and settles adiabatically into the potential wells. This idealised model predicts that at the present (or at any other epoch) the luminosity-temperature relation should be of the form $L \propto T^{7/2}$ and that the gas to dark mass ratio in the X-ray emitting region should be an increasing function of temperature. Both of these predictions seem to fit quite well with current observations.

Various qualitative evolutionary trends are predicted by this model: Clusters

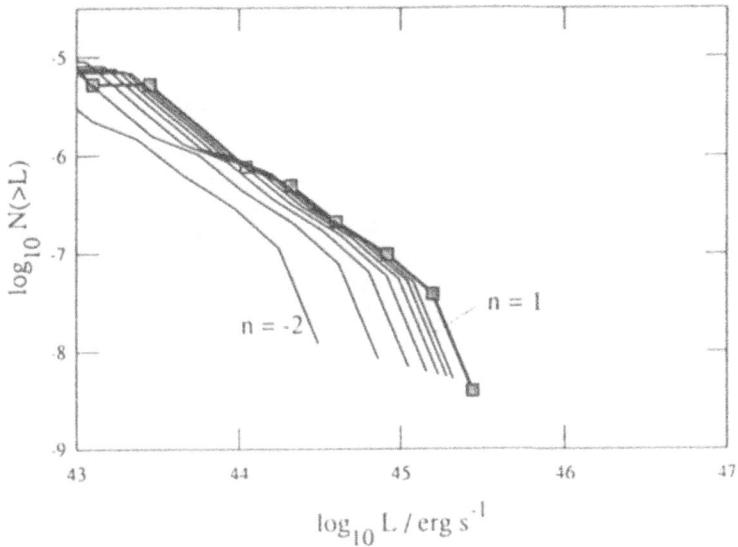

Fig. 3. Cumulative luminosity function for the preheated IGM model. The vital ingredients of this model are: i) self-similar evolution of the dark matter , ii) gas *mass set* by gravitational infall, iii) gas *density* fixed by an assumed constant entropy and hydrostatic equilibrium $\rightarrow$ $T \simeq \Phi$. The symbols are obtained from the observed current epoch temperature function, mapping temperature to luminosity according to $L \propto T^{7/2}$. The constant of proportionality in the $L - T$ relation was chosen to get the best match to the observed LF. The narrow lines are for $z = 0.2$ and are obtained similarly, but using self-similar prediction for the temperature function. For spectra with $n \gtrsim -1$ there is very little evolution at low luminosities, but the 'break' moves to lower luminosities and the evolution at the bright end can be dramatic.

should be slightly more compact in the past ($r \propto (1 + z)^{-1/2}$, but the X-ray emissivity per unit mass is smaller in the past, in stark contrast to the predictions of the self-similar model. The ratio of gas to dark matter in clusters is predicted to be lower in the past, as is the ratio of the expansion time to the cooling time. In this model, cooling flows would be less abundant in the past and should be seen primarily in the most massive clusters. Again, this is opposite to the predictions of the self-similar model.

To try to make more quantitative predictions for the evolution of the X-ray LF I have assumed that dark matter potential wells evolve self-similarly, and I have then assigned luminosities to clusters according to the predicted luminosity-temperature relation. The result is shown in figure 3. For $n \simeq -1$ there is strong negative evolution of the luminosity function. This is qualitatively compatible with the indications from existing surveys and a prediction of this model for deeper future surveys is that the evolution will be found to be less at lower luminosities.

A very different solution to the X-ray evolution problem to that proposed here is to retain the assumption of self-similarity for the gas, but adopt a redder spectrum with $n \simeq -2$ say. Such a model would not fit very well with my interpretation of the the optical observations, but this, of course, may be wrong.

From figure 2 we see that such a model would give negative evolution of the temperature function at high luminosities, and it is also a positive feature of this model that the temperature function obtained from the Press-Shechter formalism with this spectral index agrees much better with that observed. The Press-Schechter method with $n = -1$ produces too steep a temperature function, but I would argue that this may simply reflect either a failure of the approximation or perhaps non-gaussianity of the initial fluctuations.

If we discard the optical observations then the two models (pre-heated gas with $n = -1$ or self-similar with $n = -2$) would both seem to be quite compatible with the current X-ray observations. There is however, the hope of distinguishing between these models by X-ray observations alone. In the self-similar evolution of the dark matter clumps, the characteristic physical radius of the clusters varies as $r_* \propto (1 + z)^{-(5+n)/(3+n)}$ and the characteristic temperature varies as $T_* \propto (1 + z)^{(n-1)/(n+3)}$. This gives $r_* \propto (1 + z)^{-3}$ and $(1 + z)^{-2}$ and $T_* \propto (1 + z)^{-3}$ and $(1 + z)^{-1}$ for the $n = -2$ and $n = -1$ models respectively. Also of interest is the ratio of the cooling time to the Hubble time $t_c/t_H \propto \rho^{-1/2} T_*^{1/2} \propto r_*$. Relative to the $n = -1$ models then the clusters in the $n = -2$ model are cooler in the past and more able to cool. The X-ray radius should, in the self-similar model, track that characteristic radius for the dark matter clumps. In the preheated IGM model however, we saw that the radius for the X-ray emitting gas (the half-light radius for concreteness) is independent of mass and is a very slowly increasing function of redshift, so the two models are radically different in this regard. It is perhaps also worth noting that in deriving the cluster luminosities from the EMSS, Gioia et al. assumed a constant X-ray core radius, and it is not clear that the $n = -2$ model would remain compatible with the data if the luminosities were obtained with appropriate evolution of the core radii.

References

Gunn, J., 1990. in "Clusters of Galaxies" edited by Oergerle, Fitchett and Danly, Cambridge: Cambridge University Press.

Edge, A., Stewart, S., Fabian, A. and Arnaud, K.A., 1990. MN, **245**, 559

Evrard, A., 1990. Ap.J., **363**, 349.

Frenk, C, White, S, Efstathiou, G. and Davis, M., 1990. Ap.J. **351**, 10.

Press, W.H. & Schechter, P., 1974. Ap.J., **187**, 425.

Gioia, I.M., Henry, J.P., Maccacaro, T., Morris, S.L., Stocke, J.T. and Wolter, A., 1990. Ap. J. Letters, **356**, L35.

Bardeen, J.M., Bond, J.R., Kaiser, N. & Szalay, A., 1986. Ap. J., **304**, 15.

Kaiser, N., 1991. Ap. J., In press.

Tyson, A., 1991. in "After the First Three Minutes", edited by Holt, Bennett and Trimble, New York: American Institute of Physics.

Carlberg, R. and Dubinski, J., 1991. Ap.J., **369**, 13.

CLUSTER EVOLUTION

A. Cavaliere
Dip. di Fisica II Università
00173 Roma   Italy
S. Colafrancesco
Osservatorio Astronomico di Roma
00040 Monteporzio   Italy

and

N. Menci
Dip. di Fisica II Università
00173 Roma   Italy

ABSTRACT. Using a time-resolved approach, we compare the dynamical evolutions of groups and clusters of galaxies, as expected from direct hierarchical collapses and from aggregations and mergings. We find that aggregations dominate in bound environments, and in fact easily lead to a merging runaway that describes growth of cD galaxies in groups and erasure of substructures in forming clusters. In the 'open' field, instead, direct collapses set the dominant scales. Fast anti-evolution for the X-ray luminosities as indicated by the existing data may be understood in terms of the intracluster gas breaking the self-similarity associated with the dynamical evolution.

1. Introduction

Clusters and groups of galaxies are, by definition, *ill-defined* objects in space (fuzzy boundaries) and in time (metastable states), even more so than other structures dominated by non-relativistic gravity.

This is because their formation takes times $\sim$ several Gyr (actually, the average rich clusters may be settling now); the variances in times and sizes may be substantial; and not much longer than the formation stage is the average life time, after which these condensations are reshuffled into new ones.

The challenge, then, is to draw efficient cross sections through such a space-time complexity. Specific interest in cluster evolution has been recently rekindled by some apparent mismatch between optical and X-ray information.

In the optical band, the data at face value indicate that the distribution of clusters with large velocity dispersions declines only weakly, if at all, out to redshifts $z \sim 0.5$ or more (Gunn 1990, Ellis 1991). But the present surveys have to contend with small-number statistics, and with velocity measurements often limited to core galaxies and possibly contaminated by projected objects (see dicussions in previous references; and Frenk et al. 1990, Olivier et al. 1990).

A. C. Fabian (ed.), Clusters and Superclusters of Galaxies, 331–350.
© 1992 *Kluwer Academic Publishers.*

The X-ray surveys, in principle, single out physical potential wells of sizes $R \sim$ Mpc and masses $M \sim 10^{14} \div 10^{15} M_\odot$ out to $z \sim 0.5 \div 1$, taking advantage of the powerful, extended bremsstrahlung emission (Cavaliere, Gursky and Tucker 1971, Gursky et al. 1972, Solinger & Tucker 1972) common to most local clusters and many groups

$$L_x \propto \hat{n}^2 \hat{R}^3 T^{1/2} . \qquad (1.1)$$

Outputs up to $L_x \sim 10^{45}$ erg s^{-1} come from large masses $\hat{M} \sim 10^{-1} M$ of intra-cluster plasma (ICP) gathered in the wells with central densities $\hat{n} \sim 10^{-3}$ cm^{-3}, heated to virial temperatures $kT \sim G M m_p / 10 R \sim$ several keV. Moreover, the X-ray luminosities are less liable to contaminations from projection effects, being non-linear with $\hat{M} \propto \hat{n} \hat{R}^3$.

The pre-ROSAT evidence (Gioia et al. 1990, Edge et al. 1990; Henry et al. 1991) suggests a sharp decline of the *luminous* cluster sources already at $z \gtrsim 0.2$. But the X-ray evolution results from potential well dynamics combined with ICP physics.

While X-ray temperatures and galaxian velocities will constitute more direct probes of the dynamical state of these systems, at present the X-ray luminosities are the deepest and fastest probes available. They, with the related physics, constitute the thread of this review.

2. Direct Hierarchical Collapses

Within the *hierarchical* scenario, which envisages structures assembled from the bottom–up, the clustering may still proceed along *two* routes: direct collapses (DHC) from overdensities, initially with small contrast $\delta = \delta\rho/\rho \ll 1$; or aggregations and merging (AM) of units already of high contrast. N-body simulations show that both processes operate and may combine.

In the DHC mode (see Peebles 1980), the guidelines are: to relate every step of the hierarchy directly to the initial perturbations; to conserve at each epoch the mass fraction in condensations, which eventually get incorporated into a subsequent, larger condensation; to scale the internal densities in step with the external average $\rho_a(z)$. Relevant details are as follows.

In the simplest versions, the linear δs are taken to constitute a random-phase Gaussian field: power spectrum $\langle |\delta_k|^2 \rangle \propto k^n$ with $n > -3$, and ~ -1 at cluster scales; distribution on a given mass scale $p(\delta|M) \propto exp(-\delta^2/2\sigma^2)/\sigma$, with $\sigma \propto M^{-a}$ and $a = (n+3)/6$. The much discussed CDM model (see Davis & Peebles 1983; Efstathiou 1991) features instead a spectrum gently curving with decreasing mass, from $n \approx -1$ in the rich cluster range toward -2 in the galaxy range.

The overdensities are weakly gravitationally unstable even in a high-density FRW universe: $\delta(z) \propto (1+z)^{-1}$ as long as $\Omega \approx 1$ holds. On crossing the threshold of non-linearity $\delta_c \approx 1$, the perturbations detach from the Hubble flow, collapse, violently relax and settle to a virial equilibrium in a few dynamical times. To cover this interval and to end up with realistic contrasts $\delta \sim$ a few 10^2 within radii $\sim R_A = 1.5 \ h^{-1}$ Mpc, the linear regime may be extended up to $\delta_c \sim 1.3 \div 1.7$.

Spherical, homogeneous perturbations of r.m.s. amplitude in a critical universe detach as $\delta \sim \sigma \propto M^{-a} \ t^{2/3} \sim 1$, and virialize at z with a characteristic mass $M_c \propto t^{2/3a} \propto (1+z)^{-1/a}$; the present value is taken $M_c(0) \sim 5 \ 10^{14} \ h^{-1} \ M_\odot$, corresponding to a richness 1 Abell (1958) cluster. The internal densities scale like $\rho \sim 2 \, 10^2 \rho_a \propto (1+z)^3$; the sizes as $R_c \propto (M/\rho)^{1/3} \propto M_c^{(5+n)/6}$, and the specific virialized energies like $v_c^2 \propto M/R \propto M_c^{(1-n)/6}$.

3. Mass Distributions

Given the characteristic $M_c(z)$, the next step is to derive the full mass distribution functions (MFs). Consistent with the thrust of DHC clustering, these are expected (Kaiser 1986) to follow over a restricted dynamic range the self-similar comoving form

$$N(M, z) \propto M_c^{-2}(z) \, f(m), \qquad (3.1)$$

depending on the variable $m = M/M_c(z)$ with parameters set only by the initial spectrum. The evolutionary scale is given by $t_c = M_c/\dot{M}_c$.

MFs of such a form have been derived with a *quasi-static* approach, following the prescriptions set by Press & Schechter 1974: First, a general *golden rule*

$$N(M, z)M dM = -dF \qquad (3.2)$$

is taken to relate at each z the mass density of clumps in the range $M \div M + dM$ with the differential of the fractional mass F residing in the objects gone non-linear by that epoch. Second, an *ansatz* for F of the general form

$$F(M_s, z) = \int_{\nu_c}^{\infty} d\nu \; \Pi(\nu|M_s) \, M(M_s, \nu) \,, \qquad (3.3)$$

where $\nu = \delta/\sigma \propto (M_s/M_c)^a$, will specify: the site and hence the statistical weight Π of collapses beginning on the mass scale M_s, or on the size scale $R_s \propto (M_s/\rho)^{1/3}$; and the total mass involved M.

In fact, PS 1974 assumed collapse to take place at each given epoch in all independent overdense ($\delta > 1$) volumes. Then

$$\Pi \propto e^{-\nu^2/2}/R_s^3 \qquad (3.4)$$

is given by the same Gaussian $p(\delta|M_s)$. They went on to assume that all mass in the underdense fluctuations infalls onto the collapsed clumps under the equipartition condition $M \sim 2M_s$, thus forcing the normalization $F \to 1$. Their end result is

$$f(m) \propto m^{-\Gamma} e^{-\beta m^\Theta} \qquad (3.5)$$

with $\beta = 1/2$; the parameters $\Theta = 2a$, $\Gamma = 2 - a$ preserve at all times full memory of the initial field.

The PS 1974 approach carries out most curtly the thrust of the DHC view: only initial amplitudes and background density matter. It deftly skips all dynamical complexities of the collapes using something like a limit of zero dynamical and survival times (only a trace of the former remains if $\delta_c > 1$ is used). Problems of "clouds-in-cloud" remain, however, including difficulties with connectivity of the overdensity field and with mixing of time scales, except for one very special filtering as discussed by Bond et al. 1991. On the other hand, the result (3.5) is not inconsistent with the MF found from N-body simulations, although in a narrow dynamic range (Efstathiou et al. 1988; Carlberg & Couchman 1989); or with optical luminosity functions (LFs), as far as they can be inferred from still scanty and inhomogeneous data (see Bahcall 1979; Ramella, Geller and Huchra 1989).

Just beyond the PS 1974 approach, complexity increases steeply. The alternative considers collapses beginning only at high peaks of the overdensity field. Topology and geometry of the peaks against a noisy background govern the collapse statistics and the amount of infallen mass, as pioneered by Doroshkevich 1970, and analyzed in detail by Bardeen et al. 1986, and by Ryden 1988. Three main alterations result.

The weight $\Pi(\nu|R_s)$ in equation (3.3) now counts simply connected regions corresponding to well isolated peaks in the Gaussian field that have survived smoothing over the length R_s. This implies in eq. (3.4) an extra, ν- dependent factor $G(\gamma, \gamma\nu)$ as defined by Bardeen et al. 1986; the simple approximation $G \propto \nu^p$ ($p \approx 1.5$) holding in the limited but relevant ranges $\nu \sim 0.2 \div 1.5$ and $-2 < n < -1$, exhibits the expected dearth of isolated low peaks.

A bias may be consistently introduced, to account for the observed clustering patterns and to reconcile a globally critical universe with values $\Omega_0 \lesssim 0.2$ measured from condensations (Kaiser 1984). In its simplest form, biased formation starts from peaks higher than the r.m.s. value $\sigma(8h^{-1}Mpc)$ in the field; effective perturbations are renormalized to $\nu = \delta \, b/\sigma(8h^{-1}Mpc)$ with $b \sim 1.5$. Then $\beta = b^2/2$ holds, and the relevant epochs scale to $(1+z)/b$.

Finally, the infall is sensitive to the peak height ν, that governs the range of gravitational influence against the field fluctuations. The total infalling mass $M(M_s, \nu, t)$ asymptotes over long times to

$$M_{max} = M_s \, (M_s/M_c)^d \, , \tag{3.6}$$

with $1/2 \lesssim d \lesssim 1$. Actually, the upper limit could be approached only in conditions of coherent infall from a homogeneous, isotropic, high density background with noise smooth over large scales (Cavaliere, Colafrancesco and Scaramella 1991).

In any case, the infall of the mass $M(\nu, t)$ develops *in time*: building up high-ν halos takes longer than the central dynamical time $t_c \equiv M_c/\dot{M_c}$, and competes with new collapses. This is one motivation which leads to consider explicit time changes.

4. A Time-resolved Approach

The archetypal PS 1974 formalism is likely a limit of zero collapse and survival times, that recasts the whole MF at any given epoch.

CCS 1991 propose to go beyond the stage of golden rules and ansatzes for $N(M, t)$ by introducing explicit time resolution. They consider a resolution of order $t_c = M_c/\dot{M_c}$, the basic time scale for collapses, and cast the MFs of the class (3.1) (3.5) into the differential, *linear* form

$$\frac{\partial N}{\partial t} = \frac{N}{\tau_+} - \frac{N}{\tau_-} \, . \tag{4.1}$$

This contains the scale $\tau_- = t_c/(2 - \Gamma)$ insensitive to M, and the other scale $\tau_+ = 2t_c m^{-\Theta}/\Theta$ which is shorter at larger M. For example, in the PS 1974 case $\tau_- = t_c/a$ and $\tau_+ = t_c\, m^{-2a}/a$ hold; after the simple peak model, instead, τ_- is shorter by a factor $\sim 1/(p+1)$, and τ_+ may be shorter by a factor $1/b^2$ if a linear bias b is adopted.

Then shape and evolution of the MF may be viewed as the result of a *shifting unbalance* of two processes: creation of new clumps on the time scale τ_+ dominant at the large-M end; vs. destruction of previous generations on the scale τ_-, which dominates at the small-M end and may be identified with the survival time. At large masses the MF is comprised of recent structures; at small masses, at variance with the original P&S approach, the MF is represented at each z as a dynamical *superposition* of objects produced mainly with $M \gtrsim M_c(z_i)$ at previous epochs $z_i > z$, which survived reshuffling into larger clumps. The production during Δt of new objects with $M < M_c$, given by $\Delta\, t dN/dt \sim aN(M/M_c)^{2a}\Delta t/t_c$, is subdominant compared with the number of survivors, but still is comparable with the production of more massive objects than $\sim M_c$.

A shortcut to understand the shape is as follows. At small M/M_c, where random destruction prevails, $N(t)$ simply lowers at the rate given by $dN/dt \sim -2N/3t$. At large M/M_c, where creation prevails, eq. (4.1) writes $dN/M_c \sim \Theta N M^{\Theta}/2M_c^{1+\Theta}$; this yields an advancing cutoff $N \propto exp[-(M/M_c)^{\Theta}/2]$ which singles out the characteristic mass $M_c(t)$, yet allowing a range of larger masses with decreasing probabilities (Gaussian in the present case).

Note some formal properties of eq. (4.1) for comparison with those of the aggregation equation in Sect. 6 below. The r.h.s. implies a *decreasing* number of collapsed object with increasing t. A class of solutions exists, which at *all times* are of the self-similar form $N \propto M_c^{-2}(t)\, f(m)$. For these, the total mass in collapsed objects $\mathbf{M} = \int dM\, N(M)M$ is *conserved*, while the number $\mathbf{N} = \int dM\, N(M) \propto M_c^{-1}(t)$.

Guided changes of the time scales τ_-, τ_+ will modulate, and may suggest how to improve, $N(M)$. For solutions of the self-similar form (3.1) (3.5) the relationship of shape to time scale is given by

$$\frac{\partial lnN}{\partial lnM} = -2 - \frac{t_c}{t}\,\frac{\partial lnN}{\partial lnt}\,. \tag{4.2}$$

which reads $\Gamma = 2 - 3\Theta\, t/4\tau_-$ in the power-law section.

Two consequences are noteworthy. First, a *faster-steeper* theorem holds: with the evolutionary pace set by $M_c(t) \propto t^{4/3\Theta}$ from the initial spectrum, smaller values of Θ (corresponding to smaller n) yield faster evolutions and also steeper slopes. Second, enhanced destruction flattens the MF for $M < M_c$, while enhanced generation steepens it for $M > M_c$; e.g., for long survival time $\tau_- \to \infty$, the MF steepens toward $N(M) \to M^{-2}$ at the low end. Similarly, guided changes of τ_+ may account for further variance in the collapses, like collapse delays from subclustering (CC 90).

The controlled unbalance of creation and destruction expressed by eq. (4.2) is what preserves in time the self-similar shape. This control is not always effective with the other, quadratic kinetic equation that describes AM clustering; correspondingly, a merging runaway may set in. The direct comparison of DHC and AM modes in terms of their respective kinetic equations constitutes another motivation for undertaking a time-resolved approach.

5. Secondary Infall

The statistics of the halos produced by infall may be reconsidered in the time-resolved framework. The new objects produced at each epochs z_i with $M_s > M_c(z_i)$, while they survive reshuffling, accrete mass from their surroundings over long time scales.

The infall model of Gunn & Gott 1972 may be rephrased in terms of a limiting mass M_{max} to read, with the shorthand $\Delta M = M_{max} - M_s$:

$$M(t) = M_s \left[1 + \frac{\Delta M}{M_s} \frac{(t/t_i)^{2/3} - 1}{\Delta M/M_s + (t/t_i)^{2/3}} \right] . \qquad (5.1)$$

While the survivors accrete by progressive infall, they are displaced toward larger masses in the distribution. Thus, the MF is expected to steepen depending on a finite accretion rate onto the central regions that were at the leading edge with $\nu > 1$ some time back, when all numbers were larger.

The process may be represented in differential form by eq. (4.1) complemented on the l.h.s. with the term $\partial(\dot{M}N)/\partial M$ where $\dot{M}$ is computed after eq. (5.1). Such term describes the *shift* within dt toward larger masses caused by the infall, after production on the scale M_s and during survival in dt described by the r.h.s.

Numerical results for the local MF are represented in Fig. 1. Note how the steepening depends on the infall rate, with resulting shapes not far from PS's 1974. Note that the values of F including infall range from $\lesssim 0.2$ to ≈ 0.9 when d ranges from $1/2$ to 1.

From the above, local clusters are expected to be still accreting mass, at a rate $5\ 10^{13} M_\odot$ Gyr^{-1} per rich cluster. Of this amount, a fraction $\sim F_s/F$ is to be found in collapsed subclusters, which corresponds to the infall of ~ 1 group per cluster dynamical time. Regős & Geller 1989 gave direct evidence of infall flows. More is coming from ROSAT observations (see Briel et al. 1991).

Optical LFs may be derived, given the mass/luminosity ratio. There is empirical evidence that M/L_o increases toward larger scales: the trend may be represented by $M/L_o \propto M^\epsilon$ with $\epsilon \sim 1/3 \div 1/4$ (see, e.g., Hoffman et al. 1982) that connects values in solar units $\lesssim 100\ h$ for groups (with a large dispersion, cf. Ramella et al. 1989) with values $\sim 400\ h$ for clusters. In point of principle, biased formation of structure and biased light distribution (Cole and Kaiser 1989) require that M/L_o should increase toward the outskirts of large structures (Bardeen et al. 1986), and require that the trend should continue out to scales $\gtrsim 50\ h^{-1}$ Mpc to recover an overall value $\sim 1500\ \Omega_o h$ in the field. In a critical universe this is consistent with $\epsilon \approx 1/4$.

The trend $M/L_o \propto M^{1/4}$ has been used by CCS 1991 to derive optical LFs, given in their Fig. 4. The numerical computation weights each dM with the corresponding value of dL; this assumes stratification of the accreted material (see Hoffman 1988), until next reshuffling into a larger clump enforces mixing.

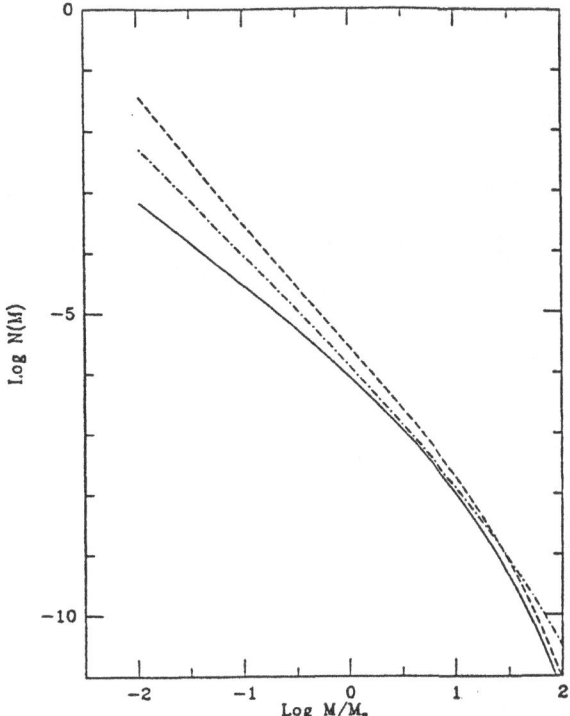

Fig. 1. The local MF from PS 1974 (dot-dashed line) is compared with those for DHC at the overdensity peaks with no (continuous line) and with maximal (d= 1, dashed line) infall. Perturbation spectrum with $n = -1.5$, $\delta_c = 1.3$, and $b = 1.5$ ($\Omega_o = 1$). Normalization to the abundance of rich clusters $\approx 10^{-6}$ Mpc^{-3} given by Scaramella et al. 1991.

6. Merging

A second route to hierarchical building of structure involves aggregation and merging processes (AM) of formed, high contrast condensations.

The analyses above assume substructure to be erased on time scales close to the minimum resolution t_d. But erasure develops over a variety of scales, from a few crossing times t_d of the structure as a whole, up to longer than a Hubble time for galaxies in clusters.

Describing such phenomena in full requires resolution on scales $< t_d$. In fact, both in N-body simulations and in the sky we discern *interactions* in the making among substructures, which involve transfers of energy to internal degrees of freedom. These play an important role during the formation and the subsequent evolution of groups and clusters, affecting not only permanence of substructure, but also inducing back reactions to delay the overall collapse, as discussed by Cavaliere & Colafrancesco 1990.

Results from N-body simulations of clusters, and discussions of subcluster interactions may be found in the papers by Cavaliere et al. 1986; Efstathiou et al. 1988; West, Oemler and Dekel 1988; Carlberg & Couchman 1988; CC 1990. Observational evidence of substructures present, or on their way to merge, stems primarily from X-ray observations, see Forman & Jones 1984, 1991; Briel et al. 1991). Additional evidence comes from optical counts associated with spectroscopic analyses, see Geller & Beers 1982; Binggeli et al. 1987; Dressler & Shectman 1988. Evidence from optical and X-ray morphologies has been pointed out by Edge 1991.

To attack the complexity of merging phenomena Cavaliere, Colafrancesco and Menci 1991 adopt the following framework: N units of individual mass M, size r, internal density ρ, compose a system with total mass M, overall size R and average mass density $\rho_a < \rho$. The system evolves mainly through random binary interactions.

The ensuing evolution of the MF may be analyzed using the classical aggregation equation of Smoluchowski 1916. Possible cosmogonic applications have been pointed out by Silk & White 1978, Lucchin 1988, Edge et al. 1991. But the rich mathematical content of this integro-differential equation has been discovered and discussed only recently, see Ernst 1986 for a review oriented toward the physical chemistry of suspensions, and CCM 1991 for phenomena relevant to cosmogony.

In continuous form the equation reads

$$\frac{\partial N}{\partial t} = \frac{1}{2} \int_0^M dM' \, K(M', M - M', t) \, N(M', t) \, N(M - M', t)$$

$$-N(M, t) \int_0^\infty dM' \, K(M, M', t) N(M', t) \, , \tag{6.1}$$

where $N(M, t)$ is the "comoving" MF, i.e., divided by the ambient mass density ρ_a and normalized to system mass M. The lower integration limits actually stand for the smallest mass in the system $M_l \ll$ M. In the intrinsically *finite* systems we shall consider, the upper infinite limit is to be replaced by $M_u <$ M.

The r.h.s. includes again, like in eq. (4.1), a construction and a destruction term. Still similarly to eq. (4.1), for small M the former vanishes while in the latter, always of the form $-N/\tau_-$, the time scale tends to become M-independent yielding a power law lowering in time. Again similarly to eq. (4.1), coefficients and symmetries with respect to M, M' are such as to yield for the first moments of the MF

$$\dot{N} < 0 \qquad \dot{M} = 0 \, , \tag{6.2}$$

i.e., mass conservation *when* the MF has an upper cutoff. The novel situation, however, arises when the fully *quadratic* construction term forces the cutoff to disappear, as discussed below; then the total mass M in the system described by the equation (6.1) is *not* conserved.

The interactions kernel $K = \rho_a \Sigma V$ includes the relative velocity and the gravitational cross section for encounters of two units with masses M and M', velocity-averaged. For encounters at relative velocities larger than the internal dispersions $V^2 >> v^2$, the efficiency $\lesssim v^2/V^2$ for energy transfers implies uninterestingly long time scales for merging.

When instead $v^2 \sim V^2$ holds (and the angular momentum is not too large, see Saslaw 1985), the cross section for merging of two units of masses $M \sim M'$ reads $\Sigma \approx \pi(r+r')^2[1+2G(M+M')/(r+r')V^2]$. The two addenda separately constitute partial cross sections scaling like M^λ with $\lambda = 2/3$ and $4/3$, respectively.

Here is the detail. The first component describes purely geometrical collisions (GC) where $\Sigma = \pi(r + r')^2$. The second describes the focused, resonant interactions (FI) with $\Sigma \approx 2\pi(r + r')^2 G(M + M')/(r + r')V^2$, which prevail in the range $V^2/v^2 \lesssim$ a few, and are most effective for merging. The radii scale as $r \propto (M/\rho)^{1/3}$.

So in terms of the characteristic mass $M_*(t) = \langle M^2 \rangle / \langle M \rangle$ and of the normalized mass $m \equiv M/M_*$, the following scalings obtain: $\Sigma V \propto M_*^{2/3} V \rho^{-2/3} (m^{1/3}+m'^{1/3})^2$ for the GC component; or $\Sigma V \propto M_*^{4/3} V^{-1} \rho^{-1/3} (m^{1/3} + m'^{1/3})(m + m')$ for the FI component.

7. Time Scales for Merging

Numerical computations given below confirm that all solutions of the integro-differential eq. (2.1) begin with a *transient* stage which still remembers the initial conditions, but rapidly go over into a *self-similar* shape where such memory is lost. Remarkably, when $\lambda > 1$ holds the memory loss may go to the extreme of a *run-away* phenomenon that remolds the whole ditribution over a few crossing times $t_d \sim 2\,R/V \sim 2\,R_{Mpc}/V_{1000}$ Gyr.

The average merging time is given by $\tau \sim 1/n\Sigma\,V \sim t_d/N(r/R)^2$. As long as the mass in normal system members M is conserved, their number density will scale like $n \propto \rho_a M_*^{-1}$. Then the merging rates effective in the kernel of eq. (6.1) read

$$\tau^{-1} \propto \frac{\rho_a\,V}{\rho^{2/3}}\,M_*^{-1/3} \quad \text{or} \quad \tau^{-1} \propto \frac{\rho_a}{\rho^{1/3}V}\,M_*^{1/3}, \tag{7.1}$$

for the GC component or for the FI component, respectively. In a heuristic look at eq. (6.1), a rate $\tau^{-1}(t)$ accelerating with time will point to a runaway process.

When GCs prevail, eq. (7.1a) implies that any attempt to accelerate the rate $\tau^{-1} \propto M_*^{-1/3}$ will be counteracted by the negative feedback provided by the very growth of $M_*(t)$. For FIs, instead, the only condition for $\tau^{-1} \propto M_*^{1/3}$ to run away is that the t-dependent coefficient in eq. (7.1b) does not decrease rapidly. If so, a *positive* feedback loop for $M_*(t)$ sets in, leading to a divergent rate, or equivalently to very short effective time scales, which suggests peculiar behaviors of the solutions of eq. (6.1).

The merging rates both scale like

$$\tau^{-1} \propto t^f\,M^{\lambda-1}\,, \tag{7.2}$$

which heuristically may be read as $dM_*/dt \propto t^f M_*^\lambda$. This suggests a diverging $M_*(t)$ under the conditions: $\lambda > 1$, physically related to the non linear, self-feeding increase of the cross section; *and* $f > -1$, related mainly to the sensitivity of all binary interaction to a decreasing ambient density. To quantify this, we parametrize $\rho_a(t) \propto t^d$, $\rho(t) \propto t^s$ and $V(t) \propto t^u$.

340

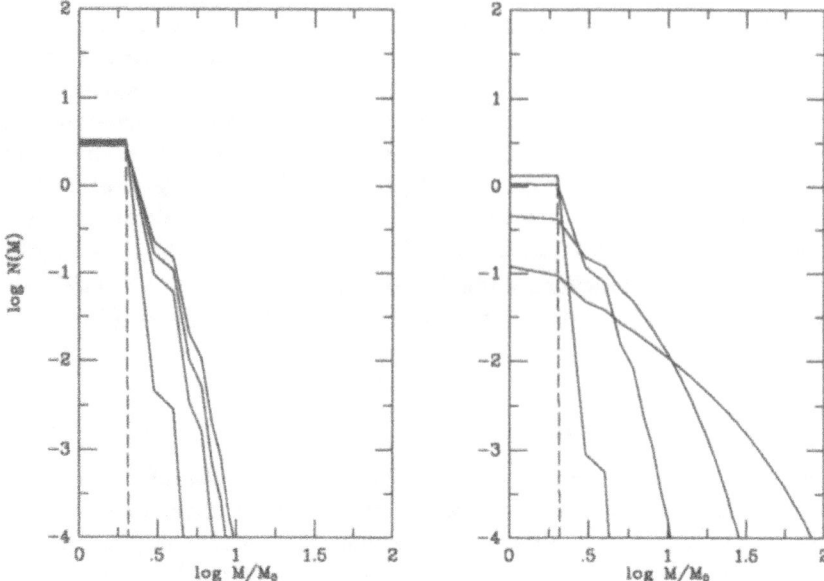

Fig. 2. Fig. 2a: The evolution of the MF under GCs ($\lambda = 2/3$) with $f = -1/3$ is computed numerically from the initial condition shown at the extreme left (dashed line). The unit of mass M_o corresponds to 10^{13} $M_\odot$ for subclusters in a cluster. The solid lines refer to the MFs evaluated at the times (from left to right) $t_1 = 0.3$, $t_2 = 0.6$, $t_3 = 1.6$, $t_4 = 3.1$ in units of t_d. Fig. 2b: As above for $f = 1/3$.

In the *field* of a FRW universe the density decreases with $d = -2 \div -3$ for $\Omega_o = 1 \div 0$. The linear velocity field grows with $u \sim 1/3$ (see Vittorio & Turner 1987). Within *bound or binding* structures $\rho_a(t)$ rises sharply during the virialization phase, and grows slowly thereafter, $d \sim 2/3$. Meanwhile, $V(t)$ undergoes only a modest increase, $u \sim 1/3$. The internal density $\rho(t)$ is poorly known and, if anything, it tends to decrease after a merging. Two limiting cases: $s \sim 0$, i.e., nearly constant internal density, relevant within structures; $s \sim d$, i.e., nearly constant contrast with the ambient, relevant for the field.

In summary, for GCs ($\lambda = 2/3$) in the field with 3-dimensional critical expansion f ranges from $\approx -5/3$ for constant internal densities, to $\approx -1/3$ for constant contrasts; inside structures instead, the corresponding range is $f \approx 5/9 \div 1$. With FIs ($\lambda = 4/3$) in the field $f \approx -7/3 \div -5/3$; inside structures $f \approx 0 \div 1/3$. It is seen that f is indeed dominated by the ambient density.

The full mathematics of eq. (6.1) is very amusing and rather subtle; we refer to CCM 1991 and Cavaliere, Colafrancesco and Menci 1992, but mention here some guidelines. Solutions, which are unique, are found with the form

$$N(M,t) \propto M_*(t)^{-\alpha}\phi(m) . \tag{7.3}$$

Then eq. (6.1) splits into a t-dependent equation for $M_*(t)$ of the form guessed above, and an m-dependent equation; the two are related by the value of the parameter α appearing in both. Asymptotic solutions are sought for large m, where construction dominates, and for small m, where destruction prevails giving rise to power-law shapes.

For $\lambda < 1$ and/or $f < -1$ the value $\alpha = 2$ holds, and the mild and gradual unbalances between construction and destruction preserve a self-similar shape

$$m \ll 1 \qquad A(t)\, m^{-\xi} \leftarrow \phi(m) \rightarrow m^{-\lambda} e^{-m} \qquad m \gg 1 \tag{7.4}$$

where $\xi \approx 1.3$ for $\lambda = 2/3$. The evolution is slow: $M_*(t)$ saturates when $f < -1$; when $f > -1$ but $\lambda < 1$ hold, it goes into a slanting asymptote $M_* \propto t^{(f+1)/(1-\lambda)}$ after a slow start.

Instead, when $f > -1$ *and* $\lambda > 1$ hold, striking phenomena appear, as expected from our heuristic look. The characteristic mass $M_* \propto (t_\infty^{f+1} - t^{f+1})^{2/(1-\lambda)}$ diverges at a *finite* time $t_\infty \sim 5 \div 2t_d$ for $f = 0 \div 1/3$. The MF splits into a *bimodal* shape

$$m \ll 1 \qquad B(t)m^{-(\lambda+3)/2} \leftarrow \phi(m) \quad \text{and a spike at } \mathbf{M}_o - \mathbf{M}(t) \tag{7.5}$$

with a steep power-law lowering in time corresponding to the residual normal galaxies, and a delta-like spike at large, increasing mass corresponding to a single forming *merger*, as illustrated by Fig. 3b. The two disjoint modes are related by an integrated mass flux from the normal galaxies to the merger. Technically, it is the requirement of a finite mass flux $\dot{\mathbf{M}}$ that fixes all the exponents, beginning with $\alpha = (\lambda+3)/2 > 2$; this is equivalent to require a consistently *time-resolved* description on a single time scale. The splitting of the MF is caused by construction of large masses overwhelming the destruction to the point of breaking away the high end, leaving behind a power law over most M-range where now $m = M/M_*(t) \ll 1$ holds. All that constitutes a rather drastic departure from self-similarity! In fact, such phenomena rather bear the hallmark of a critical point associated with a gravitational phase transition, as we discuss below.

Numerical solutions for relevant parameters are shown in Figs. 2a, 2b, 3a, and confirm the analysis in all details.

The description is stable relative to variations of the system mass. The evolutionary behavior is also robust relative to variations of the initial conditions; in fact, widely disperse initial conditions accelerate the onset of the critical phenomenon. This is also accelerated when a realistic cross section is considered, summing the GC and the FI terms. But even when GCs initially dominate, the stronger M-dependence of the FI component will cause it to prevail eventually, as numerical solutions confirm. Similarly, the t-dependence of the cross sections favors GCs, where f is increased by $\sim 1/3$ for given behaviors of $\rho(t)$, $\rho_a(t)$, and $V(t)$. But then the faster GCs drive the initial evolution of $M_*(t)$ extending $N(M)$ toward large values of M, where the stronger M-dependence of FIs takes over and drives the runaway.

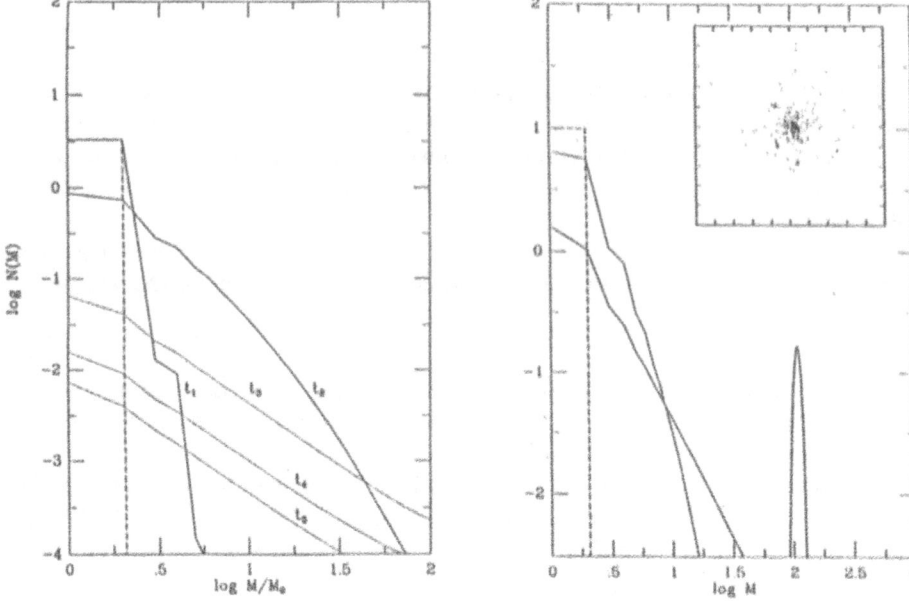

Fig. 3. Fig. 3a: The evolution of the MF under FIs ($\lambda = 4/3$) with $f = 1/3$. The times are $t_1 = 0.3$, $t_2 = 0.7$, $t_3 = 1.2$, $t_4 = 2.2$, $t_5 = 3.1$ in units of td. The initial condition is shown at the extreme left (dashed line). The solid lines refer to the MFs before the transition, with an exponential cutoff; the dotted lines refer to the MFs after the transition, with the typical power-law shape. Fig. 3b: To illustrate the transition in a group of galaxies, the full MFs in the premerger stage (thin lines) and in the postmerger stage (pure power law plus a spike, thick lines) are compared with the result from an N-body simulation (Carnevali et al. 1981).

Finally, numerical computations actually solve the discrete counterpart of eq. (6.1), with summations over a finite number of system members replacing the integrals. The agreement of numerical and analytical behaviors proves the equivalence of the two formulations.

8. From Runaway to Standstill

The general message is that in finite, self-gravitating systems the member condensations are sensitive to gravitational aggregation, depending primarily on v/V. Under such interactions shape and evolution of the member MF tend to *forget* initial conditions. For $v \sim V$ the evolution is likely to undertake a merging *runaway* capable of remolding the MF over a few dynamical times. In galaxy groups such processes correspond to aggregation of member galaxies into a large cD-like merger; in clusters, they correspond to aggregation of subclumps into a relaxed cluster configuration.

The relevant time scale t_∞ is provided by the dynamical time t_d set at system collapse, multiplied by a factor of a few to several (weakly depending on the mass ratio M/M_*). This is the case for FIs where $\Sigma \propto M^\lambda$ is more than linear, so that the feedback is positive and drives a runaway ending up in a critical phenomenon. The transition to FIs takes place at values $v^2/V^2 \gtrsim 1/3$, that in bound systems implies $N \lesssim 30$. Just below the threshold, GCs pave the way for FIs to soon take over. The threshold condition is easy to satisfy because the velocity ratio depend rather weakly on the mass ratios, $v^2/V^2 \sim (M_*/M)^{2/3}(\rho/\rho_a)^{1/3}$. Borrowing estimates from DHC clustering (as argued below), yields $v^2/V^2 \sim (M_*/M)^{(1-n)/6}$.

When $v^2/V^2 \ll 1$ the GCs prevail, and the negative feedback by the growing aggregates holds back the runaway causing the MD to settle to a mild, self-similar regime. This is liable to competition of other processes, and to an early termination by incorporation of the system into a still larger cluster. For example, in rich clusters the velocity dispersion is large enough to suppress the direct mergings of galaxies, making it difficult to build up from scratch a cD body inside rich clusters by this process (CCS 1981; Merrit 1983; Richstone & Malumuth 1983; Bothun & Schombert 1988). Slower forms of merging may instead occur, based on dynamical frition, see Richstone 1990.

A physical parameter that opposes direct galaxy merging is a large angular momentum. However, this is not statistically expected in many-body systems from the action of external tides or of statistical aggregations (see Saslaw 1985).

With all the caveats said, the *reality* of runaways gone to near completion is supported not only by the results from many N-body simulations (e.g., Carnevali, Cavaliere and Santangelo 1981, Ishizawa et al. 1983, Barnes 1989; see also Mamon 1990), but also by observations of groups dominated by a cD-like galaxy (e.g., MKW11, AWM4, AWM7) catalogued by Morgan, Kayser, and White 1975; and by Albert et. al. 1977. Indirect evidence may be provided by X-ray emission from groups (cf. Schwartz, Schwarz, and Tucker 1980; Biermann, Kromberg and Madore 1982; Kriss, Cioffi and Canizares 1983), since extensive merging induces shrinking of the overall configuration and increasing of the density of the ICP. Thus the bremsstrahlung emission is boosted, and sustained over several dynamical times.

In clusters, subclumps easily satisfy $v \sim V$, and merging occurs between sub-clusters and groups falling together at formation. The runaway leads to a rapid erasure of substructure during the collapse and virialization of the overall structure (see also White and Rees 1978). Such a processe on large scales may re-start merging of the cD galaxies already formed in subclusters and carried along into larger condensations (Ostriker & Tremaine 1975, Edge 1991).

Since rich clusters are young systems forming around the present epoch, the process is likely to be observed in action in a fair fraction of such structures, as in fact is shown by the optical and especially by the X-ray evidence mentioned in Sect. 6. In addition, N-body simulations again provide many examples of these patterns (e.g., Cavaliere et al. 1986, West et al. 1988, Carlberg & Couchman 1989).

Technically, the above pattern is closer to hold as it stands when many sub-clusters aggregate. Such conditions prevail when the spectral index of the initial perturbations $n > -1$ so that the initial units are more uniform in size, with short-range correlations. Because r/R is relatively larger in the subcluster case, the time scale $\tau \sim t_d/n_o \, \Sigma \, R$ is relatively shorter compared with galaxies in groups, $t_\infty = 2 \div 2.5 \, t_d$ for $f = 0 \div 1/3$ and with $n_o = 5 \, \mathrm{Mpc}^{-3}$.

A merging runaway in a multiple system gone to near completion may be viewed at as a *phase transition* of gravitational nature. In the perspective of the critical phenomena, the characteristic mass $M_* \propto \langle M^2 \rangle / \langle M \rangle$ plays on the mass axis a role similar to that played in ordinary space by the correlation length, whose divergence is indicative of organized behavior embracing the system. At the divergence, $N(M, t)$ becomes scale-free; then $\mathbf{M}(t)$, the decreasing mass in the system of normal galaxies, takes over in the role of the physically relevant mass. In fact, the order parameter appropriate to the present case is $1 - \mathbf{M}(t)/\mathbf{M}_o$.

While the qualitative outline of the phase transition is provided by the mean field eq. (6.1), it will constitute matter of further study – in analogy with the development of other theories of critical phenomena (cf. Domb & Lebowitz 1988) – to find how the critical indexes may be modified by proper account of realistic correlations. Here we note that a condition for eq. (6.1) to apply, namely no pairwise correlations of initial velocities, implies $(\mathbf{M}/M)^{2/3} \gg 1$. In systems constituted by a few units, the velocities may instead be correlated, and eq. (6.1) may not apply literally. But then the equation provides an upper bound to the aggregation time.

Aggregations like all 2-body interactions are very sensitive to a decreasing ambient density. We have seen that the expansion in the "field" of a FRW universe implies negative f which brakes all self-similar evolutions. Especially sensitive are the FIs with their smaller value of f, and these are easily brought to a *standstill*; but even GCs require the unlikely condition $f \gtrsim -1/3$ to really evolve, see fig. 2. In sum, in a 3-D "open" field of a FRW universe, AMs are unlikely to sustain a lasting growth of structures, and certainly will not drive a runaway.

In such conditions, DHCs easily prevail. The competition may be discussed in terms of a comparison of eqs. (4.1) and (7.1). With the runaway set aside, the evolutions of the corresponding fully self-similar regimes are driven by $M_c(t) \propto t^{4/(n+3)}$ and $M_*(t) \propto t^{(f+1)/(\lambda-1)}$. The former is faster for $n < -1$, indicating better efficiency of DHC mode at building up new structures from the field. Moreover, with DHC the slope of the MFs are steeper, indicating dominance also at masses smaller than M_c.

Actually, the two processes rather than compete play different and *complementary* roles in structure formation. Initially the DHC mode modulates the potential on the overall scale of the collapse; then the AM mode becomes effective within such structures detaching from the Hubble expansion. The share of DHC grows for $n \lesssim -1$ (the perturbation field is dominated by relatively few peaks with long range correlation and strong accretion of smaller clumps); the share of aggregations between nearly independent equals grows for $n \gtrsim 0$.

To the leading role of the canonical DHC mode for $n < -1$ there is one possible and important exception. The "open" homogeneous isotropic field may be a myth in the light of the results from deep redshift surveys (e.g., Sutherland 1988, Ramella, Geller, and Huchra 1989), large scale simulations (e.g., Efstathiou et al. 1988, Villumsen 1989, Carlberg & Couchman 1989), and quasi-linear analyses (see Shandarin & Zel'dovich 1989) which all concur in outlining cellular structures comprised of precursor ridges and filaments. Such low-dimensionality structures, looming out at each z in scaled form, may provide especially at their intersections "protected" sites with enhanced contrast and reduced expansion for possibly milder or incomplete, but more widespread merging activity both for galaxies and for groups and clusters.

9. X-ray Evolution

We now use the above dynamical considerations to discuss the X-ray evolution. The cluster X-ray luminosity by optically thin thermal bremsstrahlung emission is written in eq. (1.1) in terms of ICP parameters, and these may be related to dynamical quantities. In virial equilibrium the ICP temperature satisfies $T \propto M/R$. Density $\hat{n}$ and mass $\hat{M}$, or size $\hat{R} \propto (\hat{n}/\hat{M})^{1/3}$, are related to their dynamical counterparts, namely, average internal density ρ and virial mass M, or dynamical size $R \propto (M/\rho)^{1/3}$. Similarly, the LF should be related to the MF.

The point is that the simplest relations fail to comply with the present data. If one tries equal scalings for ICP as for dynamics, namely, $\hat{n}/\rho \propto g$ =const and $\hat{R} \propto R \propto (M/\rho)^{1/3}$, then the result of

$$L_x \propto g^2 \, M^{4/3} \rho^{7/6} \tag{9.1}$$

implies (Kaiser 1986) that the decrease at higher z (anti-evolution) of the characteristic luminosity $L_c \propto (1+z)^{(5+7n)/2(n+3)}$ is not much, since the decrease in $M_c(z)$ is nearly balanced by the increase of $\rho(z)$. On the other hand, the X-ray LFs are often derived from the MFs in eq. (3.1) (3.5) simply by the transformation $L_x \propto M^c$, with $c = 4/3$ from (9.1). The result is

$$N(L, z) \propto M_c^{-1} L_c^{-1} \ell^{-\gamma} e^{-\beta \ell^\theta} \, , \tag{9.2}$$

where $\ell \propto L/L_c(z)$, $\gamma = 1 + (\Gamma - 1)/c$ and $\theta = \Theta/c$, with two key features: flat slope $\gamma \lesssim 1.5$, and source number obviously increasing as strongly as that described by $N(M, z)$ (density evolution). Density evolution and weak luminosity anti-evolution would combine to yield an excess of the cluster counts above known uncertainties of the Einstein survey (Cavaliere & Colafrancesco 1988, Pesce et al. 1990).

The z-resolved LFs referred to in Sect. 1 and the harder, local luminosity function by Kowalski et al. 1984 instead indicate steep LFs over a wide range, and a connecting luminosity anti-evolution so *fast* in the range $z \sim 0 \div 0.3$ as to overwhelm at the bright end any density evolution. Note also that with eq. (9.2) a faster-flatter theorem would hold: any attempt to strengthen the dependence $L \propto M^c$ to speed up the evolution will flatten the power-law section of the LF, and to fit the data (over a limited dynamic range) one has to resort to the cutoff section.

So the data as they stand call for non-trivial solutions giving up at least one of the the simple scalings above. Add that the luminosities (9.1) depend strongly also on the density *at formation*, so they are liable to amplify any spread in formation epochs z_s. In DHC clustering, these depend on the initial δ because $\rho(z_s) \propto (1 + z_s)^3 \propto \delta^3$ hold, and we know the δs are distributed according to $p(\delta|M_s)$.

10. Luminosity Functions

This knowledge may be used in two ways.

From $L_x \propto M_s^{4/3} \delta^{7/2}$ and $p(\delta|M_s)$ Cavaliere, Burg and Giacconi 1991 derived the local distribution of luminosities at given M_s, namely, $p(L|M_s)$, and the corresponding distributions for derived masses like Abell's. These may be convolved a posteriori with the corresponding mass distribution to derive an unconditional LF.

Alternatively, we may consider the distributions in M and in δ on equal footing, and re-derive the luminosity function directly from the density field in a way parallel to PS 1974. Choosing L and δ as independendent variables, the number of collapsed objects per unit volume may be computed from

$$N(L, z) = \frac{\partial}{\partial L} \int_{\nu_c}^{\infty} d\nu \; \Pi(\nu|M_s) \tag{10.1}$$

where Π is given by eq. (3.4). The result includes the average spreads associated with L, and in explicit form reads

$$N(L, z) \propto M_c(z)^{-1} \; L_c(z)^{-1} [A \, \ell^u \Gamma(q, \frac{\delta_c^2}{2} \ell^s) + C \, \ell^{u+s} \, \Gamma(q+1, \frac{\delta_c^2}{2} \ell^s)] \tag{10.2}$$

where $q \approx 3.3$ and $A \sim C$. The two power-law components of the LF have steep logarithmic slopes $u \simeq -1.9$ and $u + s \simeq -1.7$. The incomplete gamma functions provide an exponential cutoff at the bright end. A somewhat flatter result obtains from Π corresponding to collapses at overdensity peaks.

Next we tackle the problem of the anti-evolution of $L_c(z)$, and discuss how the history of the ICP can give rise to dependences $g(M, z)$ in eq. (9.1) which break the invariant scaling $\hat{n} \propto \rho$.

11. The History of the ICP

The ICP history comprises two main issues (see Sarazin 1988): origin, and heating.

New points concerning the first issue include: The ratio gas/stars rises from $\hat{M}/M_* \lesssim 10^{-2}$ in X-ray galaxies (Fabbiano 1989) to ~ 1 in groups, up to values ~ 5 in rich clusters (David et al. 1990). The definitely subsolar Fe/H in rich clusters (see Edge 1989) requires abundant dilution with un-astrated baryons of the high metal yields predicted by leading stellar evolutionary models (e.g., Ciotti et al. 1991). M/L_o is believed to increase on gross average as $M^{1/4 \div 1/3}$, see Sect. 5.

If the distribution of baryons was initially uniform (and the efficiency in any pregalactic star generation was modest), all that directly implies an increasing relative content in diffuse baryons along the hierarchy, complementary to the decrease of L_o/M. The facts may be connected by the following scenario.

The initial star populations in early galaxies used up much of the initial baryons. They gave rise to strong winds from SN II and then from SN I, that swept clear of residual gas the shallow potential wells of galaxies and small groups, up to a mass scale M_1 such that $v^2(M_1) \gtrsim v_w^2$. The gas in proximity of these sites of star activity was preheated, and incidentally suppressed further star formation.

Biased structure formation implies, as discussed in Sects. 3 and 5, that collapses into progressively deeper and larger wells cause infall of matter from relatively underdense regions, out to progressively larger distances. Biased light distribution implies the outer material to be darker, i.e. containing a larger proportion of free baryons vs. those clumped into stars (including failing galaxies or delayed pregalactic perturbations), and also colder due to absence of stellar activity. These free baryons are heated up by release of gravitational energy in the large wells; they are prevented from forming stars, except in localized cooling flows, thus sustaining or amplifying light biasing from dynamic biasing. As for X-rays, the net result is to set a mass scale of order $M_1 \sim 10^{13} \, M_\odot$ where the dynamic scale invariance is maximally broken as for the gas behavior. The recovery is slow, as outlined below.

The role of the halos, invisible in X-rays, is that of temporary reservoirs of diffuse baryons, to be partly incorporated into the ICP at next reshufflings, when halos and central regions get mixed. The decrease of the relative stellar content in the central regions, or the complementary increase in available gaseous content, may be evaluated with simple, stepwise dilution models based on the relationship (3.6) that yield

$$g(M_s) \propto \frac{\hat{M}}{M_s} = \Omega_B \left[1 - \frac{M_*}{M_s} \right] \sim M_s^{0.3 \div 0.2} \qquad (11.1)$$

for dilution factors $3 \div 5$.

Another dependence $g(z)$ comes from considering how the density of the available gas, when it infalls into the forming potential wells, depends on preheating. Numerical simulations (Perrenod 1978, Evrard 1991) show that, when the infalling gas temperature (or entropy) is smaller than the virial temperature (or final entropy) associated with the potential well, a shock front forms, and moves outward leaving inside a nearly isothermal distribution at the virial value that joins to the inner boundary condition at the shock.

For an analytic evaluation, we focus on the density n_2 at the inner boundary as a function of the external density n_1 and the corresponding temperature ratio T_1/T_2. The Hugoniot adiabat provides the relationship

$$n_2/n_1 = 2 \left(1 - T_1/T_2 \right) + [4 \left(1 - T_1/T_2 \right)^2 + T_1/T_2]^{1/2} \qquad (11.2)$$

for $\gamma = 5/3$. As is well known, this reduces to the Poisson adiabat for weak shocks when $T_2 \gtrsim T_1$, and asymptotes to $n_2/n_1 \to 4$ in the limit of strong shoks when $T_2 \gg T_1$, which corresponds – as for the central density – to recovery from broken scale invariance.

The ratio T_1/T_2 is a direct function of z, hence $g(z) \propto n_2/n_1$ will be an inverse function of z, becoming ~ 1 for $M_c(z) \sim M_1$. Actually, approximate equality $T_1 \sim T_2$ is likely to be maintained down to $z \sim 1$ for the gas *within* the gravitational range for accretion by the forming structures, with the help of powerful SN I activity with slow decay $t^{-1.3}$ (see Ciotti et al. 1991). Thereafter, the external temperature drops nearly adiabatically while the virial temperature goes up following $T_2 \propto (1+z)^{(n-1)/2(n+3)}$.

Inserting the above z-dependences into eq. (11.2) we find for $g(z)$ the result plotted in Fig. 4a. We note that $g(z) \sim (1+z)^{-1.5}$ for $z \sim 0.3 \div 0.5$. The complete

348

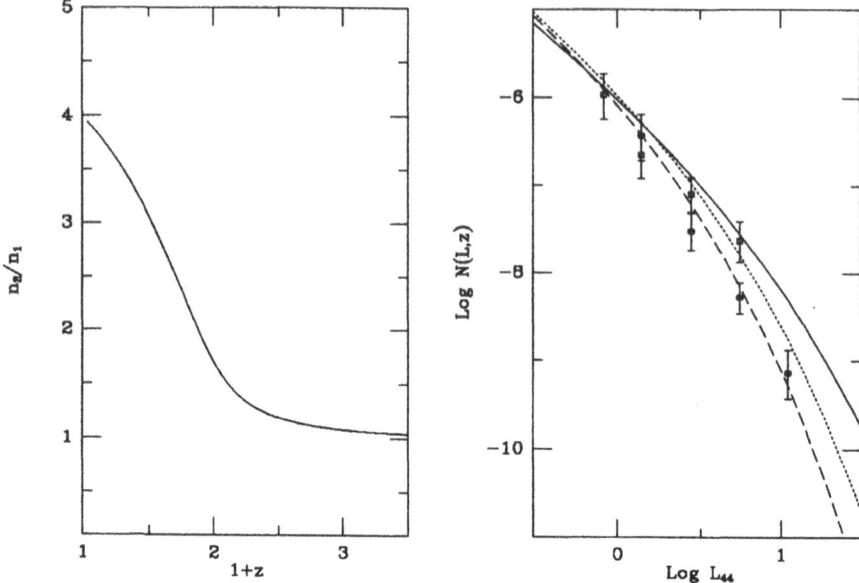

Fig. 4. Fig. 4a: The z-dependence of n_2/n_1, see text. Fig. 4b: The X-ray LFs for local and for distant clusters from eq. (10.2) with $g(M,z) \propto m^{0.2} M_c^{0.2}(z) n_2/n_1(z)$. The continuous line corresponds to $z=0$, the dotted one to $z=0.17$, and the dashed one to $z=0.35$. Data from Henry et al. 1990.

$g(M,z) = g(M)g(z) \propto m^{0.2} M_c^{0.2} n_2/n_1$ is used to compute the z-dependent LFs plotted in Fig. 4b.

12. Discussion

In the above scenario, that draws from biased DHCs, the X-luminous ICP constitutes on average a younger cluster component than the optically-luminous baryons. The dynamical invariance of the gravitational clustering is broken by nuclear energetics, with effects gradually erased by the increasing dilution with diffuse baryons. Implications include:

  i) At the bright end, strong luminosity anti-evolution to the effect that the LFs retreat rapidly with z. At given L, this will be observed as a fast decrease of the LF, namely $N(L,z) \propto L_c^{\gamma_e - 1}(z)$ on the time scale shortened by a factor $/(\gamma_e - 1)$ depending on the effective slope in the LF tail $\gamma_e = \gamma + b^2 \theta \ell^{\gamma + \theta}/2$. A bias $b \sim 1.5$ consistent with the scenario amplifies the decrease, and speeds up the evolution.

  ii) Toward the faint end the LFs remain steep, and cross due to the density evolution of the underlying MFs, with group number increasing like $M_c^{-1}(z)$.

iii) X-ray and optical evolutions differ, $L_x/L_o \propto M_c^{3/4}(z)$; cooling flows at high z are limited; the equivalent widths of high excitation lines increase at $z > 0$.

Consistent evidence is provided by changes of morphology and evolutionary rate of radio loud quasars and radio galaxies in clusters at $z \sim 0.5 \div 0.3$ (see Ellingson, Yee, and Green 1991; Hill & Lilly 1991), that may be interpreted as another effect of the rapid increase in pressure and density of the confining and stripping intracluster medium.

It is fair to say that pre-ROSAT data strain the yield of DHC scenarios. The alternative to obtain strong anti-evolution envisages instead to break the other scaling, with the assumption $\hat{R} \propto R \sim$ const. In DHC clustering (Perrenod 1980) this is not really consistent. It is more interesting in the context of clustering by AM (Edge et al. 1990), although it underplays the effects of the kinetic energy input in aggregations. The result is to slow down the dynamical evolution since the GC cross section becomes independent of M, while relatively speeding up the X-ray evolution since L_x is proportional to a high power of M. Other implications are: a fast decrease of $T(z)$, and a sharp drop of the surface brightness at all luminosities.

In this self-similar regime the balance between X-ray emission speeded up and dynamics slowed down depends sensitively on t-dependence of the ambient density; it looks unfavorable in the open field where, additionally, DHCs lead the dynamics. The initial transient regime (re-aggregation after a rather recent fragmentation episode?) may be tuned to be fast, but then it would show up also in the optical band with a sharp dearth at $z \sim 0.5$ for which there is little, or rather negative, evidence (Gunn 1990, Ellis 1991). Waiting for conclusive ROSAT data, we are investigating other, protected environments more conducive to effective aggregations.

References

Abell, G.O., 1958, ApJS, 3, 211.
Bahcall, N.A. 1979, ApJ, 232, 689.
Bahcall, N.A. 1988, Ann. Rev. Astron. Astrophys., 26, 631.
Bardeen, J.M., Bond, J.R., Kaiser, N., and Szalay, A.S. 1986, ApJ, 403, 15
Barnes, J.E. 1989, Nature, 338, 123
Biermann, P., Kromberg, P.P., and Madore, B.F. 1982, ApJ, 256, L37
Binggeli, B., Tammann, G.A., and Sandage, A., 1987, AJ, 94, 251
Bond, J.R., Cole, S., Efstathiou, G., and Kaiser, N. 1991, preprint
Briel, U.G., Henry, J.P., Schwarz, R.A., Böhringer, H., Ebeling, H., Edge, A.C., Hartner, G.D., Schindler, S., Trümper, J., Voges, W. 1991, A&A, 246, L10
Carlberg, R.G., and Couchman, H.M.P. 1989, ApJ, 340, 47
Carnevali, P., Cavaliere, A., and Santangelo, P. 1981, ApJ, 249, 449
Cavaliere, A., Santangelo, P., Tarquini, G., and Vittorio, N. 1986, ApJ, 305, 651
Cavaliere, A., and Colafrancesco, S. 1988, ApJ, 331, 660
——————— 1990, in 'Clusters of Galaxies', ed. W.R. Oegerle, M.J. Fitchett and L. Danly (Cambridge: Cambridge Univ. Press), 43 (CC 1990)
Cavaliere, A., Colafrancesco, S., and Scaramella, R. 1991, ApJ, 380, 15 (CCS 1991)
Cavaliere, A., Colafrancesco, S., and Menci, N. 1991, ApJ, 376, L37 (CCM 1991)
——————— 1992, ApJ, in press
Cavaliere, A., Burg, R., and Giacconi, R. 1991, ApJ, 366, L61
Cole, S., and Kaiser, N. 1989, MNRAS, 237, 1127.
Davis, M., and Peebles, P.J.E. 1983, ApJ, 267, 465.

350

Domb, C., and Lebowitz, J.L. eds. 1988, 'Phase Transition and Critical Phenomena' (London: Accademic Press)

Doroshkevich, A.G., 1970, Astrophysica, 6, 320

Dressler, A., and Gunn, J.I., 1988, in 'The Large Scale Structure of the Universe', IAU Symposium 130, J. Audouze ed. (Reidel: Dortrecht)

Dressler, A., and Shectman, S.A. 1988, AJ, 95, 985

Efstathiou, G. 1991, in Proc. of the NATO ASI 'Clusters and Superclusters of Galaxies', this volume

Efstathiou, G., Frenk, C.S., White, S.D.M., and Davis, M. 1988, MNRAS, 235, 715

Edge, A.C., Stewart, G.C., Fabian, A.C., and Arnaud, K.A. 1990, MNRAS, 245, 559

Edge, A.C., 1989, Cambridge University PhD Thesis

Edge A.C. 1991, MNRAS, 250, 103

Ellis, R.G. 1991, in Proc. of the NATO ASI 'Clusters and Superclusters of Galaxies', this volume

Ernst, M.H. 1986, in 'Fractals in Physics', L. Pietronero and E. Tosatti (New York: Elsevier), 289

Forman, W., and Jones, C. 1991, in Proc. of the NATO ASI 'Clusters and Superclusters of Galaxies', this volume

Gunn, J.E., and Gott, J.R. 1972, ApJ, 176, 1.

Henry, J.P., Gioia, I.M., Maccacaro, T., Morris, S.L., Stocke, J.T., and Wolter, A. 1991, preprint

Hoffman, Y. 1988, ApJ, 328, 489.

Hoffman, Y., Shaham, J., and Shaviv, G., 1982, ApJ, 262, 413.

Ishizawa, T., Matsumoto, R., Tajima, T., Kageyama, H., and Sakai, H. 1983, PASJ, 35, 61

Kaiser, N. 1984, ApJ, 284, L49.

Kaiser, N. 1986, MNRAS, 222, 323.

Kriss, G.A., Cioffi, D.F., and Canizares, C. R. 1983, ApJ, 272, 439

Lucchin, F. 1988, in Lecture Notes in Physics 332 ed. P. Flin (Berlin: Spriger), 284

Mamon, G.A. 1990, in 'Paired and Interacting Galaxies' IAU Symp. 124, J.W. Sulentic, W.C. Keel and C.M. Telesco eds., NASA CP 3098, pg. 609.

Ostriker, J.P., and Tremaine, S.D. 1975, ApJ, 202, L113

Peacock, J.A., and Heavens, A.F. 1990, MNRAS, 243, 133.

Peebles, P.J. 1980, 'The Large Scale Structure of the Universe' (Princeton: Princeton Univ. Press)

Perrenod, S.C. 1978, ApJ, 226, 566

Perrenod, S.C. 1980, ApJ, 236, 373

Pesce, J.E., Fabian, A.C., Edge, A.C., and Johnstone, R.M. 1990, MNRAS, 244, 58

Press, W.H., and Schechter, P. 1974, ApJ, 187, 425 (PS 1974)

Ramella, M., Geller, M.J., and Huchra, J.P. 1989, ApJ, 344, 57

Regös, E., and Geller, M. 1989, AJ, 98, 755.

Richstone, D. 1990, in 'Clusters of Galaxies', ed. W.R. Oegerle, M.J. Fitchett and L. Danly, STScI Symposium Series (Cambridge: Cambridge Univ. Press), 231

Richstone, D.O., and Malumuth, E.M. 1983, ApJ, 268, 30

Ryden, B.S., 1988, ApJ, 333, 78.

Sarazin, C.L. 1988, 'X-ray Emission from Clusters of Galaxies' (Cambridge: Cambridge Univ. Press)

Saslaw, W.C., 1985, 'Gravitational Physics of Stellar an Galactic Systems' (Cambridge: Cambridge Univ. Press)

Scaramella, R., Vettolani, P., Zamorani, G., and Chincarini, G. 1991, preprint

Shandarin, S.F., and Zel'dovich, Ya.B. 1989, Rev. Mod. Phys., 61, 185

Silk, J. and, White, S.D.M. 1978 , ApJ, 223, L59

Solinger, A.B., and Tucker, W.H. 1972, ApJ, 175, L107

Smoluchovski von, M. 1916, Phys. Z., 17, 557

Sutherland, W. 1988, MNRAS, 243, 159

Turner, E.L., and Gott, J.R. 1976, ApJS, 32, 409.

Villumsen, J. 1989, ApJS, 71, 407

Vittorio, N., and Turner, M.S. 1987, ApJ, 316, 475

West, M.J., Oemler, A., and Dekel, A. 1988, ApJ, 327, 1

White, S.D.M, and Rees, M.J. 1978, MNRAS, 183, 341

STRIP-MINING THE SOUTHERN SKY: Scratching the Surface

Stephen A. Shectman
The Observatories of the Carnegie Institution of Washington
Paul L. Schechter
Massachusetts Institute of Technology
Augustus A. Oemler, Jr. and Douglas Tucker
Yale University

and

Robert P. Kirshner and Huan Lin
Harvard-Smithsonian Center for Astrophysics

ABSTRACT. Recent advances in techniques for measuring galaxy redshifts have made it possible to gather large samples that reveal the structure in the universe on scales of 100 Mpc or more. Measurement of these properties is especially interesting because the mechanism for forming large structures may connect the fluctuations in the early universe with present observations. In any case, we would like to know the average properties of the Universe, such as the mass density Ω, but no determination of average properties can be reliable unless it encompasses a sufficient volume to average over fluctuations due to large voids and filaments. Conventional Wisdom (Geller and Huchra 1989) indicates that maps of galaxy distributions show inhomogeneities on scales as large as the surveys. While we expect that deeper surveys will reveal a more homogeneous pattern, the scale on which this transition takes place has not yet been reached in any published survey. This contribution reports initial results based on the first 5100 redshifts from a new survey carried out at Las Campanas Observatory in Chile. With a characteristic depth of one billion light years, the data suggest we may at last have surveyed the universe on a scale where the galaxy distribution tends toward homogeneity. The observed galaxies appear to have voids on the scale of 5000 km/s and filaments similar to those seen in the CfA work, but do not reveal obvious larger structures. Perhaps we are beginning to see the end of large scale structure.

1. Ancient History

Our collaboration has been engaged in the measurement of large scale structure for many years, inadvertently at first, and deliberately since 1981. Our first efforts were aimed at determining the luminosity function for galaxies (Schechter 1976) by obtaining redshifts for a sample of galaxies selected by apparent magnitude, and then using the observed luminosity density together with M/L derived from cluster dynamics to estimate Ω. These surveys (Kirshner, Oemler, and Schechter 1978, 1979) were hampered by large fluctuations in the properties that they set out to measure. While we developed techniques for deriving the luminosity function independent of density fluctuations, the discovery of the Bootes Void in a deeper

A. C. Fabian (ed.), Clusters and Superclusters of Galaxies, 351–363.

extension of our early work (Kirshner, Oemler, Schechter, and Shectman 1981, 1987) overshadowed the more pedestrian results on the luminosity function, luminosity density, and Ω reported in 1983 (Kirshner, Oemler, Schechter, and Shectman 1983).

The Bootes Void appeared to be a roughly spherical volume located at a redshift of 15500 km/s with a diameter of 124 Mpc (choosing h=1/2 to make the volume seem larger). No galaxies were seen in our survey within that volume, and we estimated that the probability was less than 1% that the density inside the void exceeded 1/4 of the average galaxy density. Subsequently, this interesting volume has been the subject of investigations which used other selection methods to construct galaxy samples. Emission-line galaxies selected from objective prism surveys have been studied in the Bootes Void by Tifft et al. (1986), Moody et al. (1987), Moody and Kirshner (1988), and by Weistrop and Downes (1988) and Weistrop (1989). These studies show that the Bootes Void volume is not completely empty: what is less certain is whether the proportion of emission-line galaxies is higher in this low-density region than elsewhere. The problem is that there are few appropriate samples of comparably selected galaxies for which the average space density has been established. A more straightforward sample of galaxies has been culled from the IRAS catalog and studied by Strauss and Huchra (1988) and for a fainter IRAS sample by Dey, Strauss, and Huchra (1990). They show that the density of IRAS galaxies in the center of the Bootes Void is lower than the density at the edge of the survey by a factor of about 4, which is consistent with IRAS galaxies having about the same contrast in the void as galaxies selected by apparent magnitude in the initial survey.

One of the limitations of these initial IRAS surveys is that the typical galaxy is at a redshift of 6000 km/s, much closer than the center of the Bootes Void at a redshift of 15500 km/s. To make a more effective survey of the void volume, Aldering, Bothun, Kirshner, and Marzke (1989) used the faint IRAS source catalog to construct a sample of 748 galaxies in the direction of the Bootes Void which have 60 micron flux greater than 150 mJy. At present, 396 have measured redshifts, and the rest should be completed in 1992. Eighteen galaxies are in the void volume, which corresponds to an underdensity of about a factor of 4, but a reliable result will have to await the completion of the redshift sample, since the galaxies for which redshifts are in hand do not constitute a flux limited sample.

Flushed with the success of the Bootes result, we carried out a similar survey in the diametrically opposite part of the sky in the constellation Phoenix (Kirshner, Oemler, Schechter, and Shectman 1990). As in the case of the Bootes survey, the galaxies were selected from small fields (15' x 15 ') sparsely covering a region 25° in diameter, covering just 1% of the area. The redshift distribution for the 150 galaxies we measured showed a striking resemblance to the Bootes region: dense regions at 10 000 km/s and at 20 000 km/s, but a dearth of galaxies in the range from 12 000 to 18 000 km/s. Why is the Phoenix Void less well known than its Northern partner? Because the 1% of the area surveyed did not include a large cluster of galaxies observed by Peterson et al. (1986) which is centrally located at 14 400 km/s in the direction of the Phoenix survey. While it may yet be true that this volume of space is of lower than average density, a more thorough survey is required to map the structure.

While the Bootes Void survey and its successors have helped to establish the presence of large scale fluctuations in the galaxy distribution, other, more systematic mapping efforts have been better at establishing the topology of the galaxy distribution. Notable among these are the strips surveyed at the Center for Astrophysics by John Huchra, Margaret Geller, and their colleagues ably summarized by Geller and Huchra (1989). These regions, 6° in width and 9 hours long on the sky have been completely surveyed to a limit of B = 15.5 based on the Zwicky catalog of galaxies. Four such strips have been reported, providing convincing evidence that the galaxy distribution on the scale of the survey, which has a characteristic depth of about 6000 km/s, has inhomogeneities that are as big as the survey could possibly have revealed–the "Great Wall" has a size of roughly 60/h x 170/h Mpc.

2. No More Fooling Around

A plausible next step in mapping large scale structure would be to push the scale of the survey a factor of five deeper than the CfA survey, but to retain some of its desirable features: a large filling factor and extensive coverage of the sky. Any useful survey will involve a large number of galaxies that allows the structure to be delineated. We have undertaken a redshift survey for 20 000 galaxies that covers 600 square degrees with a characteristic redshift near 30 000 km/s. This project has required some innovations in observational technique. First, since no catalogue was available to us when we began this work, we have conducted our own modest digital sky survey to construct a galaxy catalog based on CCD data down to a Gunn r magnitude of 18 (roughly corresponding to B = 20). The method used is to mount a CCD at the focus of the Swope 1 meter telescope at Las Campanas, set the telescope to the desired starting position, and turn off the telescope drives. We then clock the CCD at the sidereal rate and read each line of the data, peeling a continuous strip from the Southern sky. In practice, the effective integration time (which varies as the cosine of the declination) is about 1 minute. The length of the strips was set by the by the desire to match the photometric catalog to the size of the spectroscopic fields (1.5° x 1.5 °) used to obtain the galaxy redshifts while avoiding strips that were too short to observe efficiently. Our individual strips are 10000 pixels long at about 1"/pixel (depending on the CCD employed) resulting in a length of 3°. Seven contiguous strips are overlapped to form a "brick" that measures 1.5° by 3°.

Figures 1 and 2 show the fields which have been observed so far. Bricks with a date are those for which the photometric data are in hand. Each brick corresponds to about 110 Megabytes of data and 15 to 20 bricks must be observed and reduced in order to prepare for each spectroscopic observing run. The advent of the Exabyte tape has diminished the need for upper body strength as a requirement for this type of observational work. Careful alignment of the strips, astrometric solution based on stars from the Space Telescope Guide Star catalog, and separation of stars from galaxies has been carried out at Yale. Photometric calibration is obtained for each night by observing standard stars, and a catalog of galaxy positions and magnitudes which extends well below the limit chosen for the redshift survey emerges as the result.

354

SOUTHERN HEMISPHERE PLAN

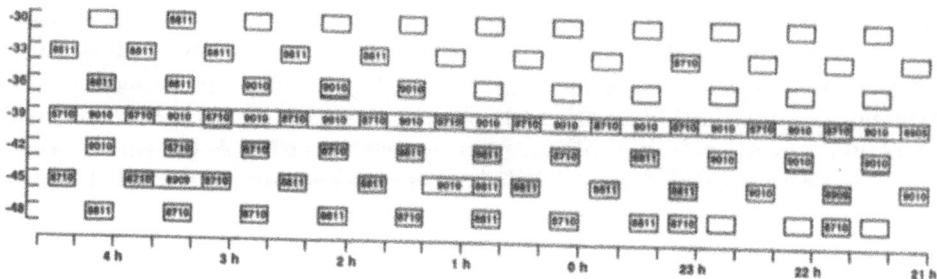

Fig. 1. Las Campanas redshift survey fields in the south galactic hemisphere as of July 1991. Numbers give dates of photometric observations. Shaded fields have been observed spectroscopically. The -39° strip will be completed next.

NORTHERN HEMISPHERE PLAN

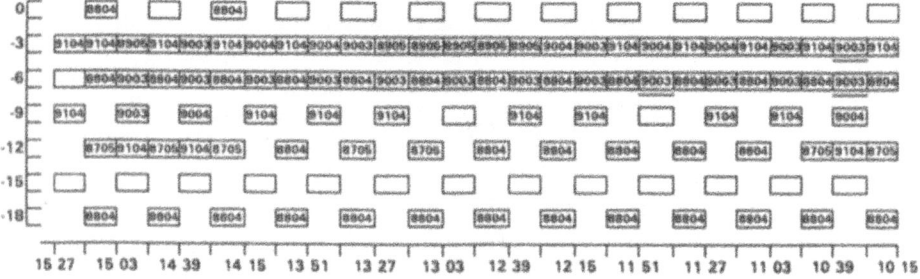

Fig. 2. Survey fields in the north galactic cap. Notation as in Figure 1. The -3° strip will be completed next.

The CfA redshifts were measured by craftsmen, one at a time, over a period of years. To carry out this ambitious survey before our decay into senescence, we had to implement a more rapid method. The 2.5 meter DuPont telescope at Las Campanas was designed in the heroic age of astronomical photography and has a very large focal plane of 2.1° diameter which was initially used for 20" x 20" (50 x 50 cm) plates. We have adapted that focal plane for fiber optic spectroscopy by building an assembly for holding large aluminum plates in a pneumatic bending fixture. Individual fibers are inserted by hand into carefully positioned holes in the plate. The fibers, encased in hypodermic tubing clad in a bicycle brake cable housing, guide the light for simultaneous spectroscopy using a two-dimensional photon-counting detector (the 2D-Frutti). The resulting "Fruit and Fiber" system has most recently been equipped with 112 fibers, each of which is 3.5" diameter at the focal plane, but all the data reported here were obtained with earlier versions of this system employed over the last 3 years, mostly with 50 object fibers per exposure.

The use of a fiber system allows a tremendous gain in observing efficiency, but imposes some constraints on the survey which are worth bearing in mind. Our basic selection criterion is an isophotal Gunn r magnitude selected to make efficient use of the fiber system by providing as many galaxies as there are fibers. For the data shown here, the isophotal limit is $r < 17.3$. In addition, since maximizing the throughput of the system by guiding on the count rate is very useful, we do not observe any galaxy brighter than $r = 16$ to avoid having a single galaxy dominate the telescope pointing. Because each fiber is small, only the central 3" of a galaxy is measured. To avoid wasting fibers on low surface-brightness systems, a second "central" magnitude corresponding to the flux in the central 3" is also used as a selection criterion. Approximately 20% of the galaxies which meet the isophotal limit are not observed because of low surface brightness. We have examined the properties of the galaxies as a function of surface brightness to assess the effects of this selection rule, and find that it is unlikely to produce a large bias in the results.

A more overt effect of the fiber system is that there are often more galaxies in an individual field than fibers. One choice would be to observe all of the galaxies by making multiple exposures in the same field. We have chosen to observe exactly the number of galaxies that the fiber system can accommodate by selecting at random from the galaxies which meet our photometric criteria. On average, 60% of galaxies that meet the photometric criteria are actually observed. Our goal of investigating structure on the largest scales is better served by observing new fields than by re-observing old ones. A careful analysis of the effect of this procedure will be carried out using simulations and sampling from complete data sets, but the obvious effect is that our redshift survey will undersample the densest regions. As an interim measure, we retain the fraction of galaxies observed in a field as part of the record for each galaxy and use this as a weighting factor in some of the further work.

Because 50 holes, each of 2.3 mm diameter, use very little of the material on a large aluminum plate, we actually drill 8 sets of 50 on the same plate. One setup can be carried out in the afternoon, but all the subsequent plugging of fibers into holes takes place while the Earth turns inexorably to the East. In practice, the exposures have an integration time of 2 hours. A comparison spectrum is obtained in the

356

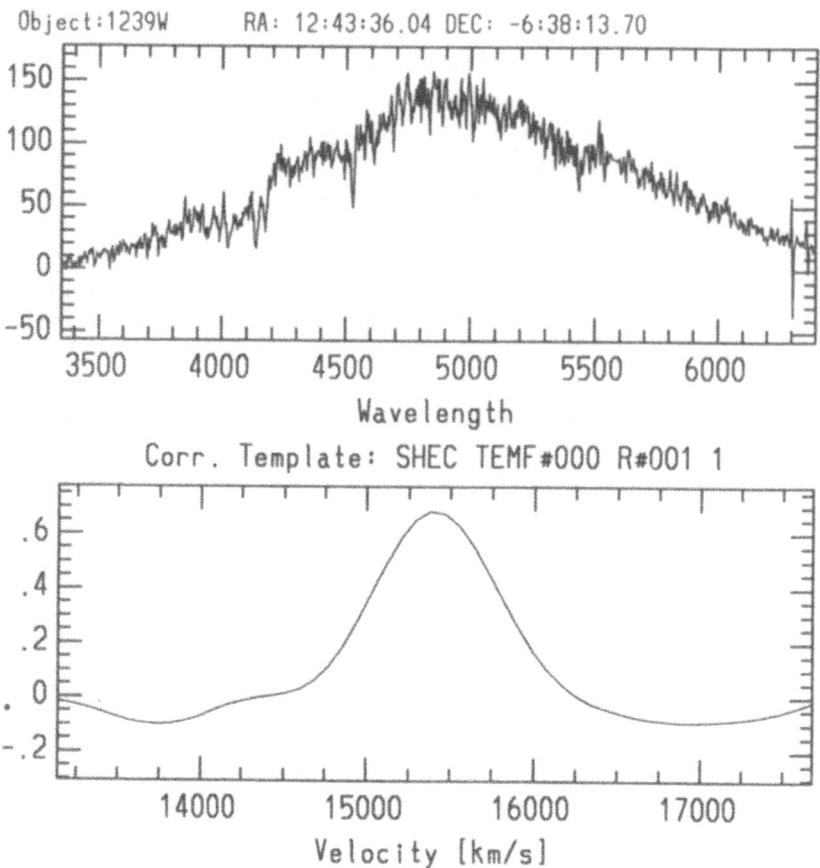

Fig. 3. Spectrum and cross-correlation peak for a galaxy with good signal-to-noise.

middle of the integration. While the spectrograph remains on the observing floor, the fiber assembly is rotated to compensate for polar misalignment and atmospheric refraction. The time taken to unplug the fibers, plug them into a new set of holes, slew to a new field and acquire the guide stars (through another set of fibers feeding the guide TV) is about 20 minutes. The plates are never changed during the night, so the entire night's observing must be carefully plotted months in advance when the plates are drilled. The advantages of an automated fiber positioning system have induced many workers to invest effort, ingenuity, and large sums of money in building fast, reliable, and versatile systems suitable for a wide range of challenging observational problems. Our approach has been more pragmatic, with the aim of developing a system which is adequate to our needs for this project and which capitalizes on 10 million years of evolution for the fiber positioning device and its control system.

The spectra cover the wavelength range from 3300 to 6000Å with an effective

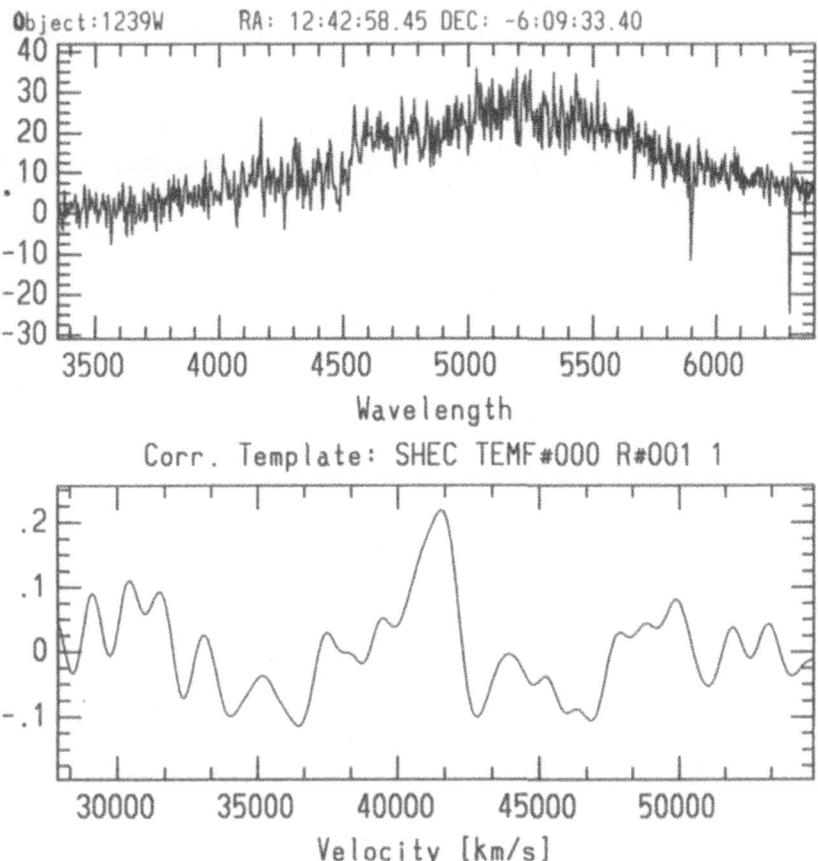

Fig. 4. Even galaxies with modest signal-to-noise can yield redshifts with an accuracy better than 75 km/s. Sky subtraction accuracy is indicated by 5577.

resolution of about 5Å. Sky subtraction is achieved by averaging the spectrum obtained from 10 sky fibers distributed over the focal plane in unused holes. Individual galaxy spectra are corrected for different fiber transmissions by using the strength of the 5577Å night sky line. The residual sky intensity after subtraction is rarely greater than 10%. It is worth noting that sky subtraction for our sample need not be very precise, since the r = 17 galaxies are comparable to the night sky brightness through the 3" fibers. The problem would become much more challenging for galaxies just a magnitude fainter. Since the integration time for all galaxies is the same, some have very nice spectra with about 100 net counts per pixel, but even at the ragged edge of the sample, we find that spectra with 20 net counts per pixel are satisfactory for velocity measurements, as shown in Figures 3 and 4.

Velocities are carefully measured with a cross-correlation program that usually gives measuring errors smaller than 75 km/s. The systematic errors in the velocity scale zero point are considerably smaller than this. We monitor the observed "velocity" of the night sky lines to check for wavelength errors. A sample of galactic stars has been inadvertently obtained when they are mistaken for high surface brightness galaxies. The mean velocity for these is satisfyingly close to zero. About 20% of the galaxies in the sample exhibit emission lines, usually [O II] 3727, but sometimes Hβ and [O III] 5007, 4959 as well. The velocities of the emission lines and the absorption lines agree in the galaxies where both can be measured. One interesting aspect of this sample is that in addition to measuring the velocities, we are developing a large and uniform set of spectra: Harvard graduate student Huan Lin is working on the spectroscopic properties of the sample.

3. Some Preliminary Results

Redshift observations have been carried out in earnest starting in November 1988. As the fiber system has improved, the speed and efficiency of the work has grown. The first three observing runs, November 1988, May 1989, and September 1989 provided 1421 redshifts. The next three accumulated 3721, and the most recent single observing run gathered 2200 so that we now have data in hand for about 7300 galaxies. The plots shown employ measurements of the first 5100. The current rate should provide more than 4000 new redshifts each year in two observing sessions: one for the north galactic cap and one for the south. Not every hole and fiber results in a measured redshift. Some stars are not separated from the highest surface-brightness galaxies and some spectra are too weak-lined or too poorly exposed to yield a measurable redshift. A few fibers are misplaced due to errors in astrometry. Overall, our success rate is 87%, so that we obtain 87 redshifts for every 100 fibers plugged. We are investigating the systematics of the errors to improve this rate.

Figure 5 shows the redshifts plotted against right ascension for the galaxies in the fields of the south galactic cap. Since the fields sampled are in a checkerboard pattern (the shaded areas in Figure 1) and the sample is not yet complete, the sampling is not uniform as a function of RA. The depth of the sample is its outstanding feature, with a typical galaxy redshift near 30 000 km/s. To set the scale, a single slice of the CfA survey is illustrated. The denser sampling of the CfA survey provides a more detailed picture of the nearby volume. Even so, the overall impression is that the observed structure in our survey resembles that of the CfA survey, but repeated many times. Less compensation for the sampling scheme is required for the data displayed in Figure 6. As the shaded areas in Figure 2 show, the strip at -6° is essentially complete, and its neighbor at -3° is well underway. Figure 6 shows just the data from the -6° strip, which is 75° long and 1.5° wide, providing an area of over 100 square degrees. Figure 7 shows the data from the -3° strip, which has a separation perpendicular to the line of sight corresponding to about 1500 km/s. As you might expect, based on the family resemblance of the CfA strips over comparable separations, the two strips in this survey share some common features, which may become more detailed as the -3° strip is completed.

In the well-sampled region from 0 to 30 000 km/s, the -6° strip shown in Figure 6 seems to exhibit a pattern of inhomogeneity that is similar in kind to that suggested from the nearby CfA sample. We do not see voids which are typically 4 times

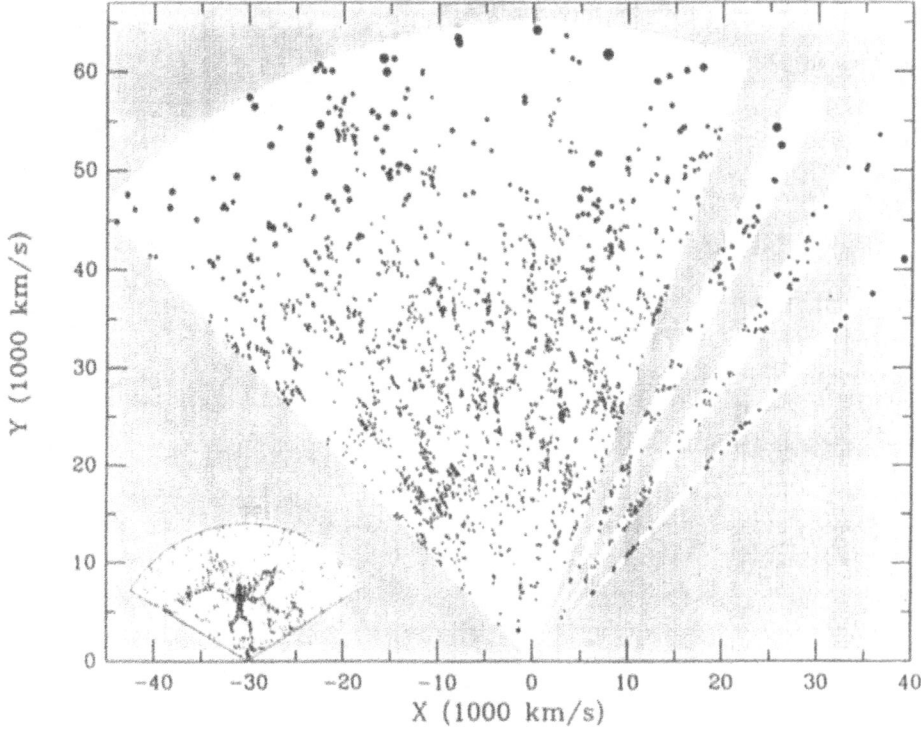

Fig. 5. Redshift distribution for galaxies in the south galactic cap as detailed in Fig.1. Note that the redshift sample extends well our beyond 30 000 km/s. Inset of a single CfA slice after Geller and Huchra (1989) sets the scale.

larger even though they could be seen in this very deep sample. The voids in our sample seem to have diameters of the order of 5000 km/s. Unless this impression is misleading, we may now be sampling the universe on a large enough scale so that it becomes more homogeneous, not less. While low amplitude structures of large extent may be revealed by subtle analysis, and we have just one strip which samples 1/400 of the sky, it is possible that we are glimpsing the end of large scale structure.

In this connection, it is interesting to note the recent work of Blumenthal et al. (1992) which uses constraints on the microwave background fluctuations and a model for the growth of voids to place a limit on the expected size of the largest voids. In their picture, the largest possible voids are expected to be 8000 km/s in diameter. More detailed analysis of our data and improved constraints on the

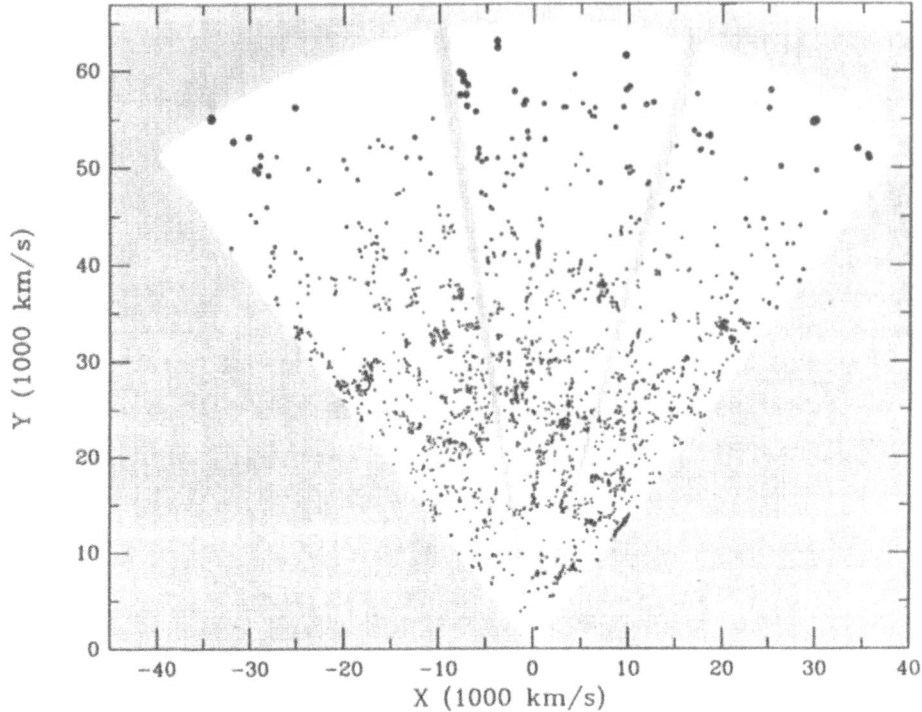

Fig. 6. Redshift distribution for 2151 galaxies in the -6° strip detailed in Fig.2. Only two spectroscopic fields are missing along the 75°length of the strip.

microwave background may help shed light on the mechanism of void formation.

4. A Little Analysis

Douglas Tucker, at Yale, has undertaken a measurement of the two-point correlation function for our sample. A preliminary result is shown in Figure 8, which indicates the usual thing: a power law with unit amplitude near 5/h Mpc, and considerable uncertainty for separations larger than 20/h Mpc. Rushing in where angels fear to tread, we also have computed the correlation function for the very large separations present in our sample, which includes galaxies in both galactic hemispheres with redshifts of 30 000 km/s.

Figure 9 illustrates that the correlation function stays small, within the errors, out to dimensions of 500/h Mpc. All of these results will become more solid as the

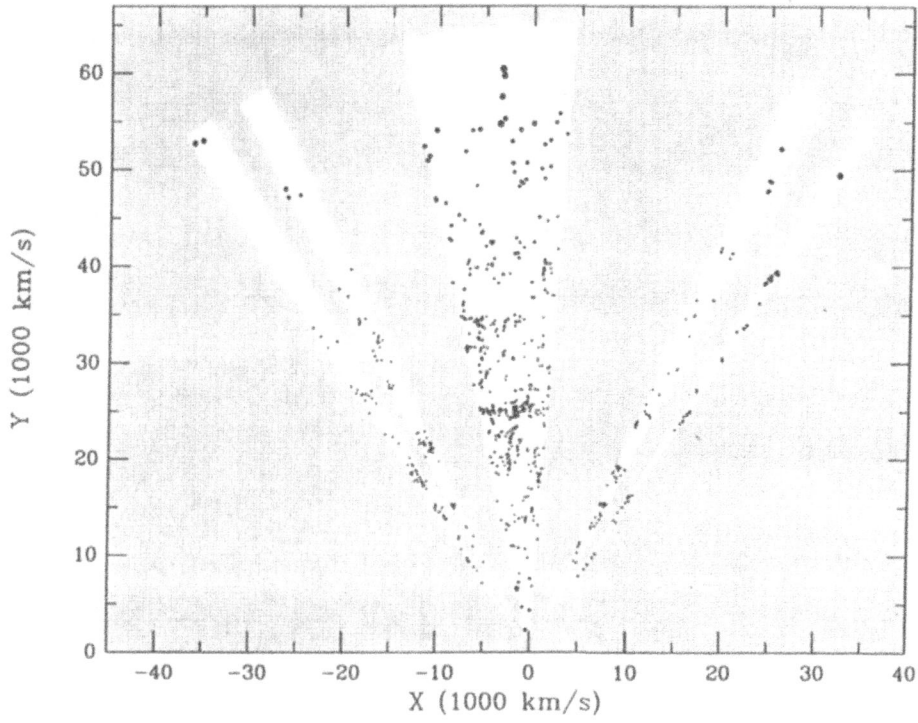

Fig. 7. Redshift distribution for 788 galaxies in the -3° strip.

sample grows and as we improve our understanding of the systematic effects which are present in the data. Since we also have spectra for the galaxies, we will be able to examine the clustering properties of the emission-line galaxies separately. Figure 8 gives a tantalizing hint that they may be more uniformly distributed than their cousins without emission from gas.

Our efforts to date have concentrated on the mind-wearying tasks of galaxy astrometry, fiber plugging, velocity measurement and list making. However, we are not completely enervated and will soon present the results of a count-in-cells analysis that resembles the work of Saunders et al. (1991), and a novel "displacement statistic" which provides a measure of the departure from homogeneity in the galaxy distribution. We expect to extract a group and cluster catalog from the photometric sample and from the redshift sample. The relation between emission lines and local galaxy density, the luminosity function for emission-line galaxies, and the spectroscopic characterization of a large galaxy sample are all under investigation.

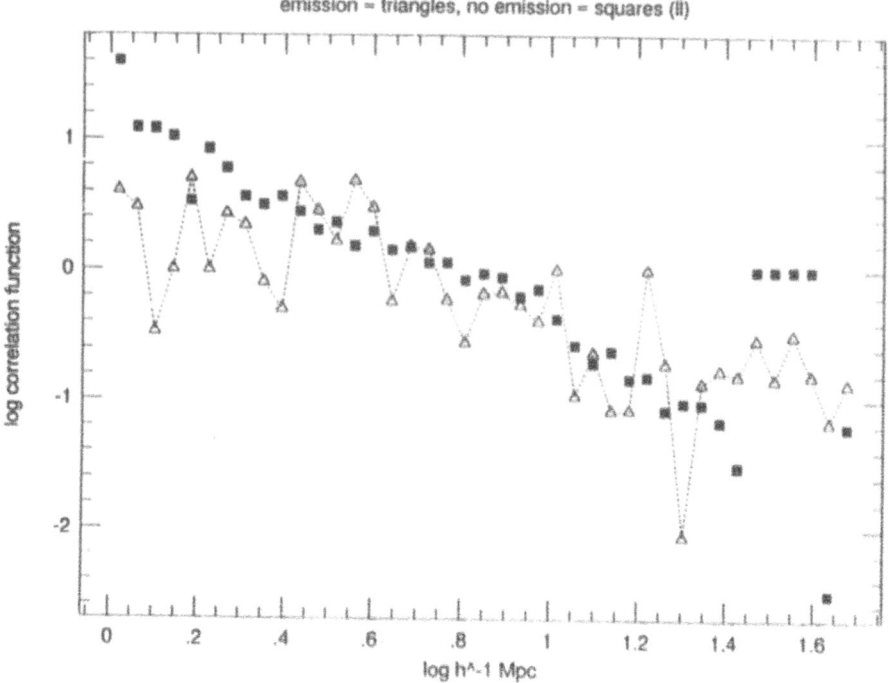

emission = triangles, no emission = squares (II)

Fig. 8. Two-point correlation function for galaxies in the sample. This is the usual log-log representation

Our aim is to make the redshift samples available at the earliest practical date to stimulate further progress in understanding the galaxy distribution. If we have, in fact, edged up to the onset of homogeneity, it may be time to return to measuring the galaxy luminosity function, the average luminosity density, and Ω.

Acknowledgements

This work is supported by the U.S. National Science Foundation through grant AST 89-21326. We use that money to support energetic graduate students Douglas Tucker at Yale and Huan Lin at Harvard. At the Carnegie Observatories, grant AST 87-17202 has enabled a number of talented workers to help bring about the technical success of this work including Chris Price, George Pauls, John Jacobs, and the inestimable John Filhaber. At the CfA, we have received valuable help with data analysis and presentation from Alberto Accomazzi, Mike Kurtz, Doug Mink, Dan Fabricant, and Emilio Falco.

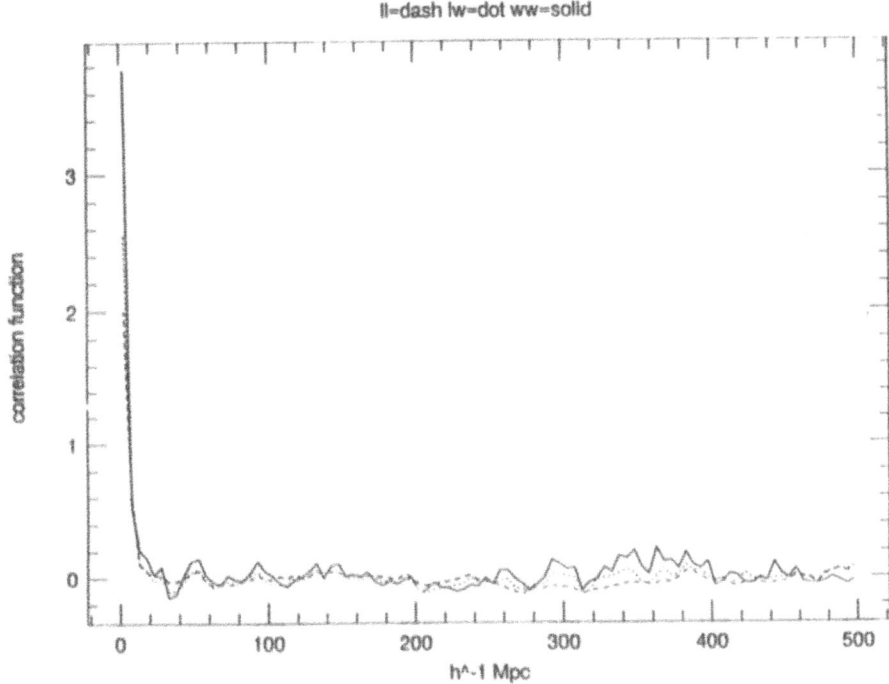

Fig. 9. Two-point correlation function for widely separated galaxies in the sample. Note that this is a linear plot.

References

Aldering, G., Bothun, G., Kirshner, R.P., and Marzke, R. 1989, B.A.A.S. 21, 1139.

Blumenthal, G., da Costa, L.N., Goldwirth, D., Lecar, M. and Piran, T. 1992, submitted to Ap.J. (CfA preprint 3259).

Dey, A., Strauss, M. and Huchra, J. 1990, A.J. 99, 463.

Geller, M.J. and Huchra, J.P 1989, Science 246, 897.

Kirshner, R.P., Oemler, A. and Schechter, P.L. 1978, A.J. 83, 1549.

Kirshner, R.P., Oemler, A. and Schechter, P.L. 1979, A.J. 84, 951.

Kirshner, R.P., Oemler, A., Schechter, P.L. and Shectman, S.A. 1981, Ap.J. Lett. 248,L57.

Kirshner, R.P., Oemler, A., Schechter, P.L. and Shectman, S.A. 1983, A.J. 88, 1285.

Kirshner, R.P., Oemler, A., Schechter, P.L. and Shectman, S.A. 1987, Ap.J. 314, 493.

Kirshner, R.P., Oemler, A., Schechter, P.L. and Shectman, S.A. 1990, A.J. 100, 1409.

Moody, J.W., Kirshner, R.P., MacAlpine, G.M., and Gregory, S.A. 1987, Ap.J. Lett. 314, L33.

Moody, J.W. and Kirshner, R.P. 1988, A.J. 95, 1629.

Peterson, B.A., Ellis, G.R., Efstathiou, G., Shanks, T., Bean, A.R., Fong, R. and Zen- Long, Z. 1986, MNRAS 221, 233.

Saunders, W. Frenk, C., Rowan-Robinson, M., Efstathiou, G., Lawrence, A., Kaiser, N., Ellis, R., Crawford, J., Xia, X., and Parry, I. 1991, Nature 349, 32.

Schechter, P.L. 1976, Ap.J. 203, 297.

Strauss, M. and Huchra, J. 1988, A.J. 95, 1602.

Tifft, W.G., Kirshner, R.P., Gregory, S.A., and Moody, J.W. 1986, Ap.J. 310, 75.

Weistrop, D. 1989, A.J. 97, 357.

Weistrop, D. and Downes, R.A. 1988, Ap.J. 331, 172.

OBJECT INDEX

SUBJECT INDEX

The manufacturer's authorised representative in the EU is Springer
Nature Customer Service Centre GmbH, Europaplatz 3, 69115 Heidelberg,
Germany. If you have any concerns regarding our products, please
contact ProductSafety@springernature.com

Printed and bound by CPI Group (UK) Ltd, Croydon, CR0 4YY
23/04/2026
02095629-0004